Commercial Mana

theory and practice

David Lowe
Senior Lecturer in Commercial Management
Manchester Business School
The University of Manchester

WILEY-BLACKWELL
A John Wiley & Sons, Ltd., Publication

This edition first published 2013

Registered Office
John Wiley & Sons, Ltd, The Atrium, Southern Gate, Chichester, West Sussex, PO19 8SQ, United Kingdom.

Editorial Offices
9600 Garsington Road, Oxford, OX4 2DQ, United Kingdom.
The Atrium, Southern Gate, Chichester, West Sussex, PO19 8SQ, United Kingdom.

For details of our global editorial offices, for customer services and for information about how to apply for permission to reuse the copyright material in this book please see our website at www.wiley.com/wiley-blackwell.

Library of Congress Cataloging-in-Publication Data

Commercial management : theory and practice / David Lowe.
p. cm.
Includes bibliographical references and index.
ISBN 978-1-4051-2468-3 (pbk.)
1. Purchasing. 2. Industrial procurement. 3. Contracting out. 4. Public contracts. 5. Project management.
I. Lowe, David
HF5437.C59 2013
658–dc23

2012037468

A catalogue record for this book is available from the British Library.

Wiley also publishes its books in a variety of electronic formats. Some content that appears in print may not be available in electronic books.

Cover design by Andy Magee
Cover image courtesy of rosevita (www.morguefile.com)

Set in 9/11.5pt Avenir by SPi Publisher Services, Pondicherry, India
Printed and bound in Singapore by Markono Print Media Pte Ltd

1 2013

To Ruth, with love
and to the memory of
Josephine Alice Lowe 1934–2009

Contents

About the Author vii
About the Contributors viii
Acknowledgements x
Preface xii
Abbreviations xiv

PART 1: INTRODUCTION 1

Introduction: Commercial Management 3

1. **Commercial Management in Project-Oriented Organisations** 23

PART 2: ELEMENTS OF COMMERCIAL PRACTICE AND THEORY 77

2. **Commercial Leadership** 79
Maria-Christina Stafylarakis

3. **Exploring Strategy** 96
Irene Roele

4. **Perspectives on Managing Risk and Uncertainty** 108
Eunice Maytorena

5. **Financial Decisions** 132
Anne Stafford

6. **Legal Issues in Contracting** 173
David Lowe and Edward Davies

PART 3: APPROACHES TO COMMERCIAL PRACTICE 237

7. **Best-Practice Management** 239

8. **Commercial Strategies and Tactics** 288
Part A: Intent 296
Part B: Deal Creation 332
Part C: Execution 359

PART 4: CASE STUDIES 379

Case Study A: Football Stadia **381**

Case Study B: Terminal 5 (T5) Heathrow **455**

Index 492

This book's companion website is at www.wiley.com/go/lowecommercialmanagement and offers invaluable resources for both students and lecturers:

- PowerPoint slides for lectures on each chapter
- sample exam questions for you to practise
- weblinks (for each chapter and for the whole book) to key journals, and to relevant academic and professional bodies

About the Author

Dr David Lowe
Manchester Business School

David Lowe is a Senior Lecturer in Commercial Management in the Executive Education Centre of Manchester Business School. He is Programme Director for the blended learning MSc in International Commercial and Contract Management and for several executive education programmes in commercial and contract management. Clients include Rolls-Royce, BAE Systems, Thales, BT, the National Skills Academy Nuclear, and the Foundation Trust Network (NHS).

David is a Fellow of the Royal Institution of Chartered Surveyors, a Fellow of the Higher Education Academy and an academic adviser to the International Association for Contract and Commercial Management (IACCM). His consultancy work includes benchmarking the engineering and project management provision of an international pharmaceutical company.

His teaching interests include: commercial management, commercial strategies and tactics, and contract management. Similarly, his research interests focus on commercial and contract management within the context of project-based industries, ranging from ICT/telecommunications to aerospace and defence. Completed research projects include: an investigation of the cost of different procurement systems and the development of a predictive model; a project to assist medium-sized construction companies develop strategic partnerships and diversify into new business opportunities offered by public and private sector clients; and an investigation into the function of commercial management in the telecommunications and construction industries. His PhD, completed at UMIST, investigated the development of professional expertise through experiential learning.

His book contributions include: 'Contract management' in *The Wiley Guide to Managing Projects* (Wiley, 2004), and *Commercial Management of Projects: Defining the Discipline* (Blackwell Publishing, 2006) which he edited with Roine Leiringer. The latter is the first book to establish a theoretical framework for commercial management.

About the Contributors

Edward Davies
Visiting Fellow, Manchester Business School/Senior Consultant, Hill Dickinson

Edward Davies specialises in the legal and commercial aspects of major infrastructure and engineering projects, with 25 years' experience in this field. He has been involved in some of the largest and most important projects in the north-west of England including the second runway at Manchester Airport, United Utilities' multibillion pound procurement programmes and Peel's £500 million waste-to-energy project at Ince Marshes in Cheshire. Further afield, his international experience includes working in conjunction with CH2M Hill as part of the Panama Canal Authority's commercial management team for the $5 billion Panama Canal Expansion Project. Prior to joining Hill Dickinson, Edward was a partner at Pinsent Masons for over 20 years and was one of the founding partners of the firm's Manchester office in 1989.

Edward lectures extensively on all aspects of engineering and construction. He is a Visiting Fellow at Manchester Business School and lectures on the MSc in International Commercial and Contract Management and MBA courses.

He is also a mediator trained by the American Arbitration Association and CEDR.

Edward wrote the chapter on drafting construction contracts in *Management of Procurement* published by Thomas Telford (edited by Denise Bower). He also co-edited *Dispute Resolution and Conflict Management in Construction – An International Review* published by E&F Spon.

In the 2011/2012 edition of the *Chambers Guide to the Legal Profession in the UK* Edward was rated in the top category of leading construction lawyers in the north-west of England for the 16th consecutive year.

Dr Eunice Maytorena
Lecturer in Construction Project Management, Manchester Business School

Eunice Maytorena is a lecturer in construction project management at Manchester Business School. She completed her PhD degree at the Bartlett School of Graduate Studies (UCL) in 2003. An architect by training, her work experience includes architectural design and consultancy, research in various aspects of the built environment and lecturing in project management. She has taught on undergraduate, masters and executive MBS programmes. Eunice has worked on several research projects investigating risk perceptions and risk management in project forms of organisations.

Irene Roele
Senior Fellow in Management, Manchester Business School

Irene Roele has worked in management, learning and development for over 25 years. For the last four years she has worked for Manchester Business School's Executive Education Centre as an organisational strategy consultant, researcher and educator primarily in the private sector. She helps individuals and organisations make constructive links between 'thinking', 'planning' and 'doing' – in other words, between theory and practice and between policy and action. Her specialism is in developing strategic thinking skills.

At MBS, Irene takes a leading role in designing and delivering a variety of bespoke executive education and strategy consulting interventions. She regularly runs strategic thinking workshops for a number of clients across the private and public sectors. Current work involves facilitating group and individual coaching for senior leadership teams to support their strategy engagement.

Irene has extensive experience developing and delivering specialist strategy modules on graduate accredited development programmes for MBS, including the MSc in Commercial Management – current clients include Rolls-Royce, BAE Systems, Thales and the Foundation Trust Network (NHS).

Prior to MBS, Irene worked at London Metropolitan University in the City of London for 10 years. There she gained extensive experience in postgraduate executive education, including curriculum design, course management and delivery. She was programme leader for the MSc in Financial Service Regulation and Compliance Management, subject leader in Strategy on four executive Masters programmes and module leader for the Chartered Institute of Marketing Postgraduate Diploma. The role included design and delivery of bespoke programmes for various organisations including Standard & Poors, Royal Bank of Scotland, Ernst and Young, BMI General Healthcare Group, the National Trainers Federation and Croydon Social Services.

Irene has a BA (honours) from the European Business School, London and an MA in Marketing from Kingston University. Her research interest is in how boards strategise – that is, in exploring the board's involvement in the strategy-making process.

Dr Anne Stafford
Senior Lecturer, Manchester Business School

Anne Stafford is a Senior Lecturer at Manchester Business School. Anne worked as a management accountant from 1985 to 1990, during which time she qualified as an accountant with the ACCA. In 1990 she was appointed as a Senior Lecturer at the University of Central England, subsequently becoming a Principal Lecturer in 1993. She taught financial accounting and reporting extensively to a wide range of students, including undergraduate, postgraduate, professional and post-experience programmes. She was Programme Director of the university's ACCA programmes from 1993 to 1999.

In 1999 Anne took a career break and worked part time as a lecturer for UMIST while completing her PhD at the University of Warwick. She was appointed to her current post within the Business School in 2004. Anne's current teaching and research interests lie in the area of financial reporting, corporate governance and business analysis.

Dr Maria-Christina Stafylarakis
Senior Fellow in Leadership, Manchester Business School

Maria-Christina Stafylarakis is a Senior Lecturer and Director of Programme Quality in Executive Education at Manchester Business School. She has designed and delivered several bespoke executive programmes and has acted as a dedicated programme director for Halliwells, UNITAS, Sita UK, UKTI and PZ Cussons.

Prior to joining MBS, Maria held teaching posts at Hull Business School and at Lancaster University where she lectured on diverse topics such as equal opportunities and diversity, organisational learning/learning organisations, managing change, organisation development, human resource management and development, leadership and qualitative research methods.

She is an AIM scholar (Advanced Institute of Management) and has also worked on various research projects. She has also worked as a training practitioner in Greece with clients such as Ford and SEAT, and has HRD consultancy experience in Vietnam.

Maria has a Masters in Human Resource Development (HRD) from the Institute for Development Policy and Management in Manchester and a PhD from Manchester Business School. Her PhD focused on leadership in the context of learning organisations

Acknowledgements

Author's acknowledgments
I would like to thank Ruth, my wife, for her love, patience and understanding during the long process of writing and editing this book. Similarly, I'd like to express my gratitude to Madeleine Metcalfe, Senior Commissioning Editor, at Wiley-Blackwell, for her encouragement, forbearance and faith that a manuscript would finally be delivered.

The book would never have been completed without the participation of Edward Davies, Eunice Maytorena, Irene Roele, Anne Stafford and Maria-Christina Stafylarakis. Thank you all for your chapter contributions, for the many exchanges and the debate we've had over the book's content, and for your inputs to the development of the commercial management programmes at Manchester Business School (MBS).

Several people have helped in the preparation of the manuscript: Lesley Gilchrist formatted the draft chapters; Vicki Mansfield drew Figure A.1; and Janine May has protected me from the realities of the 'day job', enabling me to eventually complete the project – thank you.

I would also like to thank the following colleagues at the University of Manchester: Graham Winch, Mark Winter, Eunice Maytorena and Nuno Gil (members of the Projects and Programmes affinity group at MBS) and Peter Fenn and Margaret Emsley (from the School of Mechanical, Aerospace and Civil Engineering) for their support and encouragement over the years, and Janine May, Emma Farnworth and Helen Jennings for their administrative support and good humour!

I would also like to thank the numerous students and delegates who have participated in the various 'commercial management' programmes and modules, and the representatives of the sponsoring organisations that have supported and helped develop these programmes. In particular I would like to thank Shan Morris for allowing me to 'plunder' her MSc dissertation on *Insolvency in Construction: The Collapse of Laing Construction plc* to produce the case study on the Millennium Stadium, Cardiff.

Finally, special thanks are due to Peter Fenn, without whom neither this book nor the commercial programmes that spawned it would have transpired, to Mark Winter for our many discussions on the nature of commercial management, and to Irene Roele for championing the cause of commercial management with MBS's Executive Development Centre.

Publisher's acknowledgements
We are grateful for permission to reproduce the following copyright material:

Table 6.3: EU procurement directives and the corresponding UK regulations, reprinted from European & UK Procurement Regulations, © Millstream Associates Ltd (2012); Table 8.2: Example of an approach to procurement route evaluation, reprinted from *Achieving Excellence Guide 6 – Procurement and Contract Strategies*, Office of Government Commerce © Crown Copyright (2007). Contains public sector information licensed under the Open Government Licence v1.0 www.nationalarchives.gov.uk/doc/open-government-licence/open-government-licence.htm; Table A.6: Cost breakdown and Table A.7: Breakdown of funding, reprinted from *English National Stadium Review: Final Report October 2002*, The Stationery Office, © Crown Copyright (2002). Contains public sector information licensed under the Open Government Licence v1.0 www.nationalarchives.gov.uk/doc/open-government-licence/open-government-licence.htm; Box 5.1: Extract from Vinci Annual Report 2011, reprinted from the Vinci 2011 Annual Report, Copyright (2011), with permission from Vinci, Paris, France; Box 5.3: Balfour Beatty balance sheet, Box 5.4: Balfour Beatty plc group income statement, Box 5.5: Balfour Beatty plc group statement of comprehensive

Income, and Box 5.6: Balfour Beatty cash flow statements, reprinted from the Balfour Beatty Report and Accounts 2010, Copyright (2010), Balfour Beatty plc, London; Box 6.8: EU procurement thresholds, reprinted from European Union (EU) Public Procurement, Intellectual Property Office © Crown Copyright (2012). Contains public sector information licensed under the Open Government Licence v1.0 www.nationalarchives.gov.uk/doc/open-government-licence/open-government-licence.htm; and Box B.2: The underlying assumptions of BAA's contracting approach contrasted with conventional principles (T5 contracting assumptions), reprinted from *Heathrow's T5 History in the Making*, Sharon Doherty © 2008 John Wiley & Sons Ltd.

Image 'Cardiff Millennium Stadium, May 2006' p 382 courtesy of Epaunov72; Image 'Stitched photo of the Emirates Stadium' p 399 courtesy of Ed g2s; Image 'Wembley under construction, January 2006' p 410 courtesy of ProhibitOnions at the wikipedia project; and Image 'View of the south side of the new terminal 5, July 2006' p 455 courtesy of Henrik Romby.

Photographs in Part 4: Case Studies of the Cardiff Millennium Stadium and Terminal 5 (T5) Heathrow are both licensed under the GNU Free Documentation License.

In some cases we have been unable to trace the owners of copyright material, and we would appreciate any information that would enable us to do so.

Preface

Definitions of Commercial Management:

> 'The management of contractual and commercial issues relating to projects, from project inception to completion' Lowe and Leiringer (2005, 2006)
>
> 'The identification and development of business opportunities and the profitable management of projects and contracts, from inception to completion.' International Association for Contract and Commercial Management (IACCM)/The Institute of Commercial Management (ICM) (IACCM, ND)

Commercial exists as a distinct management role in many organisations, particularly those emanating from the UK, although it is becoming more accepted globally as a valuable business activity. In particular, commercial is increasingly viewed as a dynamic capability within project-oriented organisations.

The impact of globalisation, servitisation and collaboration on business-to-business (b2b) exchanges (economic transactions) has been the formulation and management of complex interfirm contracts, agreements and relationships across the ensuing value networks. The commercial function is primarily responsible for the design, negotiation, award and management of these b2b transactions.

The aim of this book is to provide a framework for understanding commercial practice within project-oriented organisations. Additionally, it seeks to identify generic aspects of this practice, to provide a theoretical foundation to and common vocabulary for these activities (by reference to existing and emergent theories and concepts, for example, transactional cost economics and relational contracting), and to examine relevant management best practice.

The book is divided into four parts:

- **Part 1: Introduction**
  - ***Commercial Management in Project-orientated Organisations***: explores the nature of commercial practice within project environments at the buyer–seller interface, and introduces a commercial management framework
- **Part 2: Elements of Commercial Practice and Theory**
  - ***Commercial Leadership***: explores key themes and concepts from the leadership literature. It reviews established approaches to the study of leadership, for example, trait, style and contingency theories; examines contemporary theories, such as transformational and dispersed leadership; and explores leadership in relation to learning. The chapter concludes with the exposition of a conceptual framework
  - ***Exploring Strategy***: focuses on the orientation aspect of strategy, its aim being to assist commercial practitioners, identify the strategic agenda, define a clear sense of purpose and appreciate the need for a guiding policy for the firm in creating value and clearly articulate its core value-proposition
  - ***Perspectives on Managing Risk and Uncertainty***: provides an introduction to contemporary project risk management, evaluates various perspectives on risk and presents an overview of the project risk management process and its commonly used tools and techniques
  - ***Financial Decisions***: provides an introduction to financial decision making. Its aim is to equip commercial practitioners with the knowledge and ability they need to appreciate the financial information issued by their own organisation, partners, suppliers and customers, and their role in generating this data

   - ***Legal Issues in Contracting***: seeks to develop an appreciation of some of the key legal issues that influence commercial practice. It provides a selective overview of contract provisions and procedures, together with an international perspective on the legalisation and regulations that affect b2b exchanges within a project and programmes environment
- **Part 3: Approaches to Commercial Practice**
   - ***Best-Practice Management***: identifies and outlines 'best practice' in managing projects and programmes. Topics covered include governance issues associated with the Turnbull recommendations; the OGC Gateway Process; the use of process protocols, such as PRINCE2, MSP, MoP and MoV; and guidance on best practice in procurement and contract management
   - ***Commercial Strategies and Tactics***: examines commercial practice across the project life cycle from both the supply and demand perspectives, identifying the interrelationship between the demand-side procurement process (cycle) and the supply-side bidding and implementation cycles. The chapter explores how purchasers can articulate their requirements for new asset and services effectively, and then manage the process of asset/service definition and delivery. From the supply side, it addresses how suppliers can identify potential opportunities effectively, develop and communicate their proposals for the supply of asset and services, manage the deal creation stage, and ensure that commitments made are delivery profitably. It is subdivided into three sections: Part A – Intent; Part B – Deal Creation; Part C – Execution
- **Part 4: Case Studies**: Two extended case studies are provided:
  Case Study A: Football Stadia: comprising the Millennium Stadium, Cardiff; the Emirates Stadium, London; and Wembley Stadium, London
  Case Study B: Terminal 5 (T5) Heathrow

The book is informed by experience gained through developing and delivering undergraduate, postgraduate and executive education programmes in commercial management and research into commercial practice in a variety of industry sectors, including construction, ICT, aerospace and defence. Specifically, it draws on research undertaken at the University of Manchester in support of the MSc in Commercial Management (now renamed the MSc in International Commercial and Contract Management), an MBA for Commercial Executives, and the BSc in Commercial Management and Quantity Surveying. It also builds on *Commercial Management of Projects: Defining the Discipline* by Lowe with Leiringer (2006).

The intended audiences for this book are:

- Postgraduate students on MSc commercial, contract and project management, quantity surveying and supply chain management/procurement programmes, and those taking specialist modules on MBA programmes
- Final-level undergraduate students on quantity surveying, construction management, project management and supply chain management programmes

Abbreviations

4Ps	Public Private Partnerships Programme
ABA	American Bar Association
ACA	Aircraft Carrier Alliance
ACA	Association of Consulting Architects
ACCA	Association of Chartered Certified Accountants
AEC	Achieving Excellence in Construction
AHP	Analytical hierarchy process
AICPA	American Institute of Certified Public Accountants
AMA	American Marketing Association
ANN	Artificial neural networks
APM	Association for Project Management
ARR	Accounting rate of return
ASB	Accounting Standards Board (UK)
ASIC	Australian Securities and Investments Commission
b2b	Business-to-business
b2g	Business to government
BA	British Airways plc
BAA	British Airports Authority Ltd
BATNA	Best alternative to a negotiated agreement
BCIS	Building Cost Information Service
BERR	Department for Business, Enterprise and Regulatory Reform
BIM	Building information modelling
BIS	Department for Business, Innovation and Skills
BoK	Body of knowledge
BOO	Build–own–operate
BOOT	Build–own–operate–transfer
BPO	Business process outsourcing
BSF	Building Schools for the Future
BSI	British Standards Institution
C&S	Civil and structural
CAA	Civil Aviation Authority
CAT	Competition Appeal Tribunal
CBA	Canadian Bar Association
CBUK	Cleveland Bridge UK Ltd
CC	Competition Commission
CCC	Cardiff County Council
CCT	Compulsory competitive tendering
CCTA	Central Computer and Telecommunications Agency
CA	Confidentiality agreement
CDA	Confidentiality-disclosure agreement
CE	Constructing Excellence
CEDR	Centre for Effective Dispute Resolution
CFR	Cost and freight
CIB	Construction Industry Board
CIF	Cost insurance and freight
CII	Construction Industry Institute of the United States of America
CIM	Chartered Institute of Marketing

CIMA	Chartered Institute of Management Accountants
CIP	Carriage and insurance paid to
CIPFA	Chartered Institute of Public Finance and Accountancy
CIPP	Continuous improvement of the project process
CIPS	Chartered Institute of Purchasing & Supply
CLC	Colnbrook Logistics Centre
CLM	Contract life cycle management
CMM	Capability maturity model/commercial management maturity
CoE	Centres of excellence
CoPS	Complex products and services/systems
CPT	Carriage paid to
CRFC	Cardiff Rugby Football Club
CRM	Customer relationship management
CRO	Chief risk officer
CSCMP	Council of Supply Chain Management Professionals
CSFB	Credit Suisse First Boston
DAP	Delivered at place
DARS	Defense Acquisition Regulations System
DAT	Delivered at terminal
DCF	Defence commercial function
DCMS	Department for Culture, Media and Sport
DDP	Delivered duty paid
DFARS	Defense Federal Acquisition Regulation Supplement
DFBO	Design-fund-build-operate
DoD	Department of Defense
DoH	Department of Health
DRO	Debt relief order
EC	European Commission
ECC	(NEC) Engineering and Construction Contract
ECM	Enterprise contract management
EFA	Education Funding Agency
EI	Emotional intelligence
EIRM	European Institute of Risk Management
ENSDC	English National Stadium Development Company Ltd
ENST	English National Stadium Trust
EPC	Engineering, procurement and construction
ERG	Efficiency and Reform Group
ERM	Enterprise risk management
ETO	Economic, technical or organisational
EU	European Union/expected utility
EXW	Ex works
FA	Football Association
FAR	Federal Acquisition Regulations
FARS	Federal Acquisition Regulations System
FAS	Free alongside ship
FASB	Financial Accounting Standards Board
FCA	Free carrier
FIDIC	Fédération Internationale des Ingénieurs-Conseils
FM	Facilities management
FOB	Free on board
FRC	Financial Reporting Council
FV	Future value
GMP	Guaranteed maximum price
GPA	Government Procurement Agreement
HCC	Heathrow Consolidation Centre
HS&E	Health, safety and environment
IACCM	International Association for Contract and Commercial Management
IASB	International Accounting Standards Board
IBA	International Bar Association

ICAEW	Institute of Chartered Accountants in England and Wales
ICAI	Institute of Chartered Accountants in Ireland
ICAS	Institute of Chartered Accountants of Scotland
ICC	International Chamber of Commerce
ICE	Institution of Civil Engineers
ICM	Institute of Commercial Management
IFAC	International Federation of Accountants
IFPSM	International Federation of Purchasing and Supply Management
IFRSs	International Financial Reporting Standards
IIF	Incident and injury free
Incoterms®	International Commercial terms
IRM	Institute of Risk Management
IP	Intellectual property
IPMA®	International Project Management Association
IPO	Intellectual Property Office
IPR	Intellectual property rights
IPT	Integrated project teams
IRM	Institute of Risk Management
IRR	Internal rate of return
ISM	Institute for Supply Management™
IST	Integrated supply team
ITIL®	IT Infrastructure Library
ITT	Invitation to tender
IVAs	Individual voluntary arrangements
JCT	Joint Contracts Tribunal
JV	Joint venture
KPIs	Key performance indicators
KSM	Key supplier management
L&A	Liquidated and ascertained (damages)
LDA	London Development Agency
LFA	Lottery Funding Agreement
LU	London Underground
M&E	Mechanical and electrical
M_o_R®	Management of Risk
MAPE	Mean absolute percentage error
MBTI	Myers–Briggs Type Indicator
MEAT	Most economically advantageous tender
MOD	Ministry of Defence
MoP™	Management of Portfolios
MoV®	Management of Value
MPA	Major Project Authority
MS	Millennium Stadium
MSP®	Managing Successful Programmes
MVL	Members' voluntary liquidation
NAO	National Audit Office
NATS	National Air Traffic Services
NCMA	National Contract Management Association
NDAs	Non-disclosure agreements
NEC	New Engineering Contract
NGN	Next generation networks
NHS	National Health Service
NJCC	National Joint Consultative Committee for Building
NPS	National Procurement Strategy
NPV	Net present value
NRM	New Rules of Measurement
OBK	O'Brien Krietzberg
OFT	Office of Fair Trading
OGC	Office of Government Commerce
OH	Occupational health

OJEC	Official Journal of the European Community
OPM3	Organisational Project Management Maturity Model
P&L	Profit and loss
P3M3®	Portfolio, Programme and Project Management Maturity Model
P3O®	Portfolio, Programme and Project Offices
PATNA	Probable alternative to a negotiated agreement
PCM	Partnered category management
PCP	Procuring complex performance
PEP	Project execution plan
PERT	Programme evaluation and review technique
PESTEL	Political, economic, social, technological, environmental and legal
PFI	Public Finance Initiative
PfS	Partnerships for Schools
PI	Performance indicator
PID	Project initiation document
plc	Public limited company
PMI	Project Management Institute
PMM	Project management maturity
PMO	Project management office
PPE	Property, plant and equipment
PPP	Public-private partnerships
PRINCE2®	Projects in Controlled Environments
PRM	Project risk management
PSC	Project-specific criteria
PSCPs	Principal supply chain partners
PUK	Partnerships UK
PV	Present value
RBS	Risk breakdown structures
RBV	Resource-based view
RFP	Request for proposal
RFQ	Request for quotation
RFT	Request for tender
RICS	Royal Institution of Chartered Surveyors
RM	Relationship Management
RMIA	Risk Management Institution of Australasia Limited
ROA	Return on assets
ROCE	Return on capital employed
ROI	Return on investment
RPI	Retail Prices Index
RR	Risk registers
SBU	Strategic business unit
SCM	Supply chain management
SDR	Strategic Defence Review
SEC	Specialist engineering contractors/US Securities and Exchange Commission
SEI	Software Engineering Institute
SEU	Subjective expected utility
SIBET	Soft issues bid evaluation tool
SLA	Service-level agreement
SMART	Specific, measurable, agreed, realistic and timely
SPI	Smart Procurement Initiative
SRA	Solicitors Regulation Authority
SRM	Supplier relationship management
SRO	Senior responsible owner
SSM	Soft systems methodology
T5	Terminal 5, Heathrow
TA	Teaming agreement
TCE	Transaction cost economics
TfL	Transport for London
TRIPS	Trade-Related Aspects of Intellectual Property Rights Agreement

TUPE	Transfer of an Undertaking (Protection of Employment)
UCAS	Universities and Colleges Admissions Service (UK)
UCC	Universal Copyright Convention
UNCITRAL	United Nations Commission on International Trade Law
VaR	Value-at-risk
VfM	Value for money
VGPB	Victorian Government Purchasing Board
WATNA	Worst alternative to a negotiated agreement
WDF	World Duty Free
WestLB	Westdeutsche Landesbank
WIPO	World Intellectual Property Organization
WNSL	Wembley National Stadium Limited
WRU	Welsh Rugby Union
WTO	World Trade Organisation

Part 1

Introduction

Introduction Commercial Management 3
Chapter 1 Commercial Management in Project-Oriented Organisations 23

Introduction
Commercial Management

Learning outcomes

After reading this introduction you will be able to:

- Define the terms 'commercial management' and 'commercial manager'
- Identify common activities undertaken by commercial managers (practitioners)
- Describe the position of 'commercial' within global, project-oriented organisations

Additionally, this introduction seeks to provide an overview of the format and context of the book.

Introduction

Commercial exists as a distinct management role in many organisations, particularly those originating from the UK, although it is becoming more accepted globally as a valuable business activity. Despite this, while a basic internet search will generate numerous job advertisements for commercial managers, executives and directors, academic management literature is decidedly quiet on the subject. Building upon the previous publication *Commercial Management of Projects: Defining the Discipline* (Lowe with Leiringer, 2006), this text seeks to redress the situation by defining and describing in a normative way, precisely what 'commercial actors' (practitioners, managers, specialists, executives, etc.) actually do. It also provides a framework for the application of the principles and underpinning theory that support effective commercial practice.[1]

Increasingly 'commercial' is viewed as a dynamic capability within organisations. The changing nature of business-to-business (b2b) exchanges (economic transactions), resulting from the processes of globalisation, servitisation and collaboration, has necessitated the formulation and management of complex inter-firm contracts, agreements and relationships across the resulting value networks. The commercial function within organisations is primarily responsible for the design, negotiation, award and management of these b2b transactions.

What is commercial management?

The terms commercial management and commercial manager have been used for some time, predominantly, but not exclusively, in the following sectors:

- Telecommunications, electronics and ICT
- Services and outsourcing

Commercial Management: theory and practice, First Edition. David Lowe.

Table I.1 Definitions of commercial management.

Definition	Source
The process of controlling or administering the financial transactions of an organisation with the primary aim of generating a profit	Lowe *et al.* (1997)
The management of contractual and commercial issues relating to projects, from project inception to completion	Lowe and Leiringer (2005, 2006)
The identification and development of business opportunities and the profitable management of projects and contracts, from inception to completion	International Association for Contract and Commercial Management (IACCM)/The Institute of Commercial Management (ICM) (IACCM, ND)

- Energy and oil
- Defence and aerospace
- Financial services
- Pharmaceuticals and health care
- Engineering and construction

Definitions of commercial management

Despite its usage in practice, there is no universally acknowledged definition of the term **commercial management**; although a few suggestions have been proffered. These are presented in Table I.1.

Based on an investigation into the role of commercial managers in the UK construction industry, Lowe *et al.*'s definition focuses on the financial aspects of the function. Their results suggest that within the construction supply sector the role was based around *'looking after'* the profits of an organisation, which entailed minimising its costs and maximising its income, and managing 'cash flow'. This definition was subsequently modified by Lowe and Leiringer following a study of the commercial role across three, predominantly project/programme-based industry sectors: construction, telecommunications/ICT and defence/aerospace (Lowe and Leiringer, 2005, 2006). Although not explicitly stated in either definition, both Lowe *et al.* and Lowe and Leiringer acknowledged the centrality of risk management to commercial management.

The definition adopted more recently by the International Association for Contract and Commercial Management (IACCM) and The Institute of Commercial Management (ICM) reflects the application of the term beyond project/programme-oriented organisations and industry sectors, emphasising the function's involvement in recognising and realising new business opportunities. However, in doing so, the definition is open to criticism: being a rather general delineation of the commercial contribution to an organisation – that is, one that could equally be applied to a more generalist management role.

While all three definitions are concise, they each fail perhaps to convey the breadth and multifaceted nature of the function. This is not surprising. The task of framing an appropriate definition that encapsulates all aspects of the role would probably run into several pages and include numerous exclusions and provisos. It is also hampered by the use of the term to describe the use or exploitation of natural resources, for example, land (forestry and quarrying), fish stocks (fish farms) or animals (poultry, pigs, etc.). Additionally, in a global context there are issues concerning the use of the word 'commercial'; for example, the French translate the word as 'sales', which has the potential to cause confusion as 'sales' is a recognised function within organisations and an established academic discipline.

Definitions of commercial manager

Derived from the above definitions of commercial management, two definitions of the term **commercial manager** have been framed (see Table I.2). However, the reservations expressed in respect of the definitions of commercial management, equally apply to these definitions.

Table I.2 Definitions of commercial manager.

Definition	Source
... a person controlling or administering the financial transactions of an organisation with the primary aim of generating a profit	Lowe *et al.* (1997)
... is someone whose primary role is in the management or execution of such opportunities or projects[2]	International Association for Contract and Commercial Management (IACCM)/The Institute of Commercial Management (ICM) (IACCM, ND)

Table I.3 Definitions of contract management.

Definition	Source
Contract management is a niche within the procurement profession, but it has a very broad perspective in terms of the responsibilities assigned to a contract manager. The job scope ranges from the administrative skills of managing, organizing, and planning, to the excitement and challenge of negotiating a major contract	National Contract Management Association (NCMA, 2011)
Contract management is the process that enables both parties to a contract to meet their obligations in order to deliver the objectives required from the contract. It also involves building a good working relationship between customer and provider. It continues throughout the life of a contract and involves managing proactively to anticipate future needs as well as reacting to situations that arise	Office of Government Commerce (OGC, 2002)
Contract management is the process of managing and administrating the ... contract from the time it has been agreed at contract award, through to the end of the service period ...	Public Private Partnerships Programme (4Ps, 2007)
Contract management is the phase of the procurement cycle in which a supplier delivers the required goods or services in accordance with a procuring authority's specification	OGC (2009)
Ongoing monitoring and management of the provision of services in line with the agreed terms and conditions	Victorian Government Purchasing Board (VGPB, 2011)

Definitions of contract management

Alternatively, the terms **contract management** and **contract manager** have been adopted as descriptors for the 'commercial' role in organisations; however, as shown in Table I.3, there is some confusion over the precise meaning of the term contract management. For example, it is often taken to imply a 'back-room', administrative, task-orientated or procedural-focused role, and in many organisations the role is restricted to post-award activities in the procurement cycle (those carried out after a contract has been entered into); see, for example, the definitions OGC (2009) and VGPB (2011). In other organisations 'contract management' is viewed as a more dynamic capability involving the drafting and negotiation of complex b2b contracts and agreements. However, as is evident from the following review of the tasks undertaken by commercial practitioners, while contract management (in its broadest sense) forms a major area of activity, individuals with the title 'commercial' rather than 'contract' manager generally have a wider sphere of involvement both throughout the project life cycle and in the breadth of activities under their control (for example, relationship and transaction management, and regulatory and financial aspects).

Additionally, IACCM have championed the term 'commitment management', which they define as:

> '... an advanced method through which an organization applies quality principles to business terms, policies, practices and processes to drive improvement in negotiation, contract performance and

governance standards. It is a systematic way of ensuring that business requirements and capabilities are aligned to formal commitments, to ensure that opportunity selection is optimized and business relationships are fulfilled as agreed.' (IACCM, ND)

The term, however, has failed to gain the same degree of acceptance afforded to commercial and contract management, particularly within the UK context.

As mentioned earlier, the term 'commercial management' is particularly prevalent in organisations originating in the UK or regional locations that have strong historical ties with the UK. Elsewhere, for example, in the US the term 'contract management' is generally used. However, there is some evidence that 'commercial' is gaining some purchase in the Americas. Interestingly, the principle institution that supports

Table I.4 Common 'commercial' activities in project-oriented organisations.

Generally:

- Identifying and evaluating commercial opportunities
- Identifying, evaluating and mitigating risk
- Constructing novel forms of commercial arrangement (agreements and contracts)
- Reviewing and evaluating agreements and contracts
- Estimating costs
- Applying life cycle management (and its various processes and procedures) to commercial and contractual issues
- Providing commercial input to projects
- Preparing, reviewing and submitting management reports; auditing projects and undertaking commercial reviews

Pre-award

- Building business cases for business needs or requirements
- Selecting appropriate procurement strategy(ies)
- Assessing the costs and risks of a particular venture/opportunity
- Constructing a business case for a potential deal
- Dealing with contracts: drafting, negotiating and agreeing complex contracts and agreements
- Preparing bids; forming part of, or leading bid teams (panels)
- Owning the bid authorisation procedures[3]; and providing commercial authorisation
- Developing and publishing standard terms and conditions of contract for products and/or services
- Arranging 'back to back' subcontract terms to mirror main contract provisions
- Obtaining legal and regulatory sign-off for undertaking business opportunities

Post-award:

Managing and administering complex contracts and agreements (post-award contract management):

- Advising project management teams on contractual issues, providing critical business support to ensure the project is delivered in accordance with the agreed contract terms
- Managing/administering commercial/contractual issues and initiating correspondence:
  - receiving and processing contractual correspondence: variations and claims, etc.
  - valuing, preparing, submitting, assessing, negotiating and agreeing additional payment (fees) and/or extensions of time in respect of contractual changes (variation) and claims
  - producing and maintaining financial reports and forecasts
  - arranging insurance provisions
  - valuing, preparing, submitting, evaluating and agreeing interim payments and ensuring timely receipt of payment
- Procuring subcontractors and materials: establishing trading accounts with suppliers and subcontractors
  - managing/administering subcontracts and suppliers
  - receiving and processing subcontract interim payment claims and final accounts
  - liaising with the accounts section to ensure timely payment of monies due to suppliers and subcontractors
- Valuing, preparing, submitting, assessing, negotiating and agreeing final accounts

practitioners in this area of practice has assumed the name International Association for Contract and Commercial Management (IACCM),[4] an acknowledgement of the duality of terms and interpretations. Within this text, however, the term 'commercial' is used.

Commercial management activities

Commercial practitioners engage in a wide range of activities; some of these tasks are identified in Table I.4.

Lowe and Leringer (2005, 2006) established that, within the construction, telecoms/ICT and defence/aerospace sectors, the commercial function was generally responsibility for: contract negotiation, contract formulation, risk management, dispute resolution and bidding. They also found that it provided support to other functions with responsibility for: business plan development, sales, performance measurement, estimating, image/reputation management, innovation management, supply chain management, supplier evaluation, quality assessment, development appraisal and marketing. The function's contribution to cost management, price formulation, payments, cost–value reconciliation, subcontracting administration, cash flow management, value management, the creation of outline proposals, acquiring approvals and permits, procurement strategy and claim formulation, was less certain. It was either accountable for these activities or for supporting other functions responsible for carrying out these tasks.

As expected, Lowe and Leiringer (2005, 2006) found divergence in practice between industry sectors, for example, commercial practitioners in telecoms/ICT were less likely to be involved in resolving disputes than those from defence/aerospace, while cost management, cost value reconciliation, sub-contracting administration and value management were more likely to form part of the commercial role in construction than telecoms/ICT and defence/aerospace. Further, commercial staff in telecoms/ICT were less likely to be involved in payments, claims formation, cash flow management and estimating than those in construction and defence/aerospace; creating outline proposals, and acquiring approvals and permits were less prevalent in construction than telecoms/ICT and defence/aerospace; price formulation formed a greater part of the commercial role in defence/aerospace than construction and telecoms/ICT, while commercial actors in defence/aerospace were more involved in procurement strategy than their counterparts in telecoms/ICT.

Chapter 1 collates and describes the activities undertaken by commercial practitioners, within project-oriented organisations in more detail. It also presents and explains the following model of the activities, processes, capabilities and knowledge that underpin commercial practice (Figure I.1)

Commercial practice, therefore, requires an appreciation of an eclectic mix of academic disciplines and principles, which are applied to the management of projects, programmes and the organisation as a whole.

Context

This text is confined to **business-to-business** (b2b) relationships; that is, the commercial activities and transactions relating to the purchase and supply of products and services between businesses, rather than for personal use. From this perspective the buyer could be a manufacturer, service provider, consultancy, public sector or third sector (not-for-profit/non-profit-making) organisation, etc. The purchase and supply of products and/or services to public sector organisations can be further categorised as **business-to-government** (b2g) transactions (Chapman, 2004; CIM, 2011). Within this context, this book is restricted to project-oriented organisations.

Project-oriented organisations

Although there is no universally accepted definition of the term **project**, there is general agreement that a project is a temporary endeavour undertaken to generate a unique outcome. Additionally, it can be differentiated from other operational activities, by having specific objectives, a novel or distinctive scope of work, unambiguous time, quality and cost parameters and a specific project-oriented organisational form or management structure (Table I.5 lists the most commonly cited definitions). Projects commonly have a recognised life cycle; in the case of a manufacturer of equipment they may comprise the following stages – project

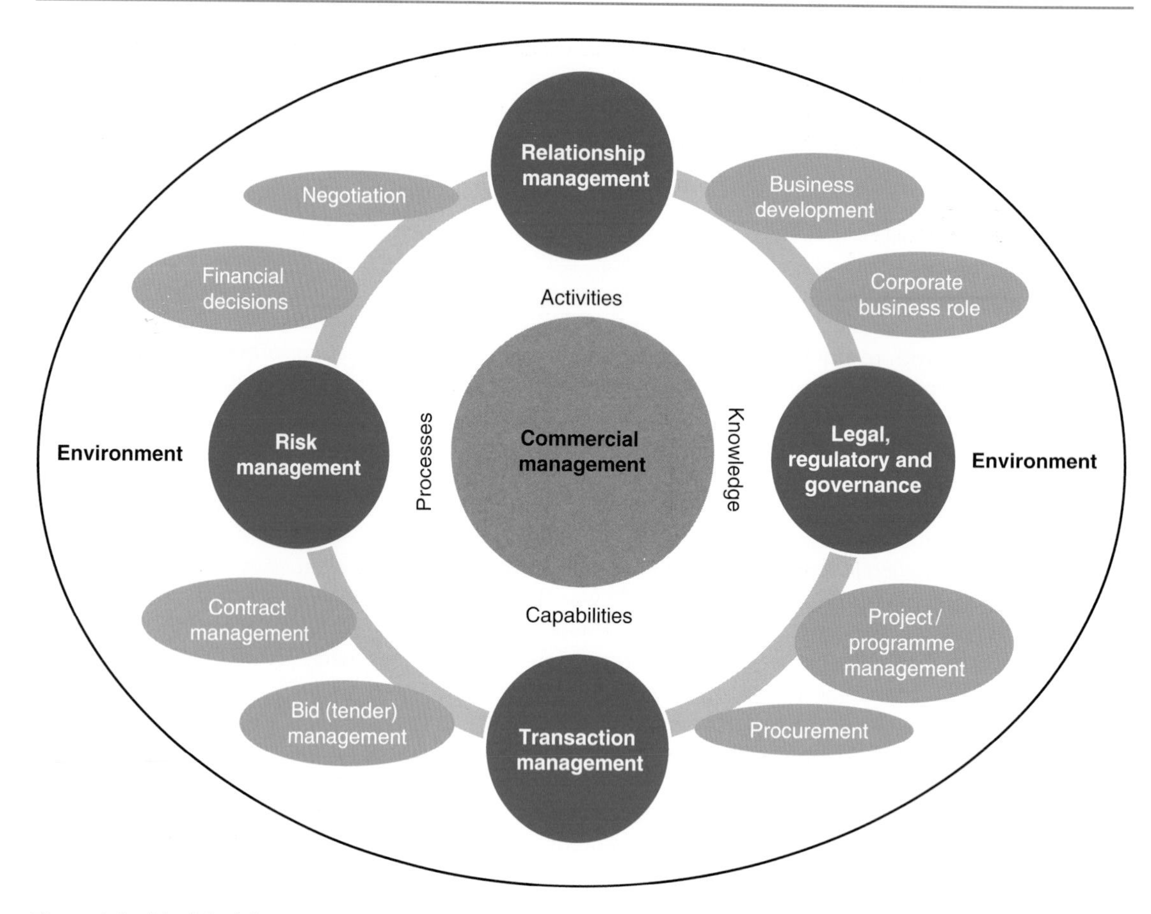

Figure I.1 Model of the activities, processes, capabilities and knowledge that underpin commercial practice.

Table I.5 Definitions of project.

Definition	Source
A project is a time and cost constrained operation to realise a set of defined deliverables (the scope to fulfil the project's objectives) up to quality standards and requirements	International Project Management Association (IPMA) (2006)
A temporary organisation which is created for the purpose of delivering one or more business products according to a specified business case	PRINCE2 (2009)
A unique, transient endeavour undertaken to achieve a desired outcome	The Association for Project Management (APM) (2010)
… unique process, consisting of a set of coordinated and controlled activities with start and finish dates, undertaken to achieve an objective conforming to specific requirements, including constraints of time, cost and resources [BS 6079-2:2000]	The British Standards Institution (BSI) (2000a)

opportunity, commitment, execution, in-service support, serial production and disposal. A project may be a stand-alone activity or a part of a larger programme (a collection of several interconnecting projects).

Frequently, projects are undertaken by multi-firm **integrated project teams** (IPT). An IPT is both an approach and a set of key capabilities assembled to ensure the delivery of a project and its objectives.

It incorporates both supply-side participants, such as suppliers, contractors, consultants, specialists and service providers, and the demand-side (customer) project team. Many IPTs co-locate for the duration of the project to facilitate deeper integration; Terminal 5 (T5) Heathrow (Case Study B) is a good example of an IPT. Alternatively, an **integrated supply team** (IST) is a similar grouping of supply chain organisations but without the integrated involvement of the customer.

Organisational structures

Organisational structures can be placed on a pure function/project-based continuum[5]; Figure I.2 illustrates potential organisational configurations at points along this range. These include:

- **Function-based organisation**: a management structure where individual functions in a business – dedicated services such as finance, legal and, marketing – are organised into specialist groups or departments. Efficiency of operation is considered to be the principal advantage gained by adopting this form of business structure
- **Function-led matrix organisation**: a management configuration where individual projects (contracts) are resourced by the separate business functions, which retain the principal reporting lines to senior management
- **Balanced functional/project matrix organisation**[6]: a management configuration where project and functional management have equal status and reporting routes to the organisation's senior management. The existence of dual reporting lines has the potential to create conflict, dividing an individual's loyalties and responsibilities between the project team and their functional base. Moreover, there is often a tendency for 'power' within a balanced matrix structure to lean towards either the project or function depending upon the strength of personalities, regardless of individuals and the particular circumstances involved
- **Project-led matrix organisation**: an arrangement where the main reporting routes to senior management are via the project (director); which is resourced by the various functional groups. Generally, under this model functional representation is maintained at business unit board level through functional directors (see Figure I.3.).
- **Project-based organisation**: is where a business's management structure is configured exclusively around its ethos: managing by projects.[7]

The key benefits of adopting a project structure are effectiveness and flexibility.

In many organisations, however, rather than selecting a single format, which is applied uniformly across the business, a variety of configurations are adopted, depending on the nature of the operations, provision of services, and types of projects, activities and tasks, etc. undertaken by its constituent parts. The term **project-oriented organisation** has been adopted in this text to reflect this.

Additionally, the term **project orientation** can be used to identify how an organisation coordinates and manages its various projects (portfolios) and also incorporates the drive of an organisation to develop its project management capability.

Examples of 'commercial' in project-oriented organisations

'Commercial' is predominantly seen as a distinct function in its own right, reporting to a commercial director (at either the group or business unit level) or alternatively to the Finance Director or General Counsel. In other instances the activity may reside within marketing and sales, business development, finance, legal and corporate affairs, procurement/purchasing, supply-chain or project/programme management.

In BAE Systems, for example, 'commercial' is acknowledged as a function and business role – an area of business management, comparable to project management, procurement, and sales and marketing. These areas are distinct from its corporate functions, such as: business development, business improvement, communications, finance, health, safety and environment (HS&E), human resources, information technology (IT), legal, quality, secretarial and administration. The function encompasses estimating, bidding, negotiating and managing complex contracts. Each business unit has its own commercial director, who reports to the finance director on BAE Systems' main group board. Similarly, within Rolls-Royce 'commercial' is viewed as a specialist area and a function of their business capability, which also includes: customer

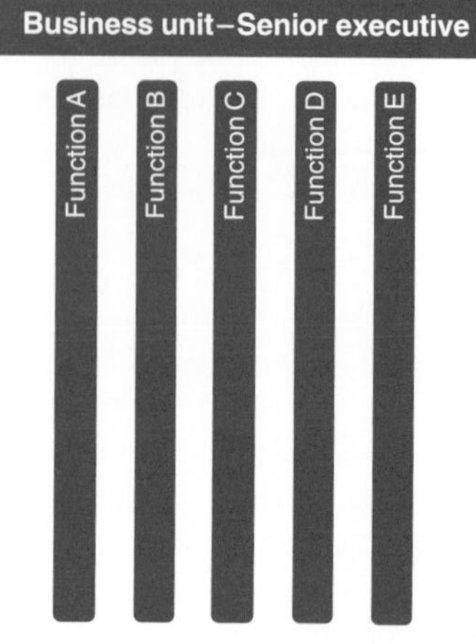

Function-based organisation

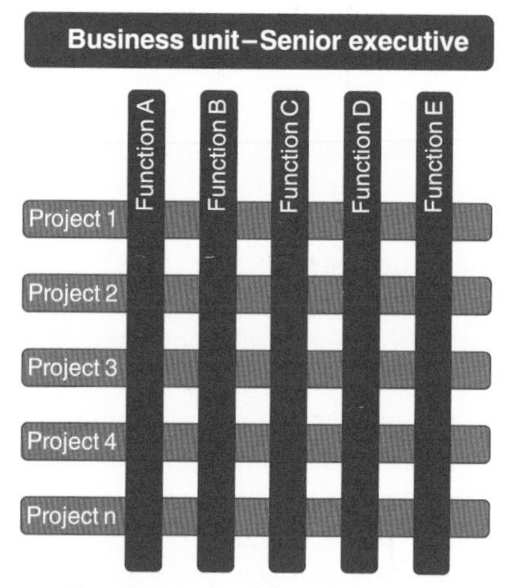

Function-led matrix organisation

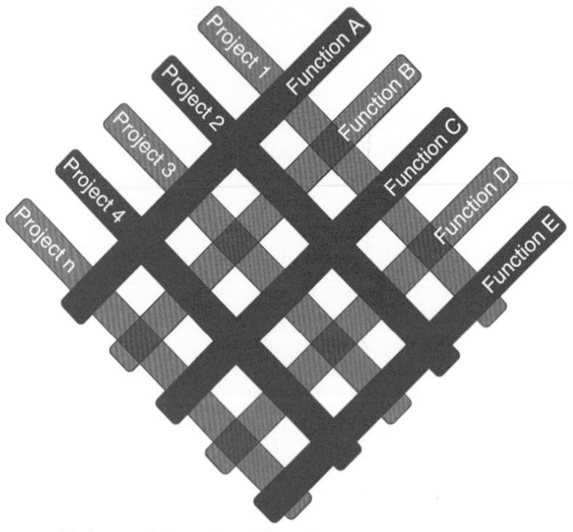

Balanced functional / project matrix organisation

Business unit–Senior executive
Project 1
Project 2
Project 3
Project 4
Project n
Function A
Function B
Function C
Function D
Function E

Project-led matrix organisation

Project-based organisation

Figure I.2 Organisational structures.

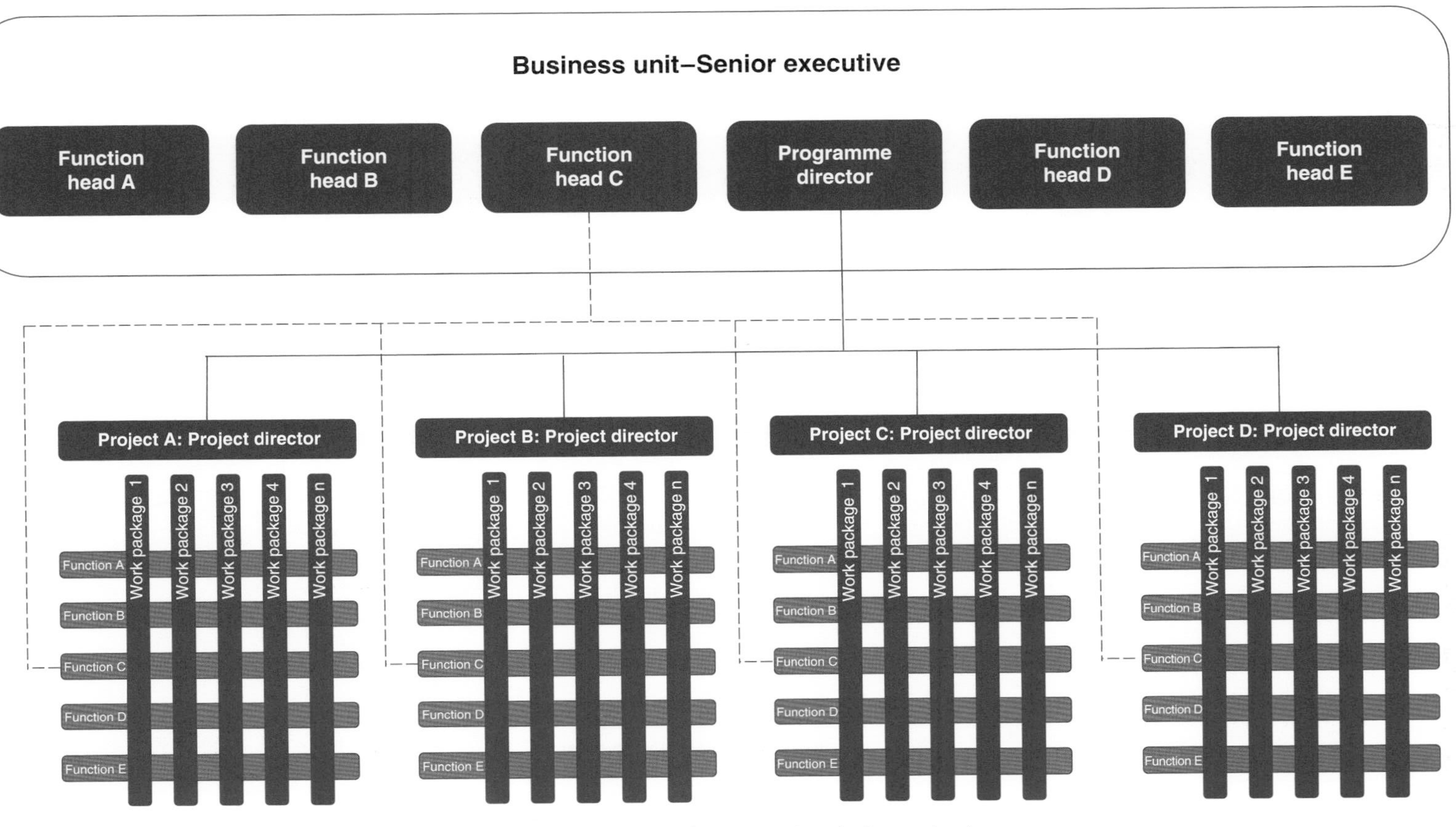

Figure I.3 Dual reporting lines in project-led organisations.

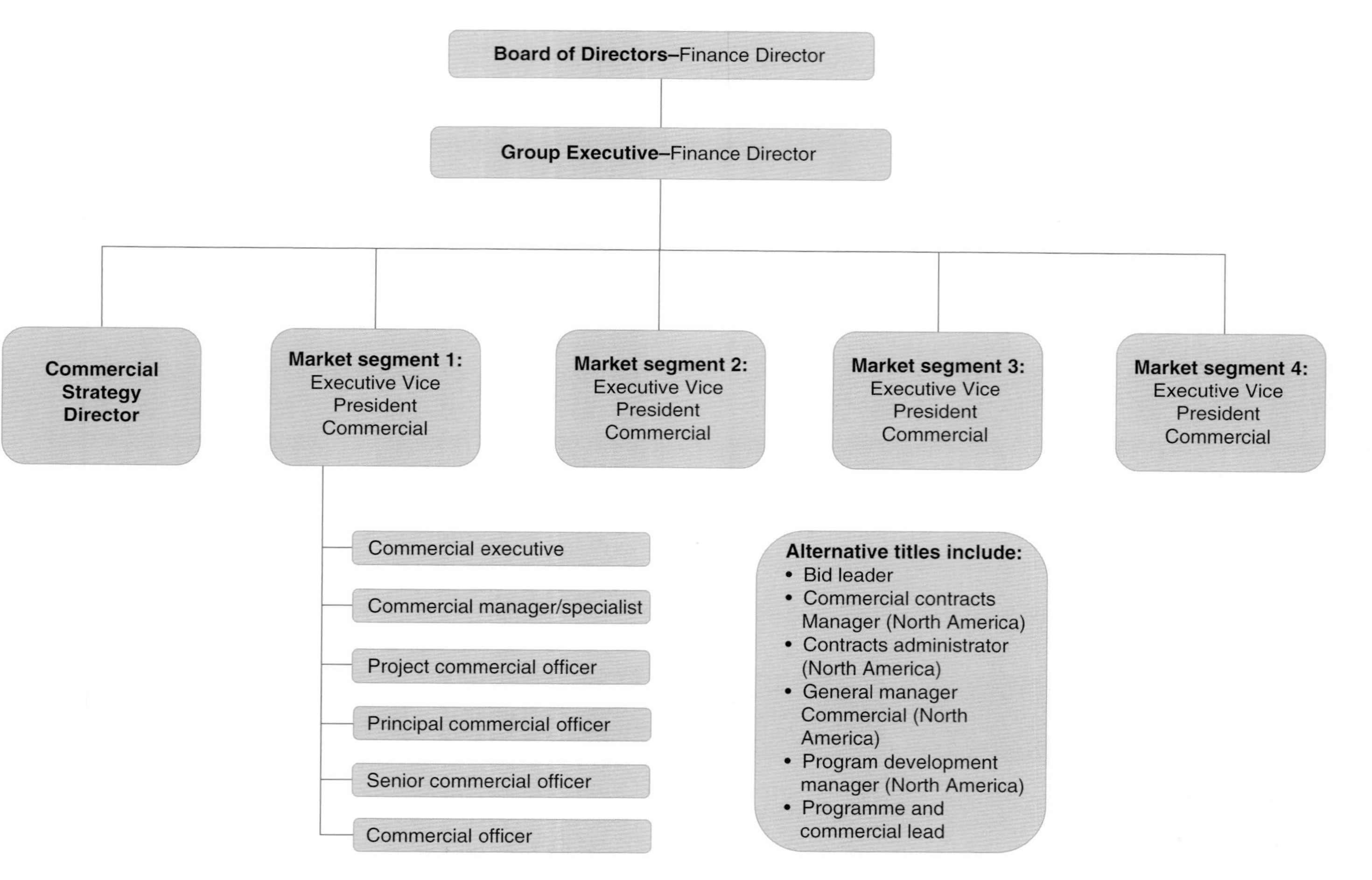

Figure I.4 Commercial job titles and reporting lines within an aerospace/defence organisation.

management, customer services, programme management, purchasing, sales and marketing, and supply chain. Each of Rolls-Royce's five business areas has its own Executive Vice President Commercial (commercial director), who provides functional leadership and direction. As with BAE Systems, 'commercial' in Rolls-Royce reports to the financial director on the main board. Figure I.4 illustrates the commercial job titles and reporting lines in a generic aerospace/defence organisation.

Within BT Global Services, a business unit of BT, 'commercial' is regarded as a function, along with strategy, marketing, propositions, and legal and regulatory. Formerly a group service, the commercial function was distributed across BT's business units in the early 2000s. Similar to BAE Systems and Rolls-Royce, each business unit has its own commercial director, who currently reports to the finance director. Previously the function was overseen by Legal. Box I.1 provides examples of how 'commercial' is configured in project-oriented organisations.

Box I.1 Examples of 'commercial' in project-oriented organisations

Aerospace/defence

BAE Systems

With operations spanning five continents and customers in over 100 countries, BAE Systems is Europe's largest defence and aerospace contractor and the second largest globally, based on revenues. A leader in the area of intelligent defence and aerospace technology, it designs, manufactures and supports military aircraft, surface ships, submarines, radar, avionics, electronic systems, guided weapons and a range of other defence products.

BAE Systems is a project-oriented organisation, divided into several business units, which are responsible for winning commissions and delivering projects. Each business unit has its own commercial capability; regarded as a business function, which operates in integrated teams in conjunction with project management, procurement, marketing and sales, and finance (project accounting, management accounting and financial accounting) personnel. Commercial is a customer-facing role responsible for:

> '... providing our Customers with innovative business solutions to meet their individual needs and to deliver competitive advantage and profitable business whilst ensuring that our contractual obligations are met.'

BAE Systems' commercial population consists of approximately 1000 people in its non-US businesses, plus a similar number within its US operations. Commercial practitioners have dual accountability through functional line management and the project(s) they are assigned to. In the UK commercial role titles, linked to business units, include:

- Commercial Officer
- Senior Commercial Officer
- Principal Commercial Officer
- Commercial Manager
- Commercial Executive
- Commercial Director
- Head of Commercial

and non-UK commercial job titles include:

- Contract(s) Administrator, BAE Systems Inc.
- Lead Contracts Administrator, BAE Systems Inc.

- Senior Lead Contracts Administrator, BAE Systems Inc.
- Contract(s) Administrator Senior, BAE Systems Inc.
- Contract(s) Administrator Principal, BAE Systems Inc.
- Commercial (Contracts) Officer, BAE Systems Australia
- Contracts Manager, BAE Systems Saudi Arabia, BAE Systems Inc.
- Contracts Management Specialists, BAE Systems Australia
- Director of Contracts, BAE Systems Inc. (an executive role: reporting to a 'sector' Vice President of Contracts)
- Vice President of Contracts, BAE Systems Inc. (an executive 'sector' role: reporting to the Inc. Vice President of Contracts)
- Inc. VP of Contracts, BAE Systems Inc.

High level support for the function (non-US business) is provided by a Head of Commercial, who is responsibility for learning and development, policy and process and corporate governance.

Rolls-Royce

Rolls-Royce is a global supplier of integrated power systems for land, sea and air-based applications with a physical presence in over 50 countries. A project-oriented organisation operating across five business areas: civil aerospace, defence aerospace, marine, energy and nuclear, its annual underlying revenues in 2010 were over £10.8 billion, of which more than half were generated from the provision of services. Rolls-Royce's customer base includes over 500 airlines, 4000 corporate and utility aircraft and helicopter operators, 160 armed forces, over 2500 marine customers (including 70 navies), and energy customers in almost 120 countries.

Each business area has its own customer focused commercial function; considered a specialist business capability along with: customer management, customer services, programme management, purchasing, sales and marketing, supply chain. Organised in teams, the aim of the function is to:

> '... enhance our reputation and ensure we provide maximum customer satisfaction. They do this by building strong relationships with our customers and third parties, and by influencing the decisions we make on everything from sales strategies to the management of company accounts.'

Commercial, therefore, is involved throughout the contract life cycle in all aspects of Rolls-Royce's business transactions with customers and suppliers, for new and existing contracts. Individual roles are typically subdivided into pre- and post-award activities. The capability has both functional responsibilities and project(s) accountabilities, reporting to executive vice presidents (formerly referred to as directors) on each of the business boards and ultimately to the finance director on the main board. In the UK, role titles include:

- Commercial Officer
- Senior Commercial Officer
- Principal Commercial Officer
- Project Commercial Officer
- Commercial Manager/Specialist
- Commercial Executive
- Executive VP Commercial (Commercial Director)

and in North America commercial job titles include:

- Bid Leader
- Commercial Contracts Manager
- Contracts Administrator, Senior Specialist
- Customer Business
- General Manager Commercial
- Program Development Manager

Reporting directly to the main board finance director, Rolls-Royce has recently appointed a Commercial Strategy Director, responsible for developing the commercial capability across its five business areas. The post holder has also been tasked with rationalising commercial processes, procedures and systems, and learning solutions across the organisation.

ICT/communications

BT

A world-leading communications services company, BT is active in 200 countries supplying integrated information and communications technology solutions to multinational corporations, domestic business and government departments. Its key activities include the provision of fixed-line services, broadband, mobile and TV products and services in addition to networked IT services.

BT is made up of four principal 'customer-facing' lines of business: BT Global Services, BT Retail, BT Wholesale and BT Openreach, which in turn are supported by two internal service units BT Innovate and Design and BT Operate. Each business has its own dedicated commercial function, which reports to the finance director.

Within BT, commercial managers provide contractual and business support and expertise to all aspects of its customer-facing activities and other channels to market. They are involved in a variety of diverse business activities: from marketing and product launches to major government and commercial outsourcing bids. This involves them in, for example, setting BT's commercial tone and policy, and providing a 'one-stop shop' for commercial issues. Additionally, the commercial function delivers innovative commercial solutions, which seeks to provide benefits to BT, furthering its group and business units' business interests, and good service to its customers. Commercial role titles include:

- Commercial Analyst
- Commercial Manager
- Commercial Contracts Manager
- Principal Commercial Manager
- Head of Commercial Management
- Head of Legal and Commercial
- Head of Commercial Management (BT Group)

Alternative commercially focused roles within BT include:

- Contract Manager
- Corporate and Government Solutions Product Manager
- Deal Architect
- Senior Deal Architect
- Senior Commercial Lawyer

Fujitsu Services Limited

Fujitsu Services (part of the global Fujitsu Group) is a leading IT services company operating in 20 countries across Europe, the Middle East and Africa. With an annual income of £2.46 billion (€3.59 billion), its activities include: consulting, applications, systems integration, managed services and product for customers. It designs, builds and operates IT systems and services for major public and private sector clients in its business markets: aviation, defence and national security, financial services, government, healthcare, local government, manufacturing and services, media, rail, retail, telecommunications and utilities. Utilising project and programme management, Fujitsu Services' business approach is based on developing enduring relationships built on success and mutual benefit.

Commercial personnel operate in teams within Fujitsu Services and are viewed as a discipline and community, together with account management, business consultancy, business services, finance, general management, global supply chain, human resources, legal, marketing, operations, procurement, project management, sales, service delivery, software and solution development, technical architecture and consulting. Reporting to commercial directors within the each business area, the discipline ultimately reports to General Council on the main board.

Engaged throughout the contract life cycle, and in both customer- and supplier-facing activities, commercial is responsible for driving the agreement of successful deals; writing and negotiating contracts; and ensuring that contracts entered into are managed professionally and profitably, in compliance with their terms and conditions, while delivering added value to Fujitsu's customers and enabling them to realise their objectives. Similarly, the discipline ensures that the organisation obtains the 'best deals' with its partners and suppliers. In Fujitsu Services commercial role titles include:

- Commercial Manager
- Senior Commercial Manager
- Head of Commercial (Specific Accounts)
- Head of Commercial at Fujitsu Services Limited
- Commercial Director

Construction

Carillion

With operations in the UK, Europe, Canada, the Middle East, North Africa and the Caribbean, Carillion plc is one of the UK's premier support services and construction companies. Its business units comprise: rail, highway maintenance, civil engineering, construction recruitment (skyblue), building, property services, defence, education, health, facilities management, planned maintenance, fleet management, TPS (specialist consultancy services), Canada and Caribbean, Middle East and North Africa, developments, private finance, and environmental and health and safety solutions (Enviros Consulting). Working across commercial property, defence, education, health, rail, regeneration, highway maintenance, facilities management, civil engineering, international, nuclear, secure (the design, construction and maintenance of prisons etc), fleet management, and utility services sectors, Carillion generates annual revenues in the region of £5 billion.

Utilising collaborative supply chains, Carillion is a project-oriented organisation, which delivers integrated, customer-focused solutions, from conception to ongoing facilities management, and support services.

In Carillion commercial role titles include:

- Quantity Surveyor
- Senior Quantity Surveyor

- Commercial Manager
- Commercial Finance Manager
- Commercial Director

Reporting to Commercial Directors in each business unit, the Commercial Manager's role is to ensure that all projects meet their contractual and commercial requirements. Also, in conjunction with other 'financial' groups in the business, they support the winning and monitoring of contracts.

Laing O'Rouke

Laing O'Rourke is the UK's largest privately owned construction solutions provider, with revenues in 2010 of £4.3 billion. Operating from two geographic bases – Europe, Middle East and Rest of World; and Australia and South East Asia – Laing O'Rourke's activities cover eight core sectors: lifestyle, business, social infrastructure, transport, power, mining and natural resources, oil and gas, and utilities and waste. The organisation comprises a number of engineering, construction and specialist services companies, which come together to offer a combined investment, development and management competency. Each of Laing O'Rouke's two international divisions is overseen by a management board, which reports to the executive board.

A project-focused organisation, Laing O'Rourke invests in, designs, engineers, constructs and manages major capital infrastructure and building assets. It possesses capabilities in all facets of implementation and project management, throughout an asset's life cycle. Adopting a collaborative approach, 'Complete Thinking', Laing O'Rourke provides integrated, multidisciplinary capabilities, services and solutions to projects from their conception to completion.

Commercial services (comprising estimating, procurement, commercial, cost and time management) in Laing O'Rourke is considered a discipline and function, along with engineering and design, construction management and project leadership. Commercial role titles include:

- Quantity Surveyor
- Senior Quantity Surveyor
- Contracts Administration Assistant
- Senior Contracts Administrator
- Assistant Commercial Manager
- Commercial Manager
- Senior Commercial Manager
- Commercial Leader
- Commercial Director

Working in integrated project teams, the commercial discipline is ultimately overseen by the Group Finance and Commercial Director, who is responsible for the organisation's finance and commercial activities, in addition to supporting business development, 'work-winning' activities and group strategy.

Sources: http://www.baesystems.com and BAE Systems Commercial Graduate Development Framework http://www.baesystems.com/BAEProd/groups/public/documents/ss_asset/bae_grad_rolelinks_com_pdf.pdf; http://www.rolls-royce.com; http://www.btplc.com/thegroup/ourcompany/index.htm; http://www.fujitsu.com and www.uk.fujitsu.com; http://www.carillionplc.com (accessed 20th March 2011)

Overview of book

The aim of this book is to provide a framework for understanding commercial practice within project-oriented organisations. Additionally, it seeks to identify generic aspects of this practice and to provide a theoretical foundation to these activities, by reference to existing and emergent theories and concepts, and to examine relevant management best practice. It adopts the following structure:

Part 1: Introduction

Chapter 1: Commercial Management in Project-Oriented Organisations

Commercial Management in Project-Oriented Organisations further explores the nature of commercial practice within project environments at the buyer–seller interface. It presents a commercial management framework, which illustrates the multiple interactions and connections between the purchaser's procurement cycle and a supplier's bidding and implementation cycles. Additionally, it outlines the principal activities undertaken by the commercial function, identifies the skills and abilities that support these activities and reviews the theories and concepts that underpin commercial practice. Finally, it identifies areas of commonality of practice with other functions found within project-oriented organisation, sources of potential conflict and misunderstanding.

Part 2: Elements of Commercial Practice and Theory

Chapter 2: Commercial Leadership

Maria-Christina Stafylarakis

Maria-Christina Stafylarakis in ***Commercial Leadership*** conceptualises leadership, and in particular commercial leadership, as a social process that is principally concerned with influencing, managing relationships, and educating multiple stakeholders: the aim being to deliver organisational goals through projects and programmes within the context of contractual commitments. The chapter explores key themes and concepts from the leadership literature. It reviews established approaches to the study of leadership, for example, trait, style and contingency theories; examines contemporary theories, such as, transformational and dispersed leadership; and explores leadership in relation to learning. The chapter concludes with the exposition of a conceptual framework that views leadership as a process involving a dynamic interaction between a task focus and learning.

Chapter 3: Exploring Strategy

Irene Roele

While acknowledging the importance of strategy implementation, in ***Exploring Strategy*** Irene Roele focuses on the orientation aspect of strategy. The chapter views strategising as an ongoing, dynamic process, and considers strategy to be a dialogue or strategic conversation, which is not solely restricted to the members of an organisation's senior leadership team. Its aim is to assist commercial practitioners: identify the strategic agenda – this requires an understanding of the complexity of the industry dynamics, and the ability to identify value; define a clear sense of purpose and appreciate the need for a guiding policy for the firm in creating value; and clearly articulate its core value-proposition.

Chapter 4: Perspectives on Managing Risk and Uncertainty

Eunice Maytorena

As identified earlier in this introduction, the commercial function, contributes significantly to the management of risk and the governance role in organisations. It is responsible for implementing risk management and ensuring the application of regulatory and control processes and procedures. In ***Perspectives on Managing Risk and Uncertainty*** Eunice Maytorena provides an introduction to contemporary project risk management. It introduces the concept of risk and uncertainty, providing definitions and descriptions. Additionally, it evaluates a number of perspectives on risk and provides an overview of the project risk management process, commonly used tools and techniques.

Chapter 5: Financial Decisions

Anne Stafford

Anne Stafford's chapter, ***Financial Decisions***, provides an introduction to financial decision-making. Its aim is to equip commercial practitioners with the knowledge and ability to appreciate the financial information issued by their own organisation, partners, suppliers and customers, and their role in generating this data.

The chapter addresses the nature and purpose of financial information, the users of financial information, and the financial environment within which companies operate. It outlines the function of the principal financial statements and the accounting conventions that lie behind their production. Crucially, it considers how managers can employ financial information to support financial decision-making, via an awareness of cost behaviour and the application of techniques, such as break-even analysis, performance measurement and cash flow management. The chapter concludes with an explanation of investment appraisal methods, for instance, discounted cash flow.

Chapter 6: Legal Issues in Contracting

David Lowe and Edward Davies

In ***Legal Issues in Contracting*** David Lowe and Edward Davies seek to develop an appreciation of some of the key legal issues that influence commercial practice. The intention is to provide a selective overview of contract provisions and procedures, together with an international perspective of the legalisation and regulations that affects b2b exchanges within a project and programmes environment. The chapter presents an outline of commercial contracts for the supply of goods and/or services in terms of generic contract provisions: the key components of contracts. To illustrate these principles, reference is made, where appropriate, to a number of industry standard form contracts. It also provides an introduction to procurement/ acquisition regulations, competition and anti-trust legislation, international trade law, intellectual property and freedom of information, insolvency and employment rights on the Transfer of an Undertaking (TUPE).

Part 3: Approaches to Commercial Practice

Chapter 7: Best-Practice Management

Major product (asset) and service (capability) acquisition projects impose significant demands on both the purchaser and supplier; however, project failure in many cases can be attributed to inadequate demand-side management. Chapter 7 identifies and outlines 'best practice' in managing projects and programmes and in particular those aspects identified in Chapter 1 as being pertinent to commercial practitioners. These includes governance issues associated with the Turnbull recommendations and the OGC Gateway Process, the use of process protocols, such as, PRINCE2, MSP, MoP, and MoV, plus guidance on best practice in procurement and contract management.

Chapter 8: Commercial Strategies and Tactics

This chapter examines commercial practice across the project life cycle from both the supply and demand perspective: identifying the interrelationship between the demand-side procurement process (cycle) and the supply-side bidding and implementation cycles. ***Commercial Strategies and Tactics*** explores how purchasers can effectively articulate their requirements for new asset and services, and then manage the process of asset/service definition and delivery successfully. From the demand side, it addresses how suppliers can effectively identify prospective opportunities, develop and communicate their proposals for the supply of asset and services, manage the deal creation stage and ensure that commitments made are delivered profitably. It is subdivided into three sections:

A. Intent
   Part A ***Intent*** from the demand side addresses **Requirement Identification**, **Requirement Specification** and **Solution Selection**, while from the supply side it covers **Opportunity Identification**, **Opportunity Development** and **Proposition Identification**.
B. Deal Creation
   Part B ***Deal Creation*** focuses on the interrelated process and actions undertaken by commercial practitioners that bring an agreement into existence. From the demand side it addresses **Asset/Service Procurement**, implementing the selected sourcing (procurement) strategy and negotiating the agreement, while from the supply side it covers **Proposition Development**, creating, preparing, submitting and negotiating a proposal in respect of the supply of a product and/or service.

C. Execution

Finally, Part C ***Execution*** addresses commercial practice, again from both the supply and demand perspective, involved in implementing, delivering and fulfilling commitments made during the deal creation stage. It outlines the principle aspects of **Contract Management**: the monitoring and control of the contract to ensure that the agreement between the purchaser and supplier operates efficiently and effectively, and that all contractual obligations are appropriately discharged. Part C concludes with an explanation of the commercial activities involved in asset disposal and service termination.

Part 4: Case Studies

Two extended case studies are provided. Using the commercial management framework presented in this introduction, the case studies illustrate commercial practice during the life cycle of four prestigious projects.

Case Study A: Football Stadia

Recently, several major stadium developments have been constructed in the UK, forming the centrepiece for urban redevelopment. The procurement and implementation of these projects exemplify elements of both 'best' and 'poor' practice, as illustrated by the following: the Millennium Stadium, Cardiff, the Emirates Stadium, London and Wembley Stadium, London.

Case Study B: Terminal 5 (T5) Heathrow

The Terminal 5 (T5) programme at Heathrow airport was at the time of its construction Europe's largest and most complex construction project. Taking over 18-and-a-half years to develop, it was delivered on time and to budget. And, despite some initial teething problems that included flight delays, technical hitches, baggage-handling issues, and protest action, T5 is deemed to have been one of the UK's most successful construction programmes.

The contracting strategy adopted by BAA for the project went beyond any previous arrangement implemented in the UK, redefining BAA's role as project client: participating fully in the delivery process; adopting a proactive approach to risk; managing the cause of risk, not its effect; and establishing a set of behaviours to facilitate innovative problem-solving. BAA's strategy embraced four key principles:

1. The client always bears and pays for the risk
2. BAA retained full liability for all project risks
3. Suppliers' profit levels were predetermined and fixed
4. Partners are worth more than suppliers

These principles were incorporated into what was branded the 'T5 Agreement': a bespoke project-specific contract developed by BAA to govern its relationships with first-tier suppliers and designed to create a framework to deliver a project success.

Summary

This introduction has presented definitions of the terms 'commercial management' and 'commercial manager', comparing the former with 'contract management' and commenting on the variance in use of these terms globally. Common activities undertaken by commercial practitioners are identified.

Further reading

Lowe D and Leiringer R (2006) Chapter 1: Commercial management – defining a discipline? In: DJ Lowe (ed.) with R Leiringer, *Commercial Management of Projects: Defining the Discipline*. Blackwell Publishing, Oxford. pp. 192–206

Adopting a b2b and project-oriented organisational perspective, it has identified five organisational structures on a function/project-based continuum and provided examples of how global, project-oriented organisations configure their 'commercial' function. Additionally, it has provided an overview of the format and context of the book.

Endnotes

1 The book is informed by experience gained through developing and delivering undergraduate, postgraduate and executive education commercial management programmes and through research into commercial practice in a variety of industry sectors, including: construction, ICT, aerospace and defence. Specifically, it is substantiated by content analysis of 90 job/function specification and competency modules.

Supporting research was undertaken at the University of Manchester in support of the MSc in Commercial Management (now renamed MSc in International Commercial and Contract Management), MBA for Commercial Executives, BSc in Commercial Management and Quantity Surveying. Investigations sought to:

1. Establish the role of commercial managers in the construction sector: reported in Lowe *et al.* (1997).
2. Compare the role of commercial managers across a range of organisations and industry sectors including: construction, telecoms/ICT and defence/aerospace: reported in Lowe and Leiringer (2005, 2006) and Lowe (2006).
3. Establish a body of knowledge for commercial management: reported in Lowe (2006, 2008b) and Lowe *et al.* (2006).
4. Establish practitioner expectations of a specialist MBA for commercial executives: reported in Lowe (2008a).
5. Analyse 90 commercial functional frameworks/job specifications: previously unreported work.

2 As outlined in IACCM's definition of Commercial Management.

3 Commercial managers often have the responsibility for approving (signing off or authorising) bids to a designated value, or for ensuring that bids are approved by the appropriate person within the business unit or company.

4 See IACCMs website: http://www.iaccm.com/

5 See, for example: Larson and Gobeli (1987, 1989), Hobday (2000).

6 I take issue here with the classification of project-based organisations proposed by Hobday (2000). His balanced matrix clearly shows the function as the primary communication route to senior management, while under his project matrix structure project managers and functional managers are deemed to have equal status, both reporting directly to senior management. I find the label project matrix unhelpful, preferring balanced functional/project matrix organisation.Additionally, hobday's functional matrix and balanced matrix have been replaced by the function-led matrix organisation: the inverse of the project-led matrix organisation.

7 See, for example: Turner and Keegan (1999, 2001); Gareis (1989); Gann and Salter (2000); Hobday (2000); Westerveld (2003); Sydow *et al.* (2004); Blindenbach-Driessen and van den Ende (2006); Hyväri (2006); Maylor *et al.* (2006); Aubry *et al.* (2007); Thiry and Deguire (2007); Lindkvist (2008); Huemann (2010); and Kujala *et al.* (2010).

References

APM (2010) *Glossary*. The Association for Project Management. http://www.apm5dimensions.com/Definitions.asp (accessed 25 March 2011)

Aubry, M, Hobbs, B and Thuillier, D (2007) A new framework for understanding organisational project management through the PMO. *International Journal of Project Management*, 25, 328–336

Blindenbach-Driessen, F and van den Ende, J (2006) Innovation in project-based firms: the context dependency of success factors. *Research Policy*, 35, 545–561

BSI (2000) *BS 6079-2:2000 (Incorporating Corrigendum No. 1): Project Management - Part 2: Vocabulary*. The British Standards Institution, London

Chapman, A (2004) *Business Contracts Legal Terms and Definitions Glossary*. Alan Chapman/Crown Copyright. http://www.businessballs.com/businesscontractstermsdefinitionsglossary.htm (accessed 31 March 2011)

CIM (2011) *The Chartered Institute of Marketing: Resource Glossary*. http://www.cim.co.uk/resources/glossary/home.aspx (accessed 31 March 2011)

4ps (2007) A guide to contract management for PFI and PPP projects, Public Private Partnerships Programme, London. http://www.localpartnerships.org.uk/UserFiles/File/Publications/4ps%20ContractManagers%20guideFINAL.pdf (accessed 22 August 2012)

Gann, DM and Salter, AJ (2000) Innovation in project-based, service-enhanced firms: the construction of complex products and systems. *Research Policy*, 29, 955–972

Gareis, R (1989) 'Management by Projects': the management approach of the future. *Project Management*, 7, 243–249

Hobday, M (2000) The project-based organisation: an ideal form for managing complex products and systems? *Research Policy*, 29, 871–893

Huemann, M (2010) Considering human resource management when developing a project-oriented company: case study of a telecommunication company. *International Journal of Project Management*, 28, 361–369

Hyväri, I (2006) Project management effectiveness in project-oriented business organizations. *International Journal of Project Management*, 24, 216–225

IACCM (ND) *What is the definition of commercial management?* IACCM, Ridgefield, USA. http://www.iaccm.com/about/faq/?questionid=15 (accessed 22 August 2012)

IPMA (2006) *ICB - IPMA Competence Baseline Version 3.0* International Project Management Association, Nijkerk, The Netherlands

Kujala, S, Artto, KI, Aaltonen, P and Turkulainen, V (2010) Business models in project-based firms – towards a typology of solution-specific business models. *International Journal of Project Management*, 28, 96–106

Larson, EW and Gobeli, DH (1987) Matrix management: contradictions and insights. *Californian Management Review*, 29, 126–138

Larson, EW and Gobeli, DH (1989) Significance of project management structure on development success. *IEEE Transactions on Engineering Management*, 36, 119–125

Lindkvist, L (2008) Project organization: exploring its adaptation properties. *International Journal of Project Management*, 26, 13–20

Lowe, DJ (2006) Establishing a body of knowledge and research agenda for commercial management. In: In: DJ Lowe (ed.) *Proceedings IACCM International Academic Symposium on Commercial Management*, Ascot, UK. University of Manchester. ISBN 0-9547918-2-7. pp. 1–13

Lowe, DJ (2008a) Commercial and contract management – practitioner expectations on the content of an MBA for commercial executives. In: DJ Lowe (ed.) *Proceedings IACCM International Academic Symposiums on Contract & Commercial Management*, London, UK and Fort McDowell, Arizona. ISBN 978-1-934697-01-6. pp. 45–72

Lowe, DJ (2008b) *Commercial and Contract Management – Underlying Theoretical Concepts & Models.* In: DJ Lowe (ed.) *Proceedings IACCM International Academic Symposiums on Contract & Commercial Management*, London, UK and Fort McDowell, Arizona. ISBN 978-1-934697-01-6. pp. 25–44

Lowe, DJ and Leiringer, R (2005) Commercial Management in Project-based Organisations. Special Edition on Commercial Management. *Journal of Financial Management of Property and Construction*, 10, 4–18

Lowe, DJ and Leiringer, R (2006) Introduction: commercial management – defining a discipline? In: DJ Lowe (ed.) with R Leiringer, *Commercial Management of Projects: Defining the Discipline*. Blackwell Publishing, Oxford. pp. 1–17

Lowe, DJ with Leiringer, R (2006) *Commercial Management of Projects: Defining the Discipline*. Blackwell Publishing, Oxford

Lowe, DJ, Fenn, P and Roberts, S (1997) Commercial management: an investigation into the role of the commercial manager within the UK Construction Industry. *CIOB Construction Papers*, 81, 1–8.

Lowe, DJ, Garcia, J and Kawamoto, K (2006) Contract and commercial management: an international practitioner survey. In: DJ Lowe (ed.) *Proceedings IACCM International Academic Symposium on Commercial Management*, Ascot, UK. University of Manchester. ISBN 0-9547918-2-7. pp. 53–57

Maylor, H, Brady, T, Cooke-Davies, T and Hodgson, D (2006) From projectification to programmification. *International Journal of Project Management*, 24, 663–674

NCMA (2011) *About the Profession*. National Contract Management Association. http://www.ncmahq.org/About/content.cfm?ItemNumber=993 (accessed 22 August 2012)

OGC (2002) *Principles for Service Contracts: Contract Management Guidelines*. The Office of Government Commerce/HMSO, London

OGC (2009) *Policy Principles: Contract Management*. The Office of Government Commerce/HMSO, London

PRINCE2 (2009) *Glossaries/Acronyms*. The Office of Government Commerce/HMSO, London. http://www.prince2.com/prince2-downloads.asp (accessed 25 March 2011)

Sydow, J, Lindkvist, L and DeFillippi, R (2004) Project-based organizations, embeddedness and repositories of knowledge: editorial. *Organization Studies*, 25,1475–1489

Thiry, M and Deguire, M (2007) Recent developments in project-based organisations. *International Journal of Project Management*, 25, 649–658

Turner, JR and Keegan, A (1999) The versatile project-based organization: governance and operational control. *European Management Journal*, 17, 296–309

Turner, JR and Keegan, A (2001) Mechanisms of governance in the project-based organization: roles of the broker and steward. *European Management Journal*, 19, 254–267

VGPB (2011) Achieving Excellence in Government Procurement: Glossary. http://www.vgpb.vic.gov.au/CA2575BA0001417C/pages/glossary (accessed 22 August 2012)

Westerveld, E (2003) The Project Excellence Model: linking success criteria and critical success factors. *International Journal of Project Management*, 21, 411–418

Chapter 1

Commercial Management in Project-Oriented Organisations

Learning outcomes

When you have completed this chapter you will be able to:

- Define the terms commercial management and commercial manager
- Identify the common activities undertaken by commercial managers
- Identify the skills and abilities that support these activities
- Identify the theories and concepts that underpin commercial practice
  - legal
  - strategy: value, activity-based theory of the firm, competitive advantage
  - managerial economics: transactional cost economics
  - marketing: make or buy decision
  - organisational behaviour
- Understand the relationship between the purchaser's procurement procedures and the supplier's bid cycle

Introduction

Commercial awareness (acumen)

All members of an organisation should be commercially aware – that is, have a basic understanding of the economics of business: the benefits and realities of an economic exchange from both the supplier's and the buyer's perspectives (UCAS, ND). They should also have an appreciation of the environment within which this transaction occurs. A key aspect is a broad understanding of its clients and/or suppliers:

> '... identifying with them, and helping them achieve their commercial objectives - their strategy. It's about understanding their culture and using their language. This requires an interest in, and liking for, business and what it's about.' (Stoakes, 2011)

We talk about people being commercially astute, minded or aware, etc. The implication is that this aspect is more than a generic skill or competency – it is something that is related to the individual's mind-set (approach, attitude, way of thinking, outlook); view point (perspective, standpoint, opinion, belief, ethos, feelings, thought); behaviour (manner, stance, position); and predisposition (inclination, propensity), which is innate (natural, inherent, or intrinsic).

Commercial Management: theory and practice, First Edition. David Lowe.

This chapter further explores the nature of commercial practice within project-oriented organisations at the buyer–seller interface. It presents a commercial management framework that illustrates the multiple interactions and connections between the purchaser's procurement cycle and a supplier's bidding and implementation cycles. Additionally, it outlines the principal activities undertaken by the commercial function, identifies the skills and abilities that support these activities and reviews the theories and concepts that underpin commercial practice.

What is commercial management?

Although an activity (discipline/function) labelled **commercial** clearly exists within many organisations, an elementary search on Google for commercial manager/director positions will generate numerous hits. However, a similar search within the academic domain will be somewhat less successful. It seems sensible, therefore, to start with a definition of the subject of this text – **commercial management** – however, the endeavour is fraught with potential difficulties.

The *Oxford Online Dictionaries*[1] define **commercial** as:

> '... concerned with or engaged in **commerce** [... the activity of **buying and selling**, especially on a large scale]... making or intended to make a **profit**: having profit rather than artistic or other value as a primary aim.'

The word 'commercial' is derived from the French *commerce* or Latin *commercium* (of trade, trading) a compound of *com-* (together) plus *mercium* (from *merx, merc-* merchandise), the root of merchant. The term is, therefore, directly related to trade: the action of buying and selling goods (assets, items or products) and services, involving an exchange of one commodity for another, which is generally referred to as a commercial transaction. Similarly, a transaction *is an instance of buying or selling something: the action of conducting business* – hence a commercial transaction.

Commercial is often used interchangeably with **business**. However, the word business, which concerns an individual's *regular occupation, profession, or trade*, has a different etymology, originating from the Old English *bisignis* meaning anxiety. Initially used to convey a *state of being busy*, its current usage denoting an *appointed task* dates from late Middle English. Likewise, trade can also be used in the sense of a person's *habitual practice of an occupation* (although historically, this was often used in a pejorative sense) or to describe a skilled artisan (tradesman). Business also relates to the activity undertaken by an individual or organisation.[2]

Etymologically, therefore, commercial is clearly associated with the activity of **buying and selling;** commercial actors supporting the associated commercial transaction (interaction, agreement [contract] and exchange process) between the parties: the buyer and seller. The term transactor, a derivative of transact, either with or without the prefix commercial, is a useful descriptor for the commercial actor; interestingly, it is rarely used. The exception to this is, within the UK public sector, where it is used to describe an individual appointed to advise procuring bodies on the commercial and contractual implications of procuring goods and services using the Public Finance Initiative (PFI) approach.

The principal participants in the exchange are generally referred to as the **buyer** and **seller**.

Buyer

A buyer is an organisation, function or individual employed within an organisation to select and purchase goods, facilities, products and/or services[3] on behalf of the organisation from suppliers and contractors. The term can be used in relation to individuals who merely requisition products and services from pre-established contracts, through to senior executives responsible for making key **value-for-money** decisions regarding major procurement projects, which are then implemented by dedicated procurement personnel (VGPB, 2011). The designations **client**, initially applied in respect to the users of professional services, **customer** and **purchaser** are frequently used interchangeably with buyer to denote an actual or prospective buying organisation (OGC, 2002) or group/individual assigned to make the purchasing decision (less frequently, the term **vendee** is also used). Although, the PRINCE2 glossary of terms defines a customer as:

> '... the person or group who commissioned the work and will benefit from the end results.' (PRINCE2, 2009)

The term **buying centre** is often used to describe the group of individuals (decision-makers), directly involved in the purchase of products or services, whereas **procuring party** refers to the division of an organisation responsible for purchasing products and/or services, frequently possessing delegated authority to sign any resulting contracts on completion of the process on behalf of the organisation (CBI, 2006). Allied terms include:

- **Intelligent customer**: a competency (capability, skills, knowledge and experience) within a buying organisation relating to an understanding of both its own and its suppliers' business aims, needs and capabilities, with the objective of facilitating a successful trading relationship and contract (OGC, 2002)
- **Sponsor**: the principal motivating influence of a project or programme, defined by (APM, 2010) as:

  > 'The individual or body for whom the project is undertaken and who is the primary risk taker. The sponsor owns the business case and is ultimately responsible for the project and for delivering the benefits.' (APM, 2010)

- **User(s)/end users**: are the intended beneficiaries (an individual or group) of the purchased product and/or services: those who will actually use or exploit the output of these products and/or services

A client, customer and user can be external or internal to the organisation.

Seller

A **seller** is a person or organisation that sells something – that is, provides or transfers something in exchange for money (or some other tangible benefit). In a b2b context, sellers are generally called suppliers, contractors or service providers. The term supplier, for example, refers to an entity (an individual, organisation, company or consortium) that provides or could provide resources (products and/or services) to a purchaser or project. The term, however, also embraces contractors, consultants and service providers:

- A **contractor** is an entity that enters into a contract to provide resources to perform a service or undertake work, either directly or through subcontractors
- A **consultant** is an entity that provides professional expert advice or services
- A **service provider** is an entity responsible for providing any required service

Additionally, the terms vendor, broker, merchant and trader can be used.

In order to respond to specific project opportunity or opportunities, individual suppliers may form joint-venture companies, consortia or alliances with other organisations:

- **Joint-venture companies**: a commercial enterprise entered into by two or more parties, while retaining their individual identities
- **Consortia**: a commercial enterprise created by two or more parties to combine complementary services or to jointly develop new products and services
- **Alliances**: involve the purchaser and their principal suppliers forming a joint organisation to implement a particular project, often becoming shareholders in a co-owned company. Personnel from the constituent organisations are then seconded to the alliance to work in an integrated team. Alliances generally have their own organisational and financial structures, with risk and reward allocated between the participants (shareholders)

In relationship to a specific project or programme, the various suppliers can be categorised in tiers; for example, a tier-1 supplier is an entity that has a direct, contractual relationship with the initial purchaser, whereas a tier-2 supplier is a supplier that has a contractual relationship with a tier-1 supplier.

The commercial activity of buying and selling products and services is termed a **market**, as in market forces. Additionally, the term is used as a collective noun for purchasers who buy goods and services.

In the following text the terms **purchaser** and **supplier** are predominantly used.

Purpose of the commercial function

Commercial activities

The commercial function is engaged in and/or undertakes an amalgam of activities that support and deliver an organisation's, business unit's and project's strategic aims and objectives and in particular their commercial objectives. Examples of these wide-ranging activities, tasks and areas of interest are provided in Table 1.1.

The value of the role is derived from this mix of activities, its integrating features, focus on value creation (profit orientation and maximisation) and the protection of an organisation's position contractually. In particular the purpose of the commercial role is one of developing and delivering innovative commercial solutions that:

- Maximise and deliver **profit** and **value** for the organisation
- Protect and enhance the organisation's **commercial and reputational interests**
- Deliver and acquire **competitive advantage**: '*The product, proposition or benefit that puts a company ahead of its competitors*' (CIM, 2011)
- Drive **growth**, improve **market penetration**, increase **market share**: by supporting the development, expansion and exploitation of the organisation's activities, products and or service offerings

Table 1.1 Examples of commercial activities, tasks and areas of interest.

- Accounts/accounting
- Acquisitions
- Analysis of suppliers' cost structures
- Analysis of cost budgets
- Annual sales, profit and overhead budgets
- Assessing costs of a venture/project
- Assessing financial viability of projects/contracts
- Bidding
- Budget monitoring and control
- Budget planning
- Budgeting
- Business case generation
- Business development
- Business planning
- Business process outsourcing (BPO)
- Business proposals
- Calculating service management costs
- Cash flow
- Cash flow forecasting
- Claims
- Communications
- Compliance
- Continuous improvement
- Contract administration
- Contract award
- Contract compliance
- Contract cost modelling
- Contract drafting
- Contract negotiation
- Contract pricing
- Contract requirements
- Contract review
- Contract selection
- Contract strategy
- Contracting
- Contractual matters
- Controlling prime costs
- Copyright issues
- Cost advice
- Cost analysis
- Cost budgets
- Cost control
- Cost engineering
- Cost forecasting/planning
- Cost modelling
- Cost reporting
- Cost value reconciliation

(*Continued*)

Table 1.1 (*Continued*).

- Costing
- Design and build
- Development of strategic partnerships
- Dispute resolution
- Due diligence
- E-procurement
- Estimating
- Ethics
- European procurement law/directives
- Final accounts
- Financial appraisal
- Financial engineering
- Financial forecasting
- Financial payments
- Financial planning
- Financial principles
- Financial processes
- Financial reporting
- Financial risk
- Financing
- Fixed price contracts
- Gainshare/painshare allocation
- Generating new clients
- Global outsourcing
- Governance (corporate and project)
- Government procurement policy, guidance and legal frameworks
- Guaranteed or agreed maximum price contracts
- Incentive pricing
- Incentivised contracts
- Insolvency
- Insurance
- Intellectual property rights (IPR)
- IT outsourcing
- Legal issues
- Life cycle costing
- Management contracting
- Market analysis
- Marketing
- Measured term/serial contracting
- Model/standard contracts
- Negotiating
- New market opportunities
- New product/service development
- Offers – preparation and submission
- Opportunity development and exploitation
- Outsourcing
- Partnering
- Performance monitoring
- Post contract financial reporting
- Price analysis
- Price benchmarking
- Price increase negotiation
- Price modelling
- Pricing policies
- Pricing proposals
- Procurement
- Procurement cycle/life cycle
- Procurement legislation and regulations
- Procurement methodologies
- Procurement routes
- Procurement strategy/policy
- Profit generation, maximisation, etc.
- Profit and loss (P&L) accounts
- Project cost management
- Project final costs
- Project finance
- Project Finance Initiative (PFI)
- Projects
- Proposals
- Public–Private Partnerships (PPP)
- Purchasing
- Quotations
- Regulatory issues
- Resource allocation
- Risk
- Route(s) to contract
- Sales
- Serial contracting
- Sourcing
- Sourcing strategy
- Strategic sourcing
- Subcontract procurement
- Subcontracts
- Supplier cost driver analysis
- Supply issues
- Sustainable procurement
- Target costs
- Tendering
- Value
- Value and cost reporting
- Variations
- Whole-life costing

Source: Analysis of 90 commercial functional frameworks/job specifications

Profit and value

Predominantly, the commercial function seeks to maximise, generate, deliver, or optimise profit, margins, and/or return on investment. These terms are all interrelated. For example, **profit** is the financial gain obtained by supplying (selling) a product and or service – that is, what remains once the costs incurred in producing, supplying, purchasing or operating something are deducted from the price obtained by selling it (the difference between the proceeds generated and the amount spent). This is also referred to as a **margin**. A **profit margin**, therefore, is this difference expressed as a percentage; for example, the gross profit margin is obtained by dividing an organisation's gross (total) income by its net sales (its operating revenues). Alternatively, their objective is to increase income or revenue. **Income** is what an organisation obtains through the sale of its products and or services to customers. Associated financial indicators include: return on investment (ROI), or the rate of return, is the value (profit) derived by an organisation from investing in an opportunity or project, expressed as a percentage; return on assets (ROA), the net income generated by assets divided by their total value (or cost); and return on capital employed (ROCE), the value gained by an organisation from investing in an asset, opportunity or project: a comparison of the revenues obtained with the funds invested.

A further aim of the function is to obtain, create, achieve and add value. In general usage, **value** is the regard (importance, usefulness, or worth) in which something is held, for example, its monetary, material or economic worth – '*the amount buyers are willing to pay for what a firm provides them*' (Porter, 1985) – and its benefit to an individual or organisation (both purchaser and supplier). Benefit is the profit or advantage gained by buying or selling the product and or service. According to Porter:

> 'Value is measured by total revenue, a reflection of the price a firm's product [or service] commands and the units it can sell. A firm is profitable if the value it commands exceeds the costs involved in creating the product [or providing the service offered]. Creating value for buyers that which exceeds the cost of doing so is the goal of any generic strategy.' (Porter, 1985, p. 38)

In relation to projects, a venture is worth undertaking if its value, expressed as benefits, can be quantified and justified in business terms (OGC, 2007a). For example, the Association for Project Management (APM) defines value as:

> 'A standard, principle or quality considered worthwhile or desirable. The size of a benefit associated with a requirement. In value management terms value is defined as the ratio of "satisfaction of needs" over "use of resources."' (APM, 2010)

They, define benefit as:

> 'The quantifiable and measurable improvement resulting from completion of project deliverables that is perceived as positive by a stakeholder. It will normally have a tangible value, expressed in monetary terms, that will justify the investment.' (APM, 2010)

Alternatively, according to BS EN 1325-1: 1997, value is:

> 'The relationship between the contribution of the function (or VA [Value Analysis] subject) to the satisfaction of the need and the cost of the function...' (BSI, 1997)

Additionally, value is also used when non-monetary features such as availability, speed of implementation, reliability and longevity are take into account. Value, therefore, is a subjective construct, which varies between the different stakeholders in a project.

Value management provides:

> '... a structured approach to defining what value means to the organisation and the project. It is a framework that allows needs, problems or opportunities to be defined and then enables review of whether the initial project objectives can be improved to determine the optimal approach and solution.' (APM, 2006)

Applicable throughout the project (contract) life cycle, value management seeks to maximise the overall performance of an organisation, ensuring that decisions taken optimise the balance of benefits with regard to cost and risk (OGC, 2007a). Its processes, methods and tools can be applied at the corporate level, encompassing stakeholder and customer value, and at the operational level to address project-oriented activities, in order to enhance the likelihood of delivering the project's objectives in terms of value for money (BSI, 2000b; OGC, 2007a).

Value for money is realised when **whole-life cost** (*'enhancing whole-life value' or 'total cost of ownership'*) and **quality** are optimally combined to meet the purchaser's/user's requirements (NAO, 2003, 2007; OGC, 2007c) and is derived from *'the effective, efficient and economic'* employment of resources:

- **Effectiveness**: a measure of the degree of success in meeting these requirements (aims, objectives, goals, etc.), occasionally labelled as *'doing the right things'*
- **Efficiency**: a comparison of the output derived with the elements of production, occasionally labelled *'doing things right'*
- **Economy**: the ability or option of acquiring comparative products and/or services at a lower cost to the organisation (price paid) (OGC, 2002)

Further, **value added** relates to the amount, excluding costs incurred, by which the value of a product or service is increased at each phase of its production or delivery. It can also be used in relation to an organisation offering enhanced, supplementary or specialised services.

Interestingly, the associated concept of **utility** – *the state of being useful, profitable, or beneficial* – and in particular its specific economic application as a measure of that which is sought to be maximised in any situation involving a choice, was not mentioned in any of the job specifications or commercial frameworks reviewed.

Generally, all projects undertaken by an organisation should be aligned to its strategy, aim and objectives, at the very least obliquely – that is, seen as a means to an end. A business may carry out a suboptimal project, for example, if it believes that by doing so it will lead to other more aligned and lucrative activity or more pertinent work with a 'key' client, increased market share or facilitate entry to a particular market sector.

The relationship between value and **competitive advantage** is explored in Chapter 3 along with the concepts of **value chain** (alternatively, value web and value constellation; Normann and Ramirez, 1993) and **supernormal profit** (Kay, 1993; Augier and Teece, 2008). Finally, value chain management has been defined as the ability:

> '... to categorise the generic value adding activities, identifying primary and support activities and analyse these specific activities to maximise value creation while minimising costs.' (MOD, 2009)

Further reading

Value management

Kelly, J (2006) Value management of complex projects. In: DJ Lowe (ed.) with R Leiringer, *Commercial Management of Projects: Defining the Discipline*. Blackwell Publishing, Oxford. pp. 298–316

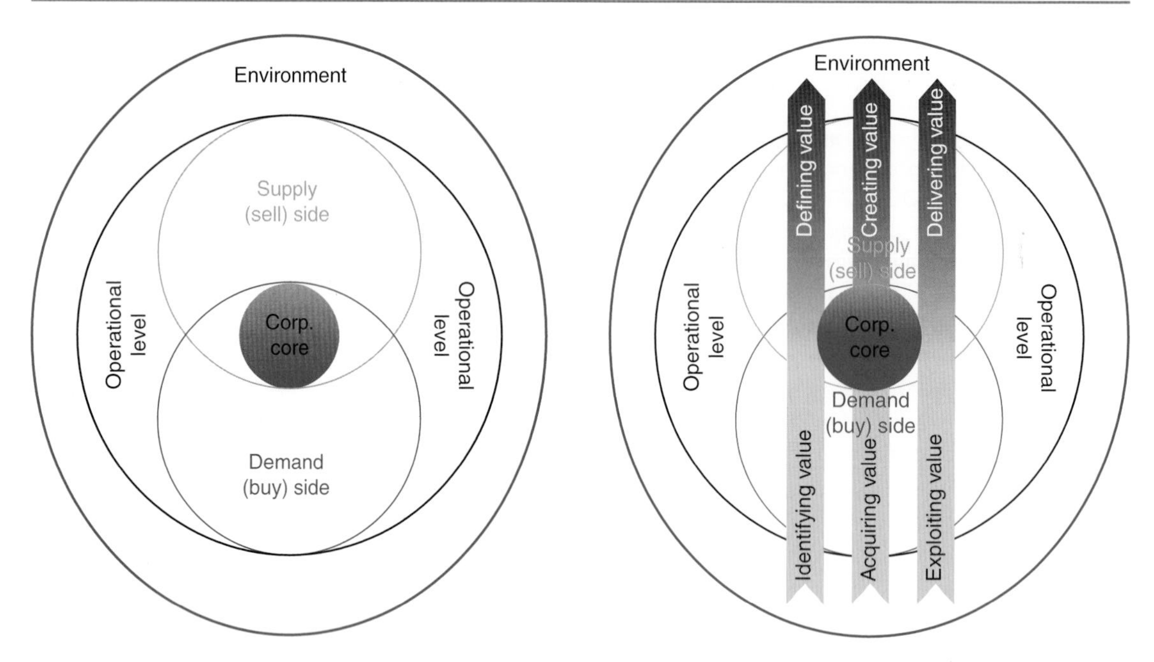

Figure 1.1 Organisational interfaces.

Figure 1.2 Transmission of value through an organisation.

Transmission of value through an organisation

The transmission of value via the supply-side (customer facing), demand-side (supplier facing) and corporate core of an organisation is represented figuratively in Figures 1.1 and 1.2.

Supply-side (customer-facing) perspective

Customer-facing supply activities focus upon defining, creating and delivering value:

- Defining value: stating or describing precisely the scope, nature, or substance of the value to be derived by the purchaser from acquiring the product and/or service offered
- Creating value: generating and producing value through the actions of the supply organisation for both itself and its customers
- Delivering value: providing and transferring the promised or anticipated value to the purchaser

Demand-side (supplier-facing) perspective

Alternatively, supplier-facing purchasing activities centre on identifying, capturing and exploiting value:

- **Identifying value**: recognising, distinguishing and establishing value in the product and/or service sought by the acquiring organisation or offered by a suitable supplier
- **Acquiring value**: seeking, purchasing or obtaining assets, products and services through which the purchaser can realise value. This aspect could also include the development and acquisition of new skills and abilities
- **Exploiting value**: in the positive sense of deriving benefit from or fully utilising a resource, asset, product or service (as opposed to the term's negative implications, such as, unfair or underhand dealing or unjustly or illegally benefiting from the work of others)

Figure 1.2 illustrates the development and transition of value through a single organisation. However, value is transferred at the point of exchange between organisations and passes 'downstream' through the entire supply/value chain.

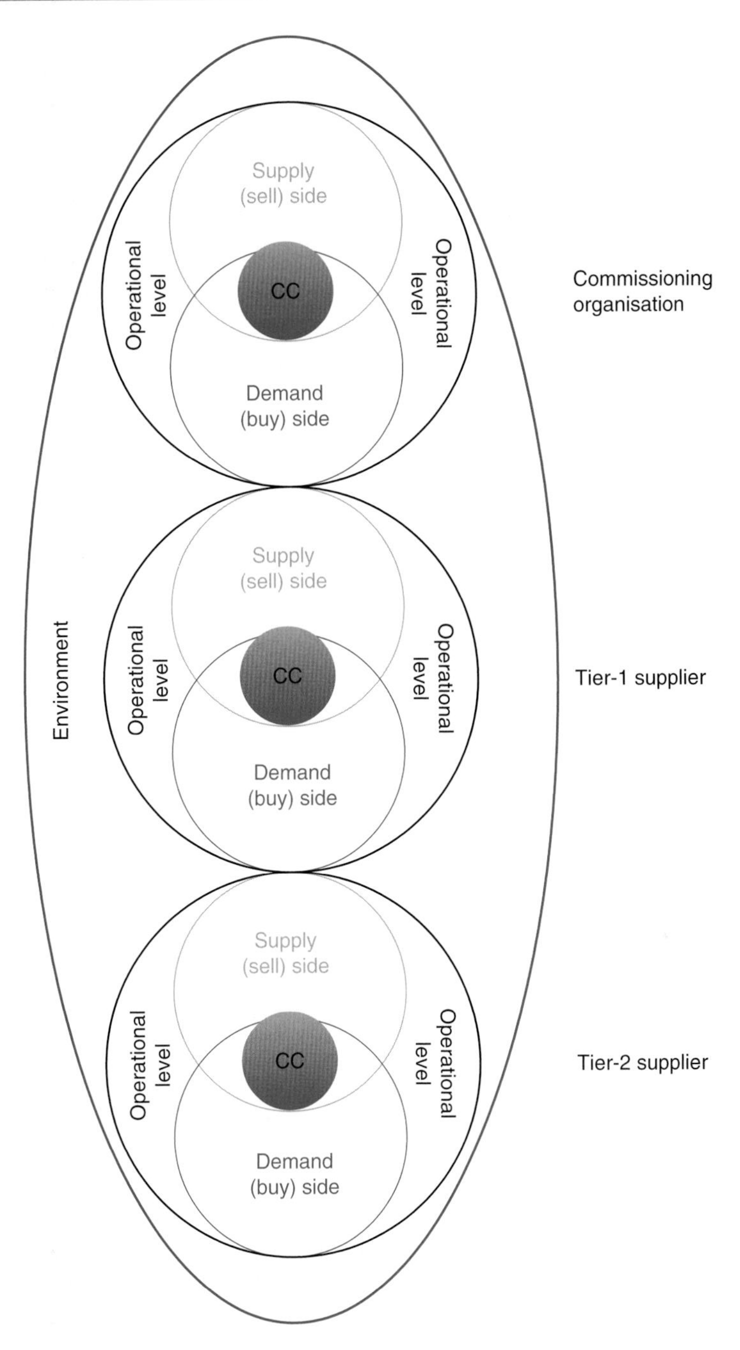

Figure 1.3 Organisational interfaces – traditional arms-length contracting. CC = corporate core.

These points of exchange are illustrated in Figures 1.3 and 1.4. Figure 1.3 represents the traditional 'arms-length' approach to contracting, while Figure 1.4 shows a collaborative partnering approach, where, for example, the purchaser's demand-side activities are often integrated and co-located with those of the principal Tier-1 suppliers' (supply-side) delivery activities. In relation to this latter arrangement,

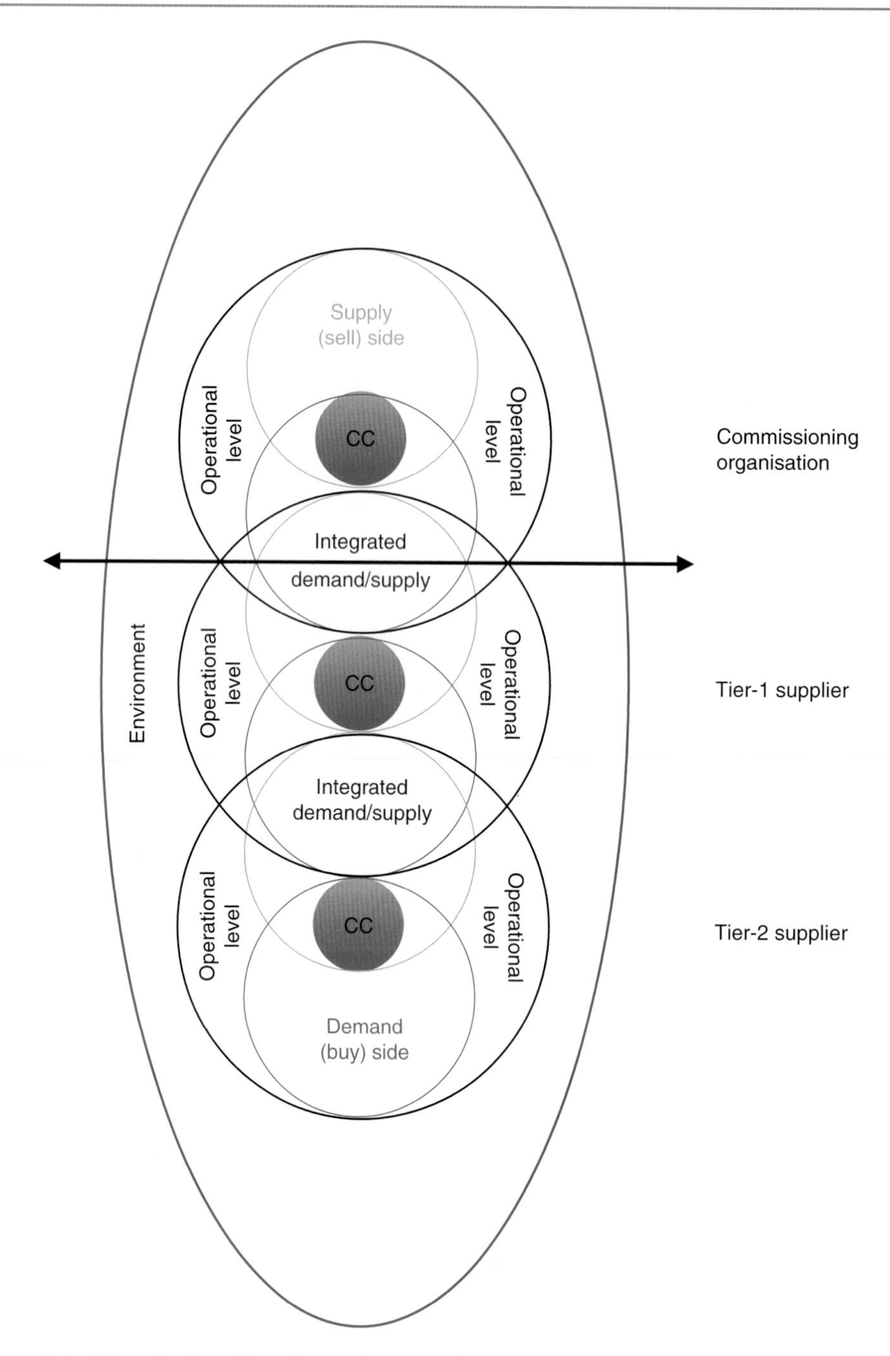

Figure 1.4 Organisational interfaces – integrated contracting (partnering). CC = corporate core.

Figure 1.5 illustrates the complexity of the internal interfaces within an integrated demand/supply team; for simplicity, only four Tier-1 suppliers are shown. However on the T5 project, for example, there were 60 plus suppliers. Similarly, Figure 1.6 illustrates the multiple commercial interfaces across the various tiers of a supply network.

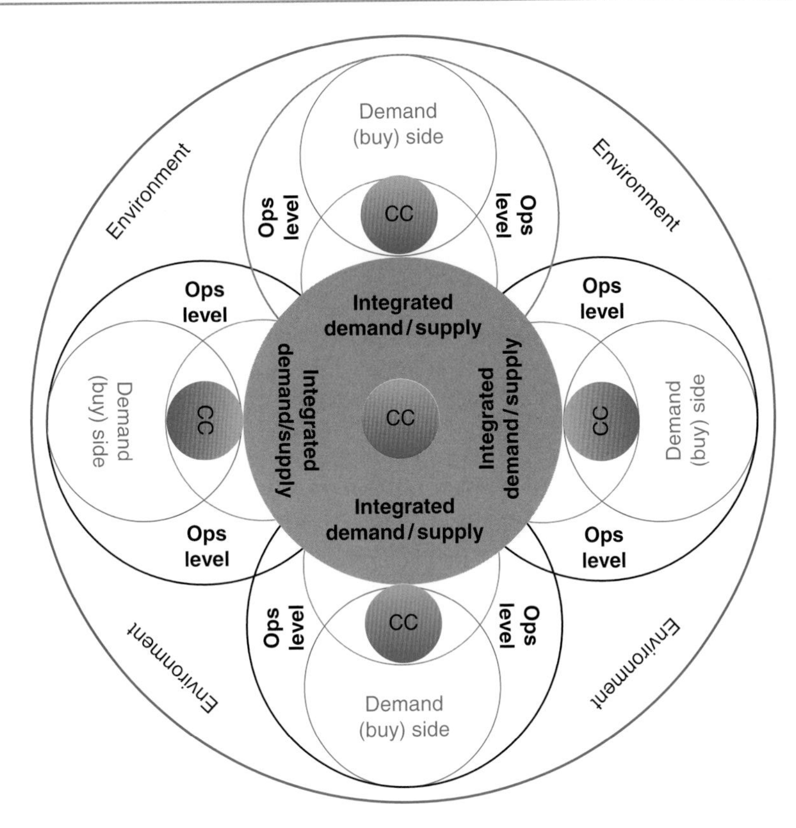

Figure 1.5 Tier-1 commercial interfaces in an integrated demand/supply network.

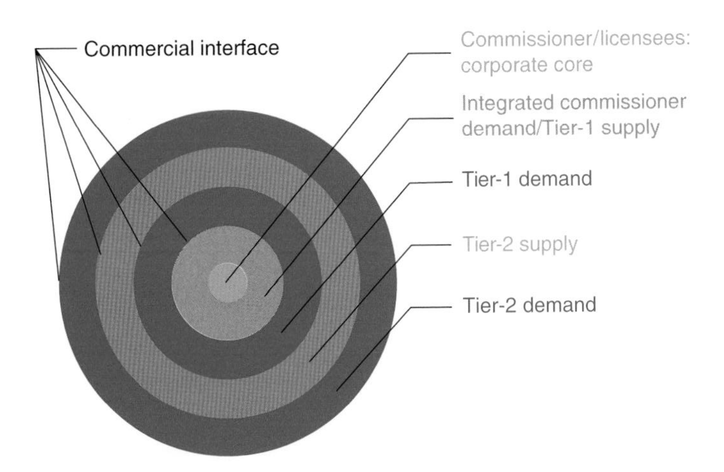

Figure 1.6 Inter-organisational commercial interfaces.

Commercial management framework

The following commercial management framework (Figure 1.7) exemplifies the multiple interactions and connections between the purchaser's procurement cycle and a supplier's bidding and implementation cycles. For simplicity it shows the interface between two organisations, although in reality the connections

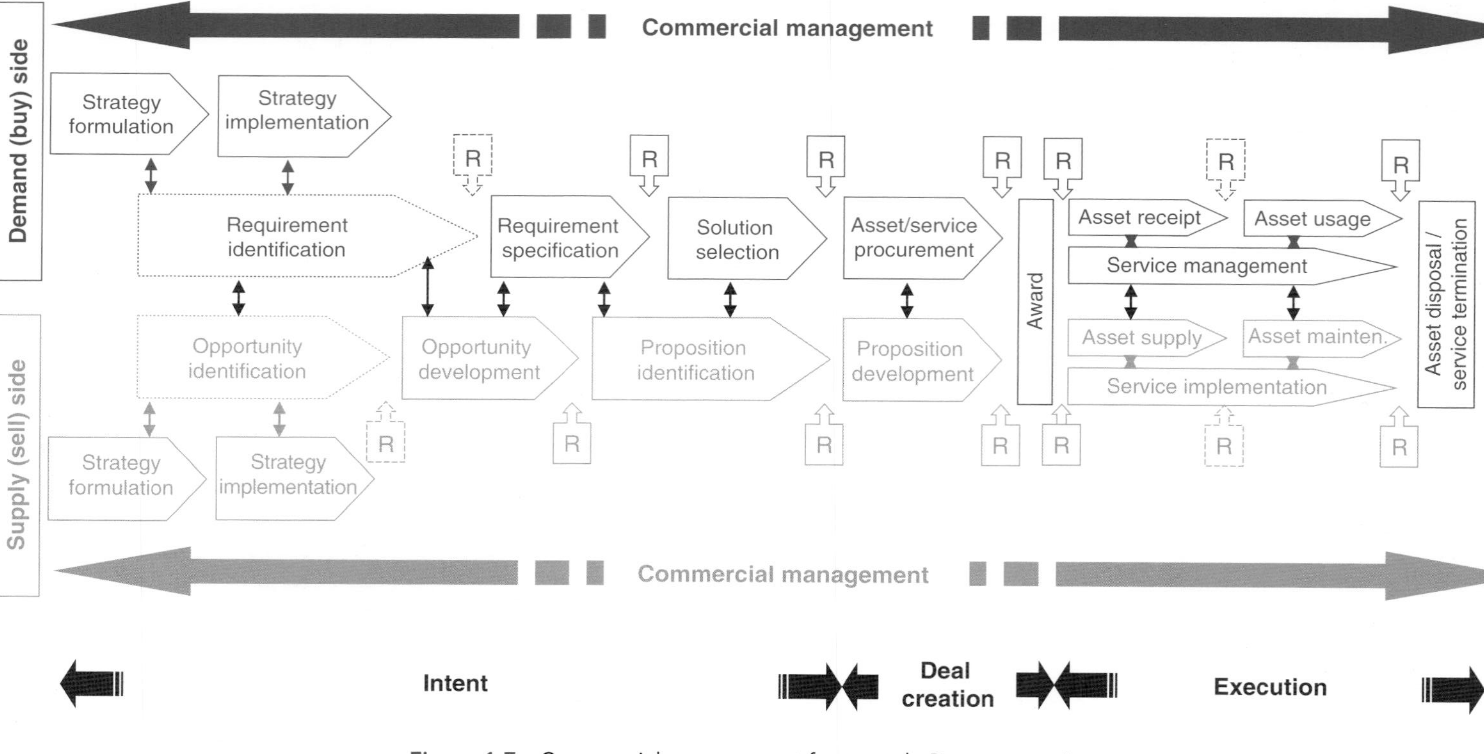

Figure 1.7 Commercial management framework. R = stage review.

are multilayered, the purchaser appointing numerous suppliers sequenced at different points across a project's life cycle. Similarly, suppliers will generally contract with several if not many customers; while in the vertical dimension there will be parallel interfaces between the numerous tiers of the value (supply) network.

Comprising a series of interrelated processes and actions, the framework is organised into three discrete stages: **intent**, which links the proposed acquisition or opportunity to supply to the strategic objectives of the organisation; **deal creation**, which culminates in the award of a contract; and **execution** – the implementation stage – which concludes with the disposal of an asset or termination of a service agreement. However, it is acknowledged that in practice contract life cycles will contain minor variations depending upon the specific industry sector or even business unit involved. The framework incorporates several stage gates (review points) to ensure that the proposed contract delivers the anticipated benefits (value) for both the purchaser and supplier. The Office of Government Commerce's (OGC) Gateway™ Project Review Process, which is outlined in Chapter 7, provides examples of the types of questions to be asked and techniques to be applied at each point of the process.

Although there is a clear boundary between the pre- and post-contract award stages, the boundaries of other activities are not as precise. Moreover, the constituent processes are likely to be iterative, containing cyclic decision loops, and with some actions being carried out simultaneously.

Stages of the framework

Intent

The **intent** stage bases and connects the proposed acquisition or identified opportunity to supply with the overall strategy (motivation, purpose, aim or objective: the *raison d'être*) of each organisation and to its implementation. The phase includes the following elements:

- **Strategy formulation**: creating or preparing a plan intended to achieve or deliver a long-term or overall aim (*a purpose or intention; a desired outcome*) of the organisation
- **Strategy implementation**: executing, effecting or enforcing the plan
- **Requirement identification**: identifying and confirming a need or want, a potential opportunity to be exploited or an issue to be resolved
- **Opportunity identification**: identifying potential customer requirements that the organisation is capable of satisfying
- **Requirement specification**: collating and comprehending the internal customer's needs, stating (specifying), invariably in a written form, a detailed description of precisely what is required. The resulting document, the specification, can include: design details, a description of or reference to the materials to be used, service levels or the standard of workmanship to be achieved, and references to acknowledged standards. The phase will also include an evaluation of the viability of the proposed endeavour and the production of a business case
- **Opportunity development**: cultivating a relationship and dialogue with potential customers, and creating, preparing and marketing potential solutions (product/and or services) to them
- **Solution selection**: identifying and deciding upon the most appropriate means of satisfying the acknowledged organisational need, which the market is capable of supplying (preferred solution), and the route to obtaining the selected product(s) or services (sourcing strategy). During the phase the selected solution will be progressed and appraised; concurrently the business case will be reviewed and revised
- **Proposition identification**: identifying and refining a potential solution to a customer's identified and specified requirement

Deal creation

The **deal creation** stage focuses on bringing an agreement (principally a contract) by two or more parties into existence for their mutual advantage. Colloquially referred to as brokering or 'striking' the deal, the phase contains the following elements:

- **Asset/service procurement**: obtaining or procuring property, products and/or services through the implementation of the selected sourcing (procurement) strategy, the tendering process, and negotiating the agreement: the precise terms of the exchange

- **Proposition development**: creating, preparing and submitting a proposal (an offer, bid or tender) in response to a client's request for proposal, quotation or tender (RFP/RFQ/RFT) in respect of the supply of property, products and/or services

The stage concludes either with a decision not to proceed with the 'purchase' or in the award of a contract:

- **Award**: assigning or granting a contract (or commission) to an individual or organisation. It also marks a transition point: the commencement of the implementation stage, requiring the delivery of the commitments made by the contracting parties during the deal creation phase

Execution

The **execution** stage involves implementing, delivering and fulfilling the commitments made during the deal creation stage and incorporated into the contract, concerning the required asset, products and/or service. The term has been adopted due to its legal connotation: an executed contract is one where the provisions have been fully performed by both parties. Depending upon the nature of the purchase (asset/product or service) and associated sourcing strategy the phase comprises the following alternative components.

The purchase of an asset or product involves the following elements:

- **Asset supply**: designing, manufacturing, constructing, providing and delivering the procured asset or product(s) in accordance with the contract
- **Asset receipt**: receiving and accepting the procured asset or product(s) in accordance with the agreed terms and conditions of the contract
- **Asset usage**: applying or utilising the procured asset or product(s)
- **Asset maintenance**: preserving and maintaining the procured asset or product(s)

The stage concludes with the disposal of the asset:

- **Asset disposal**: divesting of property and assets

Alternatively, the acquisition of a service includes the following elements:

- **Service implementation**: executing, effecting and delivering the procured services in accordance with the agreed terms and conditions of contract
- **Service management**: receiving, utilising, monitoring and management of the procured services in accordance with the agreed terms and conditions of contract

In this instance the stage concludes with the termination of the service:

- **Service termination**: bringing to an end a service contract[4]

Each of the above elements will involve **contract management**: the monitoring and control of the contract to ensure that the agreement between the purchaser and supplier operates efficiently and effectively, and that all contractual obligations are appropriately discharged. Similarly, each element (usually covered by separate contracts) will conclude with a **closeout** component, involving a final review of the exchange and the conclusion of all contract matters.

Chapter 8 adopts this structure; similarly, the case studies are also presented under the above heading.

The components of commercial management

Commercial practitioners perform and contribute to several business processes – management, operational and supporting – each involving an array of interrelated structured activities and tasks (Figure 1.8). Each component will generally be integrated within organisational policies, processes and procedures, and

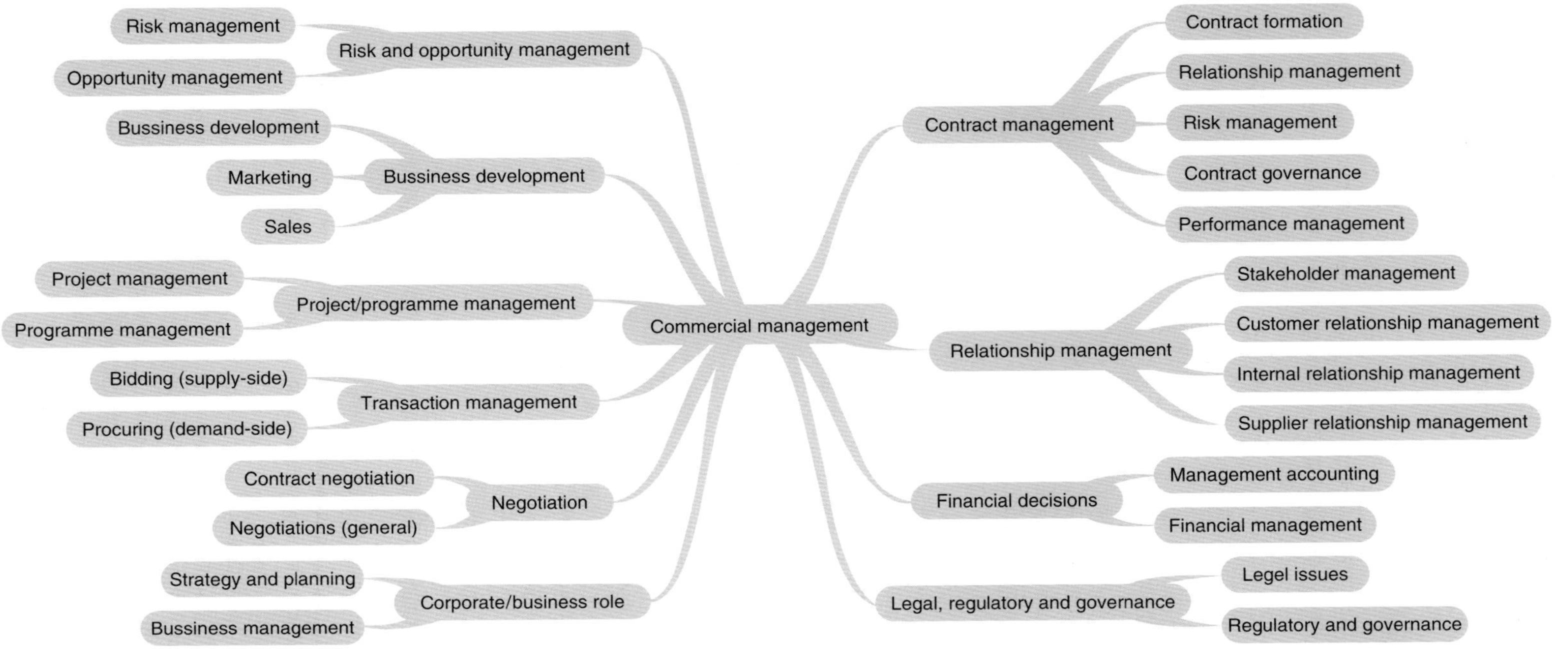

Figure 1.8 Processes frequently undertaken by the commercial function. Source: Analysis of 90 commercial functional frameworks/job specifications.

supported by underlying knowledge, theories and concepts. They will also require the practitioner to possess and/or develop specific capabilities.

The processes frequently undertaken by the commercial function are described below, in order of commonality.

Contract management

Contract management concerns the application of a variety of processes, activities and procedures aimed at ensuring the agreement (contract) between the purchaser and supplier operates efficiently and effectively, and that all contractual obligations are discharged. Commencing with the identification of the purchaser's need(s) and concluding with the completion of the contract (referred to as the contract life cycle), the dimension is undertaken by both purchasers and suppliers. It has two main characteristics:

- **Relationship management**: establishing and maintaining appropriate relationships and channels of communication between the contracting parties, their agents and suppliers, based wherever possible on openness, trust and respect. The aim is to facilitate the smooth running of the contract/project through proactive intervention: identifying and resolving issues before they turn into disputes
- **Risk management**: identifying, allocating and managing risk – balancing risks with their associated costs. This aspect is closely related to post-contract award performance/ implementation management

Both these aspects are explored in more detail in later in this chapter.

The primary aim of contract management is to deliver the requisite corporate, business and operational/ project level objectives encompassed in the contract, attain value for money (best value), and ensure the individual party's agreed contractual position is safeguarded.

Pre-award

Preliminary, pre-award components of contract management include:

- **Establishing the contract management team**: deciding upon the structure and composition of the contract management team and when it should be formed
- **Forming the contract**: interpreting the requirement through drafting, negotiating and approving an appropriate agreement (contract). Agreeing an appropriate contract is crucial to the ultimate success of the purchase/project, as it forms the basis on which the trading relationship is shaped and contractual compliance is measured and controlled

This 'transaction focused' aspect of the role is often undertaken by different personnel from those responsible for post-award contract management. However, this segregation of the role needs careful management to ensure that knowledge gain during the procurement process is utilised during the implementation stage.

Post-award

In addition to relationship and risk management, post award elements of the role include:

- **Contract governance**: administering and implementing contract procedures, including the formal communication and interface between the parties, such as notification of any variations to the contract, certification of any interim payments or the submission of a claim for additional payments
- **Performance/implementation management**: monitoring and ensuring that the asset and or service is delivered in accordance with the contract terms and conditions – to the specified performance and quality standards and to the stipulate programme – and that the allocated risk distribution is upheld. It also involves looking for any detrimental execution or service trends, applying any financial remedies for non-performance, evaluating the effectiveness of any subsequent remedial action, and implementing continuous improvement in performance and functionality over the contract life cycle.[5]

The terms **enterprise contract management** (ECM) and **contract life cycle management** (CLM) are frequently used interchangeably in relation to the provision of organisation-wide contractual services based on current best practice and IT-enabled processes, which efficiently and effectively support the management of a business' contracts and agreements.[6]

Commercial managers are accountable for the commercial acceptability, quality, legality and management of contracts. This involves:

- Acting as the principal interface with customers, partners and suppliers regarding contract management issues, and briefing and providing guidance and advice on contractual issues to senior management and project teams
- Contributing to the financial effectiveness and profitability of the business by developing and implementing organisation-wide 'whole-of-life' commercial and contract management policy, procedures, processes and techniques; developing or aiding the formulation of contractual strategies and contract management frameworks in respect of core business activities; and developing and publishing standard terms and conditions for products and services
- Drafting, evaluating, negotiating, agreeing, reviewing and monitoring/managing contracts and agreements (for example, licensing, service level, confidentiality – non-disclosure, distribution, teaming and partnership agreements and memorandums of understanding) with customers, partners and suppliers; ensuring that contract requirements are clearly specified and in accordance with the organisation's strategy; clarifying customer requirements and ensuring that the resulting contract reflects these needs
- Developing and maintaining a register of contracts and beneficial contractual relationships with customers, partners and suppliers
- Managing contracts and the commercial resources of projects throughout their life cycle, including: developing, implementing and reviewing effective commercial activity; providing support and expertise to ensure that contracts are appropriately administered, monitored, reported on and fulfilled so that all contractual obligations are performed and rights conferred by the agreement are appropriately exploited; realising business plans and project objectives, maximising profit, taking advantage of opportunities and minimising risk; and delivering continuous improvement; compiling, notifying, assessing, approving and reconciling contract variations (change management) and claims for loss and expense; negotiating and agreeing contract disputes and price increases to realise commercially acceptable solutions; agreeing final accounts; and managing contract close-out, extensions or renewals

Further reading

Contract management

Hughes, W (2006) Contract management. In: DJ Lowe (ed.) with R Leiringer, *Commercial Management of Projects: Defining the Discipline*. Blackwell Publishing, Oxford. pp. 344–355

Relationship management

Relationship management (RM) is a systematic process involving the application of various activities and procedures, plus an ability to proactively develop (establish, build, create, cultivate) contacts, manage appropriate working relationships, including communicating, negotiating, interacting with and influencing significant suppliers, customer, colleagues and stakeholders. The activity includes:

- **Stakeholder management**: the identification, analysis and planning of exchanges with stakeholders aimed at developing and maintaining their committed and active support to the implementation of the contract or project. A stakeholder is any organisation, group or individual that is involved in, affected by (or consider itself to be affected by), has a vested interest in, or can influence change in, a system, an enterprise (programme, project, activity, change or risk), or an organisation and its sphere of operation. The term includes shareholders, users and decision-makers.[7]

Commercial practitioners develop, manage, maintain and act as the primary interface with key stakeholders forming effective, strong collaborative relationships with colleagues, customers, suppliers and all interested parties. They create a network of contacts with high level influencers to shape the strategic aims of the business, deliver anticipated benefits and fulfil commitments made

- **Customer relationship management (CRM)**: the development and management of effective relationships with customers (clients, purchasers, buyers, buying centres and procuring parties, etc.).

  The commercial function identifies, develops, maintains, manages, understands, supports and has ownership of/is the prime point of contact with customers forming effective, strong, key 'customers focused' relationships. This entails understanding the business goals of customers, appreciating and clarifying their requirements (establishing the client's brief) and end-user constraints, gaining their trust and respect, and managing their expectations and satisfaction. The principal aims being to improve market share and secure profitable business
- **Internal relationship management**: the development and management effective working relationships with colleagues.

  Commercial personnel create, develop, lead and manage close, effective, strategic and collaborative relationships with other functions and teams – for instance, interfacing with colleagues from business development, engineering, finance and accounting (treasury, tax, insurance and business assurance), legal, marketing, sales, communication and information, procurement, project and programme management, operations, service delivery and production at senior management, business unit and project/programme levels. The activity is associated with demand management, the identification and analysis of anticipated future requirements and the selection of appropriate sourcing/procurement approaches to meet this demand. It also entails fostering a positive cross-function organisational culture and effective team working, creating business-wide decision-making frameworks, facilitating the development and implementation of appropriate business development, commercial, product and services strategies in accordance with corporate objectives and achieving social and corporate responsibility aspirations and requirements. The key goals are to improve the quality/efficiency of products and services, develop new offerings, ensure that offerings meet and exceed customer expectations, minimise customer churn, and secure the position of provider of choice, while at the same time delivering the maximum return on investment
- **Supplier Relationship Management (SRM)**: understanding, categorising, developing, coordinating and managing relationships with suppliers (contractors, consultants and service providers) with the aim of maximising or optimising value. An alternative term is key supplier management (KSM), which involves the provision of corporate and strategic data on the most important suppliers to the purchasing community. A key feature of the activity is supplier development, which incorporates supplier performance measurement against pre-agreed criteria, benchmarking and the concept of continuous improvement. The activity is a constituent part of supply chain management (SCM) covered later under a separate heading.[8]

  Again, the commercial function is involved in developing, managing, maintaining, understanding and coordinating suppliers and partners; forming effective relationships across the supply network. These tasks require a detailed knowledge of the supply network, the implications of working with third parties, and how to incentivise suppliers and their employees. It involves analysing supply networks – reviewing existing relationships with, sourcing, co-ordinating and optimising the performance of external contractors, suppliers and partners to ensure successful delivery, and designing and implementing appropriate SRM, procurement and supply chain strategies. The objective being to provide value for money, improve supply networks and deliver mutual benefits

These activities are frequently facilitated by IT systems; they are not, however, despite the implications of some definitions, totally reliant on them.

This component of the commercial function requires highly developed interpersonal, influencing and communication (including presentation and linguistic) skills; plus the ability to account for individual and group cultural, social, economic and political differences and to use constructive feedback to improve relationships.

Further reading

Performance measurement

Horner, M (2006) The governance of project coalitions – towards a research agenda In: DJ Lowe (ed.) with R Leiringer, *Commercial Management of Projects: Defining the Discipline*. Blackwell Publishing, Oxford. pp. 270–297

Financial decisions

The provision of current and accurate financial information is crucial to effective management planning, decision-making and control. This information is predominantly supplied by the **accounting and finance** function within businesses. As Anne Stafford explains in Chapter 5, this entails classifying, measuring and summarising transactions, then analysing and communicating this data. A significant aspect of accounting is the preparation of **financial reports**, for example, **balance sheets**, **income statements** and **cash flow statements**. Although there is some interrelationship, these reports can be classified as either **financial accounting** (data prepared for external consumption) or **management accounting** (information compiled to aid internal planning, evaluation and control).

The financial aspects of the commercial role can primarily be subdivided into management accounting and financial management.

Management accounting

Management accounting, according to Charles Tilley, Chief Executive of the Chartered Institute of Management Accountants (CIMA, 2011), is:

> '... where finance and business meet and helps corporations build more efficient and effective organisations'.

Within the project environment, the component involves the provision of effective financial control of projects/contracts via:

- **Cost management**: the management of monies to be spent; and
- **Cash management**: the management of monies received by the organisation

This entails analysing, evaluating, planning, estimating, forecasting, budgeting, coordinating, controlling, monitoring, and reporting cost and revenue data, throughout the asset/project/contract life cycle. It also involves the application of techniques, such as value management and cost-value reconciliation, and operating 'open-book accounting' on projects, where integrated project teams have access to project cost data.

Financial management

Financial management, likewise, is a key management activity, organisational function, and decision-support process that seeks to ensure an organisation's sustainability by efficiently and effectively investigating the availability of, sourcing of and utilising of financial resources (capital funds). Associated management activities include appraising, planning, balancing, controlling, reporting and budgeting/forecasting, enabling an organisation, business unit or project to realise its stated objectives.

Regarding this component, commercial practitioners are responsible for the cash/cost management and profitability of projects and contracts, liaising with and supporting the finance, accounting and tax functions. This involves:

- Contributing to business, financial and tax planning and compliance: ensuring that suitable corporate governance and financial management processes and procedures are adopted and implemented, in particular the efficient and effective operation of appropriate commercial and financial systems and techniques
- Developing and implementing business plans: leading discussions on the establishment of annual sales, profit and overhead budgets; initiating and maintaining measures of sales and financial performance
- Driving revenue growth across the organisation's business activities/projects (including bids and contracts), selecting investment opportunities, and pursuing all opportunities to maximise sales, profit and cash for the business and/or deliver value for money
- Managing funding requirements and financial structuring: ensuring that adequate financial resources are available to support commercial activity, arranging international financing and taking action to mitigate currency fluctuations
- Undertaking budgeting and financial forecasting: generating accurate cost estimates and developing contract/project-related cost models for assets, products or services; and carrying out ad hoc analysis and financial assessment
- Providing pre-award advice and guidance on the financial viability of projects/contracts: evaluating the financial standing of customers (and suppliers) including any funding, budgetary and phasing constraints; assessing capital and revenue expenditure over the whole life of an asset/project/contract; determining the financial risk associated with contractual non-compliance and obtaining appropriate financial sign off
- Providing post-award cost control and financial and management reporting to ensure the financial stability of the project/contracts: preparing project/contract budgets and cash flow forecasts for projects and individual work packages; establishing payment schedules and reviews; tracking payments and financial targets, and providing performance monitoring information; comparing planned spend against costs actually incurred, preparing monthly management accounts and cost value reconciliation reports; valuing work in progress for interim payments; instigating change management, valuing changes and calculating the final project cost (final account); managing loss and expense claims; maximising cash flows; carrying out value engineering and benchmarking exercises; and ensuring the accuracy of monthly financial key performance indicators (KPIs)

The component necessitates numeracy skills and the ability to determine, maintain and communicate apt cost information in respect of projects, products and or services.

Legal, regulatory and governance

Legal exists as a distinct function within many organisations. Generally headed by a general counsel, the traditional role of legal is to provide legal advice, plus the oversight of **governance** and **compliance management** (relating to organisational regulations and procedures), particularly in respect of reputation and risk management. The role includes aspects of **corporate governance**, the maintenance of a rigorous system of internal control, **project governance**, the internal control of project-related issues and activities, and **performance management**, assessing achievements against goals, targets, performance frameworks and external benchmarks.[9]

In relation to legal, regulatory and governance, the commercial function can be responsible and accountable for:

- Ensuring compliance with the organisation's objectives, guiding principles and systems, and adherence to commercial, contract and procurement law, financial and statutory regulations, and other relevant legal requirements
- Acting as the initial point of contact for any incoming legal-related issues; providing or obtaining legal advice, briefing and managing in-house legal expertise/counsel or interfacing with external lawyers concerning commercial issues and contracting policies that impact upon the organisation's business activities
- Determining the organisation's commercial policy: developing and ensuring that effective commercial procedures, contracting frameworks and contract management protocols are adopted and implemented that are legally sound and beneficial to the organisation, and that protect its commercial and reputational interests

- Interpreting and clarifying contract terms and conditions; ensuring that all contracts and agreements are commercially sound, are negotiated, agreed, executed, implemented and consistent with organisational objectives and policies; recommending the inclusion of terms and conditions that protect the organisation's interests and accurately document the intended business relationships; informing senior management of all legal responsibilities contained within contracts and agreements; and developing the organisation's capability to monitor and report on contract compliance; ensuring contract compliance, i.e. that commitments made by all the parties are met
- Establishing effective monitoring processes: monitoring and assessing the impact of legal issues on the commercial function and responding appropriately; monitoring and reporting on contract compliance, ensuring that the contracting parties deliver on their commitments (obligations) in accordance with the terms of the contractual framework
- Providing commercial direction and support to senior management in meeting all governance and reporting requirements; ensuring that appropriate corporate governance and robust internal and external control systems are in place (in accordance with statutory requirements and professional standards) and adhered to; and promoting standards, codes and practices aligned with corporate governance, values and ethics
- Ensuring that all commercial and financial aspects of projects are managed effectively and efficiently in accordance with business procedures and systems; that all business opportunities, sales proposals, bids/quotations and contracts are prepared in accordance and comply with the organisation's policy and procedures; that risk exposure is appropriately acknowledged, understood, accepted, minimised or mitigated; and signing off or obtaining regulatory and legal sign-off of all commercial aspects of business opportunities, proposals, bids/quotations and contracts in accordance with delegated authority

The capability requires an understanding of business, commercial, competition, contract and environmental law; legislation and legal frameworks, in particular national, EU and international procurement legislation; jurisdiction, government procurement policy and guidance; health and safety and duty of care; intellectual property (IP) and trade mark issues, including: the protection and breach of trademarks, copyright, patents, royalties, licensing, sharing information (identification and apportionment of background, foreground, arising IP); and standard forms of contract, etc.

Further reading

Project governance

Winch, G (2006) The governance of project coalitions – towards a research agenda. In: DJ Lowe (ed.) with R Leiringer, *Commercial Management of Projects: Defining the Discipline.* Blackwell Publishing, Oxford. pp. 344–355

Corporate/business management role

At senior levels, commercial actors contribute to the overall management of organisations. Having responsibility, in conjunction with other senior colleagues, for the strategic leadership, direction and development of the organisation, they participate in strategy formulation and setting business objectives. Specific activities undertaken may include:

- Driving and providing commercial support to the strategic decision-making processes in the business
- Leading and shaping the organisation's commercial and business planning activities
- Leading and supporting the development of business units, for example, identifying, developing and implementing joint ventures and strategic partnerships
- Contributing to and establishing the organisation's corporate core purpose, vision, mission and values

At other levels they may contribute to the development of the organisation's strategy and direction at the corporate, business unit, division or function level. This may involve:

- Leading, advising, driving, assisting and providing strategic input and commercial expertise to the development of corporate, commercial and service (delivery) strategy – for example, by providing insights into competitor activity
- Converting strategic corporate aims, business unit objectives and functional policy goals into commercially viable activities
- Developing, implementing and monitoring integrated business and commercial plans, policies, systems and decision-making processes to support the organisation's business objectives and strategies, and communicating these both internally and externally
- Promoting a shared vision of the organisation's values and strategic objectives

Additional responsibilities/accountabilities may include:

- Contributing to and participating in the effective general management and direction of the business; providing advice and support to senior management, business units and other management functions as appropriate; forecasting the probable impact of proposed business decisions including, for example, undertaking scenario analysis, assisting in the preparation of annual budgets, and distilling and generating 'fit for purpose' management information
- Leading, managing and developing commercial business activities and solutions with third parties and ensuring these activities support the organisation's vision, values and objectives
- Delivering operational and economic performance and continuous improvement of the commercial function against predetermined targets and KPIs

The capability requires an understanding of and the ability to analyse the business unit, the corporate debate and the organisation as a whole – its management systems and business processes; the role and responsibilities of the commercial function within the business; and the key drivers and elements of its business plans. In addition, an ability to employ commercial experience, knowledge, expertise and acumen to maximise the organisation's potential, add value and generate competitive advantage.

Further reading

Strategies for solutions

Davies, A and Hobday, M (2006) Strategies for solutions. In: DJ Lowe (ed.) with R Leiringer, *Commercial Management of Projects: Defining the Discipline*. Blackwell Publishing, Oxford. pp. 132–154

Negotiation

Negotiation is a competence, activity and 'bargaining' process, through which two or more entities interact in order to reach an agreement, consensus or compromise. Both formal and informal negotiations are undertaken throughout the project, product and service life cycle.

Negotiating with internal and external stakeholders, customers (clients and commissioners), partners and suppliers, is an integral part of the commercial role, although some commercial practitioners may predominantly focus on either the customer or the supply chain (suppliers, contractors and partners). Depending on the level of the individual within the organisation, they may be responsible for leading, managing or facilitating the process, and either leading or participating in negotiation teams at the strategic, programme or project level. The role can involve:

- Formulating appropriate negotiation processes, assembling negotiation teams, consulting internally and externally from initial enquiry to final settlement, establishing objectives and setting the negotiation strategy and ensuring that it is applied
- Negotiating, within agreed delegated limits, various standard and bespoke contracts (commercial and legal terms and conditions, contractual risks and liabilities) for the supply or purchase of goods and services and for aftermarket support, including formal, short form, and annual contracts; agreements, for example, Teaming Agreements (NDAs), Non Disclosure Agreements (NDAs), Intellectual Property Rights (IPR) and commercial exploitation agreements; tenders, deals and prices; and various opportunities and business plans – providing commercial solutions
- Negotiating and agreeing post-award contract changes, variations (additions or omissions) and price increases; disputes (conflict avoidance, management and dispute resolution); claims, liabilities and implications of changes to contract terms and conditions; high value or complex subcontracts; and final accounts and contract close-out
- Ensuring that the output of the negotiations reflect and are consistent with the corporate/business units/ programme/project's strategic plan, objectives and requirements; protect business needs and interests; are commercially acceptable; generate the optimum return for the business; meet the customer's requirements; identify and mitigate exposure to risk; support and maintain equitable and balanced business relationships; deliver savings or efficiencies; and balance the drive for sales with business risk

The capability requires an understanding of the emotional and psychological aspects of negotiation and of the principal influencing techniques and strategies. It also requires the ability to: communicate and negotiate at a senior level both within the organisation and externally; determine and deploy negotiation strategies; be forthright, tactful, and gain the trust of the other parties; and secure concessions without undermining relationships.

Transaction management

Bidding (supply-side)

The terms 'bidding' and 'tendering' are generally used interchangeably to refer to the process and activities undertaken by suppliers when preparing and submitting an 'offer' to a prospective purchaser. An offer (also termed a bid, quote or tender) is a proposal to meet the purchaser's requirements and can include an economic (price) as well as a commercial and technical element. They are predominantly submitted in response to a formal invitation (invitation to treat), such as, an invitation to tender (ITT), request for proposal (RFP), request for quotation (RFQ) or request for tender (RFT). Under English contract law, an offer to contract:

> '... must be made with the intention to create, if accepted, a legal relationship. It must be capable of being accepted (not containing any impossible conditions), must also be complete (not requiring more information to define the offer) and not merely *advertising*.' (Chapman, 2004)

An offer, therefore, must signal a desire to enter into a legally binding contract based on stated terms that requires no further negotiation or forms a further invitation to treat. The formal submitted bid document presenting the offer is also called a tender, while the terms 'bidder', 'tenderer' and sometimes 'respondent' are used in respect of the party (supplier) submitting an offer.

Bid management is an integral part of the commercial role. Generally, the function is responsibility for leading and managing the proposal (bid/tender/offer) development stage, developing and implementing bid strategies, and establishing and managing the bid production and estimating processes. The role also involves bid team management: leading, supporting and liaising with business development, marketing and sales, operations, production and project management colleagues throughout the bid process to coordinate the development and production of proposals and offers. The prime objectives are to ensure that bid opportunities meet the business's strategic objectives – for example, that they drive revenue/sales growth, maximise profit, address potential risks and opportunities, and that all products and services are offered on appropriate, competitive terms and conditions.

Commercial practitioners, therefore, work as part of bid teams to generate proposals, dealing with the commercial aspects of bid (offer/tender) production and submission, and providing commercial solutions and market information. Additionally, they can be involved in:

- Developing, monitoring and implementing processes and procedures to select and process bid opportunities, produce tender documentation and evaluate proposals, in order to comply with internal governance and improve tender success rates
- Identifying, delivering and managing (within delegated powers) the entire bidding process; responding to tenders and proposals, managing the pre-qualification process, and reviewing requests for quotations (RFQs); capturing customers' requirements, screening projects (the bid/no bid decision); formulating win themes; defining and selling the customer proposition, preparing commercial responses to customer invitations to tender (ITTs); developing the commercial shape of deals; establishing pricing and payment policy, obtaining requisite bonds/guarantees/insurance/export control licences; reviewing, drafting, compiling editing, pricing, approving, and submitting proposals; owning and implementing the bid authorisation process and providing commercial authorisation; participating in bid presentations, negotiating with the customer in respect of any offers submitted; and ensuring that effective CRM techniques are applied and maintaining the commercial integrity of relationships and the deal throughout the process
- Providing an assurance function, ensuring that bid proposals are submitted on time, in the correct format, and in accordance with company policy; that the executive commercial director (or bid sponsor) is provided with progress briefings and kept abreast of potential issues; that tenders obtain the appropriate approvals (sign-off) before submission. Commercial is also responsible for evaluating and reporting on the success of bids and the effectiveness of bid processes and procedures

Estimating and **pricing** activities include: leading and managing the estimating function, estimating, analysing and evaluating the costs and risks inherent in bid opportunities, overseeing the pricing process, constructing appropriate pricing strategies, modelling and negotiating prices, and establishing the commercial viability of all quotations. They also include working with the finance function to align pricing, margin, risk, contingency and discount provisions with organisational policy.

This component of the commercial role is underpinned by an ability to analyse customer requirements and analyse, define and plan the proposal, proactively project manage the proposal development stage. It also requires an understanding of the tendering process, bidding and proposal development best practice, the implications of Incoterms, the relationship between commercial competitiveness and profitability, and the financial aspects of bids and contracts.

Further reading

Bidding

Lowe, D and Skitmore, RM (2006) Bidding. In: DJ Lowe (ed.) with R Leiringer, *Commercial Management of Projects: Defining the Discipline*. Blackwell Publishing, Oxford. pp. 356–389

Procuring (demand-side)

Supply chain management (SCM) is a function and set of processes and activities responsible for planning, developing, coordinating, integrating and maintaining upstream and downstream collaborative relationships with customers, suppliers and other channel partners such as contractors, consultants, intermediaries, manufacturers, distributors and service providers, etc. – its supply chain.

A supply chain comprises a network of all the organisations, groups and individuals, plus the various activities, processes, resources and information, both within and external to an organisation, required to meet a specified need or implement a project. Supply chain activities, both inbound to and outbound from organisations, include the development, production, distribution, and marketing, etc. that transform resources (raw materials, components, and intellectual property) into a final finished product or service. This network, therefore, is responsible for realising and delivering a specific product and or service, including the requisite inputs, outputs and outcomes, to the ultimate customer.

> 'The supply chain conceptually covers the entire physical process from obtaining the raw materials through all process steps until the finished product reaches the end consumer. Most supply chains consist of many separate companies, each linked by virtue of their part in satisfying the specific need of the end consumer.' Chartered Institute of Purchasing and Supply (CIPS, 2012)

In the context of realising complex projects, the supply network can incorporate several specialist supply chains that coalesce into an integrated supply team (integrated project team). Although generally used to joint working, the constituent members, such as those in a construction setting, often move from project to project.

Functionally, SCM interacts with commercial, design, finance, IT, operations, marketing and sales, both within the organisation and across its supply chain, while its constituent activities include the sourcing and procurement, transformation and logistics management of capabilities and resources to meet an organisational need. Principally, SCM combines inter- and intra-organisational supply and demand management, with the aim of delivering enhanced customer value at a lower cost to the entire supply chain.[10]

Similarly, **procurement** is an organisational function and a set of business processes centred on the acquisition of assets (products, goods, resources and works) and or commissioning services from both external third-party suppliers and internal providers in response to a specific need or requirement. It encompasses the whole life cycle, commencing with the identification of a requirement and concluding either with the disposal of an uneconomic or unwanted asset or the completion of a service contract. The term **sourcing** (and in particular strategic sourcing) is also used in relation to the processes, approaches and strategies involved in engaging with the supply market to address current and potential business requirements. Procurement is generally undertaken by a purchase and supply team, either as business-wide activity or as a constituent part of a project or programme.

The terms 'procurement', 'purchasing' and 'commissioning' are frequently used interchangeably. Murray (2009), however, is of the opinion that this is a misuse of the terms, which causes some confusion as they mean 'different things to different people'. Further, he maintains that commissioning and procurement are discrete activities arguing that procurement is a subfunction of commissioning, which is driven by the commissioning cycle, whereas procurement is a broader activity than the purchasing cycle.

Commercial practitioners can be responsible for providing strategic leadership to the procurement function in organisations having accountability for establishing and directing procurement processes and procedures, developing and monitoring performance frameworks, and overseeing, directing and coordinating multi-functional procurement and acquisitions teams. Elsewhere, they may be responsible for coordinating the 'commercial' input, interfacing with and providing specialist expertise and support to a separate procurement department, and ensuring appropriate governance procedures are in place and applied.

They generally act as the organisation's lead interface and advisor on all commercial matters relating to external suppliers, partners and integrated project teams (IPTs). Aspects of the role include: collating and interpreting internal demand information, benchmarking and assessing market-wide supplier capability and performance data (supply network mapping, analysis and planning), evaluating alternative procurement methodologies, and utilising the results to inform the development of an appropriate corporate, business unit or programme specific procurement, sourcing and supplier relationship strategies that support the organisation's strategic objectives. The predominant aims are to achieve best value (value for money), ensure continuity of supply/service, deliver optimal commercial outcomes and create opportunities to leverage competitive advantage.

Specific buying (purchasing and strategic sourcing) activities undertaken include:

- Identifying, capturing, collating and appreciating internal customer and user needs (requirement capture), evaluating various solutions and options to address and meet these needs (developing

specifications to meet organisational requirements), applying value analysis, generating make or buy plans, establishing appropriate output and performance requirements to define them, selecting the preferred solution and determining the optimal procurement route to source the solution, including the development of specific supplier selection criteria based on both hard and soft issues
- Leading or liaising closely with and supporting the procurement function in the pre-qualification, tender, evaluation, selection, and award processes relating to supply, subcontract, outsourcing and work package contracts, etc:
  - pre-qualification (pre-qualifying bidders): recommending preferred suppliers and establishing framework agreements following the assessment, selection and accreditation of potential third-party organisations
  - tendering: initiating and managing the tender process in accordance with organisational procedures and applicable legislation and regulations. This can involve choosing an appropriate commercial approach, issuing invitations to tender, defining the obligations of each party, preparing tender documentation and ensuring conformity with procurement legislation
  - evaluating: assessing, reviewing and analysing the resulting bid submissions (supplier proposals), identifying potential commercial risks and compiling tender evaluation reports
  - negotiating: negotiating directly with suppliers (a key dimension of the procurement role), includes the negotiation of agreements and resolution of any subsequent disputes
  - awarding: recommending the preferred bidder, making recommendations on the contract award and concluding the 'deal'
  - administering: managing the resulting supplier contract

The role may also encompass broader aspects of supply chain management and logistics: inventory control, transportation – 'just in time' delivery, materials management – receipt and storage).

Commercial practitioners, therefore, require knowledge and experience of supply chain management, the procurement cycle (category management cycle), the various procurement options, such as, prime contracting, the Private Finance Initiative (PFI), project and strategic partnering, public–private partnerships (PPP), and tendering processes and procedures available to the organisation, and relevant procurement legislation and regulations. They also need an understanding of best value, contingency planning, cost planning, incentivised pricing, KPIs, performance-based outputs, reverse auctions, supplier pricing policies, sustainable procurement and whole-life costing.

The ability is needed to apply relevant sourcing tools and techniques, compile tender documents and reports, identify the potential risks associated with the proposed solution, to balance cost and quality, and manage and control costs during the pre-award stage factors, develop incentive pricing models, analyse complex and innovative commercial arrangements, manage and control costs during the pre-award stage, analyse supplier's proposals (the underlying costs of labour, materials, overheads and profit), and evaluate the suitability of bids.[11]

Further reading

Procurement, strategic purchasing and supply chain management

Langford, D and Murray, M (2006) Procurement in the context of commercial management. In: DJ Lowe (ed.) with R Leiringer, *Commercial Management of Projects: Defining the Discipline*. Blackwell Publishing, Oxford. pp. 71–92

Cox, A and Ireland, P (2006) Strategic purchasing and supply chain management in the project environment – theory and practice. In: DJ Lowe (ed.) with R Leiringer, *Commercial Management of Projects: Defining the Discipline*. Blackwell Publishing, Oxford. pp. 390–416

Project and/programme management

Project management

Project management is the process and activity of planning, controlling, monitoring, managing, organising, coordinating, defining, delegating, delivering and directing all aspects of a project. It includes the management of resources (materials, suppliers, contractors, consultants, etc.) and the motivation and leadership of the individuals involved throughout the project life cycle. Its aim is to achieve the project objectives safely, on time, within agreed cost, scope, performance/quality and risk criteria, deliver value and satisfy the aspirations of the stakeholder. Moreover, project management is widely acknowledged as the most efficient way of bringing about change.[12]

Programme management

Programme management is a structured and coordinated approach to the management of interrelated projects, involving the application of various processes, activities and procedures. It provides a framework for the implementation of proposals, strategy, and change, and for monitoring and evaluating whether or not a programme delivers its intended benefit to stakeholders. While the precise interpretation of what represent a programme varies between businesses and industry sectors, fundamental programme management practice exists along with an associated ability to establish and manage a programme or programmes. The underlying principles, procedures, techniques and activities are closely linked to those that support project management.[13]

Within the context of projects and programmes, the commercial function may have the responsibility for managing projects or delivering a portfolio of projects (from their inception to completion), including the oversight of their associated project teams. More often, however, commercial practitioners 'project manage' the proposition development and bid submission, solution selection and asset/service procurement, and award stages of the project life cycle. Post-award, they generally have the responsibility for the commercial (contractual, financial and possibly procurement) aspects of projects and programmes, supporting individual projects, interfacing with project implementation teams, and supporting and liaising with the project's commercial director, project manager/director and programme manager/directors. As mentioned earlier, the commercial function often has dual accountability, reporting to both the commercial executive and project/programme director or manager. Additionally, at the programme and project level the commercial role can involve:

- Managing the commercial input to programmes and individual projects under their control (in accordance with organisational systems and procedures) and having responsibility for their commercial success. As discussed earlier, commercial success is generally measured in terms of profitability and achieving value for money. Specific activities include:
  - developing strategies to support a project execution programme and the creation of an appropriate commercial strategy to help deliver a project's objectives
  - establishing and implementing project commercial operating systems and the project management office
  - ensuring that each project is commercially viable and appropriately monitored and evaluated during its implementation stage, generating and managing financial and performance monitoring information, securing due monies and fees, identifying issues and opportunities to improve profit margins and preparing business cases for board approval to ensure a return on investment and value for money
  - maintaining contractual records and documentation, reviewing and monitoring contract compliance, resolving disputes and managing conflict, and identifying, analysing and managing risk and opportunities
  - reporting on all commercial matters relating to assigned projects and on the success of individual projects to the commercial executive
- Providing commercial services to individual projects and programmes: this may include, for instance:
  - supplying commercial and contractual advice, support, guidance and expertise to project managers when preparing proposals; briefing project teams on the application of relevant contracts, agreements

and associated documentation, particularly on the organisation's key liabilities and obligations, and the implications of non-compliance; promoting project activities in accordance with the strategic plan. They can also become involved in training inexperienced project managers and team members in contracting procedures

The capability requires an awareness of the economic, legal, technological, political and environmental backdrop against which projects are executed and of the roles and responsibilities of all those involved. Specifically, it entails an understanding of the numerous project and programme management approaches, issues, principles, processes, techniques and tools. These include, for example, establishing and managing multifunctional teams, opportunity mapping, project handbook, project execution plans (PEPs), project planning, project programming, multiproject programming, activity schedules, flow diagrams, Gant charts, critical path, key milestones, progress monitoring, issue logs, cash flows, commissioning and handover procedures, and close-out reports. It also requires the ability to successfully manage commercial/procurement projects and programmes, identify and create programmes, coordinate resources within time, quality and budget constraints, track commercial activities and benefits, control performance against established targets, and identify and manage key project commercial and procurement risks.

Business development

Business development is a strategy-focused business activity associated with establishing strategic relationships and enduring partnerships with both customers and suppliers. Centred on realising new business opportunities, creating deals and delivering sustainable organisational success and increased sales, its processes and practices include designing business models, products and services, marketing and sales.

Again, depending on the structure of specific organisations, the commercial function can be responsible for overseeing business development activities. Generally, however, their involvement in this area involves leading commercial discussions with third parties and working in conjunction with and providing commercial support and expertise to colleagues in other functions, such as, business, marketing, sales, operations and programme management, and the senior management team. The objective is to ensure that robust business development, marketing and sales strategies are generated and that commercial issues are considered in associated business development, marketing and sales processes. Further, the role incorporates the identifying and exploiting of new markets and capitalising on business development opportunities to achieve business objectives, such as securing new customers and contracts, extending offerings to existing customers, expanding product and service capability, increasing market share/penetration, becoming the provider of choice in a chosen market, and delivering sales and margin targets.

Specific activities may include: leading and supporting **business development** initiatives and activities – targeting, initiating and developing business relationships; formulating and implementing corporate and business unit business development strategies and plans, and ensuring they are adhered to in order to leverage opportunities, deliver efficiencies and drive sustainable growth.

Further tasks undertaken include:

- Identifying, initiating, evaluating, advising the board on, cultivating, implementing, managing and exploiting **business opportunities** (including: acquisitions, business, commercial, development, joint venture, market, new product, partnership and service opportunities) and generating new customers; plus obtaining regulatory and legal sign-off for undertaking new business opportunities, ensuring that all potential opportunities meet the organisation's strategic objectives and investment criteria, and maximise the financial return
- Analysing, evaluating, and generating **business cases** and **proposals** (or providing commercial input to their development); particularly, generating and managing appropriate input data, for example, assessing the likely costs and risks involved in a particular venture, and ensuring that a robust review process is in place and followed, so that the organisation can make sound investment decisions

Marketing and sales

Marketing and sales, either individually or combined into a single department, are management functions, processes and activities centred on promoting and selling an organisation's products and/or services.

Marketing is responsible for anticipating, identifying and satisfying customer needs by defining, creating, communicating and delivering proposals (offerings) that possess or generate value for customers and partners profitably. Activities include maintaining key relationships and generating publicity and interest in an organisation at a reputational and corporate level.

Marketing activities carried out by commercial practitioners include: designing, developing, implementing, reviewing and updating marketing strategies; overseeing, directing or contributing to the preparation of marketing plans; providing advice, support and management to marketing campaigns, brand management and new product development; developing customer markets and cultivating customer relationships that are aligned with the organisation's overall strategic plan, maintain and enhance its reputation and contribute to its overall strategic vision. Further marketing tasks may include:

- Establishing a customer demand-focused organisational culture, identifying customers and unmet customer needs, developing appropriate product and service solutions, and ensuring that effective marketing resources, policies, protocols and procedures are in place to promote the organisation's products and services effectively
- Undertaking market research, analysis and benchmarking: developing and maintaining market intelligence systems to maintain knowledge of the market, review and analyse competitors, identify opportunities, mitigate commercial risk, inform strategy development and business and financial planning, and test the viability of new product and service offerings.

Alternatively, commercial practitioners may provide support services to a discrete marketing function, for example, in relation to possible legal and commercial risks inherent in new marketing activities and product/services.

Sales is the activity of selling (the organisation's products and/or services); it is a systematic process, focused on the exchange of value between the buyer and seller. Key aspects include cultivating relationships, seeking out, nurturing and securing opportunities then converting these prospects into actual transactions.[14]

The commercial involvement in sales may entail leading, managing and developing, or providing commercial support and expertise to, the sale function and sales teams, ensuring that commercial issues are accounted for in sales activities and that sales proposals are issued in line with company policy. This can involve contributing to and shaping the development of sales/campaign strategies and promotional plans, identifying and analysing new sales opportunities, generating and communicating strategic vision and sales stories, evaluating risks and opportunities in individual campaigns, balancing the drive for sales with business risks, defining and selling customer propositions, developing proposals and offers, and overseeing the implementation of sales plans.

This area of commercial activity requires knowledge and understanding of the various business environments/market sectors in which the organisation engages, the strengths and weakness of competitors' products and services, the capabilities and capacity of suppliers and partners, and the strength of the business's brand and reputation. Further, an ability to engage with and research the market, interpret market research, identify key decision makers in customer organisations, understand the customer's business objectives and end user needs, translate customer requirements into output specifications and apply appropriate business development tools are important requirements.

Risk and opportunity management

In some project-oriented organisations **risk management** is a discrete business function, either in its own right or in association with legal, regulatory or governance. Generally, however, it is viewed as a systematic, iterative process or group of management policies, processes or procedures – a defining feature of the culture of the organisation, which is associated with its internal control, governance and strategic management. The term also encompasses a coordinated set of activities, approaches, practices, principles, and structures that are applied to risk at the corporate, business and project level and underpin many if not all management activity. At the project level risk management occurs throughout the project/contract life cycle.

There are numerous definitions of risk and risk management. Eunice Maytorena in Chapter 4 of this book reviews and comments on several of these definitions and provides an introduction to risk, uncertainty and the process of risk management. She also outlines relevant underlying theory and provides an overview of some of the available risk management tools and techniques.

Risk is linked to an event (a decision, activity, function, or process) or series of events, the circumstances or environment in which it (they) occurs, and is bounded within a time frame. Additionally, the event or events have a potential consequence, either a positive opportunity or a negative impact, for which a probability or likelihood of occurrence can be established together with an estimation of the magnitude of its potential impact. In a business environment, this impact is commonly expressed as the sum of money the organisation may directly lose, or the adverse influence its occurrence may have both monetarily and reputationally on the business. As Eunice Maytorena discusses in Chapter 4, while some sources see risk incorporating both a positive (opportunity) and negative dimensions, the term is predominantly viewed as having only negative consequences. This view is also supported by the review of commercial job roles and frameworks with risk generally seen as a negative factor – something to be mitigated, reduced, and avoided. The commercial function, however, is engaged in risk and opportunity management.

Risk (and opportunity) management enables an organisation to identify, contextualise, analyse/assess/evaluate, quantify, understand, prioritise and communicate potential risks and opportunities. These risks, in turn, may be accepted or transferred to another party (a supplier or purchaser). Within a b2b context a degree of risk taking is expected, and is clearly associated with value (profit) creation (an opportunity): an organisation may agree to accept a specified level of risk, at a premium. A general commercial principle is that risk should be borne by the party best able, economically, to control, manage or insure against its consequences (Wearne, 1999). On identifying or accepting a potential risk, an organisation will seek to mitigate and reduce its likely impact by developing, justifying and implementing a risk response that can be coordinated, monitored and controlled.

Risk management is, therefore, a decision-support activity, which allows decision makers to make informed, proactive management decisions, striking a suitable balance between opportunity maximisation and the minimisation of potential threats and losses for each aspect of a project or for an entire project.

Specifically, the commercial function is involved in identifying, responding to and managing potential commercial, contractual and procurement risks related to satisfying an acknowledged requirement or responding to an identified opportunity. In particular, this involves drafting, negotiating, agreeing, managing and fulfilling any associated contractual commitment, with the aim of supporting the definition, creation and delivery of sustainable value (supply-side) or the identification, capture and exploitation of value (demand-side).[15]

Further reading

Risk management

Kähkönen, K (2006) Management of Uncertainty. In: DJ Lowe (ed.) with R Leiringer, *Commercial Management of Projects: Defining the Discipline*. Blackwell Publishing, Oxford. pp. 211–233

Combinations of these common components

Commercial roles across industry sectors and even within organisations can comprise different combinations of these common components. Different prominence is given to aspects of the role due in part to the dominance of other functions within the business (for example, marketing and sales, finance and accounting, legal, procurement/supply chain management and project management) and to the different 'histories' (backgrounds) of individual commercial practitioners. The role has been described as the accidental 'profession' as there is no designated entry route, unlike other more established business functions. Moreover, the function recruits from all the aforementioned areas.

Management, leadership and communityship

Management

The label '**manager**' is frequently given to commercial practitioners, irrespective of whether or not they have a line-management position, in order to confer status. However, the job descriptor commercial specialist is becoming more common, conferring equal status to those commercial actors who do not have line-management responsibilities.

Similarly, the terms '**managing**' and '**management**' occurred throughout the earlier overview of the processes undertaken by the commercial function. In fact, all the job specifications/functional frameworks included elements of managing and management. Table 1.2 lists the most frequently occurring management aspects of the commercial role.

The most common definitions of management have as their origin the description of managing provided by Henri Fayol (1916), who described managing as '*planning, organising, commanding, coordinating, and controlling.*' Latterly, **leading** is usually used in lieu of commanding. There is, however, less agreement over

Table 1.2 Aspects of management undertaken by commercial practitioners.

- Bid team management
- Brand management
- Budget management
- Business management
- Category management
- Change management
- Claims management
- Commercial management
- Commercial relationship management
- Compliance management
- Conflict management
- Contract/contractual management
- Cost/cash management
- Customer relationship management
- Data management
- Demand management
- Dispute management
- Facilities management
- Financial management
- Global account management
- Knowledge management
- Life cycle management
- Line management
- Management accounting
- Management contracting
- Management of due diligence
- Managing advisers
- Marketing and communications management
- Opportunity management
- Performance management
- Portfolio management
- Prepare management accounts
- Production management
- Programme management
- Project management
- Quality management
- Relationship management
- Reputation management
- Requirements management
- Resource management
- Risk management
- Sales management
- Service management
- Stakeholder management
- Strategic management
- Sub-contract management
- Supplier management
- Supplier relationship management
- Supply (chain/network) management
- Team (people) management
- Tender management
- Time management
- Value chain management
- Workload management

Source: Analysis of 90 commercial functional frameworks/job specifications

what is meant by the terms **leader** and **leadership**, although numerous definitions have been proposed; for example:

> 'A leader shapes and shares a vision which gives point to the work of others.' Handy (1992)

> 'Leaders are individuals who establish direction for a working group of individuals who gain commitment from these group of members to this direction and who then motivate these members to achieve the direction's outcomes.' Conger (1992, p. 18)

One area of literature differentiates between leadership and management, as illustrated by Kotter (2011):

- **Leadership**: is responsible for generating the systems that managers manage and revising these processes and structures in order to respond to risks and opportunities and allow the organisation to grow and evolve. Leadership involves:
  - creating the organisation's vision and formulating its strategy
  - establishing and communicating its direction
  - motivating action
  - aligning individuals
- **Management**: ensuring that these systems, which integrate technology and people, run consistently, efficiently and effectively. Management involves:
  - planning and budgeting
  - organising and staffing
  - controlling and problem solving

Conversely, another section, exemplified by Mintzberg (2009), argues that leadership should be seen as a component of management, which is itself embedded within what he terms **communityship**: *communities of actors who get on with things naturally*. Mintzberg views management as a combination of:

- **Art**: the exp.ression or application of human creative skill and imagination; a skill at doing a specified thing, typically one acquired through practice
- **Craft**: an activity involving skill in making things by hand – *art and craft;* or the skills in carrying out one's work
- **Science**: a systematically organised body of knowledge

Analysis of commercial functional frameworks/job specifications revealed that commercial practitioners were predominantly involved in **leading** (including directing, mentoring and overseeing) and **planning** (which included strategy development). Additionally, there were some references to **coordinating** (and administering) and **controlling**; although these were limited. Surprisingly, there were no references to **organising** tasks. Interestingly, there appears to be reluctance in some organisations to use the term administration, particularly in relationship to post-award contract administration/management. The term is unfashionable, infers (inappropriately) subservience, and its lack of use is probably driven by a desire for the commercial contribution not to be seen as a 'back-room' technical support role. Having said this, many of the job descriptions included the requirement to provide support and assistance to others, and/or be a source of advice, guidance and expertise: the terms 'lead consultant', 'trusted advisor' and 'advisor of choice' were to be found.

In addition to managing and leading, commercial actors provide an **advisory** role – a source of specialist knowledge (therefore a specialist, an artisan); a **regulatory** (policing, protecting and enforcing) role, having responsible for overseeing, monitoring and ensuring compliance and performance, and sanctioning action; and an **advocacy** role, persuading and influencing others.

Maria-Christina Stafylarakis explores these issues further in Chapter 2 and in particular the importance of personal leadership (mastery) to commercial practitioners.

Commercial management professional bodies and associations

Commercial practice and commercial practitioners are principally supported by two bodies: the International Association for Contract and Commercial Management (IACCM) and the Royal Institution of Chartered Surveyors (RICS):

- **IACCM**: with members spread across 28 countries and 8804 organisations, the IACCM supports public and private sector professionals and corporations drawn from numerous industry sectors in the area of contracting and relationship management. It is a non-profit membership organisation, offering advisory, research and benchmarking services, together with web-enabled training and certification. Its membership comprises contract and commercial managers, negotiators, legal and supply chain specialists
- **RICS**: established in London in 1868, and granted a Royal Charter in 1881, the RICS is an independent organisation that represents professionals engaged in aspects of land, property and construction, plus associated environmental matters. It has approximately 100 000 qualified members spread across 140 countries. It is divided into several professional groups, which have developed generic statements of competencies required of members. Its Quantity Surveying and Construction group supports approximately 40 000 professionals who have an interest in the cost and procurement of construction projects

However, due to the eclectic nature of the role, as defined earlier in the chapter, there is a certain degree of overlap between the activities undertaken by the commercial function and the areas of interest of other

Box 1.1 Various professional bodies with an interest in commercial activities

Contract management

- **National Contract Management Association** (NCMA): A membership-based, professional society, NCMA, based in Ashburn, Virginia USA, was established in 1959 to advance the professional development of its members. The organisation has developed a Contract Management Body of Knowledge, now in its third edition: www.ncmahq.org/ProfessionalDevelopment/NCMAProductDetail.cfm?ItemNumber=501 (accessed 25 March 2011)

Financial decisions

- **The International Federation of Accountants** (IFAC): A global body comprising 163 individual professional accountancy organisations, IFAC represents over 2.5 million accountants in public practice, education, government service, industry and commerce. Separate organisations have developed their own bodies of knowledge
- **American Institute of Certified Public Accountants** (AICPA): Based in New York, AICPA is the world's largest accounting profession association representing 370 000 members in 128 countries. Its members are active in business and industry, public practice, government, education and consulting
- **Chartered Institute of Management Accountants** (CIMA): Active in 168 countries, CIMA is the world's leading and largest professional body of management accountants. Established in 1919, it has 183 000 members working in industry, commerce, the public sector and third-sector organisations
- **Additional UK-based organisations** include: The Institute of Chartered Accountants in England and Wales (ICAEW); The Institute of Chartered Accountants of Scotland (ICAS); The Institute of Chartered Accountants in Ireland (ICAI); The Association of Chartered Certified Accountants (ACCA); and The Chartered Institute of Public Finance and Accountancy (CIPFA)

Legal

- **The Law Society**: Founded in 1825 as 'The Society of Attorneys, Solicitors, Proctors and others not being Barristers, practising in the Courts of Law and Equity of the United Kingdom', the Law Society represents in excess of 145 000 solicitors (qualified in England and Wales) that practice globally. The Law Society Group contains the Solicitors Regulation Authority (SRA), which is responsible for regulatory and disciplinary matters relating to solicitors in England and Wales. It also establishes, oversees and enforces professional standards
- **The Commerce and Industry Group**: An unincorporated association (recognised by the Law Society) established in 2001, to represent lawyers (solicitors, barristers, company secretaries, legal clerks, legal executives and paralegals) working in-house in commerce and industry
- **The International Bar Association** (IBA): With a global membership that includes 197 professional bodies, plus over 40 000 individual members, IBA is the world's principal organisation of international legal practitioners, bar associations and law societies. Established in 1947 its function is to influence the development of international law and the legal profession globally
- **American Bar Association** (ABA): With a membership approaching 400 000 ABA is the largest (voluntary) legal professional association in the world. Acting for the profession at a national level, it accredits law schools, provides continuing professional education and instigates proposals to improve the US legal system
- Further **International bodies** representing the legal professionals include the Law Council of Australia, the Canadian Bar Association (CBA) and the Law Society of Hong Kong

Transaction management

- **International Federation of Purchasing and Supply Management** (IFPSM): With a global perspective, IFPSM, based in Switzerland, is a union of 43 National and Regional Purchasing Associations, which have a total membership of approximately 200 000 professionals. The body's aim is to promote the procurement profession and aid the development and distribution of knowledge
- **Institute for Supply Management™** (ISM): ISM has a membership of over 40 000 supply management professionals supported by a network of affiliated associations. Founded in 1915, it supports research, promotes education – offering a wide range of educational products and programmes, undertakes promotional activities and develops standards of excellence
- **The Chartered Institute of Purchasing and Supply** (CIPS): An international organisation supporting the purchasing and supply profession, CIPS promotes good practice and high standards of professional skill, ability and integrity. Established in 1932, CIPS assists individuals, organisations and the overall profession. CIPS were awarded a Royal Charter in 1992

Project/programme management

- **International Project Management Association** (IPMA®): Tracing its roots to 1965, IPMA is an international umbrella organisation for over 50 national project management associations representing over 40 countries. Its aspiration is to lead the promotion, development and recognition of the project management profession. To this end it certifies project managers, publishes a number of project management publications, and provides standards and guidelines for project management personnel via the IPMA Competence Baseline (ICB) version 3. http://ipma.ch/resources/ipma-publications/ipma-competence-baseline/ (accessed 29 August 2012)

- **The Association for Project Management** (APM): Having more than 18000 individual and 500 corporate members, APM is the largest project management professional body in Europe. In addition, it has a branch in Hong Kong and an international online community. APM's objective is to raise awareness and standards in the profession by developing and promoting project and programme management. Underpinning APM's qualifications, accreditation, research, and publications, the APM Body of Knowledge (fifth edition) is a principal part of APM's 'Five Dimensions of Professionalism'. http://www.apm.org.uk/APM5Dimensions (accessed 29 August 2012)
  - representing the UK, APM is the largest member of the IPMA; it is currently seeking a Royal Charter
- **Project Management Institute** (PMI): Based in Newtown Square, PA, USA and with regional offices in EMEA, India, Asia Pacific and China, PMI is a global project management profession association. Formed as a not-for-profit organisation, it has over half a million members and aspiring members across 185 countries. APM promotes project management through its standards and qualifications, and supports research programmes and professional development. Its Project Management Body of Knowledge (PMBOK®) is now in its fourth edition. http://www.pmi.org/PMBOK-Guide-and-Standards.aspx (accessed 29 August 2012)

Marketing and sales

- **The Chartered Institute of Marketing** (CIM): CIM is the world's largest organisation for professional marketers, with bases in eleven countries beyond the UK. It is involved in training, developing and representing the marketing profession
- **American Marketing Association** (AMA): AMA is a USA-based global professional association for marketers, both individuals and organisations, involved in the training and certification of individuals, and the development of the marketing practice

Risk and opportunity management

- **The European Institute of Risk Management** (EIRM): EIRM is a knowledge network, which generates, collates and disseminates risk management information to the private and public sectors
- **The Institute of Risk Management** (IRM): IRM is a global enterprise-wide education institute with members and aspirant members in over 50 countries; an advocate for the risk profession, it supports risk professionals by offering qualifications and development programmes. IRM's membership comes from a variety of risk-related disciplines across variety of industry sectors
- **The Risk Management Institution of Australasia Limited** (RMIA): Within the Asia-Pacific region RMIA is the largest professional association for risk management. Its membership is derived from numerous diverse disciplines, industry and government sectors. RMIA provides networking opportunities for its members through local chapters and special interest groups. It also provides educational programmes, professional accreditation and publications

Sources: IACCM http://www.iaccm.com; RICS http://www.rics.org; NCMA http://www.ncmahq.org/; IFAC http://www.ifac.org/; AICPA www.aicpa.org; CIMA http://www.cimaglobal.com; http://www.lawsociety.org.uk/home.law; http://www.cigroup.org.uk/; IBA http://www.ibanet.org; ABA http://www.americanbar.org/aba.html; CBA http://www.cba.org/; http://www.lawcouncil.asn.au/; IFPSM http://www.ifpmm.org; ISM http://www.ism.ws; CIPS http://www.cips.org; IPMA® http://www.ipma.ch; APM http://www.apm.org.uk; PMI http://www.pmi.org; CIM http://www.cim.co.uk; AMA http://www.marketingpower.com; EIRM http://www.eirm.com; IRM http://www.theirm.org; RMIA http://www.rmia.org.au (accessed 29 August 2012)

Box 1.2 Extract from the APM Body of Knowledge

5.0 Business and commercial
- Business case
- Marketing and sales
- Project financing and funding
- Procurement (includes contract management)
- Legal awareness

2.3 Value management
4.3 Estimating
4.5 Value engineering
7.0 People and profession
- Communication
- Teamwork
- Leadership
- Conflict management
- Negotiation
- Behavioural characteristics

Source: APM (2006)

business functions. Box 1.1 includes various professional bodies that represent other business functions with a common interest in commercial activities.

Within most businesses there is a certain amount of competition between the various business functions; for example, between project management, supply chain management, and commercial management for the position of 'lead consultant' or 'trusted advisor' within the organisation in respect of managing the exchange interface (transaction management). Similarly, each of the professional organisations listed in Box 1.1 has its own agenda and remit to promote and support the capabilities of its members.

Many of these organisations have developed specific **bodies of knowledge** (BoK), which identifies the underlying principles of the area of its specialisation and guides to practice. They may also include glossaries of common terms, lists of acronyms and suggested readings (see, for example, Morris *et al.*, 2006a, 2006b). An example of a BoK is the APM Body of Knowledge (APM, 2006) developed to support the development and certification of project managers.

A **project manager** is the individual (or entity) given the overall responsibility, authority and accountability for managing a project, administrating the contract (or contracts) and leading the project team in facilitating the successful delivery of the project, the required products or the realisation of specific objectives within pre-established constraints.[16]

It is interesting to note that one of APM BoK's seven sections is entitled 'Business and commercial', which includes many of the key activities undertaken by commercial practitioners (see Box 1.2). Other sections of the APM BoK include further activities undertaken by commercial practitioners – these include value engineering and management, and estimating. It is unclear whether the authors of the BoK claim 'ownership' of these activities for the project management function or classify them as a constituent part of the broader and strategic management of projects for which project managers require a working knowledge of commercial awareness.

Developing the commercial function

In the absence of universally accepted cross-industry sector professional standards, major organisations such as Rolls-Royce, BT and BAE Systems have invested in the development of comprehensive commercial competence matrices, process excellence and functional development programmes, and university award-bearing qualifications. Similarly, many organisations have appointed senior-level function champions and/or established **Centres of Excellence** (CoE) to rationalise the function across business units (divisions) and to provide a focus for supporting and coordinating their commercial function, providing strategic leadership and championing efforts to improve the function's resources and competence by introducing consistent processes, techniques and standards, assurance mechanisms, training and development opportunities, knowledge management capabilities and learning solutions, plus benchmarking functional performance against comparable organisations. Additionally, this intervention sought to raise the status of the commercial function both within the organisation and externally.

From a public sector, demand-side perspective, various UK government reports have revealed major weaknesses within central government departments in a number of the commercial skills viewed as crucial to the delivery of complex projects. For example, the NAO report: *Commercial Skills for Complex Government Projects* (NAO, 2009) found that departments were not utilising their limited commercial skills and experience to the '*best effect*'; areas of weakness included contract management, the commissioning and management of advisers, risk identification and management, and business acumen. The report established that the main obstacles to developing commercial skills were the demands to reduce public spending and the frequent movement of commercial personnel. It also presented a '*Commercial Skills for Complex Projects Framework*' comprising a set of key commercial skills and behaviours. As a result several initiatives have been instigated to improve the commercial capability within central government. For example, the MoD has developed the *Defence Commercial Function Skills Framework* (MoD, 2009) and *The Commercial Toolkit* (MoD, 2008). The former provides a set of competences for the defence commercial function, comprising market knowledge, commercial operations, whole supply network management, and regulatory and legal requirements. It also includes a glossary of terms, a set of appropriate skills, a commercial 'road map' and a competence map.

Commercial capabilities and activities

The core capabilities ('soft' skills: individual and interpersonal processes and behaviours) and activity-based (task-specific) competencies frequently sought by employers in their commercial staff are listed, in order of preference, in Table 1.3.

Although considered to underpin the commercial role, these skills are in the main generic and pertinent to most professional and managerial practice. However, a survey of commercial practitioners (Lowe, 2008a) established the importance attached to various capabilities (Table 1.4) and task-based competencies

Table 1.3 Top 20 'commercial' capabilities and competencies.

Rank	Capability	Rank	Capability
1	Communication and presentation skills	11	Contract management skills
2=	Commercial acumen/awareness	12=	Decision-making skills
	Relationship/people management skills		Staff development skills
4=	Leadership skills	14=	Creative/innovative thinking
	Negotiating skills		IT/IS skills
6	Team/collaborative working skills	16	Strategic thinking
7	Influencing skills	17=	Problem solving skills
8=	Analytical skills		Time management skills
	Project management skills	19=	Change management skills
10	Financial acumen/awareness		Cost management skills

Source: Analysis of 90 commercial functional frameworks/job specifications

Table 1.4 Individual and interpersonal processes and behaviours.

Primary commercial 'soft' skills	Secondary commercial 'soft' skills
1 Negotiation	11 Leadership
2 Communication	12 Creativity/ innovation
3 Problem solving	13 Evaluation
4 Commercial acumen	14 Team building
5 Logical thinking	15 Cultural awareness
6 Influencing others	16 Organisational
7 Analytical	17 Financial acumen
8 Broad perspective	18 Numerical
9 Collaborative working	19 Political awareness
10 Team working	20 IT/IS

Source: Lowe (2008a)

Table 1.5 Activity-based competencies.

Primary commercial task-specific competencies	Secondary commercial task-specific competencies
1 Negotiate contracts	8 Manage workloads
2 Analyse and manage risk	9 Manage knowledge
3 Understand business objectives	10 Manage change
4 Contract management	11 Project management
5 Manage relationships	12 Value management
6 Manage conflict	13 Bidding
7 Draft contracts	14 Performance management
	15 Manage documents
	16 Manage supply chains
	17 Develop new business

Source: Lowe (2008a)

(Table 1.5), subdividing these items into core and secondary skills/competencies. These results are generally consistent with previous IACCM surveys, although the list of potential skills and abilities used in Lowe's survey was considerably more extensive.

The main area of divergence, however, is the relatively low importance given to the abilities to manage documents, subdivide work into packages and manage cash flow, and numerical and IT/IS skills. These skills and abilities had previously been found to be important to commercial managers working within the UK construction sector (Lowe *et al.*, 1997). Further, competencies considered to be of moderate importance included developing new business, the use CM/sourcing software and marketing. Additionally, it is interesting to note that 'leadership' was only rated as a secondary skill, considering commercial managers' aspirations to seek a more strategic role within organisations. In order to achieve this, commercial practitioners need to be seen as 'leaders'. This dichotomy is addressed by Maria-Christina Stafylarakis in Chapter 2.

Lowe found a relatively high degree of homogeneity in the practitioners' responses, irrespective of their regional location, function/type of commercial/contract professional, level of academic qualification, and length of experience. Although, the results indicate that those with the least experience (particularly those with less than five years' experience) consider the skills/abilities associated with bidding, team building, project management, marketing, IT, managing documents, use of CM/sourcing software and thinking logically to be more important than those with longer experience; these tasks being more pertinent, perhaps, to their current commercial role.

Will your company help you acquire the skills for the future?

A previous IACCM survey (Lowe *et al.*, 2006) established a clear positive relationship for both job satisfaction and career path with support from the organisation in the acquisition of skills and knowledge, and in providing clear and meaningful rewards for high performance. However, it also found that while 42% of the respondents thought that their company would help them acquire the skills they need for the future, a further 42% were unsure that their company would and 16% considered that their company would not help them. Organisations, therefore, should consider introducing (or enhancing existing) mechanisms that recognise and reward high performance and enable commercial staff to acquire new skills and knowledge.

Capability maturity model

According to Ibbs *et al.* (2004), project management maturity (PMM) is an indication of an organisation's present project management sophistication and capability: for example, its managers know which technique to apply, appropriate to a project's requirements. By inference, therefore, commercial management maturity (CMM) is a measure of a firm's commercial management aptitude and erudition.

Several models designed to measure PMM have been developed, these include:

- The Berkeley Project Management Process Maturity Model
- ESI International's Project Framework
- Software Engineering Institute's (SEI) Capability Maturity Model Integration
- Project Management Institute's (PMI) Organisational Project Management Maturity Model (OPM3)
- Office of Government Commerce's (OGC) Project Management Maturity Model

However, relatively little development has been undertaken in respect of CMM. The exception is IACCM's Capability Maturity Model (David *et al.*, 2008; IACCM, ND), which addresses the following nine business dimensions:

1. Leadership
2. Customer/supplier experience
3. Execution and delivery
4. Solution requirements management
5. Financial
6. Information systems/knowledge management
7. Risk management
8. Strategy
9. People development

The tool enables an organisation to evaluate both its demand and supply side process performance against each dimension. Exemplar descriptors are provided for each of the following phases:

Phase 1: start up
Phase 2: disciplines under development
Phase 3: discipline is functional
Phase 4: continuous development
Phase 5: world class (best-in-business)

Box 1.3 provides example descriptors for the people development dimension.

Within the project environment, Ibbs *et al.* (2004) illustrate that organisations with developed project management practices exhibit superior project performance. Improvements in PMM, therefore, can result in:

- Improved cost and schedule performance and reliability
- Lower direct project management costs

The fundamental argument they make is that improved project performance is associated with enhanced PMM and capability, and can, as a result, stimulate corporate success. Similarly, IACCM claim that by benchmarking their contracting capability maturity, using the CMM model, an organisation can emulate the success of 'best-in-business' companies: generating additional savings and cost reductions, and thereby becoming more profitable than the market norm.

There are, however, dissenting views. Cooke-Davies (2004), for example, states that the simplistic process perspective inherent within maturity models does not embrace the complexity of project processes. Further, he argues that they are unlikely to be the panacea that some commentators suggest, as they:

- Lack a theoretical underpinning
- They assume there is a single 'ideal' path towards maturity

However, despite these comments, he asserts that project management maturity models make a positive contribution to the domain.

Box 1.3 Example descriptors for the people development dimension

Phase 1: Start up

- Organisation desires best-in-business competencies to support business
- Skills, abilities and competencies do not exist to support commitment management
- Job family +/or function does not exist
- People lack leadership, vision and meaning in work lives

Phase 2: Disciplines under development

- Commitment made to develop right competencies, hire right resources
- Initial assessment of contracting capabilities and limitations completed
- Immediate and irresolvable skills gaps addressed
- A vision for the future is in place and articulated to the entire organisation

Phase 3: Discipline is functional

- Team is 're-chartered' with new vision and mission upfront
- Behavioural absolutes are clearly defined and accepted
- Change management is given high priority and attention
- Job family and profiles (descriptions) are developed
- New hires reflect a mix in talent and perspective
- Assessment of competencies against standards is completed

Phase 4: Continuous Improvement

- Team is performing to new expectations and delivering value to the business
- Skills assessment has driven meaningful development plans
- Gaps in performance are aggressively managed
- Performance expectations have been raised significantly
- Both results and process metrics are in place to track performance

Phase 5: World class (best-in-business)

- Team members viewed as trusted business advisors by the organisation
- Skills gaps have been closed and higher order skills being developed
- Team is performing at high levels with great collaboration; excellence in execution is a given
- Metrics reflect stable, repeatable performance
- Team members being tapped to take on bigger roles in organisation

Source: adapted from IACCM Capability Maturity Model

Is commercial management a profession?

The question centres around the issues of status and the ability of the commercial function to interface on equal terms with other functions within and external to the organisation. For example, the engineering and legal functions have well-established professional status and representation. Similarly, the Chartered Institute of Purchasing and Supply (CIPS), Chartered Institute of Marketing (CIM) and Chartered Institute of

Management Accountants (CIMA) have all recently acquired their Royal Charters, a sign of achieving professional status within the UK. While these bodies originated in the UK they all have global reach.

While there is disagreement over what constitutes a profession, there is generally agreement on the importance of its knowledge base, for example:

> '... the power and status of professional personnel is considerably influenced by the degree to which they can lay claim to unique forms of expertise and the value placed on that expertise'. (Eraut, 1994)

Therefore, to appropriate Morris (2004), if there really is a distinct commercial discipline:

> '... then there needs to be some knowledge about it that can be articulated with a reasonable degree of robustness'. (Morris, 2004)

However, as is evident from the foregoing discussion of the key activities undertaken by commercial practitioners, the commercial function cannot lay claim to a discrete knowledge base. There is much that they share with other business functions, such as supply chain management and project management. Both these areas have been subject to considerable research. Examples include the theory and practice of supply chain management (see for example, Burgess *et al.*, 2006; Cousins *et al.*, 2006a, 2006b; Giunipero *et al.*, 2006; Harland *et al.*, 2006; Storey *et al.*, 2006) and the development of a conceptual framework for supply chain management (Croom *et al.* 2002; Giannakis and Croom, 2004). Similarly, work within the project management field includes: project contract management and a theory of organisation (Turner and Simister, 2001) and developing a theory of project management (Söderlund, 2004; Turner, 2006a, 2006b, 2006c), although the later has received some criticism (see for example, Morris, 2002; Sauer and Reich, 2007).

Returning to the question – can commercial management be viewed as a profession? The function has elements of codified knowledge derived from the legal, regulatory, governance and financial dimensions of the role. Moreover, a subset of the function – commercial managers within the construction sector in the UK context – are frequently members of the RICS. However, Mintzberg (2009) regards management as an art or craft not a science, and as such cannot be a profession; he considers management to be incompatible with the notion of 'profession'. Similarly, in respect of project management Morris (2002) argues that, while we are able to differentiate good project management practice, '... *there will never be an overall theory of project management*'. In fact, he considers the idea to be mistaken.

One can sympathise with commercial practitioners in their desire to attain professional status. They are widely considered to be subject matter experts – a source of specific knowledge, expertise and advice – and are often responsible for interfacing with externally sourced experts. In addition to the RICS, the IACCM supports commercial, contracting and commitment practitioners, providing a source of communityship (community of practice), a repository of knowledge and best practice, and a potential means by which to raise the status of the commercial (and contract) management function.

Theoretical background

The commercial function draws on an eclectic mix of theories, concepts and academic disciplines as illustrated by Table 1.6 and Figure 1.9.

Transaction cost economics (TCE) is an interdisciplinary subject that combines aspects of economics, contract law and organisation theory. As a result, it is highly relevant to commercial management, providing both a theoretical framework and common vocabulary by which to explore and explain commercial practice.

Transaction cost economics

Oliver Williamson, drawing on the work of Ronald Coase, Chester Barnard and Herbert Simon,[17] is considered to be the key proponent of TCE. His research focuses on economic **governance** and particularly the **boundaries of the firm.**[18] The **governance structure** is the institutional framework within which economic

Table 1.6 Underlying theoretical concepts and models.

Awareness			
1	Leadership	14	Trust
2	Best value/value for money	15	Negotiation analysis
3	Competitive advantage	16	Decision analysis
4	Governance	17	Corporate control
5	Strategy	18	Transaction costs
6	Globalisation	19	Contract theory
7	Core competence	20	Group dynamics
8	Commitment	21	Bidding theory
9	Creativity	22	Organisational behaviour
10	Negotiation theory	23	Problem-solving interest-based negotiation
11	Supply chains	24	Strategic purchasing
12	Value chain	25	Discounted cash flow
13	Contract law model		

Source: Lowe (2008b)

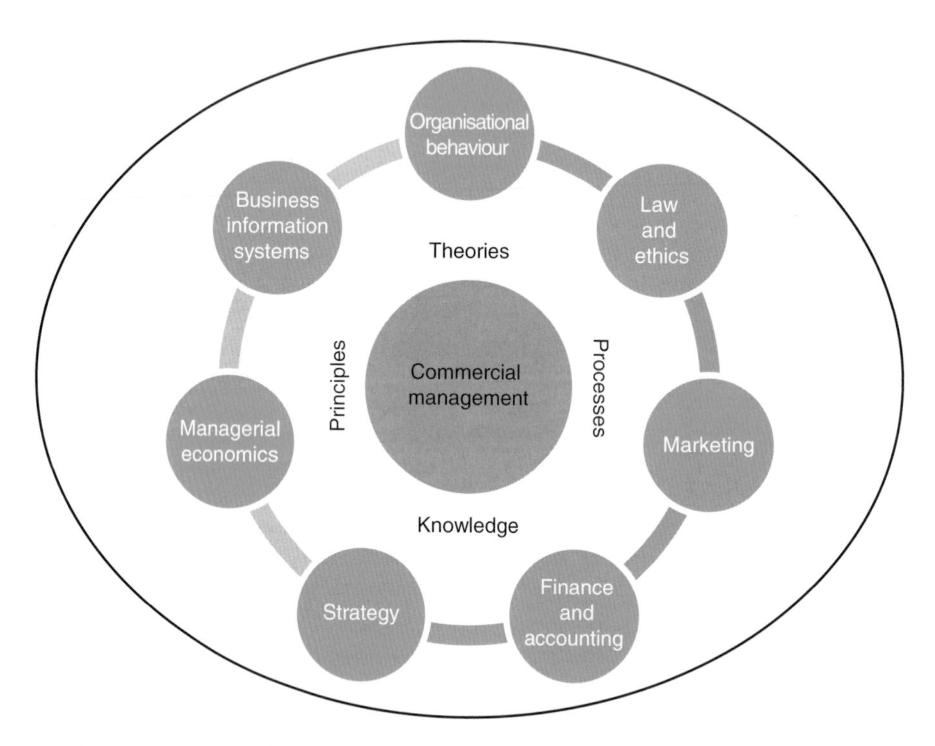

Figure 1.9 Principal academic disciplines underpinning the commercial function.

exchanges (**transactions**) are conducted. There are principally two choices: **markets** (purchasing products and services external to the organisation) and **hierarchies** (manufacturing the product or supplying the service from within the organisation) the classic '**make or buy**' decision. Although in certain circumstances an alternative **hybrid** approach is appropriate which seeks to combine the advantages of markets and hierarchies.

Employing the **lens of contract**, TCE adopts the **transaction** as its unit of analysis and **governance** as the instrument through which order is introduced to the exchange, consequently mitigating conflict and attaining reciprocal benefit. A **transaction cost** is incurred, according to Williamson (1981), whenever a product or service is '... *transferred across a technologically separable interface*', necessitating the development of skills, abilities and the means by which to produce the product or deliver a service. It is, therefore, the cost of acquiring a product or service by means of the market as opposed to producing or supplying them from within the firm: the costs incurred in an exchange. Transaction costs can be categorised under the following activities:

- **Selecting**: for example, the costs involved in collecting and collating information in order to identify and evaluate potential trading partners
- **Contracting**: for example, the costs associated with drafting and negotiating agreements
- **Monitoring**: for example, the costs associated with post-award (**ex post**) contract management in ensuring that the commitments made by both parties are fulfilled
- **Enforcing**: for example, the costs incurred as a result of ex post haggling (negotiation) and implementing sanctions when the other party fails to carry out the terms of the agreement (Williamson, 1985; North, 1990)

These activities, as identified earlier in the chapter, are predominantly undertaken by commercial practitioners: the commercial function is, therefore, essentially a transaction cost.

Although TCE generally assumes a demand-side perspective, transaction costs are incurred by both the purchaser and supplier. While the demand side incurs the direct cost involved in undertaking the activities of selecting, contracting, monitoring and enforcing, they will, more than likely, also pay the costs incurred by suppliers in complying with and responding to these activities. Suppliers usually include these costs as an overhead, which is recovered on contracts undertaken.

In identifying appropriate governance structures (**safeguards**), TCE emphasises the issue of post-award (ex post) **governance**, and in particular the provision of control mechanisms to prevent or moderate the opportunistic behaviour of an exchange partner. Williamson defines opportunism as '*self-interest seeking with guile*' and, while he acknowledges that many trading partners will not behave opportunistically, he highlights the difficultly in determining pre-award (**ex ante**) those who will, although, it is considered that most '*have their price*' – a point at which the rewards for behaving opportunistically are too high to ignore.

The main focus of attention in TCE is the inability to provide appropriate or adequate adjustment (**maladaptation**) during the contract execution phase. For example, in most projects change is inevitable due to unforeseen circumstances and technical advancement. The objective, therefore, is to predict potential contractual hazards during the pre-award phase, identify appropriate control mechanisms and incorporate these into the governance structure (contractual arrangement) adopted.

Williamson notes that Western exchange transactions are predominantly safeguarded by means of legal contracts. These contracts lie on a continuum ranging from simple discrete to complex bespoke; the cost of drafting, negotiating and agreeing these contracts increases significantly as one progresses from simple to complex. Ultimately, a point is reached where the cost of creating appropriate contracts is prohibitive and it becomes more cost-effective to integrate the activity vertically. Additionally, while the identity of the parties is considered relatively unimportant in discrete contracts used to purchase generic goods and services, they are significant in complex contracts (particularly those that are recurrent).

TCE utilises three attributes to describe exchange transactions:

- **Asset specificity**: concerns the extent to which trading parties can redeploy specialised assets acquired to either manufacture a product or deliver a service, and the importance of these assets to the relationship. Williamson identifies the following specific assets: physical, human, site, dedicated, brand name, and temporal. Examples can include the purchase of specialised plant or equipment, undertaking extensive project specific personnel induction activities and the co-location of integrated project teams. Where a party is unable to reutilise these assets or recoup their investment in them bilateral dependency is created, which it is assumed can increase the potential for opportunistic behaviour
- **Uncertainty**: long-term contracts are subject to change and variations due to unforeseen circumstances – for example, as a result of technological change. In order to accommodate these changes contracts

incorporate mechanisms to coordinate and reconcile these adaptations. Uncertainty can give rise to ex ante environmental cost and ex post behavioural costs
- **Frequency**: regular and recurrent transactions encourage the preservation of relationships due to the potential value to be derived from future exchanges. Frequent exchange generates **self-interest trust** and incentivises the parties to invest in specialised governance

Additionally, Williamson identifies two behavioural characteristics associated with these attributes:

- **Opportunism**: as defined above, involves a party going beyond the pursuit of self-interest to include dishonesty, double-dealing, deception and falsifying information in order to achieve its objectives
- **Bounded rationality**: individuals when making decisions are constrained by the availability of information, their cognitive skills to process and analyse this information, and a finite amount of time in which to reach a decision. Bounded rationality, when faced with complex and uncertain environments, will lead (it is theorised) to hierarchy

Winch (2006) has suggested the addition of a third behavioural characteristic:

- **Learning**: linked to the frequency of transactions and the length of the relationship between the parties, the more often and longer two parties transact the more they discover of the other's requirements, aims, processes, etc. and their motivation to transact

Asset specificity is the key feature of TCE. The theory asserts that transaction costs increase as the exchange parties (**transactors**) invest more in specialised assets, due to the need to safeguard themselves against the potential opportunistic behaviour of the other party. According to Williamson (1991), '*asset specificity increases the transaction costs of all forms of governance*'.

- **Market governance (classical contracting)**: this is the predominant governance structure for both repeat and one-off contract exchanges, where the product or service is standardised or commoditised. Under this structure the identities of the parties are generally unimportant with minimal effort expended on relationship development and management, as an on-going relationship is not sufficiently valued by either side. Scope and performance is determined by discrete contract terms and conditions, opportunistic behaviour is tempered by the availability of alternatives within the market and litigation is used as the prime dispute resolution mechanism. Features of this approach include: transactional, arms-length and competitive relationships regulated by contracts, standards and rules
- **Trilateral governance (neo-classical contracting)**: this is seen as an appropriate governance structure for both mixed and highly idiosyncratic infrequent transactions. The main feature of the structure is the use of third-party intervention (arbitration) to resolve post-award disputes, rather than relying on the legal system (litigation). As both parties are generally motivated to complete the contract, due to their investment in specialised assets and the potentially high switching/termination costs,[19] the focus is on facilitating the maintenance of the relationship. An example is the use of the **Dispute Adjudication Board** under the FIDIC Construction Contract (described in Chapter 6)
- **Transaction-specific governance (relational contracting)**: these are specialised governance structures that are frequently utilised on mixed and highly idiosyncratic frequent non-standard transactions. In the case of 'intermediate-production market transactions', Williamson identifies two approaches:
  - **bilateral governance (obligational contracting)**: here bespoke governance arrangements are negotiated by the transacting parties to meet the specific circumstances of the exchange. It is a hybrid, networked, third-way approach positioned between markets and hierarchies. Examples include collaborative arrangements, such as, project and strategic partnering, framework arrangements and alliances, which incorporate risk and revenue-sharing mechanisms, and relational approaches based on trust[20]
  - **vertical integration**: here the manufacturer of the product or supplier of the service is brought (directed) within the boundaries 'acquiring' organisation

TCE has been criticised for overemphasising the appropriateness of either the protection afforded by contracts or vertical integration (Poppo and Zenger, 2002), while Ghoshal and Moran (1996) argue that organisations are not direct '*substitutes for structuring efficient transactions when markets fail*'. Moreover, they assert that TCE fails to allow for the value creating components of organisations when compared with markets. Nor does it address the limitations of the hierarchies and in particular availability of or the ability to develop the required capabilities (Cousins *et al.*, 2008). Additionally, Zajac and Olsen (1993) suggest that, in certain circumstances, greater value may be generated by adopting an apparently inefficient exchange approach, when viewed from a TCE perspective.

Case study A

Football Stadia

The three football stadia projects outlined in case study A (Millenium Stadium, Cardiff, Emirates Stadium, London and Wembley Stadium, London) provide examples of a more traditional arms-length approach to contracting, although the Emirates case illustrates the benefits of a well-established and integrated supply chain.

Further reading

Transaction cost economics and project governance

Winch, G (2006) The governance of project coalitions – towards a research agenda. In: DJ Lowe (ed.) with R Leiringer, *Commercial Management of Projects: Defining the Discipline*. Blackwell Publishing, Oxford. pp. 344–355

Müller, R (2011) Project governance. In: PWG Morris, JK Pinto and J Söderlund (eds), *The Oxford Handbook of Project Management*. Oxford University Press, Oxford. pp. 297–320

Trust

Alternative mechanism to contracts have been identified. These include, for example, the concepts of 'trust' and 'self-enforcing' agreements. Sako (1992), for example, categorises three forms of trust involved in transactions and associated with the development of long-term relationships:

- **Contractual trust**: derived from Western classical contracting
- **Competence trust**: as illustrated by the trust placed in professional advisors
- **Goodwill trust**: the exemplification of established Japanese values of 'open commitment' and 'obligational contractual' relationships

Additionally, Lyons (1995) identifies two further types of trust:

- **Socially orientated trust**: reflecting the connection between experience and past relationships
- **Self-interest trust**: which links the party's assessment of potential future opportunities to social orientation

Formal contracts are widely held to essentially undermine trust and, therefore, promote opportunistic behaviour (Poppo and Zenger, 2002).

Further reading

Trust

Swan W, McDermott P and Khalfan M. (2006) Trust and commercial managers: influences and impacts. In: DJ Lowe (ed.) with R Leiringer, *Commercial Management of Projects: Defining the Discipline.* Blackwell Publishing, Oxford. pp. 172–191

Gil N, Pinto J and Smyth H (2011) Trust in relational contracting and as a critical organizational attribute. In: PWG Morris, JK Pinto and J Söderlund (eds), *The Oxford Handbook of Project Management.* Oxford University Press, Oxford. pp. 438–460

Relational contracting

Due to criticism of classic contracting as a means of structuring transactions, Ian Macneil was instrumental in conceiving an alternative field of study, one that emphasises the value of informal mechanisms.[21] Macneil sought to describe behaviour within an exchange, rather than label these exchanges by their governance structure. He advocated that transactions could be placed on a continuum – discrete to relational exchanges, depending on the intensity of the relationship between the parties (Macneil, 1983). Initially adopting the term '**relational contract theory**', he subsequently adopted the designation '**essential contract theory**' to label his approach (Macneil, 2000a,b). Macneil (2000a,b) identified ten common contract behavioural patterns and norms (see Chapter 6), which he asserts exist in all contracts, of which he deems the power norm to have the most influence on exchange relationships.

These relational norms and behavioural patterns are viewed by some as an alternative to explicit, complex contracts or vertical integration.[22] Moreover, relational contracts – '*informal agreements sustained by the value of future relationships*' – are widely used in ongoing, collaborative b2b exchanges (Baker *et al.*, 2002). Examples from the Japanese car manufacturing sector are frequently used to illustrate how relational contracting may be operationalised,[23] while Gil (2009) reports on the introduction of the approach within project-based environments, specifically the Heathrow Terminal 5 project.

Relational contract theory has been criticised for its failure to appreciate the importance of contractual consent (Barnett, 1992), while Eisenberg (2000) asserts that relational contracts are not a distinct class of contract, as all or practically all contracts are relational. Moreover, the findings of Poppo and Zenger (2002) and Arranz *et al.* (2011) support the view that transactional and relational mechanisms are complementary. Although, Poppo and Zenger have established that contractual complexity and relational governance have separate origins and have different roles in fostering transaction performance.

Case study B

Heathrow T5

The T5 case study is provided to illustrate the application of the collaborative, relational approach to contracting on a major infrastructure project.

The power perspective

The central tenet of the power perspective, as outlined by Cox and Ireland (2006), is that operational and commercial outcomes in any exchange transaction are determined by the relative power of the purchaser and supplier. Moreover, it challenges the current received wisdom (best practice) in procurement and supply management, that is, the championing of long-term collaborative approaches sustained by trust. Cox (2001) asserts that this approach is at odds with the sound logic of economic theory. Further, he suggests:

> '... the best defence of the buyer's position, and the one that ensures that suppliers innovate and pass value to buyers, is the maintenance of perfectly competitive (or highly contested) supply markets, with low barriers to entry, low switching costs and limited information asymmetries.' (Cox, 2001)

Cox (1997) refutes the concept of a single 'best practice', suggesting that purchasers' and suppliers' competence is to be found in their ability to make appropriate choices in respect of how they construct and behave in different trading relationships. Superior procurement and supply chain management performance requires an appreciation of the '*power and leverage*' circumstances in which they trade. The Power Matrix developed by Cox *et al.* (2002) provides a mechanism to assess the relative power of buyers and suppliers:

> 'The matrix is constructed based on the idea that all buyer and supplier relationships are predicated on the relative utility and the relative scarcity of the resources that are exchanged between the two parties'. (Cox *et al.*, 2002)

Buyer and supplier power circumstances can be classified as:

- **Buyer dominance**: where the buyer has relative power over the supplier
- **Interdependence**: where there is '*no relative power imbalance between the parties. Both the buyer and supplier have significant leverage opportunities over the other and they must accept the prevailing price and quality levels*'
- **Independence**: where there is '*no relative power imbalance between the parties. Neither the buyer nor the supplier has significant leverage opportunities over the other because neither party possesses key supply chain resources*'
- **Supplier dominance**: where the supplier has relative power over the buyer (Cox *et al.*, 2000)

Cox and Ireland (2006) illustrate the power perspective by reference to the UK construction sector (see further reading).

Further reading

Power perspective

Cox, A and Ireland, P (2006) Strategic purchasing and supply chain management in the project environment – theory and practice. In: DJ Lowe (ed.) with R Leiringer, *Commercial Management of Projects: Defining the Discipline*. Blackwell Publishing, Oxford. pp. 390–416

Environment

Exchange transactions take place within a context – the **environment** – which TCE theory and essential contract theory refer to as the concept of 'atmosphere'. The environment includes factors both external to and from within the organisation. Externally, for example, this would incorporate: home and host government systems and intervention; cultural values; legal systems and frameworks, such as, civil codes and precedent based legal models, procurement legislation/directives, competition/anti-trust laws, corporate governance legislation and regulatory obligations; organisation of labour; market structures, and financial system, for example, constraints imposed by the World Bank and other funding agencies. Internally, influences could include the predisposition of senior managers, power and politics, and interfunctional and business unit competition for resources.

Although not covered in detail in this text, **culture** has a significant impact on trading relationships, particularly with the increasing globalisation of markets. For a discussion of the influence of culture on commercial practice see Fellows (2006).

Further reading

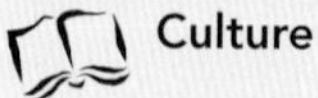

Culture

Fellows, R (2006) Culture. In: DJ Lowe (ed.) with R Leiringer, *Commercial Management of Projects: Defining the Discipline*. Blackwell Publishing, Oxford. pp. 172–191

Subsequent chapters will present further theoretical concepts and frameworks including, for example:

- Organisational behaviour and in particular leadership (Chapter 2)
- Competitive advantage and the resource based view of the firm (Chapter 3)
- Risk, uncertainty and opportunity management (Chapter 4)
- Transactional cost economics, relational contracting (essential contract theory), the power balance between parties and negotiation theory (Chapters 6 and 8)

Summary

This chapter has explored the nature of commercial practice within project-oriented organisations at the buyer–seller interface. It has presented a commercial management framework, which illustrates the multiple interactions and connections between the purchaser's procurement cycle and a supplier's bidding

Exercise **Commercial management in project-oriented organisations**

Think about the organisation you work for (or have previously worked for):

- How does your organisation position itself in the supply/value network?
  - illustrate the position of your organisation in its supply/value network and show how it configures its relationships and interacts in order to achieve its stated aims, etc,
- How does your organisation configure itself to support and deliver its aim and strategy?
  - illustrate its organisational structure to include business unit/functional and project programme relationships/interactions
- How does commercial management support and deliver the organisation's aim and strategy?
  - at the corporate, business unit and function level
- Consider the following, within the organisation:
  - is commercial and contract management considered to be a dynamic capability?
  - is the commercial management function perceived to be a distinctive capability/'trusted adviser'?
  - where do challenges to the commercial management role come from?
- Do you consider commercial management to be a profession?
- Evaluate the current commercial management maturity of your business unit

and implementation cycles. Additionally, it has outlined the principle activities undertaken by the commercial function and identifies the skills and abilities that support these activities. Areas of commonality of practice with other functions found within project-oriented organisations, sources of potential conflict and misunderstanding are identified. It considers the question: is commercial management a profession? The chapter concludes by viewing the theories and concepts that underpin commercial practice and in particular transaction cost economics.

Endnotes

1 Unless noted all definitions are from the *Oxford Dictionaries Online*, Oxford University Press, Oxford. http://oxforddictionaries.com/page/contactus/contact-us (accessed 29 August 2012)
2 For a more detailed discussion of the etymology of 'commercial' plus an exposition of the historical context of the job title 'commercial manager' see Stewart (2010)
3 Generally referred to as products and services in this text
4 Termination also refers to the premature end of a supply contract (*termination of a contract*). It occurs when the contracting parties are released from their contractual obligations due to performance of their duties (an executed contract); agreement of the parties; breach of contract; or frustration.
5 See, for example: 4ps (2007); Department of Health (2010a); IPMA (2006); OGC (2002,2007b,2009); VGPB (2011)
6 See, for example: Saxena (2008)
7 See, for example, glossaries in the following documents: APM (2010); BSI (2000a); BSI (2002); Chapman (2004); MOD (2003); OGC (2006); PMI (2012); PRINCE2 (2009).
8 See, for example, glossaries in the following documents: CIM (2011); CIPS (2007), MOD (2003).
9 For a more detailed description of the current role of legal within organisations see Nabarro (2010)
10 See, for example, glossaries in the following documents: APM (2010); Chapman (2004); Christopher (2005); CIM (2011); CIPS (2007); Council of Supply Chain Management Professionals (CSCMP, ND); Institute for Supply Management (Cavinato, 2010); OGC (2005, 2007b); Rolls-Royce (2000).
11 See, for example: APM (2006); Department of Health (2010a, 2010b); IFPSM; Institute for Supply Management (Cavinato, 2010); IPMA (2006); National Procurement Strategy (NPS), Office of the Deputy Prime Minister/Local Government Association (2003); Rolls-Royce (2000); VGPB (2011)
12 See, for example: APM (2010); BSI (2000a); Chapman (2004); MOD (2009) PMI (2012); PRINCE2 (2009)
13 See, for example, the following definitions: APM (2006); IPMA (2006); MOD (2009)
14 See, for example: American Marketing Association (AMA) (2007); APM (2006); Chapman (2004); The Chartered Institute of Marketing (CIM) (2011)
15 See, for example, glossaries and definitions in the following documents: APM (2004, 2010); BSI (2000a, 2002); Chapman (2004); CIPS (2007); Institute of Risk Management (2002); IPMA (2006); ISO (2009); MOD (2009); OGC (2002,2007a); Orange, ND; PMI (2012); PRINCE2 (2009); VGPB (2011)
16 See for example, definitions in the following documents: APM (2010); BSI (2000a); NHS (2011); PMI (2012); PRINCE2 (2009); VGPB (2011)
17 See, for example: Barnard (1938); Coase (1937, 1960); and Simon (1951,1978)
18 See, for example: Williamson (1971, 1973, 1975, 1979, 1981, 1983, 1985, 1991, 1993, 2000, 2002, 2005a, 2005b). Also, David and Han (2004); Dixit (2009); Gibbons (2005); Klein *et al.* (1978); Lacity *et al.* (2011); Shelanski and Klein (1995)
19 See Monteverde and Teece (1982)
20 See Powell (1990)
21 See, for example: Macneil (1978, 1983, 2000a, 2000b) and Barnett (1992) and Ivens and Blois (2004)
22 See, for example, Arranz *et al* (2011); Baker *et al.* (2002); Blomqvist *et al.* (2005); Dyer, (1997); Feinman (1990); Eisenberg (2000); Hagedoorn and Hesen (2007); Schwartz and Scott (2003); Zheng *et al.* (2008)
23 See, for example, Doornik (2006)

References

APM (2004) *Project Risk Analysis and Management (PRAM) Guide*. 2nd edition. Association for Project Management (APM), High Wycombe

APM (2006) *Body of Knowledge: Definitions*. The Association for Project Management. http://www.apm.org.uk/sites/default/files/Bok%20Definitions.pdf (accessed 25 March 2011)

APM (2010) *Glossary.*, The Association for Project Management .http://www.apm5dimensions.com/Definitions.asp (accessed 25 March 2011)

Arranz, N and Fdez. de Arroyabe, JC (2011) Effect of formal contracts, relational norms and trust on performance of joint research and development projects. *British Journal of Management*, DOI: 10.1111/j.1467-8551.2011.00791.x

Augier, M and Teece, DJ (2008) strategy as evolution with design: the foundations of dynamic capabilities and the role of managers in the economic system. *Organization Studies*, 29, 1187–1208

Baker, G, Gibbons, R and Murphy, KJ (2002) Relational contracts and the theory of the firm. *The Quarterly Journal of Economics*, 117, 39–84

Barnard, C (1938) *The Functions of the Executive.* Harvard University Press, Cambridge, MA

Barnett, RE (1992) Conflicting visions: a critique of Ian Macneil's Relational Theory of Contract. *Virginia Law Review*, 78, 1175–1206

Blomqvist, K, Hurmelinna, P and Seppänen, R (2005) Playing the collaboration game right-balancing trust and contracting. *Technovation*, 25, 497–504

BSI (1997) *BS EN 1325-1: 1997: Value Management, Value Analysis, Functional Analysis Vocabulary - Part 1. Value Analysis and Functional Analysis.* British Standards Institution, London

BSI (2000a) *BS 6079-2:2000 (Incorporating Corrigendum No. 1): Project Management – Part 2: Vocabulary.* British Standards Institution, London

BSI (2000b) *BS EN 12973:2000: Value management.* British Standards Institution, London

BSI (2002) *PD ISO/IEC Guide 73:2002: Risk management – Vocabulary – Guidelines for use in standards.* The British Standards Institution, London

Burgess, K, Singh, PJ and Koroglu, R (2006) Supply chain management: a structured literature review and implications for future research. *International Journal of Operations & Production Management*, 26, 703–729

Cavinato, JL (2010) *Supply Management Defined.* Institute for Supply Management. http://www.ism.ws/tools/content.cfm?ItemNumber=5558&navItemNumber=20096 (accessed 29th December 2010)

CBI (2006) *Buying the Best for the NHS: Ensuring Smarter Capital Procurement.* CBI, London. http://www.cbi.org.uk/pdf/healthreport0406.pdf (accessed 31 March 2011)

Chapman, A (2004) *Business Contracts Legal Terms and Definitions Glossary.* Alan Chapman/Crown Copyright. http://www.businessballs.com/businesscontractstermsdefinitionsglossary.htm (accessed 31 March 2011)

Christopher, M (2005) *Logistics and Supply Chain Management: Creating Value-added Networks.* Pearson Education Ltd, Harlow

CIM (2011) The Chartered Institute of Marketing: Resource glossary. http://www.cim.co.uk/resources/glossary/home.aspx (accessed 31 March 2011)

CIMA (2011) CIMA and AICPA embark on international management accounting initiative. Press Release, CIMA, London. http://www.cimaglobal.com/About-us/Press-office/Press-releases/2011/March-2011/cima-aicpa-jv/ (accessed 29 August 2012)

CIPS (2007) *Taking the Lead: A guide to More Responsible Procurement Practices* The Chartered Institute of Purchasing and Supply. http://www.cips.org/Documents/Resources/Knowledge Summary/Taking the lead – a guide to more responsible procurement practices.pdf (accessed 29 December 2010)

CIPS (2012) *CIPS Position on Practice Purchasing and Supply Management: Supply Chain Management.* CIPS in Association with KnowledgeBrief, Stamford. http://www.kbresearch.com/cips-files/Supply%20chain%20Management.%20PoP.pdf (accessed 29 August 2012)

Coase, R (1937) The nature of the firm. *Economica*, 4, 386–405

Coase, R (1960) The problem of social cost. *Journal of Law and Economics*, 3, 1–44

Conger, JA (1992) *Learning to Lead.* Jossey-Bass, San Francisco

Cooke-Davies T (2004) Project management maturity models. In: PWG Morris and JK Pinto (eds) *The Wiley Guide to Managing Projects.* John Wiley & Sons, Hoboken, NJ. pp. 1234–1255

Council of Supply Chain Management Professionals (CSCMP) CSCMP Supply Chain Management Definitions. http://cscmp.org/aboutcscmp/definitions.asp (accessed 29th December 2010)

Cousins, PD, Lawson, B and Squire, B (2006a) An empirical taxonomy of purchasing functions. *International Journal of Operations & Production Management*, 26, 775–794

Cousins, PD, Lawson, B and Squire, B (2006b) Supply chain management: theory and practice – the emergence of an academic discipline? *International Journal of Operations & Production Management*, 26, 697–702

Cousins, PD, Lamming, RC, Lawson, B and Squire, B (2008) *Strategic Supply Management: Principles, Theories and Practice.* Pearson Education, London

Cox, A (1997) *Business Success.* Earlsgate Press, Boston, UK

Cox, A (2001) The power perspective in procurement and supply management. *Journal of Supply Chain Management*, 37, 4–7

Cox, A and Ireland, P (2006) Strategic purchasing and supply chain management in the project environment – theory and practice. In: DJ Lowe (ed.) with R Leiringer, *Commercial Management of Projects: Defining the Discipline*. Blackwell Publishing, Oxford. pp. 390–416

Cox, A, Ireland, P, Lonsdale, C, Sanderson, J and Watson, G (2002) *Supply Chains, Markets and Power: Mapping Buyer and Supplier Power Regimes*. Routledge, London

Cox, A, Sanderson, J and Watson, G (2000) *Power Regimes: Mapping the DNA of Business and Supply Chain Relationships*. Earlsgate Press, Boston, UK

Cox, A, Sanderson, J and Watson, G (2001) Supply chains and power regimes: toward an analytical framework for managing extended networks of buyer and supplier relationships. *Journal of Supply Chain Management*, 37, 28–35

Croom, SR, Romano, P and Giannakis, M (2002) Supply chain management: an analytical framework for critical literature review. *European Journal of Purchasing & Supply Management*, 6, 67–83

David, M, Kawamoto, K and McCarthy, T (2008) Capability Maturity Model – a vehicle to understand and drive commercial excellence. *Contracting Excellence*, IACCM, February. http://www.iaccm.com/news/contractingexcellence/?storyid=509 (accessed 2 January 2012)

David, R and Han, S-K (2004) A systematic assessment of the empirical support for transaction cost economics. *Strategic Management Journal*, 25, 39–58

Department of Health (2010a) *Commercial Skills for the NHS*. Department of Health, London. http://www.dh.gov.uk/prod_consum_dh/groups/dh_digitalassets/@dh/@en/@ps/documents/digitalasset/dh_114573.pdf (accessed 31 March 2011)

Department of Health (2010b) *Procurement Guide for Commissioners of NHS-Funded Services*. Department of Health, London http://www.dh.gov.uk/prod_consum_dh/groups/dh_digitalassets/@dh/@en/@ps/documents/digitalasset/dh_118219.pdf (accessed 21 February 2012)

Dixit, A (2009) Governance Institutions and Economic Activity, *American Economic Review*, 99, 5–24

Doornik, K (2006) Relational contracting in partnerships. *Journal of Economics and Management Strategy*, 15, 517–548

Dyer, JH (1997) Effective interfirm collaboration: how firms minimize transaction costs and maximize transaction value. *Strategic Management Journal*, 18, 535–556

Eisenberg, MA (2000) Why there is no law of relational contracts, Northwestern University Law Review, 94, 805–821

Eraut M (1994) *Developing Professional Knowledge and Competence*. The Falmer Press, London

Fayol H (1916) *Administration Industrielle et Générale* (English translation: Industrial and General Administration, 1949

Feinman, JM (1990) The significance of contract theory. *Cincinnati Law Review*, 58, 1283–1318

Fellows, R (2006) Culture. In: DJ Lowe (ed.) with R Leiringer, *Commercial Management of Projects: Defining the Discipline*. Blackwell Publishing, Oxford. pp. 172–191

4ps (2007) *A Guide to Contract Management for PFI and PPP Projects*. Public Private Partnerships Programme, London. http://www.localpartnerships.org.uk/UserFiles/File/Publications/4ps%20ContractManagers%20guideFINAL.pdf (accessed 31 March 2011)

Ghoshal, S and Moran P (1996) Bad for practice: a critique of the transaction cost theory. *The Academy of Management Review*, 21, 13–47

Giannakis, M and Croom, SR (2004) Toward the development of a supply chain management paradigm: a conceptual framework. *Journal of Supply Chain Management*, 40, 27–37

Gibbons, R (2005) Four formal(izable) theories of the firm? *Journal of Economic Behavior and Organization*, 58, 202–247

Gil, N (2009) Developing project client-supplier cooperative relationships: how much to expect from relational contracts? *California Management Review*, Winter, 144–169

Giunipero, L, Handfield, RB and Eltantawy, R (2006) Supply management's evolution: key skill sets for the supply manager of the future. *International Journal of Operations & Production Management*, 26, 822–844

Hagedoorn, J and Hesen, G (2007) Contract law and the governance of inter-firm technology partnerships – an analysis of different modes of partnering and their contractual implications. *Journal of Management Studies*, 44, 342–366

Handy, C (1992) The language of leadership. In: M Syrett and C Hogg (eds) *Frontiers of Leadership*. Blackwell, Oxford

Harland, CM, Lamming, RC, Walker, H, Phillips, WE, Caldwell, ND, Johnsen, TE, Knight, LA and Zheng, J (2006) Supply management: is it a discipline? *International Journal of Operations & Production Management*, 26, 730–753

IACCM (ND) Benchmarking: The IACCM Capability Maturity Model, IACCM. http://www.iaccm.com/research/benchmarking/ (accessed 2 January 2012)

Ibbs CW, Reginato JM and Kwak, YH (2004) Developing project management capability: benchmarking, maturity, modeling, gap analysis, and ROI STUDIES. In: PWG Morris and JK Pinto (eds), *The Wiley Guide to Managing Projects*. John Wiley & Sons, Hoboken, NJ. pp. 1214–1233

Institute of Risk Management (IRM) (2002) *A Risk Management Standard*, IRM-AIRMIC-ALARM RM STANDARD. http://www.theirm.org/publications/documents/Risk_Management_Standard_030820.pdf (accessed 25 March 2011)

IPMA (2006) *ICB - IPMA Competence Baseline Version 3.0*. International Project Management Association (IPMA), Nijkerk. http://www.ipma.ch/Documents/ICB_V._3.0.pdf (accessed 25 March 2011)

ISO (2009) ISO *31000:2009 Risk management – Principles and Guidelines*. International Organization for Standardization, Geneva, Switzerland

Ivens, BS and Blois, KJ (2004) Relational exchange norms in marketing: a critical review of Macneil's contribution. *Marketing Theory*, 4, 239–263

Kay J (1993) *The Foundations of Corporate Success*. Oxford University Press, Oxford

Klein, B, Crawford, R and Alchian, A (1978) Vertical integration, appropriable rents and the competitive contracting process. *Journal of Law and Economics*, 21, 297–326

Kotter, JP (2011) *Management vs Leadership: What is the Difference between Management and Leadership?* Kotter International. http://kotterinternational.com/KotterPrinciples/ManagementVsLeadership.aspx (accessed 20 February 2012)

Lacity, M, Willcocks, LP and Khan, S (2011) Beyond transaction cost economics: towards an endogenous theory of information technology outsourcing. *Journal of Strategic Information Systems*, 20, 139–157

Lowe, DJ (2008a) Commercial and contract management – practitioner expectations on the content of an MBA for commercial executives. In: DJ Lowe (ed.) *Proceedings IACCM International Academic Symposiums on Contract & Commercial Management*, London, UK and Fort McDowell, Arizona. ISBN 978-1-934697-01-6. pp. 45–72

Lowe, DJ (2008b) Commercial and contract management – underlying theoretical concepts and models. In: DJ Lowe (ed.) *Proceedings IACCM International Academic Symposiums on Contract & Commercial Management*, London, UK and Fort McDowell, Arizona. ISBN 978-1-934697-01-6. pp. 25–44

Lowe, DJ, Fenn, P and Roberts, S (1997) Commercial management: an investigation into the role of the commercial manager within the UK construction industry. *CIOB Construction Papers*, 81, 1–8

Lowe, DJ, Garcia, J and Kawamoto, K (2006) Contract and commercial management: an international practitioner survey. In: DJ Lowe (ed.) *Proceedings IACCM International Academic Symposium on Commercial Management*, Ascot, UK

Lyons BR (1995) Specific investment, economies of scale, and the make-or-buy decision: a test of transaction cost theory. *Journal of Economic Behavior and Organization*, 26, 431–443

Macneil, IR (1978) Contracts: adjustment of long-term economic relations under classical, neoclassical, and relational contract law. *Northwestern University Law Review*, 72, 854–905

Macneil, IR (1983) Values in contract: internal and external. *Northwestern University Law Review*, 78, 340–418

Macneil, IR (2000a) Contracting worlds and essential contract theory. *Social and Legal Studies*, 9, 431–438

Macneil, IR (2000b) Relational contract theory: challenges and queries. *Northwestern University Law Review*, 94, 877–907

Mintzberg, H (2009) *Managing*. Berrett-Koehler Publishers Inc., San Francisco, California

MOD (2003) *Commercial Awareness: A Beginner's Guide for Those New to MOD Acquisition Work*. Commercial Services Group, DSAD (PC), Keynsham, Bristol. http://www.metasums.co.uk/uploads/asset_file/Commercial%20awareness%20-%20a%20beginners%20guide%202003.pdf(accessed 1 March 2012)

MOD (2008) *Commercial Toolkit*. http://www.aof.mod.uk/aofcontent/tactical/toolkit/content/topics/postcost.htm (accessed 28 June 2010)

MOD (2009) *Defence Commercial Function Skills Framework* Version1.1. Ministry of Defence (MOD) Acquisition Operating Framework. http://www.aof.mod.uk/aofcontent/tactical/toolkit/downloads/people/Competences.pdf (accessed 25 March 2011)

Monteverde, K and Teece, D (1982) Supplier switching costs and vertical integration in the automobile industry. *Bell Journal of Economics*, 13, 206–213

Morris, PWG (2002) Science, objective knowledge and theory of project management. *Proceedings of ICE, Civil Engineering*, 150, 82–90

Morris PWG (2004) The validity of knowledge in project management and the challenge of learning and competency development. In: PWG Morris and JK Pinto (eds), *The Wiley Guide to Managing Projects*. John Wiley & Sons, Hoboken, NJ. pp. 1137–1149.

Morris, PWG, Crawford, L, Hodgson, D, Shepherd, MM and Thomas, J (2006a) Exploring the role of formal bodies of knowledge in defining a profession – the case of project management. *International Journal of Project Management*, 24, 710–721

Morris, PWG, Jamieson, A and Shepherd, MM (2006b) Research updating the APM Body of Knowledge. 4th edition. *International Journal of Project Management*, 24, 461–473

Murray, JG (2009) Towards a common understanding of the differences between purchasing, procurement and commissioning in the UK public sector. *Journal of Purchasing and Supply Management*, 15, 198–202

Nabarro (2010) *From In-House Lawyer to Business Counsel: A Survey and Discussion Paper*. Nabarro LLP, London. http://www.nabarro.com/downloads/From-in-house-lawyer-to-business-counsel.pdf (accessed 18 September 2011)

NAO (2003) *PFI: Construction Performance*. National Audit Office, The Stationery Office, London. http://www.nao.org.uk/publications/0203/pfi_construction_performance.aspx?alreadysearchfor=yes (accessed 31 March 2011)

NAO (2007) *Benchmarking and Market Testing the Ongoing Services Component of PFI Projects*. National Audit Office, The Stationery Office, London. http://www.nao.org.uk/publications/0607/benchmarking_and_market_testin.aspx (accessed 31 March 2011)

NAO (2009) *Commercial Skills for Complex Government Projects*. The Stationery Office, London. http://www.nao.org.uk/publications/0809/commercial_skills.aspx (accessed 1 March 2012)

NHS (2011) *The ProCure21+ Guide: Achieving Excellence in NHS Construction*. The Department of Health, Leeds. http://www.procure21plus.nhs.uk/resources/downloads/ProCure21Plus%20Guide%20v2.2%202011.pdf (accessed 29 August 2012)

Normann, R and Ramirez, R (1993) From value chain to value constellation: designing interactive strategy. *Harvard Business Review*, July/August, 65–67

North, D (1990) A transaction cost theory of politics. *Journal of Theoretical Politics*, 2, 355–367

OGC (2002) *Principles for Service Contracts: Contract Management Guidelines*. The Office of Government Commerce/HMSO, London. http://www.ogc.gov.uk/documents/Contract_Management.pdf (accessed 31 March 2011)

OGC (2005) *Supply Chain Management in Public Sector Procurement: A Guide*. Office of Government Commerce, London. http://www.ogc.gov.uk/documents/scm_final_june05.pdf (accessed 31 March 2011)

OGC (2006) Category Management Toolkit: Stakeholder Management Plan

OGC (2007a) *Achieving Excellence in Construction Procurement Guide 4 – Risk and Value Management*. Office of Government Commerce, London. http://www.ogc.gov.uk/documents/CP0064AEGuide4.pdf (accessed 31 March 2011)

OGC (2007b) *Achieving Excellence in Construction Procurement Guide 5 – OGC: The Integrated Project Team Team-Working and Partnering*. Office of Government Commerce, London. http://www.ogc.gov.uk/documents/CP0065AEGuide5.pdf (accessed 31 March 2011)

OGC (2007c) *Achieving Excellence in Construction Procurement Guide 6 – Procurement and Contract Strategies*. Office of Government Commerce, London. http://www.ogc.gov.uk/documents/CP0066AEGuide6.pdf (accessed 31 March 2011)

OGC (2009) *Managing contracts and service performance*, Office of Government Commerce, London. http://www.ogc.gov.uk/delivery_lifecycle_managing_contracts_and_service_performance.asp (accessed 31 March 2011)

Orange (ND) *Glossary of Terms for Large Projects Management*. http://www.orange-business.com/en/mnc2/large-projects-management/glossary/(accessed 29 August 2012)

PMI (2012) *PMI Lexicon of Project Management Terms*. Project Management Institute Inc., Newton Square, PA. http://www.pmi.org/PMBOK-Guide-and-Standards/~/media/Registered/PMI_Lexicon_Final.ashx (accessed 29 August 2012).

Poppo, L and Zenger, T (2002) Do formal contracts and relational governance function as substitutes or compliments? *Strategic Management Journal*, 23, 707–725

Porter M (1985) *Competitive Advantage: Creating and Sustaining Superior Performance*. The Free Press, New York

Powell, W (1990) Neither market nor hierarchy: network forms of organization. *Research in Organizational Behavior*, 12, 295–336

PRINCE2 (2009) *Glossaries/Acronyms*. http://www.prince2.com/prince2-downloads.asp. The Office of Government Commerce/HMSO, London (accessed 25 March 2011)

Sako, M (1992) *Prices, Quality and Trust: Inter-firm Relations in Britain and Japan*. Cambridge University Press, Cambridge

Sauer, C and Reich, BH (2007) What do we want from a theory of project management? A response to Rodney Turner. *International Journal of Project Management*, 25, 1–2

Saxena, A (2008) *Enterprise Contract Management: A Practical Guide to Successfully Implementing an ECM Solution*. J Ross Publishing, Fort Lauderdale, Florida

Schwartz, A and Scott, RE (2003) Contract theory and the limits of contract law. *The Yale Law Journal*, 113, 541–619

Shelanski, H and Klein, P (1995) Empirical research in transaction cost economics: a review and assessment. *Journal of Law, Economics, and Organization*, 11, 335–361

Simon, H (1951) A formal theory of the employment relationship. *Econometrica*, 19, 293–305

Simon, H (1978) Rationality as process and as product of thought. *American Economic Review*, 68, 1–16

Söderlund, J (2004) Building theories of project management: past research, questions for the future. *International Journal of Project Management*, 22, 183–191

Stewart, I (2010) *Exploring the Experience of Historical and Contemporary Commercial Managers*. Unpublished PhD thesis. The University of Manchester, Manchester

Stoakes, C (2011) *All You Need to Know About Commercial Awareness 2011/2012: What it is and Why You Need it to Become a Successful Professional*. Longtail Publishing Limited, London

Storey, J, Emberson, C, Godsell, J and Harrison, A (2006) Supply chain management: theory, practice and future challenges. *International Journal of Operations and Production Management*, 26, 754–774

Turner, JR (2006a) Towards a theory of project management: the nature of the project. *International Journal of Project Management*, 24,1–3

Turner, JR (2006b) Towards a theory of project management: the nature of the project governance and project management. *International Journal of Project Management*, 24, 93–95

Turner, JR (2006c) Towards a theory of project management: the functions of project management. *International Journal of Project Management*, 24,187–189

Turner, JR and Simister, SJ (2001) Project contract management and a theory of organisation. *International Journal of Project Management*, 19, 457–464

UCAS (ND) *Glossary of Competencies*. http://www.ucas.ac.uk/seps/glossary (accessed 31 March 2011)

VGPB (2011) *Victorian Government Purchasing Board: Glossary*. http://www.vgpb.vic.gov.au/CA2575BA0001417C/pages/glossary (accessed 31 March 2011)

Wearne SH (1999) Contracts for goods and services. In: G Lawson, S Wearne and P Iles-Smith (eds), *Project Management for the Process Industries*. Institution of Chemical Engineers, Rugby. pp. 261–283

Williamson, O (1971) The vertical integration of production: market failure considerations. *American Economic Review*, 61, 112–123

Williamson, O (1973) Markets and hierarchies: some elementary considerations. *American Economic Review*, 63, 316–325

Williamson, O (1975) *Markets and Hierarchies: Analysis and Antitrust Implications*. Free Press, New York

Williamson, O (1979) Transaction cost economics: the governance of contractual relations. *Journal of Law and Economics*, 22, 233–261

Williamson, O (1981) The economics of organization: the transaction cost approach. *American Journal of Sociology*, 87, 548–577

Williamson, O (1983) Credible commitments: using hostages to support exchange. *American Economic Review*, 73, 519–540

Williamson, O (1985) *The Economic Institutions of Capitalism*. Free Press, New York

Williamson, O (1991) Comparative economic organization: the analysis of discrete structural alternatives. *Administrative Science Quarterly*, 36, 269–296

Williamson, O (1993) The evolving science of organization. *Journal of Institutional and Theoretical Economics*, 149, 36–63

Williamson, O (2000) The new institutional economics: taking stock, looking ahead. *Journal of Economic Literature*, 38, 595–613

Williamson, O (2002) The theory of the firm as governance structure: from choice to contract. *The Journal of Economic Perspectives*, 16, 171–195

Williamson, O (2005a) Transaction cost economics and business administration. *Scandinavian Journal of Management*, 21, 19–40

Williamson, O (2005b) Why law, economics, and organization? *Annual Review of Law and Social Science*, 1, 369–396

Winch GM (2006) The governance of project coalitions – towards a research agenda. In: DJ Lowe (ed.) with R Leiringer, *Commercial Management of Projects: Defining the Discipline*. Blackwell Publishing, Oxford. pp. 324–343

Zajac EJ and Olsen CP (1993) From transaction cost to transactional value analysis: implications for the study of interorganizational strategies. *Journal of Management Studies*, 30, 131–145

Zheng, J, Roehrich, JK and Lewis MA (2008) The dynamics of contractual and relational governance: evidence from long-term public-private procurement arrangements. *Journal of Purchasing and Supply Management*, 14, 43–54

Part 2

Elements of Commercial Practice and Theory

Chapter 2	Commercial Leadership *Maria-Christina Stafylarakis*	79
Chapter 3	Exploring Strategy *Irene Roele*	96
Chapter 4	Perspectives on Managing Risk and Uncertainty *Eunice Maytorena*	108
Chapter 5	Financial Decisions *Anne Stafford*	132
Chapter 6	Legal Issues in Contracting *David Lowe and Edward Davies*	173

Chapter 2

Commercial Leadership

Maria-Christina Stafylarakis

Learning outcomes

When you have completed this chapter you should be able to:

- Distinguish between different viewpoints of leadership
- Understand the relevance of reflectivity and critical reflectivity to effective leadership
- Identify some core issues in leadership studies
- Distinguish between leadership and management
- Understand the difference between leadership and power
- Identify the main trends in the leadership literature
- Identify desirable qualities in a leader
- Envision leadership as a task and learning focused process

Introduction

In spite of its universal appeal, extensive research and many scholarly journals and popular books devoted to the subject, leadership remains an ambiguous concept. Rather than ask '*What is leadership?*' Alvesson and Deetz (2000, p.52) suggest that a more useful question might be '*What can we see, think, or talk about if we think of leadership as this or that?*' What then can we see, think or talk about when we think of **commercial leadership**, that is, the application of leadership in the context of commercial and contract management – itself an ambiguous concept with wide practical applicability? Is leadership in this context any different from leadership in other contexts?

Although the absence of literature in this field suggests that it is not, there are other possible explanations for this paucity of research. For a start, most leadership studies tend to focus on the study of formally appointed leaders suggesting that leadership is hierarchical and position-based. This view is further reinforced by notions of **followership** and the implied power ascribed to leaders contained in most durable definitions of leadership. These construe leadership as a social process that takes place in a *group* context, where the leader *influences* followers to behave in a certain way so as to achieve organisational *goals* (Bryman, 1992; Shackleton, 1995, Northouse, 2010). As a result, people who typically occupy technical support roles in an organisation are excluded from being thought of as leaders even though the activity of '*leading*' may be an integral part of what they do.

This situation is further exacerbated by disagreements over the perceived significance of leadership to the commercial manager's role and the fact that practitioners themselves frequently do not think of themselves as leaders. A 2008 survey by Manchester Business School and IACCM which surveyed IACCM's worldwide membership, for instance, revealed that leadership was rated as a secondary skill in terms of importance (Lowe, 2008). Yet paradoxically, managing relationships and influencing, both central features of leadership,

Commercial Management: theory and practice, First Edition. David Lowe.

were seen as core skills. In contrast, a more recent analysis of commercial functional frameworks/job specifications revealed that commercial practitioners are predominantly involved in leading (see Chapter 1).

This chapter adopts the latter view to conceptualise **commercial leadership** (used interchangeably with leadership) as a social process that is primarily about:

- Influencing
- Managing relationships
- Educating multiple stakeholders (as opposed to followers) in order to achieve project goals within the remit of contractual obligations

It is with these meta-goals in mind that key themes and concepts in the leadership literature are explored. A review of traditional approaches to the study of leadership (for example, trait, style and contingency theories) as well as newer theories (for example, transformational and dispersed leadership) is offered in addition to engaging with the literature that explores leadership in relation to learning. Arguing that any one view of leadership is just that, one approach and therefore partial, the chapter concludes with the derivation of a conceptual framework that encourages a more holistic multifaceted view of leadership as a process that involves a dynamic interplay between a task-focus and learning.

Point of departure

The emphasis on learning is a particular point of departure for this chapter. It draws on the idea that learning occurs through a continuous dialectical process of action and reflection (Marsick and Watkins, 1990, p. 8), or what Schön (1983) terms reflective practice. The latter entails reflection-in-action, a process that consists of '*on-the-spot*' [as opposed to retrospective] '*... surfacing, criticising, restructuring, and testing of intuitive understandings of experienced phenomena*' and is fundamental to the way in which problems are framed and understood (Schön, 1983: 243). The assumption is that practitioners engage in a reflective dialogue with a situation, drawing on their existing experience to shape or impose coherence upon the situation in terms of their acquired repertoire of examples, images, understandings and actions. Experimental actions are subsequently taken or hypothesised, yielding consequences and new discoveries on the basis of which the situation is reframed and retested in a process that '*... spirals through stages of appreciation, action and reappreciation*' (Schön, 1983, p. 132).

The more advanced notion of critical reflectivity has also been explored by various writers (for example, Brookfield, 1985a, 1985b; Marsick and Watkins, 1990; Mezirow, 1991, 1994;Garrison, 1992; Tennant, 1997; Reynolds, 1998). It differs from reflection in that the latter concentrates on the current details of a task or problem. Conversely, critical reflectivity, according to Reynolds (1998), focuses on an analysis of power and control and an examination of the tacit, taken-for-granted assumptions and beliefs within which the problem or task is situated. He argues further that the objective here should be social given the socially situated nature of experience and the role of communication and dialogue in the creation of meaning (Reynolds, 1998). In this way he echoes Freire's notion of **conscientisation**, a process whereby people come to understand how socio-historical forces shape their own consciousness and can oppose their own interests (in Tennant, 1997). This, in turn, leads to an awareness of the self as an agent who is able to act on the environment in order to transform it. For Reynolds (1998), therefore, the characteristics that encapsulate critical reflectivity are that it involves questioning assumptions, it has a social focus that pays particular attention to the analysis of power relations and it is emancipatory in nature in terms of challenging distorted frames of reference. However, daily activity can involve significant learning where learners are conscious of how their assumptions have been culturally influenced and assimilated without necessitating a critique of social factors (Mezirow, 1994, p. 22).

In this way, leadership is framed beyond the performative task of leading, as a learning process that takes into account the intra- and interpersonal contexts through which leadership is constructed as well as the sociocultural and environmental forces that shape it. This further encourages a conscious awareness of leading as a necessarily politicised process and highlights the significance of skills like reflectivity and critical reflectivity for effective leadership practice.

Some issues in leadership

For a fuller understanding, it is important to first discuss some additional issues pertaining to the nature of leadership studies and how leadership differs from management and concepts of power.

The nature of leadership studies

When reading the literature, it is important to keep in mind that there are several issues that somewhat limit its usefulness.

The introduction to this chapter has already hinted at some of the many different viewpoints and approaches to leadership. Grint (2005) distils these into a typology of four different, sometimes overlapping ways of understanding leadership: leadership as **person** suggests that it resides in select people with innate qualities and skills; as **process** in the context of the interactions between leaders and followers; as **results** in the outcomes that leaders achieve; and as **position** in the formal position they hold in the organisation. Constituting a valuable piece of the puzzle called leadership, this spate of theories and approaches has in turn generated a flurry of critical reviews and contradictory or inconclusive empirical results that have prevented the building of a more cumulative theory of leadership. Adding to this difficulty is the fact that different leadership theories focus on different levels of analysis (that is, individual, dyadic, group or organisational level) and also that the terms or categories used to describe various attributes or characteristics of a leader vary in their level of abstraction (Yukl, 2002, 2006). This plurality in dialogue, though desirable, is not especially helpful when seeking to unpack what leadership might mean in a largely unexplored context like commercial management.

An additional issue pertains to the leader/follower differentiation that pervades much of the literature. As shown earlier, this view privileges the leader and places him or her in a position of dominance. However, the idea of leading from the front may have limited relevance in the context of commercial practice which requires greater collaboration and interdependence between the various parties involved and where the boundaries of ownership and responsibility for delivering a project are likely to be blurred.

Much of the leadership literature is also strongly dominated by positivist/normative assumptions (Alvesson and Deetz, 2000) that flow from the researcher's point of view. Such approaches tend to assume a static reality and employ quantitative approaches to ascertain what leadership is, thus largely neglecting the interdependent, fluid and context-bound experience of leading. This clearly suggests the need for additional research that flows from the insider reality of commercial practitioners and that uses qualitative approaches, such as interviewing, storytelling and ethnography in line with a more interactional, context-focused line of inquiry.

It is against this backdrop that this chapter explores the mainstream leadership literature.

Leadership versus management

The notion of managers as leaders has been hotly debated. Arguing that the distinction between them revolves around personality and their orientation to change, in a reprint of his classic article, Zaleznik (1992, p. 127) asserts that managers and leaders are '*... different in motivation, personal history and in how they think and act*'. Likewise, Bennis and Nanus (1985) pigeonhole managers and leaders into two groups: those who do things the right way and those who do the right things respectively. However, as Yukl (2002) rightly counters, such assertions lack empirical support.

A more useful distinction is made by Kotter (2011). As we saw in Chapter 1, he posits that leadership is about creating the organisation's vision and formulating its strategy, establishing and communicating its direction, motivating action and aligning individuals around the vision, whereas management is about planning and budgeting, organising and staffing, controlling and problem solving. A similar view and one which the author agrees with is promulgated by Rost (1991). He construes leadership as a multidirectional influence process that is focused on developing mutual purposes whereas management is seen as a unidirectional relationship based on authority that is geared around coordinating activities to complete a job.

For our purposes, the strength of this view is that it recognises the fluidity of the leadership process and lends itself to understanding leadership at a predominantly intra-personal and relations-oriented level of analysis. This is the emphasis in this chapter.

However, it is important to keep in mind that effective leadership and management are both necessary to the successful attainment of project goals. Leadership without strong management and rigorously executed processes can lead to massive failures and misdirected efforts. Strong management of processes implemented by a team of people who are committed to mutual purposes are critical success factors. Management and leadership, although distinct in their focus of application, are thus overlapping and complementary constructs.

Leadership and power

The concept of power is closely related to leadership. It refers to the largely unilateral ability to influence and overcome resistance on the part of others in order to produce desired results. Often linked to transactional theories of leadership, discussed shortly, the most common view treats power as a relational construct between the leader and followers. Generally, this view asserts that the use of power is effective only if followers perceive that the leader has something of value to offer or, in the case of coercive power, can withhold something of value.

The most widely cited framework of power, proposed by social psychologists French and Raven (1962), identifies five power bases:

1. **Legitimate power** is the most important kind of power and is associated with status and the formal authority vested in the position one holds
2. **Reward power** depends on the ability of the leader to confer valued material rewards, such as benefits, time off, desired gifts, promotions or increases in pay or responsibility. It is an obvious source of power but can be ineffective if abused
3. **Coercive power** derives from the ability to impose negative consequences, such as demoting someone or withholding rewards. It is the most obvious of all the power bases but least effective as it builds resentment and resistance in those on whom it is exercised
4. **Expert power** is based on followers' perception of the leader's competence and derives from the specialist skills and expertise of the leader in their particular domain and to the task at hand. Information power is a sub-element of expert power and refers to the extent that the person is well informed, up to date and has the ability to persuade others
5. **Referent power** refers to the ability of individuals to attract others and build loyalty. Based on the interpersonal skills of the power holder, it is often referred to as charisma

A key feature of the power bases is that they are dynamically interrelated, meaning that the use of one power base can impact on the utility of another. The application of coercive power, for instance, can lead to the loss of referent power, whereas the use of expert power can lead to the individual gaining referent power (Huczynski and Buchanan, 2001). The theory also postulates that individuals can operate from several bases of power in different combinations with the same person in different contexts and at different times. Certainly, the application of the power bases will change as situations change. In the past, for example, leadership largely depended on coercion and legitimate power. As work has become more specialised, however, expert power and referent power have become more prevalent (Huczynski and Buchanan, 2001).

For the contract and commercial manager, the most obvious source of power is clearly expert power in terms of the financial, legal and contractual expertise they bring to the task. Information power is also an important power base. Used successfully, these power bases can potentially generate referent power.

Trends in the leadership literature

This section offers an illustrative rather than exhaustive review of the mainstream leadership literature with a view to highlighting different approaches to understanding the topic.[1] Table 2.1 summarises the core themes that underscore six different approaches to leadership that are considered in this chapter. A brief discussion of each follows.

Table 2.1 Trends in the leadership theory and research.

Approach	Core theme
Trait approach	Leadership ability is innate
Style approach	Leadership effectiveness is to do with how the leader behaves
Contingency approach	Effective leadership is dependent on the situation
Transformational leadership	The charismatic and affective aspects of leadership define leadership effectiveness
Dispersed leadership	Leadership is shared or distributed between the various stakeholders
Learning leadership	The role of the leader is to design and manage learning processes in the organisation

Source: adapted from Bryman (1992)

Trait theories

The trait approach, sometimes referred to as the 'great man' or 'great woman' theory, is premised on the argument that leaders are born not made and focuses on the study of physical traits (e.g. height), abilities (e.g. intelligence) and personality characteristics (e.g. introversion–extroversion) to distinguish leaders from non-leaders or effective from less effective leaders (Bryman, 1992, 1999; Shackleton, 1995).

Although this body of literature has generated an extensive list of desirable traits that one might wish to develop in order to be perceived as a leader, a definitive set of leadership traits has yet to be identified. The failure to produce consistent results has thus led to an initial rejection of the approach. Additional problems pertain to the fact that the same person can be both effective and ineffective in the same role due to contextual variables thereby challenging the idea that traits and effective leadership are linked; that traits are difficult to define and can therefore lead to inaccurate interpretations and questionnaire ratings; and that trait approaches do not provide an adequate explanation of how leadership traits are interrelated, nor how they interact to influence leadership behaviour and effectiveness, given that most trait studies test for simple, linear relationships (Yukl, 2002, 2006).

Recent years, however, have seen a resurgence of interest in traits in the popular management literature and in academic research. Many organisations, for example, still use Myers–Briggs Type Indicator (MBTI) personality tests to screen job applicants for leadership potential.

A more recent line of enquiry has focused on the concept of emotional intelligence (EI), defined as the ability to perceive and express emotion, assimilate emotion in thought, understand and reason with emotion, and regulate emotion within oneself and in relationships with others (Mayer *et al.*, 2000). Often credited with popularising the concept, Goleman (1995) believes that EI consists of a set of personal and social competencies that are especially important for leadership roles.

His conceptualisation identifies five dimensions: self-awareness, the ability to regulate feelings, motivation, empathy and social skills, which enable one to effectively manage relationships. In later work, Goleman *et al.* (2002) collapse the competencies of EI into four clusters:

1. **Self-awareness** (emotional self-awareness; accurate self-assessment; self-confidence)
2. **Self-management** (self-control; transparency; adaptability; achievement; initiative; optimism)
3. **Social awareness** (empathy; organisational awareness; service)
4. **Relationship management** (inspiration; influence; developing others; change catalyst; conflict management and teamwork; and collaboration)

Although this conceptualisation has been heavily criticised, among other things, for a lack of evidence to support the notion that EI can predict leadership outcomes, its focus on managing relationships gives it an intuitive appeal that has particular relevance for commercial leadership which is necessarily concerned with the effective management of relationships with multiple stakeholders. Its significance is that it promotes the view that leadership influences and is influenced by the socio-emotional context in which it occurs.

Style/behavioural theories

In contrast to trait approaches, behavioural or style leadership theories are concerned with what leaders do rather than what attributes set them apart. The underlying premise is that leaders can be nurtured once the behaviour that comprises effective leadership is known (Bryman, 1992, 1999; Shackleton, 1995; Northouse, 2010).

One of the best-known studies by a group of Ohio State University researchers identifies two independent clusters of behaviour. **Consideration** refers to the extent that leaders and their subordinates have a relationship built on mutual trust, liking and respect whereas **initiating structure** denotes the extent to which leaders organise work, structure the work context and clarify roles and responsibilities (Bryman, 1992; Northouse, 2010). Relating these dimensions of behaviour to various outcome measures such as job satisfaction and group performance, early findings report that consideration (i.e. a people orientation) is associated with increased morale and job satisfaction, whereas initiating structure (i.e. a task focus) tends to be associated with poorer morale but better performance Later findings have revealed that high levels of both dimensions are the best leadership style (Bryman, 1992, 1999).

Blake and Mouton's (1964) managerial grid (later renamed the leadership grid) is another well-known model that typifies the style approach, although it also spans contingency theory. The grid measures on a nine-point (*x*, *y*) scale how a concern for results or the task (*x* axis) and a concern for people (*y* axis) are related. Briefly, a 1, 1 score on the *x*, *y* axes refers to **impoverished management** where the leader is typically disengaged and shows little concern for both the task and for maintaining interpersonal relationships. A 1, 9 score refers to a **country-club management** style where the leader has a low concern for the task and is primarily concerned with maintaining interpersonal relationships whereas the converse is true for a 9, 1 **authority-compliance** style of leadership. Here the leader is primarily focused on the task at the expense of relationships. A 5, 5 score refers to a **middle-of-the-road management** style where the leader seeks to achieve balance between a task and people focus. The most effective style according to the authors, **team management**, scores a 9, 9 on the scale and emphasises a shared commitment to the task as well as the maintenance of good relationships.

However, there is only limited support that a high–high style is effective in all situations (Yukl, 2002, 2006). It is not too difficult to imagine, for example, that at times the commercial manager will need to rely on a more authoritative, task-focused style of leadership to ensure compliance to legal and financial requirements.

More general criticisms of the style approach pertain to the fact that like trait theories, it has failed to produce a universally agreed upon set of behaviours that defines effective leadership. Other problems are that: it ignores situational factors; it disregards informal leadership; it assumes causality between leadership style and performance when the opposite might be true; it relies on combined measures of subordinate perceptions and thus may not show inconsistent behaviours of a leader towards a particular subordinate; and it measures perceptions of leadership behaviour rather than the behaviour itself (Bryman, 1992).

In spite of these limitations, however, there is considerable research to support the viability of this approach to understanding leadership. Its significance lies in the fact that it encourages leaders to consider their impact along both task and relationship dimensions and thereby promotes a more holistic assessment of the leadership experience.

Contingency theories

Rejecting the view that there is one best way of leadership, contingency theories seek to offset the failure of previous approaches to take contextual variables into account. The central premise is that the style of leadership or behaviours adopted by a leader depends on the situation. Different theories highlight different situational factors.

The path-goal theory of leadership identifies four leadership styles:

1. **Directive** (focuses on clear task assignments)
2. **Supportive** (shows concern for employee well-being and creates a good working environment)
3. **Achievement-oriented** (sets high expectations of employees)
4. **Participative** (invites employee input into decisions) (House, 1971)

The theory is that the effectiveness of each leadership style depends on the characteristics of subordinates (skills, motivations, locus of control and expectations), and on the nature of the task and its context (goals, job design, resources and time available). Of special interest is the notion **locus of control** which refers to an individual's beliefs about who controls their lives. It can be inferred that someone with an internal locus of control is more likely to be self-directed and will probably respond better to a participative leadership style whereas someone with an external locus of control is likely to be other-directed and may respond better to a directive style. Such assumptions, however, are tentative given that an empirical relationship between locus of control and self-direction has yet to be proven. A large number of studies that have tested the theory have also yielded inconclusive results (Yukl, 2002, 2006).

One of the better known and most widely applied theories similar to this approach has been proposed by Hersey and Blanchard (1988). This model juxtaposes a task (guidance/directive) and relationship (support) orientation to identify four modes of leading that depend on the ability and willingness of the subordinate to perform a task. These include:

- **Telling** (involving close supervision and low relationship behaviours)
- **Selling** (where the leader still makes the decision but allows the opportunity for clarification)
- **Participating** (where the leader shares ideas and seeks consensus)
- **Delegating** (involving very low supervision and low support behaviours)

The central premise is that the leader can reduce task and relationship behaviours as followers mature in their ability and technical competence (job maturity) and self-confidence (psychological maturity) (Hersey and Blanchard, 1988). A later conceptualisation of this theory reframes the high directive, high support *selling* style as a *coaching* approach, although the leader still makes the final decision. Similarly, the high directive, low support *telling* style is renamed *directing*, whereas the low directive, high support *participating* style is renamed *supporting* in the revised version of the theory (Northouse, 2010).[2]

Although criticised for not having an adequate research grounding, this model is popular in practitioner circles because it incorporates a developmental focus by identifying follower readiness, that is, ability and willingness, as the main influencing factors in the choice of leadership style. Moreover, its central tenets are easily understood and it serves as a useful guideline or blueprint for engaging others whose ability and willingness are likely to change when faced with different tasks or even during the life of a single project. The flexibility it offers, as with path–goal theory, is thus a key part of its appeal.

Table 2.2 summarises the key similarities between the style and contingency approaches to leadership studied thus far.

Table 2.2 Comparing style and contingency approaches to leadership.

Theory	Task focus	People focus
Ohio State Leadership	Initiating structure	Consideration
Blake and Mouton's Managerial/Leadership grid	Concern for production	Concern for people
Path–Goal Theory	Nature and context of the job (task support)	Characteristics of subordinates (psychological support)
Hersey and Blanchard Situational Leadership	Task behaviour (guidance)	Relationship behaviour (support)

Transformational leadership

A wide assortment of frameworks, collectively referred to as the New Leadership by Bryman (1992, 1999) emerged in the 1980s. Employing terms such as *charismatic*, *transformational* and *visionary* leadership, these theories marked a shift from previous conceptualisations that focus on a task and person orientation to a view of leadership as '*the active promotion of values which provide shared meaning about the nature of the organization*' (Bryman, 1999, p. 27).

Of these, perhaps the most widely known is that of **transformational leadership**. Originating in the work of Burns (1978) who first distinguished between *transactional* and *transforming* leaders, transformational leadership is concerned with binding people around a common purpose (Bryman, 1992, 1999). Often incorporating visionary and charismatic leadership, it involves influencing and motivating followers to achieve more than what is expected of them (Northouse, 2010). This is facilitated through self-reinforcing behaviours that followers gain from successfully achieving a task and from a reliance on intrinsic rewards (Avolio and Bass, 1988). Transformational leaders also actively construct change, facilitate and teach followers and foster cultures of creative change and growth (Bass and Avolio, 1990, 1993).

By contrast, transactional leaders operate within the existing culture of the organisation to maintain the status quo (Bass and Avolio, 1993). They limit themselves to the exchange of extrinsic rewards that satisfy followers' needs in return for compliance and conformity with the leader's wishes (Bryman, 1992, 1999). This has prompted many to assert that transactional behaviours typify the manager whereas transformational behaviours capture the leader. Avolio and Bass (1988), however, view transactional leadership as a necessary feature of the transformational leader's role because (s)he must also 'transact' with followers as part of organisational reality. The key aspects of transformational and transactional leadership as described by Bass and Avolio (1994) are summarised in Table 2.3.

The appeal of this particular model of transformational leadership is that it is considered effective in any situation or culture and it has been replicated at different levels of authority, in different organisations and in several different countries (Yukl, 2002, 2006; cf. Avolio *et al.*, 1991; Hartog *et al.*, 1999).

Table 2.3 Transformational and transactional leadership behaviours.

Key factor		Behaviours
Transformational behaviours	Idealised influence	Leaders act as role models, are admired, respected and trusted, consider the needs of others over their own; are consistent in their behaviours; share risks with others and conduct themselves ethically
	Inspirational motivation	Leaders motivate and inspire others by providing meaning and challenge; they rouse team spirit; are enthusiastic and optimistic; communicate expectations and demonstrate commitment to shared visions
	Intellectual stimulation	Leaders encourage innovation and creativity through questioning assumptions and reframing problems. They avoid public criticism
	Individualised consideration	Leaders attend to individual needs for achievement and growth, engage in coaching and mentoring, create new learning opportunities, value diversity and avoid close supervision
Transactional behaviours	Contingent rewards	Leaders provide rewards on the condition that followers conform with performance targets
	Management by exception	Leaders take action when task related activity is not going according to plan

Source: Bass and Avolio (1994)

Other points in its favour are that transformational leadership has been shown to impact positively on followers' development as well as on indirect followers' performance (Dvir *et al.*, 2002). It has also been shown to cascade from one organisational level to the next in a falling dominoes effect (Avolio *et al.*, 1991), suggesting that transformational leaders either select and/or develop transformational followers or that followers imitate their leaders. It thus has particular relevance to a learning orientation given its targeted aims of follower development, improved performance and its emphasis on mutuality.

In spite of the existence of a substantial body of evidence that transformational leadership is effective, the above model is also subject to several conceptual weaknesses. In his critique of the theory, Yukl (1999) points to issues with ambiguous or overlapping constructs; an insufficient description of explanatory processes as to how the constructs were derived and how they interrelate; a narrow focus on dyadic processes at the expense of group or organisational level processes; the omission of some important behaviours; an inadequate specification of contextual variables that might moderate the effects of transformational leadership and an overemphasis on heroic conceptions of leadership. Nevertheless, transformational leadership remains one of the most well-researched areas of leadership.

Notably, many expositions on the new leadership tend to align themselves with earlier trait and behavioural approaches. Goleman's (1995) theory of emotional intelligence, discussed earlier in this chapter under trait theories, is a case in point. Bennis's (1989) exposition offers another example of the durability of trait and behavioural approaches. He identifies having a guiding vision, passion, integrity, trust, curiosity and daring as the basic ingredients of leadership whereas O'toole (1996) argues for values-based leadership based on integrity, trust, listening and respect for followers. Based on his literature review and converging research findings, Yukl (2002) proposes a comparable set of guidelines where transformational leaders:

- Articulate a clear and appealing vision
- Explain how the vision can be attained
- Act confidently and optimistically
- Express confidence in followers
- Use dramatic, symbolic actions to emphasise key values
- Lead by example
- Empower people to achieve the vision

Clearly, a key theme running through most of this work is the ability to translate vision into reality and create meaning, with two resultant implications. First, it implies that anyone directly involved in creating and interpreting meaning can be a leader, suggesting that leadership is more widely dispersed than most theories discussed until now typically acknowledge. Second, it ties in with the notion of 'sensemaking', thereby implicating 'leading' as a constituent factor in processes of learning at the organisational level. Significantly, this means that leading can be thought of as a process of responding to as well as 'enacting' or shaping the environment – something that previous leadership theories have not really addressed. A third key theme is the idea that leaders should lead from a strong set of values. Table 2.4 summarises some of the core themes in the new leadership.

Distributed leadership

Recent years have also seen the growth of **dispersed** or **distributed leadership** theories which recognise that leadership can be found at all organisational levels (Bryman, 1999). Chiefly including the notions of team-based leadership related to self-led or self-managed work teams, these approaches advocate a greater sharing of power between leaders and followers (Bryman, 1999; Gordon, 2002). Katzenbach and Smith (1993), for example, believe that the role of the team leader is to develop leadership in others within the team. Belasen (2000, p. 148) similarly observes that self-managed teams '... *seek to create a critical mass of shared leadership*'. Operating as quasi-autonomous groups, the team's members thus take on multiple leadership roles that include **enacting the team's environment**, **managing influence channels**,

Table 2.4 Core themes in the new leadership.

- Vision/mission
- Infusing vision
- Motivating and inspiring
- Creating change and innovation
- Empowerment of others
- Creating commitment
- Stimulating extra effort
- Interest in others and intuition on the part of the leader
- Proactive approach to the environment

Source: Bryman (1992)

horizontal networking and **handling external relations**, all of which demand strong interpersonal, negotiation and presentation skills (Belasen, 2000).

In his model of 'distributed' leadership, Barry (1991) identifies four broad leadership roles that are essential to project-based, problem-solving and policy-making self-managed teams although the relative importance of each role is likely to vary depending on the type of team and the stage of the team's development. These roles are **envisioning**, **organising**, **social integrating** and **external spanning**. Envisioning builds shared vision, organising structures the group's tasks, social integrating maintains cohesion within the team and external spanning links the team's efforts to its outer environment. Each role is thus critical to maintaining team dynamics, although they tend to be mutually exclusive and mastery of one role may hinder mastery of the others. The different roles also vary in their level of importance at different stages in the team's or group's development. For example, envisioning is especially important when the group is forming, whereas organising is more important once the group has agreed on an objective (Barry, 1991). In this way, distributed leadership is not unlike contingency theory. A key difference, however, is that in distributed models, leadership is viewed as an emergent property of a group of interacting individuals rather than the property of one individual alone.

An interesting model of cross-functional teams that appears to have particular relevance to contract and commercial managers is offered by Mumford *et al.* (2002). This model identifies **technical expertise, cognitive skills, interpersonal skills, project management skills** and **political skills** as being important to the leaders in these teams. Being able to communicate one's expertise to team members from diverse backgrounds is critical to building a shared understanding and gaining commitment to the task. Cognitive skills entail the ability to solve complex problems. This requires creativity, systems thinking and an understanding of the contribution the various functions can make to the role. Good interpersonal skills are important to help the leader influence the team by understanding their values and needs, resolving conflicts and building cohesion within the team. Project management skills entails things like the ability to plan and organise project activities and handle financial responsibilities, whereas political skills involves developing coalitions, gaining resources and also securing assistance and influence from important stakeholders (Mumford *et al.*, 2002).

However, it is questionable whether leadership can ever be truly shared because of the way in which traditional power relations that are hierarchically based can become embedded in the organisation. These can invariably inhibit people from enacting leadership or cause them to behave in ways that comply with what they think their superiors might expect. Nevertheless, the idea of distributed leadership is attractive because it widens the boundary of leadership to many instead of a select few. As stated earlier, it views leadership as a property of a group rather than the property of one individual. It also implies greater collaboration and integration in leadership activity than previous conceptualisations.

Leadership and learning

Earlier in this chapter 'educating others' was identified as a key meta-goal of commercial leadership, for example, in regard to the procedures, rules and regulations that need to be adhered to in order to get the 'job done'. This section thus explores some of the new leadership literature that is specifically linked to engaging others in processes of learning.

Table 2.5 Senge's leadership skills.

Building shared visions	Encouraging personal vision Communicating and asking for support Visioning as an ongoing process Blending extrinsic visions (relative to a competitor) with intrinsic visions (such as setting new standards) Distinguishing between positive and negative visions
Surfacing and testing mental models	Seeing leaps of abstraction and avoiding premature conclusions Balancing inquiry and advocacy Distinguishing espoused theory from actual theory in use Recognising and defusing defensive routines
Systems thinking	Seeing interrelationships, not things, and processes, not snapshots Moving beyond a blame culture Distinguishing detail complexity from dynamic complexity Focusing on areas of high leverage where a change requiring minimum effort is likely to result in enduring improvements Avoiding symptomatic solutions

Source: adapted from Senge (1996)

Although they are largely based on conjecture and specifically focused on leadership in the context of **learning organisations**, two models that link leadership and learning are promulgated by Senge (1990, 1996) and Marquardt (1996). Briefly, a learning organisation is one that purposefully constructs structures and strategies to make the most of organisational learning (Dodgson, 1993) and that actively seeks to shape its future through the concerted effort to engage in transformation by facilitating the learning and development of all its members (Pedler *et al.*, 1991; Watkins and Marsick, 1993, 1996; Mumford, 1995; cf. Senge, 1990) or through the pursuit of a constantly enhanced knowledge base (Nevis *et al.*, 1997).

Senge (1990, p. 345) believes that the leader's task is to design the learning processes that enable people to deal productively with the critical issues they face, and develop their mastery in the learning disciplines. These new leaders are **designers**, **teachers** and **stewards** that require a new range of skills (Senge, 1990; cf. Watkins and Marsick, 1993). Senge (1990, 1996) explains that as designers of the **social architecture** of the organisation, the new leaders are responsible for the governing ideas that underpin the policies, strategies and structures that inform business decisions and actions and which allow them to build shared vision. As teachers, they coach, guide or facilitate people towards achieving a more accurate, insightful and empowering view of reality through surfacing and challenging prevailing mental models or tacit assumptions about how things work. Finally, as stewards, the subtlest of the leader's roles, they harbour an attitude of wanting to serve the people they lead and hold a personal commitment to the organisation's mission (Senge, 1990). The three critical areas of skill that are necessary to undertake these roles include building shared visions, surfacing and challenging mental models and engaging in systems thinking (Senge, 1996). These are further broken down into a subset of skills summarised in Table 2.5.

Notably, Senge conceptualises the teacher, not in the authoritarian sense of teaching people the one correct view of reality, but rather in the sense of helping everyone, including oneself, gain greater insights.

Building on this work, Maquardt (1996) produces his own list of similar, sometimes identical roles that are important for the leader. The roles of **instructor**, **coach** and **mentor** although closely related, recognise that managers will need to vary their approach to cater to the versatile learning needs of the people they are dealing with. Given the time spans of project life cycles, this would suggest that the role of instructor and coach are more likely to apply to contract and commercial managers with their focus on achieving the task and delivering the contract.

In their second role as **knowledge managers**, Marquardt (1996) asserts that leaders need to encourage the collection, storage and distribution of pertinent knowledge both internal and external to the unit and to ensure that the necessary mechanisms to support this process are in place. In their role as **co-learner** and **model for learning**, leaders also need to engage visibly in learning practices and to actively encourage

Table 2.6 Senge's (1990, 1996) and Marquardt's (1996) leadership roles.

Senge's leadership roles		Marquardt's leadership roles
Designer	➔	Architect and designer Knowledge manager
Steward	➔	Co-learner and model for learning Coordinator
Teacher	➔	Instructor, coach and mentor Advocate and champion for learning processes and projects

Source: Denton (1998)

others to learn through action learning, risk-taking and questioning that is oriented towards innovation. The fourth role of **architect** and **designer**, like Senge's proposition, involves designing and integrating new technologies, structures and processes into systems that support learning. It also includes revamping human resource practices such as selecting, training and rewarding so as to encourage people to participate in the new learning system. Finally, the role of **coordinator** involves coordinating people's work so that they can perform at their best, while in an extension of the co-learner role, leaders must act as **advocates** and **champions** for learning processes and projects to ensure that learning is more widely dispersed across the organisation (Maquardt, 1996).

Marquardt then goes on to identify the same broad areas of leadership skills as Senge but he extends his list to include the ability to coordinate multiple, task-focused teams, to encourage creativity, innovation and risk taking, and lastly, to inspire learning and action through transformational leadership (Maquardt, 1996). The two models are thus comparable with the exception that Marquardt offers a more refined set of roles and acknowledges the need for transformational leaders. Senge's (1996) leader, by contrast, is based on the notion of **servant leadership** and is much less individualistic. Table 2.6 compares the two frameworks.

Clearly, the chief value of both these models lies in their explicit emphasis on learning. But, that aside, it is difficult to see how they differ radically from previous accounts of leadership. Marquardt's instructor, coach and mentor, for example, correspond to Bass and Avolio's (1994) individualised consideration. These roles also utilise the two key factors of earlier style and contingency theorists. For instance, it could be argued that the instructor is task-oriented and initiates structure, the coach combines a task orientation with a concern for people, whereas the mentor leans towards a people orientation. Similarly, the role of the coordinator can be summed up as initiating structure.

Coining the term **learning leader** Schein (1992) takes a slightly different tack. He speculates that the role of the learning leader is to be a perpetual learner and to promote the kinds of assumptions upon which learning cultures are based. He conceives of learning leadership in terms of several propositions where the learning leader:

- Assumes that the environment can be managed
- Regards oneself and others as proactive problem solvers and learners
- Accepts that (s)he won't have all the answers and teaches others to do the same
- Holds to Theory Y assumptions about people but recognises that human nature vacillates (Theory Y assumes that people are motivated to do well and prefer to be self-directed rather than led)
- Varies his or her leadership style according to the nature of the task at hand
- Selects time units that allow proposed solutions to be tested in a timely way
- Communicates task-relevant information openly and establishes a fully connected network
- Values diversity
- Achieves a healthy balance between task and relationship-oriented behaviours
- Holds to a systemic view of the world (Schein, 1992)

Desirable qualities of a leader

So far, a number of leadership models have been considered and a whole host of traits, behaviours, attitudes and attributes have been identified that are important to a leader. At this point, it is instructive to

Table 2.7 The qualities/attributes of successful managers.

Basic knowledge and information	1 Command of basic facts 2 Relevant professional understanding
Skills and attributes	3 Continuing sensitivity to events 4 Analytical, problem solving and decision making skills 5 Social skills and abilities 6 Emotional resilience 7 Proactivity – inclination to respond purposefully to events
Meta-qualities	8 Creativity 9 Mental agility 10 Balanced learning habits and skills 11 Self-knowledge

Source: Pedler *et al.* (2007)

consider an additional model advanced by Pedler *et al.* (2007) because of its durability and its universal applicability. Although the authors do not discuss leadership as such, they identify 11 qualities of successful managers that are still relevant today, more than 30 years after the initial research endeavour was carried out.[3] These are grouped into three overarching categories as shown in the Table 2.7.

The authors further highlight the interconnection between these qualities. The meta-qualities (categories 8 to 11) allow managers to develop and utilise the skills in categories 3 to 5 and to develop additional situation-specific skills as necessary. The skills and attributes in categories 3 to 7 directly impact on behaviours and performance and allow managers to obtain basic knowledge and information (1 and 2) which in turn enable them to make informed decisions (Pedler *et al.*, 2007). The appeal of this model lies in its simplicity and resonance with contemporary managerial practice.

Towards a new conceptual framework of leadership

This chapter has illustrated that there is no shortage of existing models or theories of leadership. It is a complex, multifaceted process. In this section, a conceptual framework is advanced that portrays commercial leadership as a twofold task and learning-focused process. Leadership is understood as a social, diffused process whereby more than one person can perform the role at any one time. It can thus be alternated or shared with others involved in a commercial transaction and is not the sole purview of the contract or commercial manager. Leadership in this framework is also envisioned as a multilevel concept that applies at intrapersonal, dyadic and group levels.

The process of commercial leadership begins with the underlying beliefs and characteristics that allow commercial managers to function as leaders. This includes their overarching theories and key person-related competences or qualities. The term 'overarching theories' derives from the work of Schön (1983) and refers to the beliefs, principles, philosophies, meaning schemes and meaning perspectives by which managers make sense of the situations they manage through. These are held to guide reasoning and action and ultimately play an important role in shaping behaviour. The term **competency** was popularised by Boyatzis (1982) and refers to motives, traits, skills, job-related knowledge and even features of one's self-image or social role. Here, the term is used rather broadly to refer to the generic motives, traits, skills and behavioural strategies managers bring into play in order to lead at individual, dyadic and group levels.

Figure 2.1 portrays leadership as a process of inputs that are converted into outputs in the course of doing work. As stated earlier, it conceives of leadership as a learning as well as task-focused process whereby managers are presumed to be able to advance their own learning and that of the team or multiple stakeholders, surface and facilitate the integration of individual mental models into shared interpretations and create and reinforce appropriate conditions that are conducive to learning and to attaining shared goals. Learning and the variables of getting the job done thus blend and run into one another to the point where they become indistinguishable.

In managing through a task, it is assumed that contract and commercial managers bring their personal inputs (Node B) into play depending on situational and other intervening variables which they can sometimes

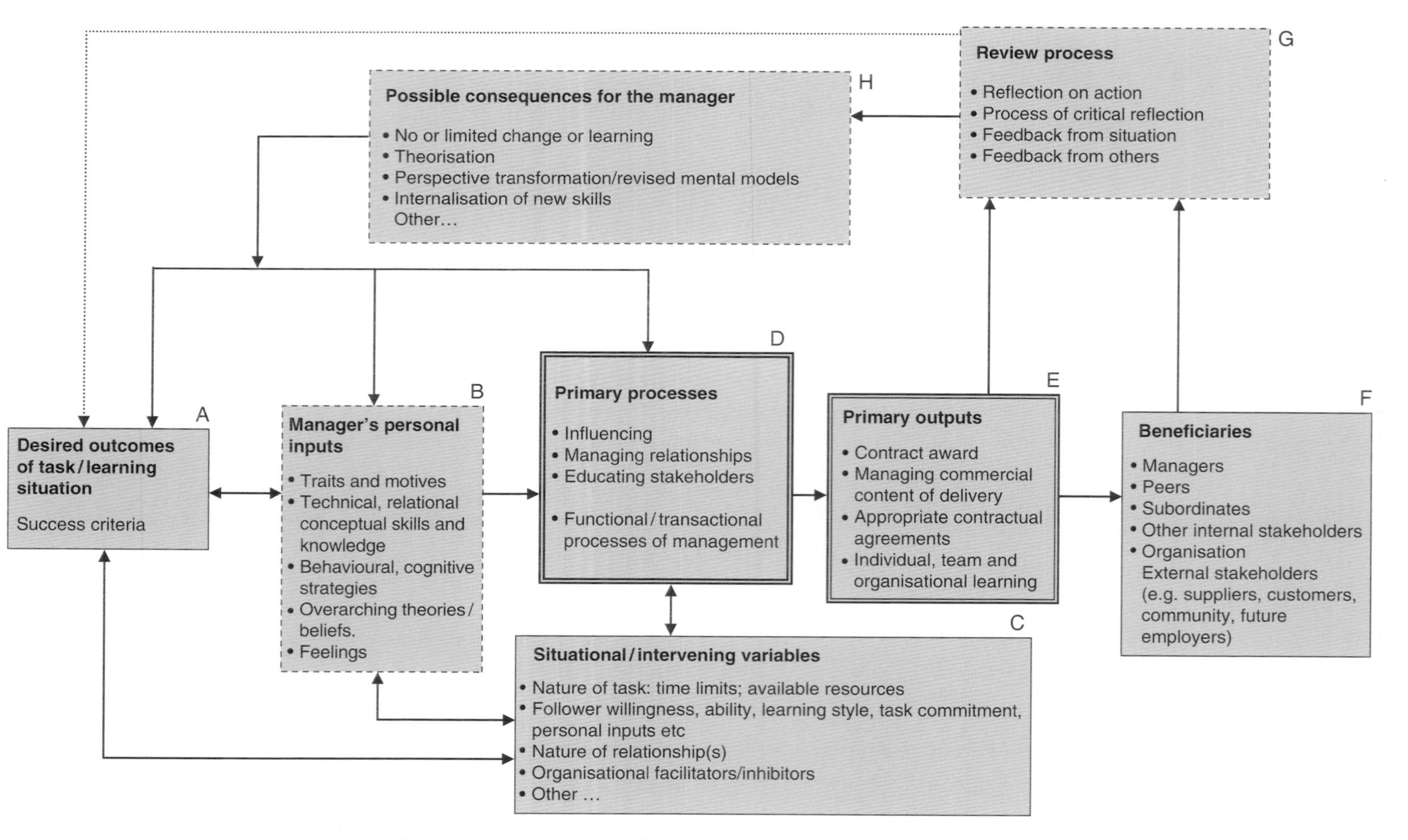

Figure 2.1 A conceptual framework of commercial leadership.

influence (Node C) in order to achieve the desired outcomes or success criteria (Node A). The double borders indicate that the primary learning processes and outputs (Nodes D and E) run parallel to other task-related processes and outputs whereas the shadows indicate that the contents of these boxes can be further broken down into sub-processes and sub-indicators. The perforated boxes largely represent a manager's inner world and incorporate the basic precepts of learning to show the interaction between managers' assumptions of leading and learning and their workplace experiences. In other words, the experience of the process represented by Nodes A to E once reflected upon (Node G) allows managers to theorise and draw abstract generalisations (Node H) which are then tested out in future situations at both the task and learning level. A weakness in this framework is that it simplifies the complexity of the environment in which managers operate and does not capture important processes, such as processes of reflection-in-action and the multiple-learning conversations that are likely to be going on inside managers' heads or in interaction with others at each node. These conversations are, however, implied in the arrows that link the nodes.

Summary and implications for the commercial management function

This chapter has conceptualised leadership, and in particular commercial leadership, as a social process, that is, principally concerned with:

- Influencing both internal and external stakeholders at the multiple interfaces of exchange transactions
- Managing these relationships during the pre- and post-award stages of a project's life cycle
- Educating the multiple stakeholders engaged in an exchange – for example, coaching project managers on the content of a specific contract or agreements generally

The overall aim of the process is the delivery of organisational objectives via projects and programmes within the context of contractual commitments.

The chapter has explored key themes and concepts from the leadership literature. It has reviewed established approaches to the study of leadership, such as trait, style and contingency theories; examines contemporary theories, such as transformational and dispersed leadership; and explored leadership in relation to learning. The chapter concluded with the exposition of a conceptual framework that views leadership as a process involving a dynamic interaction between a task focus and learning.

It therefore provides an introduction to the leadership literature, enabling commercial practitioners to appreciate the importance of the process to the execution of their day-to-day activities and to their future progression within the organisation. Moreover, at a functional level it provides evidence to support the need for the commercial function to develop further its leadership capability so that it can be truly accepted as a key participant within the senior management of organisations.

Further reading

Commercial leadership

The Results Driven Manager (2005) *Becoming an Effective Leader*. Harvard Business School Press, Boston

Mumford, M. and Connelly, MS (1991) Leaders as creators: leader performance and problem-solving in ill-defined domains. *Leadership Quarterly*, 2, 289–315

Mumford, MD, Zaccaro, SJ, Connelly, MS and Marks, MA (2000) Leadership skills: conclusions and future directions. *Leadership Quarterly*, 11, 155–170

Mumford, TV, Campion, MA and Morgeson, FP (2007) The leadership skills strataplex: leadership skill requirements across organisational levels. *Leadership Quarterly*, 18, 154–166

Yukl, G (2002) *Leadership in Organizations*. Prentice Hall, Upper Saddle River

Exercise Commercial leadership

Compare and contrast the commercial leadership exhibited on the Wembley Stadium project with that adopted on the Heathrow T5 project.

Endnotes

1 For a detailed, comprehensive review of the leadership literature, the reader is referred to texts such as Bryman (1992), Northouse (2010), Shackleton (1995), and Yukl (2002). Full references are available in the reference list.

2 A detailed description and diagram of the model can be found in Blanchard *et al.* (1985). A review of the model is also offered in Northouse (2010).

3 The initial framework can be found in Burgoyne and Stuart (1976).

References

Alvesson, M and Deetz, S (2000) *Doing Critical Management Research*. Sage, London

Avolio, BJ and Bass, BM (1988) Transformational leadership, charisma and beyond. In: JG Hunt, BR Baliga, HP Dachler and CA Schriesheim (eds) *Emerging Leadership Vistas*. DC Heath and Company, Lexington. pp. 29–49

Avolio, BJ, Waldman, DA and Yammarino, FJ (1991) Leading in the 1990s: the four Is of transformational leadership. *Journal of European Industrial Training*, 15, 9–16

Barry, D (1991) Managing the bossless team: lessons in distributed leadership. *Organization Dynamics*, 20, 31–47

Bass, BM and Avolio, BJ (1990) Developing transformational leadership: 1992 and beyond. *Journal of European Industrial Training*, 14, 21–27

Bass, BM and Avolio, J (1993) Transformational leadership and organizational culture. *Public Administration Quarterly*, 17, 112–121

Bass, BM and Avolio, BJ (1994) *Improving Organizational Effectiveness through Transformational Leadership*. Sage, Thousand Oaks

Belasen, AT (2000) *Leading the Learning Organization: Communication and Competencies for Managing Chang*. State University of New York, New York

Bennis W (1989) *On Becoming a Leader*. Addison Wesley, New York

Bennis, W and Nanus, B (1985) *Leaders: The Strategies for Taking Charge*. Harper Collins, New York

Blake, RR and Mouton, JS (1964) *The Managerial Grid*. Gulf Publishing Company, Houston

Blanchard, K, Zigarmi, P and Zigarmi, D (1985) *Leadership and the One Minute Manager: Increasing Effectiveness through Situational Leadership*. William Morrow, New York

Boyatzis, RE (1982) *The Competent Manager: A Model for Effective Performance*. Wiley, New York

Brookfield, SD (1985a) Self-directed learning: a conceptual and methodological exploration. *Studies in the Education of Adults*, 17, 19–321

Brookfield, SD (1985b) A critical definition of adult education. *Adult Education Quarterly*, 36, 44–49

Bryman, A (1992) *Charisma and Leadership in Organizations*. Sage, London

Bryman, A (1999) Leadership in organizations. In: SR Clegg, C Hardy, and WR Nord (eds) *Managing Organizations: Current Issues*. Sage, London. pp. 26–62

Burgoyne, JG and Stuart, R (1976) The nature, use and acquisition of managerial and other attributes. *Personnel Review*, 5, 19–29

Burns, JM (1978) *Leadership*. Harper Row, New York

Denton, J (1998) *Organisational Learning and Effectiveness*. Routledge, London

Dodgson, M (1993) Organizational learning: a review of some literatures. *Organization Studies*, 14, 375–394

Dvir, T, Eden, D, Avolio, BJ and Shamir, B (2002) Impact of transformational leadership on follower development and performance: a field experiment. *Academy of Management Journal*, 45, 735–744

French, JR and Raven, B (1962) The bases of social power. In: D Cartwright (ed.) *Group Dynamics: Research and Theory*. Harper and Row, New York. pp. 259–269

Garrison, R (1992) Critical thinking and self-directed learning. *Adult Education Learning*, 42, 136–148

Goleman, D (1995) *Emotional Intelligence: Why It Can Matter More Than IQ*. Bloomsbury, London
Goleman, D, Boyatzis, R and McKee, A (2002) *The New Leaders: Transforming the Art of Leadership into the Science of Results*. Time Warner Books, London
Gordon, RD (2002) Conceptualising leadership with respect to its historical contextual antecedents to power. *The Leadership Quarterly*, 13, 151–167
Grint, K (2005) *Leadership: Limits and Possibilities*. Palgrave Macmillan, Basingstoke
Hartog, D, Deanne, N, House, RJ, Hanges, PJ, Ruiz-Quintanilla, SA and Dorfman, PW (1999) Culture specific and cross-culturally generalizable implicit leadership theories: are attributes of charismatic/transformational leadership universally endorsed? *Leadership Quarterly*, 10, 219–255
Hersey, P and Blanchard, KH (1988) *Management of Organizational Behaviour: Utilising Human Resources*. Prentice Hall, Englewood Cliff
House, RJ (1971) A path goal theory of leadership effectiveness. *Administrative Science Quarterly*, 16, 321–328
Huczynski, A and Buchanan, D (2001) *Organizational Behaviour: An Introductory Text*. Financial Times/Prentice Hall, Harlow
Katzenbach, JR and Smith, DK (1993) *The Wisdom of Teams: Creating the High Performance Organization*. Harvard Business School, Boston
Kotter, JP (2011) *Management vs Leadership: What is the Difference Between Management and Leadership?* Kotter International. http://kotterinternational.com/KotterPrinciples/ManagementVsLeadership.aspx (accessed 20 February 2012)
Lowe, DJ (2008) Commercial and contract management – practitioner expectations on the content of an MBA for commercial executives. In: DJ Lowe (ed.) *Proceedings IACCM International Academic Symposiums on Contract & Commercial Management*, London, UK and Fort McDowell, Arizona. ISBN 978-1-934697-01-6. pp. 45–72
Marquardt, MJ (1996) *Building the Learning Organization*. McGraw Hill and ASTD, New York
Marsick, VJ and Watkins, KE (1990) *Informal and Incidental Learning in the Workplace*. Routledge, New York
Mayer, JD, Salovey, P and Caruso, DR (2000) Models of emotional intelligence. In: RJ Sternberg (ed.) *Handbook of Intelligence*. Cambridge University Press, Cambridge. pp. 396–420
Mezirow, J (1991) *Transformative Dimensions in Adult Learning*. Jossey Bass, San Francisco
Mezirow, J (1994) Understanding transformation theory. *Adult Education Quarterly*, 44, 222–232
Mumford, A (1995) *Learning at the Top*. McGraw Hill, Maidenhead
Mumford, MD, Scott, GM, Baddis, B and Strange, JM (2002) Leading creative people: orchestrating expertise and relationships. *Leadership Quarterly*, 13, 705–750
Nevis, EC, Dibella, AJ and Gould, JM (1997) Understanding organizations as learning systems. In: D. Russ-Eft, H Preskill, and C Sleezer (eds) *Human Resource Development Review: Research and Implications*. Sage, London. pp. 274–298
Northouse, PG (2010) *Leadership: Theory and Practice*. Fifth edition. Sage, Thousand Oaks
O'toole, J (1996) *The Argument for Values-Based Leadership*. Ballantine, New York
Pedler, M, Burgoyne, J and Boydell, T (1991) *The Learning Company: A Strategy for Sustainable Development*. McGraw Hill, London
Pedler M, Burgoyne J and Boydell T (2007) *A Manager's Guide to Self-Development*, Fifth edition. McGraw-Hill Professional, Maidenhead
Reynolds, M (1998) Reflection and critical reflection in management learning. *Management Learning*, 29, 163–200
Rost, JC (1991) *Leadership for the Twenty First Century*. Praeger, New York
Schein, EH (1992) *Organizational Culture and Leadership*. Josey Bass, San Fransisco
Schön, D (1983) *The Reflective Practitioner: How Professionals Think in Action*. Basic Books, London
Senge, PM (1990) *The Fifth Discipline: The Art and Practise of the Learning Organizatio*. Century Business, London
Senge, PM (1996) The leader's new work: building learning organizations. In: K. Starkey (ed.) *How Organizations Learn*. International Thomson Business Press, London. pp. 288–315
Shackleton, V (1995) *Business Leadership*. Routledge, London
Tennant, M (1997) *Psychology and Adult Learning*. Routledge, London
Watkins K E and Marsick V J (1993) *Sculpting the Learning Organization: Lessons in the Art and Science of Systemic Change*. Jossey-Bass, San Francisco
Yukl, G (1999) An evaluation of conceptual weaknesses in transformational and charismatic leadership theories. *Leadership Quarterly*, 10, 285–305
Yukl, G (2002) *Leadership in Organizations*. Prentice Hall, Upper Saddle River
Yukl, G (2006) *Leadership in Organizations*. Prentice Hall, Upper Saddle River
Zaleznik, A (1992) Managers and leaders: are they different? *Harvard Business Review*, March–April, 126–135

Chapter 3

Exploring Strategy

Irene Roele

Learning outcomes

When you have completed this chapter you should be able to:

- Explore the difference between strategic thinking and strategic planning
- Explore key change drivers and factors that influence the attractiveness of an industry and the firm competitive position
- Explore the potential sources of competitive advantage for the firm.
- Identify and explore strategy at the corporate and business/competitive levels
- Understand the core premise of the resource-based view of the firm
- Explore the concept of dynamic capabilities

Introduction

> 'By strategy, I mean a cohesive response to a challenge. A real strategy is neither a document nor a forecast but rather an overall approach based on a diagnosis of a challenge. The most important element of a strategy is a coherent viewpoint about the forces at work, not a plan.' (Rumelt, 2008)

Strategy is about how to achieve and sustain competitive advantage in the long term. It is about having a clear sense of purpose and the organisational capacity to make difficult choices about the direction and evolution of the firm. Rather than being a tactical, support activity responsible for the implementation of strategy created by others within the firm, the commercial management function can play a significant role in informing strategic thinking and action.

Today's fast pace of change makes it increasingly challenging, yet even more of an imperative, for senior management to think and act strategically, as they must both protect current activity while planning for the future:

> 'The basic problem confronting an organization is to engage in sufficient exploitation to ensure its current viability and, at the same time, devote enough energy to exploration to ensure its future viability.' (March, 1991)

This dilemma is the one that the US corporate Boeing faced in the second half of 2011 with its hugely successful 737 family of aircraft in the short-haul market.[1] After many years, Boeing's dominance in this sector had slipped, in particular since the launch of the Airbus 320 family of narrow-body aircraft (Airbus is part of EADS, the European conglomerate) in the 1990s.

Commercial Management: theory and practice, First Edition. David Lowe.

Before committing to a final decision, Boeing assigned separate teams to assess the two options of *exploitation* – to continuously improve its current product and re-engine its 737 family at an estimated development cost of $3 billion or *exploration* – designing an entirely new type of aircraft at an estimated development cost of $10 billion, '*a revolutionary approach that could lead to potentially twice the efficiency savings and allow it to dominate the market for decades*' (*Financial Times*, 24 July 2011).

Market and industry dynamics, notably the successful launch of the A320neo by Airbus,[2] forced Boeing into making a quick decision. It opted for exploitation: the re-engined 737max model is scheduled to enter service with American Airlines, one of its core customers, in 2018. However, with other new entrants, such as the CSeries from Canada's Bombardier and, in particular, the fledgling Chinese aircraft maker Comac, the status quo has altered: 'the days of the duopoly … are over' announced Jim Albaugh, CEO of Boeing's civil aircraft division (Odell, 2011).

There is no single definition of strategy but a multiplicity of perspectives, ranging from the airport kiosk toolkit approach to abstract conceptual pieces that bear little relation to commercial practice. '*There are now so many varied views of strategy it has become hard to be sure what we mean when we use the term*' (Cummings and Daellenbach, 2009). Some of the strategy models '*… have become diluted with familiarity*' (Pettigrew and Whipp, 1993) and are so overused as to lose all meaning, while much of the academic language of strategy remains impenetrable to most practitioners (Cummings and Wilson, 2003; Jones, 2008).

The field of strategy obviously has military origins, and draws on a variety of disciplines, most notably economics, the theory of the firm, psychology and sociology. '*The problems addressed in the practice of strategic management require insights from multiple disciplines*' (Augier and Teece, 2008, original emphasis). It could be argued that there aren't really any theories of strategy, and as Derman has said: '*Models are reductions in dimensionality that always simplify and sweep dirt under the carpet. Theories tell you what something is. Models tell you merely what something is partially like*' (Derman, 2011). Strategy models should not, therefore, be used indiscriminately.

Collis and Rukstad (2008) suggest there are three critical components in articulating a strategy: *objective, scope*, and *advantage*:

1. A definition of the *objective(s)* that the strategy is designed to achieve – which should '*… include not only an end point but also a time frame for reaching it*'
2. A definition of the '*scope*', or '*domain of the business*', the territory in which the firm will compete, importantly, this should include what it will *not* do, a much emphasised point made by Porter[3] (1996)
3. '*How*' the firm will achieve its objectives – that is, the basis on which it will compete to achieve and sustain a competitive advantage

> *The '[strategic management]* field studies the choices that enterprises must make in order to survive and prosper in an environment where it is assumed that there is competition for customers, technology, people, financial capital, and other inputs.' (Augier and Teece, 2008)

Cummings and Wilson (2003) suggest that strategy can be perceived as a combination of '*orientation*' (the cognitive components of diagnosis and choice) and '*animation*' (the affective components of commitment and action).[4] The emphasis in this chapter is more on 'orientation', as some of the 'animation' was covered in Chapter 2.

Understanding key change drivers

The forces at work in the external environment make strategic choice particularly complex for managers. These change drivers include, for example:

- Increased globalisation, which heightens competition
- Technology, notably '*the unseen domain that is strictly digital*' (Arthur, 2011), which is blurring market and industry boundaries and changing their dynamics in a non-linear way. This results in opportunities for new business models, but also threats as '*disruptive* technologies' race ahead, leaving incumbents struggling to stay in the game

- Consolidation of industries and the consolidation/fragmentation paradox where, in many industries, we are seeing '*... a move to building ever bigger scale and capacity to bid for bigger contracts* ... [and at the same time]... *opportunities for rivals with smaller operations, lower overheads, local knowledge and greater flexibility*' (Gray, 2011)
- Increasingly complex regulation of, for example, the utility sectors
- Political change, such as regime change and legislation
- The 'eco-factor', as firms have to grapple with the sustainability of every aspect of their operations

It is crucial, therefore, to have a proper diagnosis of the impact these change drivers could have on the firm as it will affect its capability to adapt, and put its strategic thinking into action.

Porter's Five Forces model[5] (1980, 2008) assesses the key determinants of profitability in an industry and helps to analyse the firm's competitive position within it. The five forces are:

1. The bargaining power of buyers
2. The bargaining power of suppliers
3. Barriers to entry
4. The threat of substitutes
5. Degree of competitive rivalry

Porter stressed the importance of looking at the impact of the combination of all the forces, not just one or two of them. Augier and Teece observe that:

> 'His Five Forces framework implicitly advised that the way to escape the zero profit condition and earn supernormal profits was (1) to pick an attractive industry (e.g. one that is growing, and faces limited competition, and isn't exposed to a squeeze from buyers or suppliers) and (2) to enter or expand output in that industry while (3) building defenses to shield oneself from competitors who will undoubtedly try to compete away supernormal profits and leave the enterprise with zero economic profit. Shields available to defend from competition include product differentiation or achieving the lowest cost.' (Augier and Teece, 2008)

Exploring competitive advantage

What do we mean by 'competitive advantage'?

> 'If strategy is to have any meaning at all, it must link directly to a company's results. Anything short of that is just talk.' (Magretta, 2012)

There is no single definition of competitive advantage, as it is obviously a relative term, but it can be summarised as follows:

> Competitive advantage occurs when the firm has the capability to differentiate itself from its rivals, when it is able to create more economic value than competing firms by providing a product or service to its customers that is in some way superior to that of its competitors. Importantly, this advantage should result in consistent superior performance, often referred to as '**supernormal profit**'[6] (Kay, 1993; Augier and Teece, 2008)

As Porter (1980, 1985) argued, competitive advantage can only be sustained by either pursuing differentiation or cost leadership, endeavouring to avoid the middle-ground. A differentiation strategy involves configuring the firm's activities to create products or services:

- The characteristics of which cannot be matched by the competition
- Where the customer is prepared to pay a superior price, giving the potential for profit derived from that premium price, or where the perceived added value to the customer yields increased market share

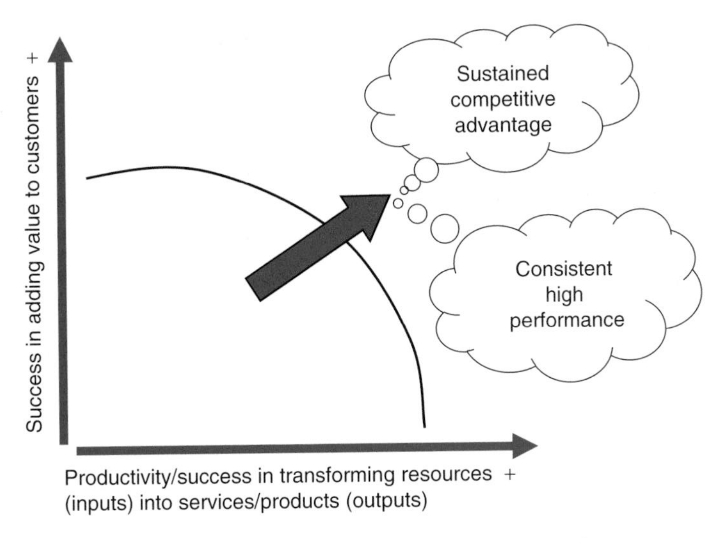

Figure 3.1 Where is the firm on 'the curve'?

In a cost leadership strategy the firm's competences are used to generate a unique low-cost way of creating or delivering to market a competitively priced alternative, with some additional margin left to create a profitable operation.

Subsequently, Porter's generic strategies have been much debated. On the whole the market and industry dynamics pre-1990 meant that firms could only compete effectively on one of these two areas. What we then saw was firms seemingly being able to do both, but they were actually just competing on the curve of operational efficiency (Porter, 1996), rather than making a strategic breakthrough.

Where is the firm on the 'curve'?

Porter's influential publication 'What is strategy?' (1996):

> '... lays out the characteristics of strategy in a conceptual fashion, conveying the essence of strategic choices and distinguishing them from the relentless but competitively fruitless search for operational efficiency.' (Collis and Rukstad, 2008)

Figure 3.1 encapsulates Porter's generic strategy concept: the vertical axis depicting differentiation, and the horizontal axis cost leadership. The macro-environmental dynamics, in particular technological developments and hyper-competition, result in the most successful firms competing on the notional curve of best practice, known as the productivity frontier, in the relentless search for operational efficiency. Porter argues that firms need to do this just to survive, and that strategy is about '*breaking through*' this productivity frontier and about '*playing the game differently*' (Porter, 1996).

Quite a few contributors to the field have developed his thinking further, notably in the concept of **blue ocean strategies** (Kim and Mauborgne, 2008). Their work clearly articulates how the relentless pursuit of operational efficiency often results in a zero sum game: the firm's focus on jockeying for position comes at the cost of innovation, as it is left vulnerable to potential new entrants disrupting the rules of a game with an entirely new business model.

The strategic importance of trying to break though into the blue ocean, where the firm aims to set new rules, and capture supernormal profit is illustrated in Figure 3.2.

Market-led innovation

> '... fundamental strategic innovation is achieved by the creation of new competencies and new business models that break the rules of their industry.' (Jacobs and Heracleous, 2005)

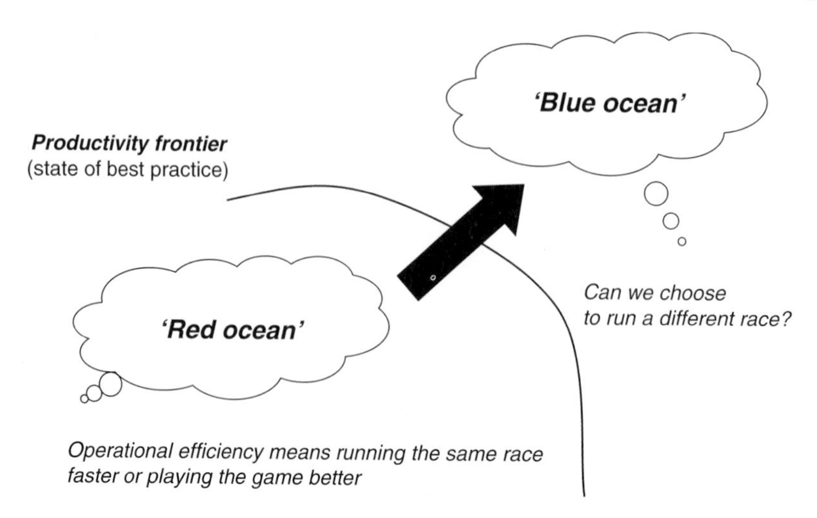

Figure 3.2 Where are we on 'the curve', are we ambidextrous? Adapted from Porter (1996), Markides (2001), Kim and Mauborgne (2008), Smith, Binns and Tushman (2011).

Innovation has become part of the strategy lexicon. It is not only about product or service breakthroughs, but can also be about process innovation. Reilly and Tushman (2008) argue that innovation occurs in roughly three distinct ways:

- *'Incremental innovation in which an existing product or service is made better, faster, or cheaper'* (Nelson and Winter, 1982), *using existing competencies*
- Major improvements, typically through an advance in technology, which require different competencies or skills from those the incumbent has
- *'Seemingly minor improvements in which existing technologies or components are integrated to dramatically enhance the performance of existing products or services. These architectural innovations, while not based on significant technological advances, often disrupt the existing offerings'* (O'Reilly and Tushman, 2008)

The pace of change, particularly in technology, is resulting in the emergence of new business models. These arise when companies, through effective market sensing, have identified existing or latent customer needs which they have then been able to translate into an appropriate value-proposition, based on a unique configuration of the firm's activity system and the structure of the firm. This type of innovation goes back to some fundamentals of strategy: defining the market, and really understanding the 'who, what and how' aspects of one's business (Chandler, 1962; Abell, 1980; Markides, 1999).

Corporate level strategy

> 'Corporate strategy is the way a company creates value through the configuration and co-ordination of its multi-market activities.' (Collis and Montgomery, 2004)

At the corporate level, the strategic agenda is concerned with two critical interrelated issues:

1. **Domain selection**: deciding which business(es) to be in and identifying the added value created through the portfolio of activities of the 'corporate parent' (Goold *et al.*, 1994). This involves determining the logic of the portfolio and the nature and extent of diversification.[7] *'To make sense from an economics/financial perspective, any multidivisional business must have a greater ongoing value than the sum of its component stand-alone businesses. This is the age-old corporate dynamic of* ***synergy****'* (Angwin et al., 2011)

2. **Domain management**: creating and capturing the value of the businesses/activities across the portfolio. This is concerned with the role of the corporate centre and its management style/approach – how the corporate centre adds and captures the values to the businesses in the portfolio

Corporate-level strategy includes much analysis regarding the types of diversification, which are defined as related or unrelated, with the latter obviously likely to incur greater risk.

Related diversification typically refers to:

- **Vertical integration** – bringing upstream and/or downstream activities into their business activity with the aim of reducing transaction costs or
- Seeking to exploit economies of scale and scope through **horizontal integration** by bringing in activities with products and/or services that are complementary

> 'Despite nearly 30 years of academic research on the benefits of related diversification, there is still considerable disagreement about precisely how and when diversification can be used to build long-run competitive advantage.' (Markides and Williamson, 1994)

Unrelated diversification refers to businesses or activities within the portfolio that do not share any common factors or characteristics – hence the increased risk.

The emergence of the multibusiness corporation, which aimed to spread risk '... *gave rise to portfolio management tools, the most famous of which was the growth-share matrix, or more colloquially, the Boston Boxes (from its origins with the Boston Consulting Group*' (Angwin *et al.*, 2011).

Generally, **portfolio analysis** seeks to measure the diversity of a firm's activities, with the aim of achieving and sustaining a '*balanced*' portfolio of businesses, often referred to as SBUs (**strategic business units**), each at a different stage of development: some for current profitability, some for medium term and some for long term. Portfolio analyses generate '... *strategic imperatives such as "invest", "divest", "harvest", and "manage for cash"*' (Angwin *et al.*, 2011).

There are a plethora of portfolio tools, most notably the GE/McKinsey Business Screen (Wensley, 1982). These tools are multifactor matrices, incorporating more industry and market factors, but are less used today than in their heyday in the 1970s.

The limitations of portfolio planning gave rise to what is known as **parenting advantage** (Goold *et al.*, 1994),[8] which deals not only with the terrain but particularly the role of the corporate parent in managing the portfolio, i.e. **how** the corporate centre adds value to the individual business units. The corporate parent is defined as '*all those levels of management that are not part of customer-facing, profit-responsible business units*' (Goold *et al.*, 1994). A key premise behind this concept is '*that many of the business units in multibusiness companies could be viable as stand-alone entities: to justify its existence, the corporate parent must influence the business collectively to perform better than they would as stand-alone entities*' (Goold *et al.*, 1994). The parent must therefore either carry out functions that the subsidiaries or business units would be unable to perform as cost effectively for themselves or it must influence the subsidiaries to make better decisions than they would have made on their own.

An example of a classic multibusiness corporation is Ferrovial, the Spanish construction and infrastructure group, which seeks to exploit synergies across its portfolio of related activities, or 'business lines' (Ferrovial, 2012). These comprise:

- **The Toll Roads Division**: which manages more than 3000 km of toll road in Spain, Chile, Canada and elsewhere across the globe
- **The Airports Division**: which owns BAA, and currently operates five airports in the UK including Heathrow, Stansted and Glasgow, following the sale of Edinburgh airport to GIP in April 2012
- **The Construction Division**: which is involved in '*all areas of construction, including civil works and buildings in Spain and across the world*'. Core markets include the UK, Ireland, Italy, Portugal, Chile, Puerto Rico, Greece and the US. For example, working in a consortium comprising Ferrovial Agroman, BAM

Nuttall and Kier Constuction (BFK), it is involved in three major programmes for the UK's Crossrail project, including the building of Farringdon Station in London

- **The Services Division**: which is mainly involved in infrastructure maintenance and management in Spain and the UK. Activities include '*infrastructure upkeep, facilities management, municipal and waste management services*'. Its Ferroser subsidiary, for example, is the leading provider of 'end-to-end hospital maintenance services' in Spain (Ferrovial, 2012)

Business and/or corporate level strategy

The concept of generic strategies was initially conceived at the business level, rather than the corporate level:

> 'Porter focuses on business-unit strategy because his early research showed that overall corporate return in a diversified corporation is best understood as the sum of the returns of each of its businesses. So for Porter, "strategy" always means "competitive strategy" within a business.' (Allio and Fahey, 2012)

When Porter was talking about achieving a competitive advantage only through differentiation or cost leadership, he was essentially distinguishing between corporate-level and business-level strategies. In other words, at the corporate level you could be enjoying a competitive advantage, while having one division pursuing differentiation and another cost leadership. For example, Volkswagen pursues competitive advantage at the corporate level by aiming to be the number one global auto manufacturer, with all the economies of scale that this brings. At the business level it pursues a cost-leadership strategy with its Seat and Skoda brands, while pursuing differentiation through, among others, premium brands Audi, Bugatti and Bentley.

The Porterian approach is often seen as an '*outside-in*' perspective, as the focus is on looking outside the firm to identify the external forces which a firm can harness.[9] While immensely valuable, this approach did not help to understand the idiosyncratic drivers of success within firms in any great depth;[10] the resource-based view of the firm provides a useful complement.

The resource-based view of the firm

> 'Competitive advantage of firms lies with its managerial and organizational processes, shaped by its (specific) asset position, and the paths available to it.' (Teece *et al.*, 1997)

The **resource-based-view** (RBV) adopts an '*inside-out*' perspective and argues that the key determinants of superior performance reside *within* the firm. To achieve a competitive advantage, not only must the firm identify a gap or '*market-space*' (Kim and Mauborgne, 2008) , it also needs to '*... anchor differentiation upstream, basically noting that if a firm is going to be able to differentiate its products, it must be different in its capabilities and/or business model*' (Augier and Teece, 2008).

From this perspective, sources of competitive advantage derive from the firm's **activity system** (Porter, 1996[11]) or **business model** (Zott and Amit, 2010[12]). It is the firm's distinct configuration of activities alone that allows it to identify, create and deliver superior customer value. These strategically valuable resources and capabilities have the following properties:[13]

- They create superior value for customers
- They are scarce
- They are hard to imitate
- They are appropriable, and enable the firm to capture the profit
- They enable the firm to learn and unlearn, making the organisation more innovative and adaptive to change

The RBV approach comes from disciplines within economics and, although it has only gained traction in the last 20 years, it originated in the work of Edith Penrose in 1958. It is relevant at both the corporate and the business level.

In essence the RBV seeks to provide a coherent structure to help discern the capabilities of a firm and how they can deliver competitive advantage. This can be challenging, as while these capabilities are '*theoretically differentiable*' they are '*largely inseparable*' (Liedtka and Rosenblum, 1996). Further, their effectiveness often relies on interconnected, overlapping managerial and organisational processes. Other academics have described them as '*invisible*' (Itami, 1987), or '*rarely observable directly*' (McGee and Thomas, 2007), making identification of '*... the exact relationship between the firm's bundle of capabilities and its performance unclear*'(Amit and Schoemaker, 1993). This so-called '**causal ambiguity**' (Lippmann and Rumelt, 1982) can therefore be a significant source of competitive advantage.

This is not dissimilar to the commercial management function, with its interconnected managerial processes across the firm's value creation activities (from inception to post-award [sales] – 'through life' project management). Thus commercial management can sit at the heart of a firm's dynamic capabilities.

The terminology used in RBV is not consistent across the extensive literature on the subject; for the purposes of this chapter, however, the terms defined by Amit and Schoemaker (1993) will be used:

- 'The firm's **Resources** will be defined as stocks of available factors that are owned or controlled by the firm. Resources are converted into final products or services by using a wide range of other firm assets and bonding mechanisms such as technology, management information systems, incentive systems, trust between management and labor, and more. These resources consist inter alia, of knowhow that can be traded (e.g. patents and licenses), financial or physical assets (e.g. property, plant and equipment), human capital etc.'
- '**Capabilities** in contrast, refer to a firm's capacity to deploy Resources, usually in combination, using organisational processes, to effect a desired end. They are information-based, tangible or intangible processes that are firm-specific and are developed over time through complex interactions among the firm's Resources. .... Unlike Resources, Capabilities are based on developing, carrying and exchanging information through the firm's human capital' (Amit and Schoemaker, 1993)

As the RBV area has grown, so the terminology has evolved, although some ambiguity regarding the various terms remains (O'Reilly and Tushman, 2008).

The next significant shift involved the introduction of the notions **meta-capabilities** (Liedtka and Rosenblum, 1996), and **dynamic capabilities** (Teece *et al.*, 1997). These concepts aim to describe the interrelationship of the processes that enable a firm to identify, create and deliver value and, critically, embrace the importance of continuous adaptation.

> '[M]meta-capabilities are composed of a set of distinct, yet inter-related skills. Learning is one. In fact, the ability to learn new sets of skill on an ongoing basis has been argued by some to represent the only sustainable source of advantage for the future.' (Liedtka and Rosenblum, 1996)

As a consequence, there has been much more focus on the role of senior management and leadership teams that enable learning within the organisation: '*... senior team capability may be a key discriminator between those firms that thrive as environments that shift versus those that do not*' (O'Reilly and Tushman, 2008).

As the commercial management function straddles a range of activities within the firm, its suppliers, and customers, at the corporate and/or business level, it is particularly well-placed to '*link and leverage*' (Normann and Ramirez, 1993) their capabilities, as described by the RBV perspective.

Conclusion

The focus in this chapter has been on the orientation aspect of strategy; it is important to note that implementation is equally, some would argue significantly more, important (Mintzberg, 1994). Strategising is an ongoing, dynamic process rather than a '*set solution*' (Montgomery, 2008) in the form of a plan. Here, strategy is seen more as a dialogue or strategic conversation (Liedtka and Rosenblum, 1996; Van der Heijden, 2005; Weick and Sutcliffe, 2007) and not exclusively one among the senior leadership team, after all', *... conversation is the very medium through which change occurs*' (Ford and Ford, 1995).[14]

The process of exploration and analysis outlined in this chapter seeks to help practitioners:

- Identify the strategic agenda: the fundamental issues that concern the firm's development in the long term, which involves market sensing, really understanding the complexity of the industry dynamics, and identifying value
- Define a clear sense of purpose and understand the need for a guiding policy for the firm in creating value: '*... an overall approach chosen to cope with the obstacles identified in the diagnosis*' (Rumelt, 2011)
- Clearly articulate the core value proposition, with a '*specifically tailored value chain*' (Magretta, 2012) or delivery system

Exercises **Exploring strategy**

Further data on these organisations can be found on company websites and sources such as ft.com or Hoovers.com. It is recommended that you carry out some additional reading, as indicated in the end notes to each question, prior to carrying out this exercise.

1. What is the strategic agenda for Boeing's civil aircraft division? Identify the key forces shaping the industry[15]
   - Identify the main change drivers
   - Critically comment on the industry's attractiveness
   - Identify the critical success factors in the industry
2. Does Boeing's civil aircraft division enjoy a competitive advantage? Compare Boeing with EADS' Airbus to explore each of the firms' business level strategies[16]
   - Identify Boeing's competitive position: does it have a competitive advantage in its short-haul or long-haul activities?
   - If so, what is the nature of the competitive advantage and is it sustainable?
   - On what basis is each firm competing?
3. Use Ferrovial group to explore corporate level strategy[17]
   - What is the logic of their portfolio and the nature and extent of their diversification? How does the collection of activities add value?
   - Does Ferrovial have a parenting advantage?
   - Explore the role of the corporate parent and critically comment on its parenting style. What are the implications for commercial management function?
4. Use Boeing to explore the RBV perspective[18] and dynamic capabilities
   - Using Amit and Schoemaker's (1993) categorisation, identify Boeing's core resources and capabilities
   - Does Boeing have the dynamic capabilities to both exploit and explore?
   - How can the dynamic capability construct be applied to the commercial management function?
5. How do the issues raised in these cases relate to your own organisation?

Further reading

Exploring Strategy

Angwin, D, Cummings, S and Smith, C (2011) *The Strategy Pathfinder Core Concepts and Live Cases.* John Wiley & Sons, Chichester

Chandler, AD (1962) *Strategy and Structure. Chapters in the History of the American Enterprise.* MIT Press, Cambridge, MA

Collis, DJ and Montgomery, CA (2004) *Corporate Strategy – A Resource-Based Approach*. McGraw Hill/ Irwin, Boston

Goold, M, Campbell, A and Alexander, M (1994) *Corporate-Level Strategy: Creating Value in the Multi-Business Company*. John Wiley & Sons, New York

Grant, RM (2013) *Contemporary Strategy Analysis*. 8th edition. John Wiley & Sons, Chichester. Chapters 3, 4, 5, and 8, 9 and 10

Lafley, AG and Martin, R (2013) *Playing to Win: How Strategy Really Works*. Harvard Business School Press, Boston

Magretta, J (2012) *Understanding Michael Porter: The Essential Guide to Competition and Strategy*, Harvard Business School Press, Boston, MA. Chapters 1, 2, 3 and 4

Rumelt, R (1974) *Strategy, Structure and Economic Performance*. Harvard Business School Press, Boston, MA

Endnotes

1 '*Narrow-body aircraft like the 737 are the workhorses of the global airline industry and analysts predict airlines will order about 22000 worth almost $2000bn over the next 20 years*' (Financial Times, 2011).

2 Airbus launched the A320neo, a re-engined version of the 320 range, in 2010: '*the A320neo has racked up well over 1000 orders, most recently the sizeable commitment from American Airlines. The American order was particularly painful for Boeing. Since 1996 the group has had an exclusive supply arrangement with the airline*' (Financial Times, 24 July 2012).

3 See Magretta (2012) for an in depth discussion on '*making clear what the organization will* **not** *do*'.

4 Cummings and Wilson (2003) posit that a good strategy would '*give focus, direction, and purpose to an organization (orientation), and encourage and move people to seek to achieve expectations for an organization or surpass and recreate these expectations (animation).*' Further research led to the authors concluding that '*with the benefit of hindsight, however, we would add another definitional element of what a good strategy should do, and it is an assumption that underlies and thus unifies all of the definitions we have described here. Strategy is about orientation, and animation, and integration*' (Angwin et al, 2011).

5 Recommended reading: Magretta (2012).

6 Supernormal profit, also known as abnormal profit, occurs when revenue exceeds the opportunity cost of inputs.

7 Recommended reading: Chandler (1962); Rumelt (1974); Prahalad and Bettis (1986); Porter (1987); Goold *et al.* (1994); Markides and Williamson (1994); Collis and Montgomery (2004).

8 See also Piercy (2009).

9 However, Porter's value chain analysis of 1985 did focus on the firm and was a major contribution. See Magretta (2012) for an excellent analysis and discussion of this.

10 Value-chain analysis makes a significant attempt at doing so, although it is prescriptive in approach.

11 Although Porter is rarely categorised within the RBV perspective, it can be argued that his activity system is in essence an evolution of his value-chain concept, and embraces RBV criteria.

12 Zott and Amit (2008) define a business model '*... as "the structure, content and governance of transaction" between the focal form and its exchange partners*' (for example, clients/customers, suppliers, third-party intermediaries, licensors/licensees).

13 See Barney (1991); Wernefelt (1984); Peteraf (1993).

14 See also Jacobs and Heracleous (2005); van der Heijden (2005); Weick and Sutcliffe (2007).

15 Recommended reading: Porter (2008); Grant (2013) Chapters 3 and 4; Magretta (2012) Chapters 1 and 2.

16 Recommended reading: Porter (1996); Grant (2013) Chapter 7–10; Magretta (2012) Chapters 3 and 4.

17 Recommended reading: Goold *et al* (1994); Angwin *et al.* (2011).

18 Recommended reading: Amit and Schoemaker (1993); Augier and Teece (2008); O'Reilly and Tushman, (2008); Grant (2013) Chapter 5.

References

Abell, DF (1980) *Defining the Business: The Starting Point of Strategic Planning*. Prentice Hall, Englewood Cliffs, NJ

Allio, RJ and Fahey L (2012) Joan Magretta: what executives can learn from revisiting Michael Porter. *Strategy and Leadership*, 40, 5–10

Amit, R and Schoemake,r P JH (1993) Strategic assets and organizational rent. *Strategic Management Journal*, 14, 33–46
Angwin, D, Cummings, S and Smith, C (2011) *The Strategy Pathfinder: Core Concepts and Live Cases*. 2nd edition. John Wiley & Sons Ltd, Chichester
Arthur, WB (2011) The second economy. *McKinsey Quarterly*, October, 1–9
Augier, M and Teece, DJ (2008) Strategy as evolution with design: the foundations of dynamic capabilities and the role of managers in the economic system. *Organization Studies*, 29, 1187–1208
Barney, J (1991) Firm resources and sustained competitive advantage. *Journal of Management*, 17, 99–120
Chandler, AD (1962) *Strategy and Structure: Chapters in the History of the American Enterprise*. MIT Press, Cambridge, MA
Collis, DJ and Montgomery, CA (2004) *Corporate Strategy – A Resource-Based Approach*. McGraw Hill/Irwin, Boston
Collis, DJ and Rukstad, MG (2008) *Can you say what your strategy is? Harvard Business Review*, April, 82–90
Cummings, S and Daellenbach, U (2009) A guide to the future of strategy? – The history of long range planning. *Long Range Planning*, 42, 234–263
Cummings, S and Wilson, D (2003) *Images of Strategy*. Blackwell, Oxford
Derman, E (2011) *Metaphors, Models & Theories*.www.edge.org
Ferrovial (2012) *Ferrovial Intelligent Infrastructure – Business-Lines*. http://www.ferrovial.com/en/Business-Lines (accessed 14 February 2012)
Ford, JD and Ford, LW (1995) The role of conversations in producing intentional change in organizations. *Academy of Management Review*, 20, 541–570
Goold, M, Campbell, A and Alexander, M (1994) *Corporate-Level Strategy: Creating Value in the Multi-Business Company*. Wiley, New York
Grant, RM (2013) *Contemporary Strategy Analysis*, 8th edition. John Wiley & Sons Ltd, Chichester
Gray, A (2011) Groups have public contracts in their sights, *Financial Times*, 20 March. http://www.ft.com/cms/s/0/c4d1bd18-532c-11e0-86e6-00144feab49a.html#axzz24p93AVhR (accessed 21 August 2012)
Itami, H (1987) *Mobilizing Invisible Assets*. Harvard University Press, Boston
Jacobs, CD and Heracleous, LT (2005) Answers for questions to come: reflective dialogue as an enabler of strategic innovation. *Journal of Organizational Change Management*, 18, 338–352
Jones, P (2008) *Communicating Strategy*. Gower Press, Aldershot
Kay, J (1993) *The Foundations of Corporate Success*. Oxford University Press, Oxford
Kim, WC and Mauborgne, R (2005) *Blue Ocean Strategy: How To Create Uncontested Market Space And Make The Competition Irrelevant*. Harvard Business School Press, Boston, MA
Liedtka, JM and Rosenblum, JW (1996) Shaping conversations: making strategy, managing change. *California Management Review*, 39,141–157
Lippman, S and Rumelt, R(1982) Uncertain imitability: an analysis of interfirm difference in efficiency under competition. *Bell Journal of Economics*, 13, 418–438
Magretta, J (2012) *Understanding Michael Porter: The Essential Guide to Competition and Strategy*. Harvard Business School Press, Boston, MA
March, JG (1991) Exploration and exploitation in organizational learning. *Organization Science*, 2, 71–87
Markides, CC (1999) Six principles of breakthrough strategy. *Business Strategy Review*, 10, 1–10
Markides, CC and Williamson, PJ (1994) Related diversification, core competences and corporate performance. *Strategic Management Journal*, 15, 149–165
McGee, J and Thomas, H (2007) Knowledge as a lens on the jigsaw puzzle of strategy. *Management Decision*, 45, 539–563
Mintzberg, H (1994) The fall and rise of strategic planning. *Harvard Business Review*, January/February, 107–114
Montgomery, CA (2008) Putting leadership back into strategy. *Harvard Business Review*, 86, 54–60
Nelson, R and Winter, S (1982) *An Evolutionary Theory of Economic Change*. Harvard University Press, Boston, MA
Normann, R and Ramirez, R (1993) From value chain to value constellation: designing interactive strategy. *Harvard Business Review*, July/August, 65–67
Odell, M (2011) Boeing and Airbus call time on duopoly. *Financial Times*, 21 June
O'Reilly, CA and Tushman, ML (2008) Ambidexterity as a dynamic capability: resolving the innovator's dilemma. *Research in Organizational Behavior*, 28, 185–206
Peteraf , MA (1993) The cornerstones of competitive advantage: a resource-based view. *Strategic Management Journal*, 14, 179–191
Pettigrew, A and Whipp, R (1993) *Managing Change for Competitive Success*. Blackwell, Oxford
Porter, M (1980) *Competitive Strategy – Techniques for Analyzing Industries and Competitors*. The Free Press, New York
Porter, M (1985) *Competitive Advantage – Creating and Sustaining Superior Performance*. The Free Press, New York
Porter, M (1987) From Competitive Advantage to Corporate Strategy. *Harvard Business Review*, 65, 43–59
Porter, M (1996) What is strategy? *Harvard Business Review*, November–December, 61–78
Porter, M (2008) The five competitive forces that shape strategy. *Harvard Business Review*, January, 78–93

Prahalad, CK and Bettis, RA (1986) The dominant logic: a new linkage between diversity and performance. *Strategic Management Journal*, 7, 485–502

Rumelt, R (1974) *Strategy, Structure and Economic Performance*. Harvard Business School Press, Boston, MA

Rumelt, RP (2008) Strategy in a 'structural break', *The McKinsey Quarterly*, December, 1–9 *http://www.mckinseyquarterly.com/strategy_in_a_structural_break_2257* (accessed 21 August 2012)

Rumelt, R (2011) The perils of bad strategy. *McKinsey Quarterly*, June

Teece, DJ, Pisano, G and Shuen, A (1997) Dynamic capabilities and strategic management. *Strategic Management Journal*, 18, 509–533

Van der Heijden, K (2005) *Scenarios: The Art of Strategic Conversation*, 2nd edition. John Wiley & Sons Ltd, Chichester

Weick, KE and Sutcliffe, KM (2007) *Managing the Unexpected*. Jossey Bass, San Francisco

Wensley, R (1982) Strategic marketing: betas, boxes or basics. *The Journal of Marketing*, 45, 173–182

Wernerfelt, B (1984) A resource-based view of the firm. *Strategic Management Journal*, 5, 170–180

Zott, C and Amit, R (2008) The fit between product market strategy and business model: implications for firm performance. *Strategic Management Journal*, 29, 1–26

Zott, C and Amit, R (2010) Business model design: an activity system perspective. *Long Range Planning*, 43, 216–226

Chapter 4

Perspectives on Managing Risk and Uncertainty[1]

Eunice Maytorena

Learning outcomes

When you have completed this chapter you should have:

- An understanding of the concepts of risk and uncertainty
- An awareness of the relevant theories associated with the concept of risk
- An appreciation of the broader perspectives of risk and its management
- Knowledge of the risk management process in projects
- An awareness of some of the risk management tools available, their benefits and limitations
- An understanding of the behavioural aspects associated with risk and uncertainty management

Introduction

The consideration of **risk** and **uncertainty** and its management continues to be a growing area of concern for organisations. The rapidly changing environment in which companies operate drives organisations to pay more attention to both managing risk and uncertainty, and to how they govern themselves. Indeed the Turnbull report (ICAEW, 1999) in the UK and the passing of the Sarbanes–Oxley Act of 2002 in the USA emphasised the importance of **governance** and, within this, the role of risk management for organisations. In this context, effective management of risk is crucial for improving organisational performance. The commercial function, as outlined in the Introduction and Chapter 1, plays a significant role in both the management of risk and in ensuring the application of regulatory and control processes and procedures.

The aim of this chapter is to provide an overview of the state of the art in project risk management. It will do this by providing an introduction to the concept of risk and uncertainty, including definitions and descriptions; a review of a number of perspectives on risk; and an overview of the project risk management process, and commonly used tools and techniques.

Risk

One only has to look at newspaper headlines – from global warming and genetically modified foods, to dieting and health – to appreciate that the concept of risk is an important issue in society today. It has become part of current scientific and political debates and over the past few decades the notion of risk

Commercial Management: theory and practice, First Edition. David Lowe.

has become central to our way of life. Authors such as Beck (1994, 1999) and Giddens (1991) explore how and why this has come about.

Drivers such as 'global interconnectivity', 'the atomisation of society', 'the ability to measure and monitor', and the influence of 'technology' are transforming the context in which we live and work as well as our awareness of that context; this inevitably influences our perception, understanding and approach to risk (Wilkinson *et al.*, 2003, pp. 299–301).

We have always been able to perceive **threats** and **opportunities** and establish a balance between them. However, the concept of risk did not become apparent until the eighteenth century with the expansion of industrial society when, with the advances in science and technology, people started to question long-standing beliefs and looked at humanity's capacity to shape the future (Giddens, 1999). In this sense the notion of risk is about time, the future and how we perceive and think about the future.

Definitions and descriptions

What is risk?

There is no one definition of risk. Various guides and standards provide their own definitions further confusing an already 'slippery' concept (see Table 4.1). Raz and Hillson (2005) review nine risk management standards and point out that one of the main differences in definitions is the inclusion or exclusion of positive aspects – in other words – opportunities. However, the negative view of risk, as threats, is prevalent in most organisations. Hillson (2003), therefore, has suggested extending the traditional threat-focused risk management process to take into consideration positive aspects to ensure opportunities are maximised and threats minimised to increase the potential of achieving an organisation's objectives. Although there are various statements of meaning, it is clear that risk definitions comprise three basic elements: a possible event, its possible consequence and the probability of the event occurring. Edwards and Bowen (2005) add a fourth element that is also important for managing risk: time. Therefore, when thinking about risk the following four elements need to be considered (Edwards and Bowen, 2005):

1. The possible future event and its source(s)
2. The possible consequence(s) of the event
3. The probability of the event occurring and its severity
4. The length of time vulnerable to the event and consequence

Table 4.1 Some definitions of risk.

Source	Definition
Oxford English Dictionary	*A situation involving exposure to danger; the possibility that something unpleasant or unwelcome will happen; expose (someone or something valued) to danger, harm, or loss*
BS/ISO 31000	*Effect of uncertainty on objectives. Note: effect is a deviation from the expected; uncertainty is the state of deficiency of information related to understanding or knowledge of an event, its consequence or likelihood; objectives can apply at different levels and aspects*
ISO/IEC Guide 73	*Combination of the probability of an event and its consequence*
BS6079:3	*Uncertainty inherent in plans and the possibility of something happening (i.e. a contingency) that can affect the prospects of achieving business or project goals*
PMBok® 2008 Guide	*An uncertain event or condition that if it occurs, has a positive or negative effect on a project's objectives*
Association for Project Management (APM)	*Combination of the probability or frequency of occurrence of a defined threat or opportunity and the magnitude of the consequences of occurrence*

What is uncertainty?

Although the concept of uncertainty is also present in many of the suggested definitions, this is not always adequately defined. Clearly, however, there is an association between risk and uncertainty and Edwards and Bowen (2005) attempt to address this relationship by reviewing a number of standards and risk analysis literature. From this they indicate that **uncertainty** is '*... inadequate or incomplete knowledge about the subject at issue*' (p.16) and in order to deal with this uncertainty during risk analysis one has to make assumptions or gather additional information if available. From a risk analysis perspective the level of uncertainty is 'bounded' by a probability, and so for analysis purposes, a probability is considered a measure of uncertainty.

However, Courtney *et al.* (1997) in a strategic context and De Meyer *et al.* (2000) in a project context propose different typologies of uncertainty and suggest that each type of uncertainty requires a different analytical tool and management approach. In a similar manner, Winch and Maytorena (2011) present a cognitive model of risk and uncertainty (see Figure 4.2), making a clear distinction between these two concepts. Risk is where the probability of an event occurring and its consequences can be calculated from historical data. Uncertainty is where there is a lack of knowledge: about the probability of incidence due to the lack of reliable data (known unknown), or about the threat or opportunity itself (unknown unknown). These distinctions are helpful for identifying and developing appropriate managerial actions.

Risk descriptions

We should also consider how risk is described. Risk descriptions vary across contexts (see, for example, Aven, 2010 for a thorough review) and organisational level. In a financial context for example, risk is described as variability from the central tendency of a probability distribution and is referred to as '**value-at-risk**' (VaR). VaR captures a worst-case scenario of capital loss for a given probability. By contrast, in a safety context, risk can be described as a probability, as captured in F–N (frequency–number of fatalities) curves used to describe the level of risk of fatal accidents in an organisation; or as expected values in terms of expected loss of life, or likely damage as a consequence of an event. In the context of investment appraisals, risk can be described as a probability distribution, where the potential of an investment is assessed against a probability distribution of returns – for example, a 5% probability of obtaining up to £100 000 return. In the context of projects, risk tends to be described as expected utility or disutility (Aven, 2010), where the expected utility is a function of an outcome and associated probability. This is the foundation for expected utility (EU) theory and subjective expected utility (SEU) theory which lie at the heart of project risk assessments. These are discussed later in this chapter.

The role of commercial management embraces these aspects. Therefore, when thinking and communicating about risk and uncertainty we need to be very clear of our meaning and concept. The importance of this is highlighted, for example, by research which found that the way in which information about investment options was presented influenced the investors' expectations about asset risk, returns and volatility, and consequently asset choice (Weber *et al.*, 2005).

Risk and uncertainty need to be understood in context and in the next section we shall look at some factors that influence our response to risk and uncertainty.

Influencing factors

Organisational functions must deal with risk and uncertainty constantly. Three factors interact and play a role in shaping our approaches and behaviour towards managing risk and uncertainty: cognitive, organisational (social), and situational (see Figure 4.1). Cognitive aspects relate to an individual's knowledge, preferences, attitudes and beliefs about possible futures. Our understanding of risk develops mainly from social (organisational) contexts. Organisational structures, strategies and culture provide a framework for action – for example, through governance structures, processes and procedures, vision setting, and value sharing. The influence of the organisational context can be appreciated by looking at where the risk management function sits; for example, in many organisations the focus and approaches taken towards risk are dependent on the location of the function. Likewise the array of situations that we encounter, or may encounter in the future, influence our perceptions and contribute to our growing knowledge and understanding. This understanding of different situations – a series of events, ideas and or circumstances, be they be simple, complicated, complex, or chaotic (Snowden and Boone, 2007) – is vital for developing appropriate managerial responses. Box 4.1 provides an illustration of this interaction.

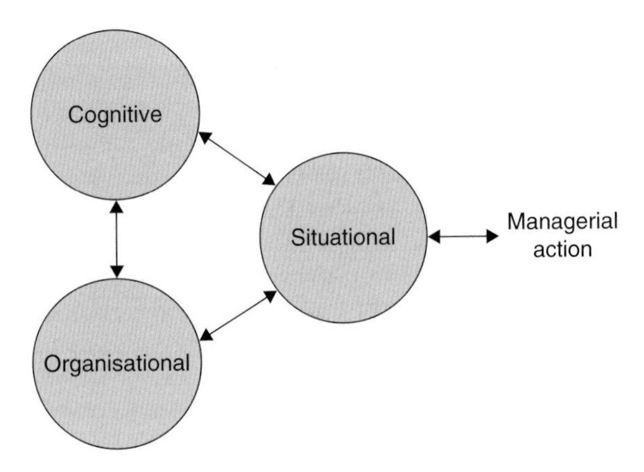

Figure 4.1 Factors that influence risk management behaviours.

Box 4.1 *Columbia* accident

On the 1 February 2003 NASA's space shuttle *Columbia* broke-up on its re-entry to earth. All seven crew members died. The technical cause of the accident was damage to the thermal protection system on the underside of the left wing of the shuttle caused by a piece of insulating foam shed by the external tank during the launch. During re-entry into the earth's atmosphere, the damage hole allowed superheated air to penetrate the wing structure, weakening it and eventually breaking the shuttle up.

The final report (CAIB, 2003) investigating the accident went beyond understanding the technical causes and looked at NASA's history and programme development; and the organisational and cultural issues that led to the accident. The investigation found problems associated with decision making, risk assessment, communication, organisational structure, processes and culture.

Managers underestimated the risk posed by the damage to the shuttle, despite the existence of evidence of damage from previous launches. The threat of damage from debris during launch was known and referred to as 'foam shedding'. Engineers raised concerns about the potential damage to the shuttle's thermal protection system shortly after the launch and requested an inspection process to gather more information in order to make an informed assessment. Orbit pictures of *Columbia* could have been obtained with the support of the Department of Defense. However, without imagery the assessment team could only develop mathematical models, which indicated overheating was most likely but could not state accurately what structural damage would result. These results were presented through a verbal summary to the mission management team who did not consider there to be an issue and who did not follow through the engineers' request for imagery. NASA's management went through a 'what if' scenario development exercise to determine the probability of future events rather than collect the information that would permit an assessment of the damage. NASA's management was used to these events, which had occurred previously without serious consequences, and, as a result of these previous successful missions, had become overconfident.

This overconfidence was not particularly new in NASA; indeed, little had changed in the 17 years since the Challenger accident in 1986. Although it is true that some changes were made due to the Challenger accident, institutional culture and practices were only slightly modified, and there remained a resistance to criticism and proposed changes from external parties. This led to poor decision making, self-deception and overconfidence in NASA's knowledge and abilities to launch people safely into space – NASA's culture began to accept escalated risks and managers became overconfident. The official investigation found that cultural resistance was a fundamental barrier to NASA's effective organisational performance.

The problem was that NASA's culture had never really adapted to the shift from designing new space craft at any cost (the lunar missions) to repeatedly flying a reusable vehicle (the space shuttle programme) on a significantly reduced budget. During the 1990s NASA was operating numerous programmes under a no-growth budget. The developmental space shuttle programme (their most expensive programme at the time) was progressing on a tight budget and schedule, competing for resources, incurring increasingly onerous maintenance costs and had a decreasing second- and third-tier contractor support base. To make matters even worse, it had a deteriorating infrastructure.

At the same time, a 'torrent of changes' was introduced. These changes were radical and discontinuous, with the aim of doing things 'faster, cheaper, better', and despite considerable concern about the impact these changes would have on safety, these protests were ignored. There was an increased reliance on a contractor workforce and the transfer of significant operations to the private sector. This increased dependence on contracting needed more effective communication and safety oversight processes, but these turned out to be flawed. The management structure changed, shifting all programme responsibility from NASA headquarters to agency field centres, reverting to the flawed structure that had been in place before the 1986 Challenger accident, and which would once again be found wanting.

Source: adapted from Columbia Accident Investigation Board (CAIB) (2003)

Perspectives of risk

To understand the concepts of risk and the management of risk we must consider how it has been studied. Risk is an important element in decision making (March and Shapira, 1987), and its significance has been confirmed by its consideration in decision theory, economics, behavioural and cognitive psychology and management. Over the years, decision-making studies of risk have varied in their approach, but four perspectives are dominant: the rationalist, behavioural, cultural and cognitive. These are summarised in the following sections.

The rationalist perspective

The concept of risk comes into play in decisions or 'choice under uncertainty'. From this perspective the development of probability, mathematical and economic theory has been highly influential in decision-making studies: see Bernstein (1998) for a thorough review of the development of the concept of risk and its relationship to these theories. In summary, concepts such as, **expected value** introduced by Blaise Pascal in 1670 (Pascal, 1670), **expected utility** introduced by Daniel Bernoulli in 1738 (Bernoulli, trans.1954), **modern utility theory** (Von Neumann and Morgenstern, 1944), and **subjective probability** and **subjective expected utility theory** (Ramsey, 1931; de Finetti, 1937 trans 1964; Savage, 1954) set the scene for understanding decision making. It was generally assumed, particularly in the area of economics, that individuals behave rationally and so **expected utility theory** was considered by decision scientists and decision analysts as either a prescriptive or descriptive model of decision making, and by economists as a predictive model of decision-making behaviour. Within the context of project risk management Von Neumann and Morgenstern's (1944) theory suggests how individuals should make decisions when the probabilities of outcomes are known (i.e. it is viewed as a normative model). Development in this area pushed science and business processes into the modern world (Bernstein, 1998).

EU theory lies at the core of risk assessments in projects and is exemplified in the application of the probability/impact matrix. The basis of this theory is that individuals are rational and act to maximise utility (defined in Chapter 1). EU theory makes three assumptions:

1. The preferences of rational individuals are clearly defined and can rationally choose between two alternatives
2. Rational individuals can decide consistently
3. The rank order of individual preferences stay the same when those preferences are mixed

The application of EU theory in a project context requires knowledge about the probabilities of various future outcomes. However, this is not always known in practice. The solution to this problem came with the development of SEU theory (Savage, 1954). This theory indicates that all probabilities are a state of the individual rather than a state of nature, and that as long as personal (subjective) probabilities are ranked there are no problems with applying the mathematics of objective probabilities (Winch and Maytorena, 2011). In this way the problem of lack of data about the future is addressed by articulating probabilities as 'degrees of belief' held by an individual and provides the foundation for turning uncertainty (lack of knowledge) into risk. However, these theories have not gone uncriticised (see Schoemaker, 1982). For example, both theories have been criticised for not describing how individuals actually make decisions. However, the effective application of SEU theory requires training in the process of elicitation (a prescribed process), the identification of all future states or events, and the consideration of individual cognitive capacity and biases. It is to these last limitations that we now turn.

The behavioural perspective

As a response to the rationalist perspective, behavioural decision theorists started to consider individual behaviour when making decisions. Simon's (1947) introduction of the concept of 'bounded rationality', where rationality is limited to the information processing capacity of an individual, and March and Simon's (1958) and Cyert and March's (1963) work on decision making in organisations contributed to the development of this area. The role of values and beliefs in individual judgements began to be considered.

Perhaps one of the most influential pieces of research related to risk and decision making is Tversky and Kahneman's (1974) experimental work which showed that individuals use a range of rules-of-thumb 'heuristics', when making many intuitive judgements (Gilovich *et al.*, 2002). The use of these heuristics reduces the requirement for processing information but it can also lead to biased judgements (see Table 4.2). This body of research demonstrated that the elicitation of subjective probabilities was indeed problematic. Based on empirical research Kahneman and Tversky (1979) proposed 'prospect theory' as an explanation of how individuals evaluate gains and losses. They found that individuals presented with a well-defined risk problem tend to assign values and subjective weights, rather than probabilities, to gains and losses. In essence this theory is a development of EU theory, which suggests that we tend to value a certain gain (a gain that is guaranteed) more than a gain that is less certain, even if the expected value is the same.

This body of work changed the focus of research towards human judgement. However, it has been criticised both on theoretical and methodological grounds (Gigerenzer, 1991; Huber, 1997). For these authors the heuristics and biases studies analysed the ways in which decision makers tended to misuse information available to them to assess probabilities. Despite this criticism, this area of study provides an explanation for poor probability judgements, which has implications for the assessment phase and communication activities of the risk management process.

The work of Flyvbjerg *et al.* (2003) provides an example of the role of biases in projects. In their review of the performance of infrastructure projects from around the globe, specifically at the feasibility/appraisal stage, the authors identified the poor consideration of risk and over-optimistic estimates as the main

Table 4.2 Heuristics and biases.

Heuristic	Description
Anchoring and adjustment	We tend to use a reference point or 'anchor' as a starting point for estimating values, gains or losses. However, we find it difficult to adjust our estimates significantly from the initial reference point even in light of new information. For example, in a project context our tendency is to anchor on the initial budget estimate regardless of new and more accurate information contradicting the initial forecast
Availability	We tend to estimate the probability of an event based on how easily instances of the event can be recalled. For example, we will base our assessment of the probability of the success of a project on our recollection of recent and similar projects' successes and failures
Representativeness	We tend to estimate probabilities of events by assessing how representative they are of a category. Consequently we ignore base-rate frequencies when making assessments, assuming that we can predict future outcomes from past outcomes, and tend to draw considerable inferences from a small number of cases. For example, we may predict a supplier's performance based on the category of supplier they represent, as was the case with the selection of Laing Construction for the Millennium Stadium project
Bias	**Description**
Conjunctive and disjunctive event bias	This is the systematic tendency to overestimate the probability of events that must occur in conjunction and underestimate the probability of events that occur independently. This helps explain the problems encountered in multistage project planning
Overconfidence bias	This is the systematic tendency to be overly confident in our estimating abilities and beliefs. This is evident in many business decisions, for example, in negotiation, business case approvals, new product development, market entry, etc.
Illusion of control	This is the systematic tendency to believe that we control or influence the outcome of future events. This is exemplified by poor risk response planning due to the belief that the risk event and its consequences can be controlled. Tony Evan's response belief, for example, was that he could control the risk associated with the dispute with CRFC (see the Millennium Stadium project case)
Optimism bias	This is the systematic tendency to underestimate the probability of negative events happening to us and overestimate the probability of good events happening to us. This is exemplified by over-optimistic estimates of passenger numbers and economic benefits of a range of transport infrastructure projects
Framing effect	This indicates that the way in which information is presented – positively framed (gains) or negatively framed (losses) – affects our decisions. Framing a situation positively can lead to risk-averse behaviour, whereas framing the same situation negatively can lead to risk-taking behaviour

reasons for poor performance. The importance of the role of optimism bias in the management of project risks has been acknowledged by government policy (HM Treasury, 2003). For instance, the London 2012 Olympic and Paralympic Games took this guidance on board with the consequence of large contingencies being allocated to projects.

The cognitive perspective

The cognitive perspective builds on the critique of the rationalist perspective and the limitations of the behaviouralist perspective. It draws on research in the area of managerial and organisational cognition (Lant and Shapira, 2001), which aims to understand how managers makes sense of and model their surrounding

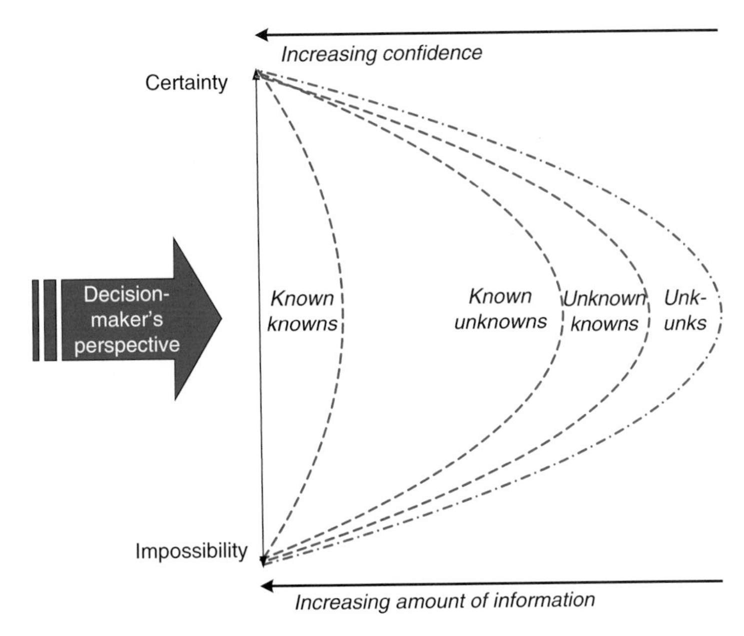

Figure 4.2 A cognitive model of risk and uncertainty on projects. Source: Winch (2010), Figure 13.2.

environment, how they process information, and how this influences behaviour. Research in this area attempts to bring individual and social processes together to enhance our understanding of organisational decision making. The cognitive perspective recognises the importance of individual mental states (beliefs, desires, motivations); it acknowledges a decision maker's limited information processing capacity; and considers how decision makers, in their organisational context, process information, and then interpret the information in order to take action (see Weick, 1995; Weick and Sutcliffe, 2001). Winch and Maytorena (2011) suggest how to put the cognitive perspective into practice using a cognitive model of risk and uncertainty in projects (Figure 4.2). This model, as indicated earlier, makes a distinction between risk (known knowns) and uncertainty (lack of knowledge: known unknowns, unknown knowns, unknown unknowns). It is cognitive because the perception of the possibility of future events taking place lies within the individual decision maker. It is therefore also subjective. It is the decision maker who makes sense of their surrounding environment, available information and mentally constructs possible future states. The future-orientated nature of projects indicates that it is common for the decision maker to navigate the 'information space' (on the right-hand side of Figure 4.2), and in so doing perceives possible future events and its sources and possible consequences; they can thus makes sense of the information available. This information space, as perceived by the decision maker, will therefore be central to how they make sense of the situation in focus, and how they make their decisions. The fact that the process is so subjective does of course have consequences for projects. The significance of this perspective for managing risk and uncertainty is highlighted in Box 4.2 which illustrates how decision makers navigating the information space can interpret and make different inferences from the same information and experiences leading to serious consequences.

The cultural perspective

This position takes into consideration the social and cultural context of the decision process and issues affecting the individuals' responses to risk. Studies in this area have tended to focus on understanding how power, social status and political views influence an individual's, group's or organisation's view of risk (Lupton, 1999). Whereas the behavioural and to some extent the cognitive perspective have focused on individual processes of choice, the sociocultural perspective focuses on how social and cultural factors

Box 4.2 Navigating the information space

In both instances of the *Challenger* and *Columbia* accidents, NASA's engineers brought to the fore issues of safety over cost and schedule. However, managers were more concerned with keeping within budget and programme. Investigations found that engineers and managers made very different inferences from years of launch experiences. Engineers inferred from increasing evidence of damage to O-rings in the solid rocket boosters (in the case of *Challenger*) and from shedding foam damage (in the case of *Columbia*) that an accident was highly likely to occur: these instances were merely near misses. Managers, however, based on increasingly successful launches, considered the shuttle to be more robust and resilient. Consequently, managers became overconfident and decreased their levels of what risks were acceptable.

Source: Starbuck and Miliken (1998); Vaughan (1987); Columbia Accident Investigation Board (CAIB) (2003)

influence judgements on risk (see Douglas and Wildasky, 1982; Douglas, 1992). For example, in the case described in Box 4.1, NASA's organisational culture had as much to do with the *Columbia* disaster as did the technical failure.

These four perspectives are different attempts to develop a more sophisticated understanding of how we make sense of risk, and they all derive, either directly, or indirectly, from the rationalist perspective. The rationalist perspective on risk *perception*, as we have just seen, can trace its intellectual origins back to the eighteenth century. So too does the topic of risk *management*: the subject of the next section.

Risk management: Origins and development

From the eighteenth century to the mid-twentieth century, the assessment of risk was primarily a concern for insurance and banking institutions, and to a limited extent for government departments concerned with public health (Hubbard, 2009). By the 1970s, however, risk management activities, driven by insurance executives, were responding to changes in the economic system and considering the 'cost-of-risk', while organisations focused on developing contingency plans.

During the 1980s, organisations were concerned about business continuity planning and the impact of natural disasters on business operations. The focus of risk management, therefore, shifted to the minimisation of corporate risk through 'disaster recovery planning' (Hopkin, 2010). Corporate governance practices were high on government agendas during the 1990s. A number of UK government reports, including for example the Cadbury Report (The Committee on the Financial Aspects of Corporate Governance, 1992) and the Turnbull Report (ICAEW, 1999), were published containing recommendations for corporate governance practices that were later integrated into the UK Corporate Governance Code 2010; similarly the Sarbanes–Oxley Act 2002 was passed in the USA, legally requiring all organisations to comply with defined financial practice and corporate governance standards. The role of 'chief risk officer' (CRO) was created to ensure compliance with regulations and to consider risk management across the organisation. Thus, the organisational, human and system aspects (see Kahkonen, 2006 for an explanation) were encouraging a more integrated approach to risk management. Or to put it another way, enterprise risk management (ERM) was raising its profile (Austin *et al.*, 2003).

The rise in public awareness and concern about risks have put pressure on organisations to demonstrate that effective risk management policies and strategies have been established (EIU, 2007). This interest has led many professional bodies and international institutes to develop risk management standards (IRM *et al.*, 2002; ISO, 2002, 2009a, 2009b; BSI, 2009, 2008) and good practice guides (APM, 2004; Lewin, 2002; OGC, 2007). These have been adopted and modified by firms across a range of sectors.

At all levels of the organisation, from corporate to project, and across a range of business areas, people will face and deal with different types of risk and uncertainty: strategic, financial, operational, commercial, technological, environmental and social. Therefore it becomes essential that organisations manage these risks effectively so that they do not impact negatively on the organisation's ability to achieve its mission and or strategic objectives. As a result, risk management has become an important area of an organisation's business as inadequate management of risk can result in serious losses.

Some of the perceived benefits of effective risk management include:

- Enhanced thinking about the future
- Improvement in planning
- Improvement in communication
- Improvement in resource allocation
- Promotion of continuous improvement
- Protection and enhancement of reputation
- Competitive advantage (see for example Clarke and Varma, 1999; BSI, 2000; Voetsch *et al.*, 2005; EIU, 2007; Salomo *et al.*, 2007)

Even so, more research on the benefits of effective risk management is required.

Projects and risk

As mentioned in Chapter 1, organisations are increasingly employing projects as a mechanism for achieving their strategic aims. Although risks are not new to projects, the rapidly changing context in which they are delivered influences both the way projects develop and their risks. According to Hartman (1997), risks seem to be more intense as new technologies, stricter regulations and changing business practices are introduced. For example, the introduction of new technologies has resulted in the drive for knowledge and the allocation of adequate resources to develop, apply and operate them. Further, the increasing need to outsource, along with the dependence on the expertise of suppliers, has created new risk areas, whereas the specific nature of regulations has lead to more focus being put on how they might change and how they might impact the project. The introduction of new forms of contract, alternative procurement routes and the pressure to remain competitive has brought a focus on understanding current business practices, technological developments and society. With project size and complexity on the rise and as competition increases, the management of risks early in the project life cycle becomes an ever more important challenge (Maytorena *et al.*, 2007).

Perhaps the first widely understood formal project risk management technique was the programme evaluation and review technique (PERT) developed in the 1950s for the *Polaris* programme. This technique introduced probabilistic rather than deterministic task duration estimates. However, this was, in essence, limited to schedule delays against plan and embodied the 'negative events only' concept of risk. This view of risks, as things that can go wrong, tends to dominate project management.

Project risk management has been an area of academic interest since the 1950s, with project management practitioners adopting formal risk management practices during the early 1980s. Artto (1997) provides a historical overview of its application from the 1980s to the late 1990s and concludes that the focus of project risk management has progressed from the development of tools to a better understanding of the underlying process and an increasing advocacy of risk management maturity models (Hillson, 2003). Risk maturity models are employed by organisations to benchmark their risk management capability and help to identify areas for improvement. The organisation's capability is captured in a level/criteria matrix. The levels indicate a degree of maturity and describe the quality in the risk management process – for example, from ad-hoc to optimised. The criteria attempt to capture the risk management practices that the organisation needs to address. Competencies are then developed to provide a description of each criterion across each of the maturity levels.

Winch and Maytorena (2011) outline developments during the past decade, identifying four areas of innovation:

1. The focus on the governance of projects (ICAEW, 1999; OGC, 2007)
2. The focus on the front-end definition of projects as a way of reducing risks at later stages of the project life cycle (Miller and Lessard, 2000; Saputelli *et al.*, 2008; Williams *et al.*, 2009)
3. The consideration and application of real options thinking (which considers the value in making an investment now to keep the option of a future investment opportunity open) as a way of generating opportunities (Gil, 2009; Jacob and Kwak, 2003)
4. The use of systems dynamics to improve our understanding of the interaction and interdependencies of risks in projects (Williams, 2002; Eden *et al.*, 2005; Ackermann *et al.*, 2006; Lyneis and Ford, 2007)

Project risk management: Process, tools and techniques

Project risk management process

Project risk management (PRM) is the systematic process of identifying, analysing, responding, monitoring and controlling project risk to improve project performance. It is an iterative process bounded by the organisational mission. The basic intent is to make the approach to managing risks more robust.

The literature on project risk management is now well developed with standards (see Raz and Hillson, 2005 for a review), guides (APM, 2004; OGC, 2007) and literature (Kahkonen and Artto, 1997; Chapman and Ward, 2003; Cooper *et al.*, 2005; Edwards and Bowen, 2005; Kahkonen, 2006) on the topic. These break down the risk management process into a number of phases which can be summarised into four basic subprocesseses (see Figure 4.3):

1. Identify and categorise the risk
2. Analyse the risk
3. Respond to the risk
4. Monitor and control the risk

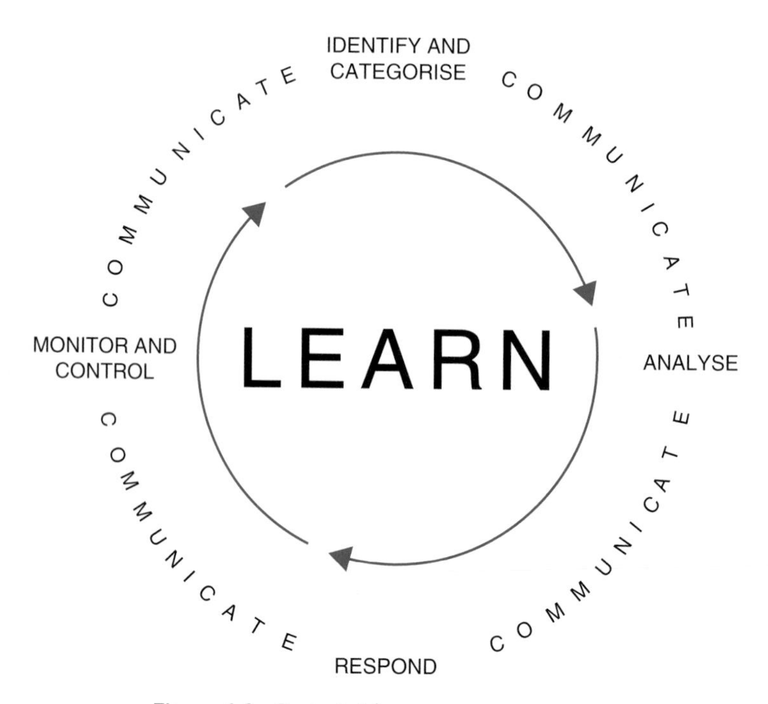

Figure 4.3 Project risk management process.

The progression from one subprocess to the next is equally important, as this is about active and meaningful communication and reporting. At the centre of the risk management process is learning and effectively building on that learning in subsequent projects.

There are a number of tools and techniques that are used during the different phases of the risk management process. This section provides an overview of the phases and most commonly used techniques, along with directed further readings.

Phase 1: identification and categorisation

In line with the cognitive perspective, the identification of risks is a cognitive problem. It is a process of identifying and categorising risks by imagining future states, and in this sense it is a creative process, seeking to capture the unusual. Risk identification should be carried out in a structured and systematic way. However, one should be careful not to allow such processes to become '*routinised*' to the point where organisational experience is inappropriately transferred to particular situations, as '*... routines are like a two-edged sword. They allow efficient coordinated action, but also introduce the risk of highly inappropriate responses*' (Cohen and Bacdayan, 1994, p. 555).

Once a risk has been identified it should be validated as far as possible in terms of information. The view is that risks are identified based on judgement, knowledge, experience and historical data. The assumption, at least in PRM, has been that experience is the key to identification. However, research (Maytorena *et al.*, 2007) has identified that experience in terms of age or number of years of working in industry are not necessarily good indicators of risk identification performance. It is risk management training and an enquiring approach that contribute to better risk identification performance, as this seems to encourage a more conscious, rather than habitual, approach to the practice of risk identification. NASA's approach to *Columbia*'s damaged wing is a good reminder of the importance of undertaking an information gathering and validation process (see Box 4.1).

There are several techniques used to aid the identification process and these can be categorised as:

- *Lists* generation in groups, through brainstorming or through interview (checklists, risk registers, risk breakdown structures, risk taxonomy)
- *Soft techniques* (causal mapping, soft systems methodology)
- *Strategic techniques* (scenario planning, future and horizon scanning)
- *Technical/engineered system techniques* (hazard and operability [HAZOP], failure mode effects analysis [FMEA]) mainly based on structured brainstorming sessions

The information generated by these activities is usually captured in risk registers (RR) and or risk breakdown structures (RBS). Risk registers (see Figure 4.4) are widely used comprising a list of all the risks that have been identified in relation to the specific project. However, there is little research on how they are constructed or developed (Patterson and Neailey, 2002), how reliable the results are and how it is used as a knowledge management tool for capturing organisational learning (Ayas and Zenuik, 2001). RBSs (see Figure 4.5) are more elaborate risk registers, which provide a hierarchical structure of potential risk sources (Hillson, 2002) from which a list of risks can be drawn through a brainstorming session. The higher level categories of RBSs can be used as prompt lists to ensure comprehensive coverage. Both the RR and RBS provide a number of insights such as understanding the type of risk exposure, uncovering the most significant sources of risk, revealing root causes of risks, indicating areas of dependency, and focusing risk response development on high-risk areas. Brainstorming sessions according to Chapman and Ward (2003) are undertaken in relation to a specific project and require a group of experienced practitioners to consider creatively possible risk sources. The list of risks should be carefully considered and key risks identified. Several issues have been raised concerning brainstorming, including the identification and selection of appropriate practitioners (participants), the frequency of brainstorming sessions throughout the project life cycle, and the avoidance of 'groupthink' dynamics, which can lead to rapid consensus without a critical assessment of all the options (Janis, 1972).

ID number	Risk category	Risk description			Probability	Impact	Confidence		Risk score	Response		Target date	Risk owner	Status of response
Identification number	Technical, contextual, HR, financial, commercial, etc.	Describe the cause of the risk and the effect of the risk occurring on your project mission			Probability of occurrence of the risk	Impact of the risk occurring	Level of confidence in ratings 90% CI		Based on probability x impact	Proactive	Reactive	The target date for the completion for each mitigation action	Name the person responsible for managing the risk	**Red**: unlikely to be met **Amber**: may not be achieved **Green**: according to plan
		Risk source	Risk event	Risk effect			P	I						
1														
2														
3														

Figure 4.4 Risk register.

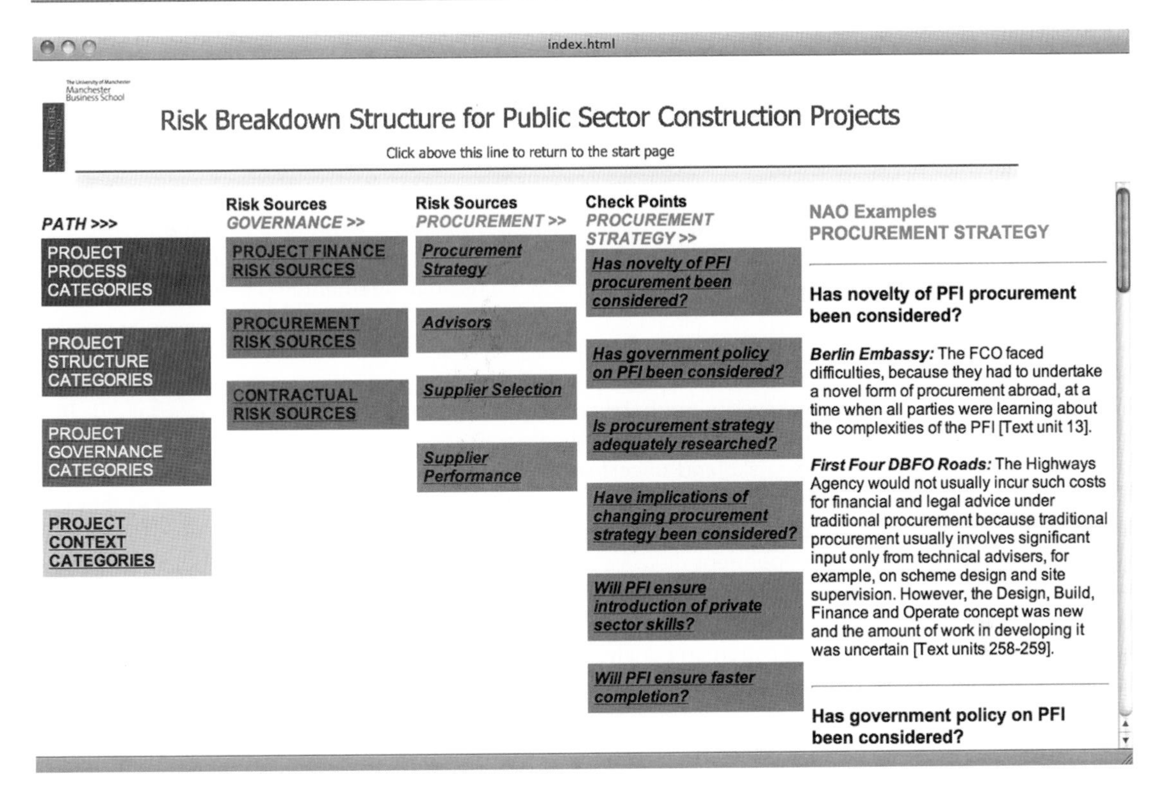

Figure 4.5 Risk breakdown structure. Source: Dalton (2007).

Soft techniques refer to a number of approaches developed in the management science area (Rosenhead and Mingers, 2001), and are also known as problem-structuring methods. Causal mapping (Eden and Ackermann, 1998) can be used to provide a better understanding of the wider project context – risks, risk sources, risk effects, risk interrelation and interdependence – and aid in the development of a risk management strategy (Williams *et al.*, 1997). Soft systems methodology (SSM) can also be used to provide a better appreciation of the project context and associated issues. It facilitates a better understanding of potential threats and opportunities, particularly at the front-end of projects (Winter, 2009).

Strategic techniques developed in the area of strategic management have increased in popularity as a means of identifying future threats and opportunities.[2] Horizon and future scanning are concerned with identifying and understanding possible future events and their potential effects on the organisation's strategic planning by using the PESTEL (political, economic, social, technological, environmental and legal) framework (Johnson *et al.*, 2006). This then enables the organisation to develop their capabilities and resilience in responding to new threats and opportunities. Scenarios (Schwartz, 1997; van der Heijden, 2004) describe in detail possible futures of how the business and organisational environment could operate given certain events, individuals, motivations and key drivers for which there is a high level of uncertainty. Scenarios enable organisations to think about and discover uncertain future aspects; they provide an understanding of the implications and facilitate conversations on important strategic features.

Technical/engineered system techniques are used to identify potential hazards, failures or deviations in a system where there is knowledge and information of the system, its design and components. They are quite different from projects, which are not known systems and have distinct qualities of their own. HAZOP studies (BSI, 2001) are commonly used in processing facilities, such as oil and gas, and for environmental and industrial health and safety assessments to identify potential hazards in a system (facility, or equipment) and processes. It is a systematic process that enables the identification of deviations from design, the causes of deviation and their potential consequences through the use of 'guide words' to facilitate the identification of key aspects of design. FMEA (BSI, 2006) is used to identify and analyse potential modes of failure, errors or defects in a process or design. It is commonly used in new product development and operations management activities. The structured process aids the identification of potential failures of individual components of the system sequentially, their severity, frequency of occurrence, and the ease of failure detection, and it helps to anticipate the effects of the failure on the whole system.

The selection of an appropriate approach will depend on the level of participant expertise and familiarity with the technique, the time horizon (short, medium, long term) to be explored, the time available to undertake the process, the quantity and quality of information or data available, and the relation to the risk management and project strategy, analysis approach, and the risk management implementation process.

Phase 2: analysis

The analysis phase is where the severity, probability and manageability of threats are ascertained. **Risk analysis** takes the outputs from the identification and categorisation phase as inputs for analysis. During this phase a richer picture of the risks and uncertainties can be established to support future decision making. The analysis and identification subprocesses are jointly considered the most important as they can have the biggest impact on the precision of the risk assessment exercise (Cooper and Chapman, 1987). It is well recognised that analysis is dependent on risks being accurately identified in the first place. If risks are not appropriately identified they cannot be analysed and managed. We must not forget the subjective nature of this exercise, the problems with eliciting subjective probabilities and the role of cognitive biases. The consequences of poor identification and analysis can be appreciated in the NASA case (Box 4.1) and in the Millennium Stadium project (Case Study A), where the risks associated with the dispute between WRU and CRFC were not adequately assessed resulting in significant post-award cost increases for Laing. The analysis phase has been well supported by research, with both qualitative and quantitative techniques being developed (see Vose, 2000; APM, 2004; Aven, 2008, for further details).

The **probability/impact matrix** is the most commonly used qualitative tool. Probability/impact matrices are used to plot the risk identified on a probability (likelihood, frequency) and impact (severity, magnitude) scale (see Figure 4.6). Scoring matrices can come in a variety of formats (3 × 3; 4 × 4; 5 × 5) and use numeric and or descriptive scales such as 0–1, 0%–90%, 1–5, very low-very high probability and impact. These, however, need to be clearly defined and explained by the designer before use. Scoring matrices are helpful for visualising the risk levels or categories, prioritising the risks and for considering the allocation of appropriate resources during the response phase. Some of the issues with their application, addressed in the perspective section, are associated with the elicitation of subjective probabilities, with the role of individual biases in estimation of probability and impact, and interpretation of outputs, and (sometimes) with the ill-defined scales employed. Cox (2008) examines the 'mathematical properties' of risk matrices and finds a number of limitations such as problems with the compression of a range of values under one category – the lack of understanding of interdependence between risks and the assumption that the intervals between scales represent the same level of magnitude. He concludes that matrices '*... should be used with caution, and only with careful explanation of embedded judgements*' (Cox, 2008).

The probability/impact matrix can be developed further by taking the risk rating (probability × impact) and plotting it against a 'manageability' scale (see Figure 4.6). This helps to visualise and consider the most and least challenging risks to manage. However, when conducting further analysis, both dimensions of risk (likelihood and impact) need to be considered, as relying only on risk ratings can be misleading (Williams, 1996).

Quantitative analysis aims to produce a mathematical model of the situation under consideration based on a number of inputs and assumptions. The outputs need to be interpreted with the quality of the inputs and parameter settings in mind. Sensitivity analysis, Monte Carlo analysis and probability trees are some of the most commonly used quantitative techniques (see APM, 2004 for further details).

Sensitivity analysis (deterministic approach) looks at the effects of varying the inputs or model parameters on the model output. For example, it can be used to estimate the effect of fluctuating material or labour costs on project cost and price.

Monte Carlo analysis (probabilistic approach) uses a range of values of a distribution as inputs to determine the effect of uncertainty on the system being modelled. The output can be visualised in graphical

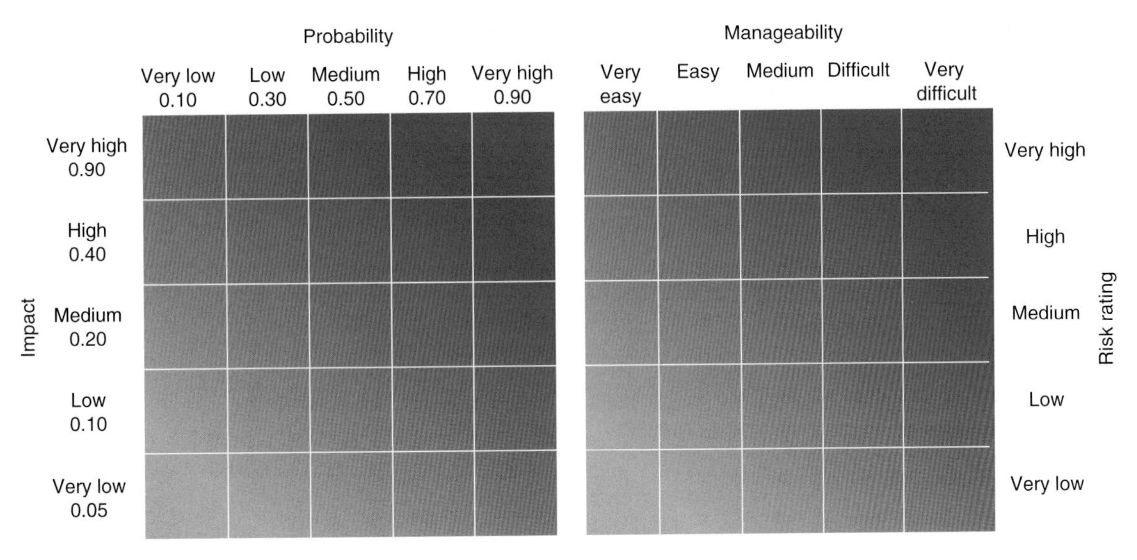

Figure 4.6 Probability/impact/manageability matrix.

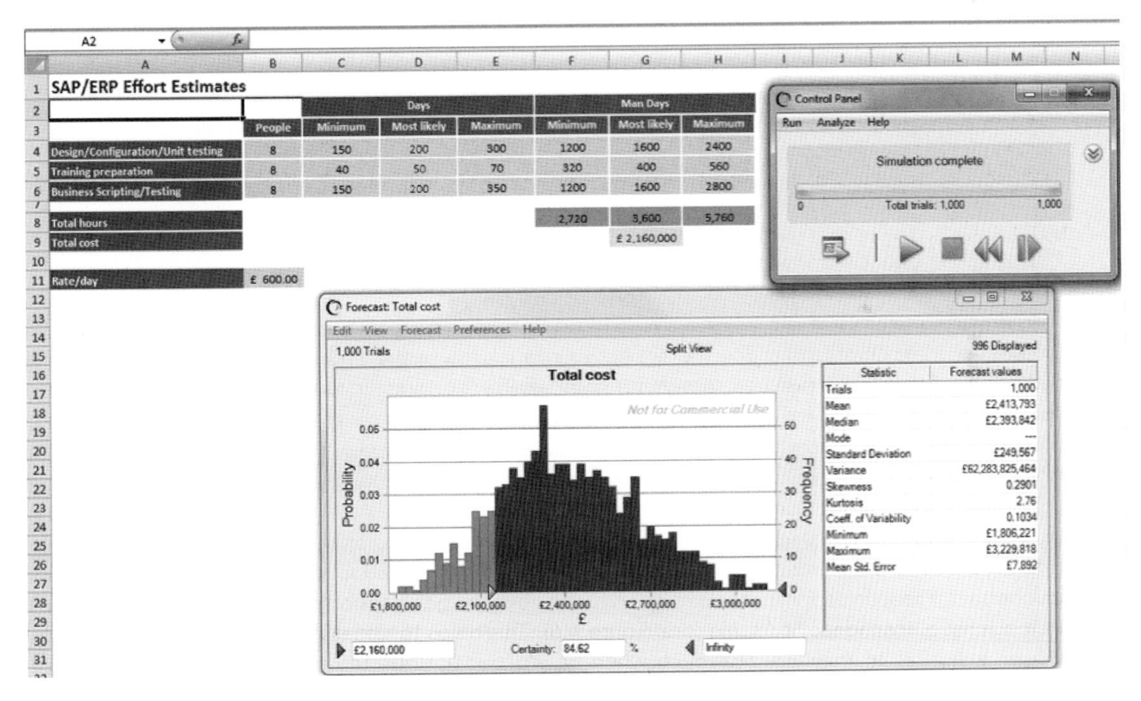

Figure 4.7 Output of Monte Carlo simulation.[3]

form as histograms, confidence intervals, percentiles and cumulative distributions. Its application is wide ranging and includes: capital investment, competitive bidding, and project cost and schedule forecasting. Figure 4.7 shows a Monte Carlo simulation output. It illustrates a range of possible project costs using three-point estimates (optimistic, pessimistic and most likely) and provides an indication of the likelihood that the total project cost will exceed £2 160 000.

Probability trees or **decision trees** illustrate graphically the possible effects of a risk event. The outcome of each event is designated a probability, so that the most probable outcome can be established.

Box 4.3 What was known?

'Shuttle reliability is uncertain, but has been estimated to range between 97 and 99 percent. If the shuttle reliability is 98 percent, there would be a 50–50 chance of losing an Orbiter within 34 flights. At a rate of 11 flights per year, there would be a 50 percent probability of losing an Orbiter in a period of just over three years. The probability of maintaining at least three Orbiters in the Shuttle fleet declines to less than 50 percent after flight 113' (The Office of Technology Assessment, 1989, p. 6)

and

'... although it is a subject that meets with reluctance to open discussion, and has therefore too often been relegated to silence, the statistical evidence indicates that we are likely to lose another Space Shuttle in the next several years...probably before the planned Space Station is completely established on orbit. This would seem to be the weak link of civil space program - unpleasant to recognize, involving all the uncertainties of statistics and difficult to resolve.' (The Augustine Committee, 1990)

Source: adapted from Columbia Accident Investigation Board (CAIB) (2003), p. 103

As with the qualitative approaches, the issues with these approaches lie with the estimation of subjective probabilities and the interpretation of outputs (as informed by the behavioural and cognitive perspectives). Box 4.3 highlights these two points concisely; they relate to NASA's statements made after the *Challenger* accident in 1986.

Phase 3: response

Commercial management has a significant role in this phase, which is where a number of managerial interventions are developed to reduce the impact or probability of the event identified and analysed. It also aims to maximise opportunities, assuming they are recognised in the first place. Two response approaches can be taken: proactive or reactive. A proactive approach aims to reduce the possibility of the event occurring, whereas a reactive approach aims to reduce the impact of the event should it occur (Winch, 2010). Proactive approaches might include: changes in design, scope, processes and procedures to avoid the possibility of a future event occurring; deciding not to proceed with a project to remove the possibility of the event happening; contractually transferring the responsibility of the risk to another party; or sharing the gain and the pain with another party specified in contractual clauses, such as in the Millennium Stadium contract. Proactive approaches can provide the basis for better-informed decisions, designing out weaknesses through resilience, and enabling vulnerabilities to be identified and addressed before they occur (Bonabeau, 2007). Reactive approaches might include: allocating contingencies (self-insurance); insuring against the cost of any loss; flexibility planning and provision to deal with residual risks (Loch *et al.*, 2006); training provision such as in the health and safety area; or mobilisation of emergency response services to reduce the impact of the event should it occur.

The response phase is one of the most challenging to manage. Usually a combination of risk responses will be used. However, high impact events rarely occur from a single source, but from interrelated factors. The 'systemicity' of risks (Williams *et al.*, 1997) and the selected response can create the possibility of new risk events, known as the 'ripple or portfolio effect' (Williams, 2002). For example, the transfer of responsibility of a risk to a supplier could lead to its impact contributing to the supplier's failure; consequently the risk falls back on the transferor (see the Wembley Stadium case for an example of the ripple effect and the T5 case for an example of a proactive response approach to avoid such an effect).

Phase 4: monitor and control

This phase is often overlooked, but nevertheless is as important as the other phases in facilitating the effective management of risk. Monitoring involves a continual review of the risks and their status, while controlling ensures that appropriate actions are taken based on the outputs of the response phase. Responsibility for each risk identified should be defined and assigned to an appropriate risk owner. In some instances responsibility will not necessarily lie with one individual but a group of individuals. Nonetheless, accountability for the outcome should also be clearly defined and allocated. The risk owner should be given authority to act and support should be provided in terms of current and accurate information and resources. The continual review process aims to identify if any of the assumptions made during the previous phases has become invalid, or if any of the risks identified are no longer possible, or if a new threat or opportunity has been spotted. Therefore, during this phase, information is gathered on status, captured on the risk register and reported via team meetings and risk reports across different management levels.

Communicate and learn

Communication is an activity that is undertaken throughout the process of managing risk and uncertainty and is a key factor in the successful management of risk. Effective communication is central to:

- The stakeholder's understanding of the benefits of effective risk management and the consequences of poor risk management implementation
- The stakeholder's understanding of quantitative assessments and their limitations
- The identification of changes in the risk status and identification of new threats and opportunities
- Ensuring risk ownership details are unambiguous in contracts and agreements
- Capturing, disseminating and building on lessons learnt

However, effective communication of risk is a challenge. There is an extensive body of literature on risk communication covering public engagement, communication strategies, and the role of trust, for example (Keeney and von Winterfeldt, 1986; Kasperson and Stallen, 1991; Renn and Levine 1991; Fischhoff, 1995; Bier, 2001; Morgan *et al.*, 2002). Maule (2008) provides a comprehensive review of research in this area and identifies three aspects that must be considered during the communication process:

1. The notion of the social amplification of risk – this considers the way in which individual and social processes can transform how risk information is received and interpreted (Kasperson *et al.*, 2003)
2. The role and nature of individual values and their influence on how the receivers of the risk communication will take action (French *et al.*, 2004)
3. The aim (the why), processes (the how) and effectiveness in communicating risks (Covello *et al.*, 1986)

Although we have provided an overview of the processes and most commonly used techniques, we should not ignore the important behavioural and cognitive issues of managing risk and uncertainty. The effective management of risk and uncertainty is critical for project and organisational performance. We must understand that having processes and tools, although useful, are not enough, the challenge is the effective implementation and this will come through leadership actions of individuals and teams and by effectively embedding risk management practice into the organisational culture (see the T5 case study for further insights).

Summary

Risk is an important issue in society today and its management is a key feature of the commercial role. Although we have always faced threats, the concept of risk is relatively new. It is concerned with two things: time and the belief in the control of the future. There is no single definition of risk, but most definitions typically include four main elements: source (of the event), consequence, probability and time. We make a distinction between risk and uncertainty:

- *Risk* is where the probability of an event occurring and its consequences can be estimated from historical data
- *Uncertainty* is where there is a lack of knowledge about the probability of incidence, the threat or the opportunity

Descriptions of risk vary greatly across contexts, so when thinking and communicating about risk we must be very clear about our precise meaning.

Our responses to risk and uncertainty are influenced by a number of factors. These should be kept in mind when undertaking the risk management process. Individual behaviour is influenced by the interaction of three factors:

1. Cognitive factors such as perceptions, attitudes, beliefs, and motivations
2. Organisational/social factors such as the organisational structure, strategy, culture, group dynamics
3. Situational factors such as our understanding of the evolving series of events during a project life cycle

Our increasing understanding of these can help us develop appropriate managerial action.

We explored four perspectives on risk to develop our understanding of risk and some of the implications for its management. The *rationalist perspective* developed from theories of economics, probability and mathematics, which were highly influential in the development of science and business – in particular EU and SEU theory – which lie at the heart of risk assessments and rely on the effective elicitation of subjective probabilities. The *behavioural perspective* critiques the rationalist perspective and demonstrates the problems with the elicitation of subjective probabilities and explains why we tend to make poor probability judgements. The *cognitive perspective* broadens its focus to the organisational and social context, by trying to understand how individuals make sense of their surrounding environment/particular situations, how they process information, interpret it and take action in a continuous 'sense-making' process. The *cultural perspective* is strongly linked to the area of risk communication and considers the social and cultural issues associated with risk responses looking beyond organisational boundaries.

We explored the process of risk identification, analysis, response, control, and communication, plus the main techniques for supporting the risk management process. We also looked more briefly at some recent developments in the area such as integrated risk management approaches and enterprise risk management (ERM), which both shift the emphasis from risk management at the project level to risk management at the portfolio level. Finally, we touched on the role of governance and the focus on front-end definition to reduce risk in later stages of the project, as well as the application of real options thinking and the use of systems dynamics to improve our understanding of risk interdependences.

Exercises Perspectives on Managing Risk and Uncertainty

Discussion questions

1. What is the difference between risk and uncertainty; why it is important to make a distinction?
2. Why are different perspectives on risk relevant for managing risk and uncertainty?
3. Why has the management of risk and uncertainty become an important area for many organisations?
4. Outline additional benefits (direct or indirect) of effective risk management implementation. How could you measure these benefits?
5. What are the challenges encountered during the risk identification stage and how would you address these?
6. What are the benefits and limitations of using quantitative and qualitative analysis techniques? When would they be appropriate to use?

7. What are the benefits and limitations of both proactive and reactive response approaches?
8. What are the key challenges of effective risk communication within your organisation, and how would you address them?
9. Your organisation is thinking of developing a risk maturity model. What would you suggest be the criteria used as a basis for developing the organisation's risk management capabilities?
10. You have been tasked to look at integrating risk management into the organisational culture. Discuss how you would go about doing this and what indicators you would suggest be used to demonstrate its effectiveness

The Millennium Stadium case discussion questions

1. From the contractor's perspective, how would you address the risk of dispute between WRU and CRFC?
2. Identify and discuss the organisational and cognitive factors that influenced the key project decisions
3. What are the key lessons to be learned from this project?

The Emirates Stadium case discussion questions

1. Discuss the role of front-end definition for reducing risks in the project
2. Discuss the role of an integrated supply chain for reducing risks on the project

Wembley Stadium case discussion questions

1. Identify and discuss the way in which the issues associated with the complexity of the organisation and environment influenced risk perceptions and decision making in this project
2. What were the main sociocultural issues which influenced risk perceptions and decision making on this project?

T5 case study discussion questions

1. What are the advantages and limitations of relational procurement or relational contracting strategy for managing risk and uncertainty in projects?
2. Identify and discuss the opportunities BAA identified and took advantage of during the project?
3. What are the main organisational factors that influenced the perceptions of risk?

Further reading

Perspectives on Managing Risk and Uncertainty

Aven, T (2010) *Misconceptions of Risk*. John Wiley & Sons Ltd, Chichester

Flyvbjerg, B, Bruzelius, N and Rothengatter, W (2003) *Megaprojects and Risk: The Anatomy of Ambition*. Cambridge University Press, Cambridge

Flyvbjerg, B, Garbuio, M and Lovallo, D (2009) Delusion and deception in large infrastructure projects: two models for explaining and preventing disaster. *California Management Review*, 51, 170–193

Kahkonen, K (2006) Management of uncertainty. In: DJ Lowe (ed.) with R Leiringer, *Commercial Management of Projects: Defining the Discipline*. Blackwell Publishing, Oxford
Kahneman, D, Slovic, P and Tversky, A (1982) *Judgement under Uncertainty; Heuristics and Biases.* Cambridge University Press, Cambridge
Loch, CH, DeMeyer, A and Pich, MT (2006) *Managing the Unknown: A New Approach to Managing High Uncertainty and Risk In Projects.* John Wiley & Sons, New York
Williams, TM, Samset, K and Sunnevag, KJ (2009) *Making Essential Choices With Scant Information, Front-End Decision Making in Major Projects.* Palgrave Macmillan, Basingstoke

Endnotes

1 This chapter draws on research carried out in EPSRC funded project grants (GR/R51452/01 and D505461/1). The author would like to acknowledge Professor Graham Winch who contributed to the development of many of the ideas presented here.
2 See the Foresight website for examples http://hsctoolkit.tribalhosting.net/The-tools.html
3 Simulation was prepared by Dr Cliff Mitchell, Manchester Business School

References

Ackermann, F, Eden, C, Williams, T and Howick, S (2006) Systemic risk assessment: a case study. *Journal of Operational Research Society*, 58, 39–51
Advisory Committee (1990) Report of the Advisory Committee on the Future of the US Space Program. Washington, D.C. http://history-nasa.gov.augustine/racfup1.htm (accessed 1 March 2011)
Artto, KA (1997) Fifteen years of project risk management applications: where are we going? In: K Kahkonen and KA Artto (eds) *Managing Risks in Projects.*E & FN Spon, London
Association for Project Management (2004) *Project Risk Analysis and Management (PRAM) Guide*, 2nd edition. Association for Project Management (APM), High Wycombe
Association for Project Management (2006) *Association for Project Management, Project Management Body of Knowledge*. 5th edition. The Association for Project Management (APM), High Wycombe
Austin, CH, Gillespie, RS, Hunter, RJ and Namek, PF (2003) Five decades of RM. *Risk Management Magazine*, 50, 38–39
Aven, T (2008) *Risk Analysis.* John Wiley & Sons Ltd, Chichester
Aven, T (2010) *Misconceptions of Risk*. John Wiley & Sons, Ltd, Chichester
Ayas, K and Zeniuk, N (2001) Project-based learning: building communities of reflective practitioners. *Management Learning*, 32, 61–76
Beck, U (1994) *Risk Society*. Sage, London
Beck, U (1999) *World Risk Society*. Polity Press, Cambridge
Bernstein, P (1998) *Against The Gods: The Remarkable Story of Risk.* John Wiley & Sons Ltd, Chichester
Bernoulli, D (trans.1954) Exposition of a new theory on the measurement of risk. *Econometrica*, 22, 22–36
Bier, VM (2001) On the state of the art: risk communication to the public. *Reliability Engineering and Systems Safety*, 71, 139–150
Bonabeau, E (2007) Understanding and managing complexity risk. *MIT Sloan Management Review*, 48, 62–68
British Standards Institution (2000) *BS 6079-3: Project Management – Part 3: Guide to the Management of Business Related Project Risk*. BSI, London
British Standards Institution (2001) *BS 61882: Hazard and Operability Studies Application Guide.* BSI, London
British Standards Institution (2006) *BSEN 60812: Analysis Technique for System Reliability. Procedure for Failure Mode and Effects Analysis (FMEA).* BSI, London
British Standards Institution (2008) *BS 31100: Risk Management: Code of Practice.* BSI, London
British Standards Institution (2009) *31000: Risk Management Principles and Guidelines.* BSI, London
Chapman, C and Ward, S (2003) *Project Risk Management: Processes, Techniques and Insights.* 2nd edition. John Wiley & Sons, Ltd, Chichester
Clarke, CJ and Varma, S (1999) Strategic risk management: the new competitive edge. *Long Range Planning*, 32, 414–424

Cohen, MD and Bacdayan, P (1994) Organizational routines are stored as procedural memory: evidence from a laboratory study. *Organization Science*, 5, 554–568

Columbia Accident Investigation Board (CAIB) (2003) Report of Columbia Accident Investigation Board, Volume 1. http://caib.nasa.gov/news/report/volume/default.html (accessed 1 April 2011)

Committee on the Financial Aspects of Corporate Governance, T.C. (1992) *The Financial Aspects of Corporate Governance (The Cadbury Report)*. London

Cooper, DF and Chapman, CB (1987) *Risk Analysis for Large Projects: Models, Methods and Cases*. Wiley, New York

Cooper, DF, Grey, S, Raymond, G and Walker, P (2005) *Project Risk Management Guidelines*. John Wiley & Sons, Ltd, Chichester

Courtney, H, Kirkland, J and Viguerie, P (1997) Strategy under uncertainty. *Harvard Business Review*, November–December, 67–79

Covello, VT, Winterfeldt, D and Slovic, P (1986) Risk communication: a review of the literature. *Risk Abstracts*, 3, 171–182

Cox, LA (2008) What's wrong with risk matrices? *Risk Analysis*, 28, 497–512

Cyert, RM and March, JG (1963) *A Behavioral Theory of the Firm*. Prentice Hall, Englewood Cliffs, NJ

Dalton, M (2007) *A Risk Breakdown Structure for Public Sector Construction Projects*. PhD Thesis. The University of Manchester, Manchester

de Finetti, B (1937, trans. 1964) Foresight: its logical laws, its subjective sources. In: HE Kyburg and HE Smokler, *Studies in Subjective Probability*. Wiley, New York

De Meyer, A, Loch, CH and Pich, MT (2002) Managing project uncertainty: from variation to chaos. *MIT Sloan Management Review*, 59–67

Douglas M and Wildavsky AB (1982) *Risk and Culture: An Essay on the Selection of Technical and Environmental Dangers*. University of California Press, Berkeley

Douglas M (1992) *Risk and Blame: Essays in Cultural Theory*. Routledge, London

Economist Intelligence Unit (2007) *Best Practice in Risk Management: A Function Comes of Age*. London

Eden, C and Ackermann, F (1998) *Making Strategy: The Journey of Strategic Management*. Sage, London

Eden, C, Ackermann, F and Williams, T (2005) The amoebic growth of project costs. *Project Management Journal*, 36, 15–27

Edwards, PJ and Bowen, PA (2005) *Risk Management in Project Organisations*. Elsevier Butterworth Heinemann, Oxford

Fischhoff, B (1995) Risk perception and communication unplugged: twenty years of process. *Risk Analysis*, 15, 137–145

Flyvbjerg, B, Bruzelius, N and Rothengatter, W (2003) *Megaprojects and Risk: An Anatomy of Ambition*. Cambridge University Press, Cambridge

French, DP, Sutton, SR, Marteau, TM and Kinmonth, AL (2004) The impact of personal and social comparison information about health risk. *British Journal of Health Psychology*, 9, 187–200

Giddens, A (1991) *Modernity and Self. Self and Society in the Late Modern Age*. Polity Press, Cambridge

Giddens, A (1999) *Runaway World: Risk*. www.lse.ac.uk/Giddens/pdf/17-nov-99.pdf (accessed 5 January 2004)

Gigerenzer, G (1991) From tools to theories: a heuristic of discovery in cognitive psychology. *Psychological Review*, 98, 254–267

Gil, N (2009) Project safeguards: operationalizing optionlike strategic thinking in infrastructure development. *IEEE Transactions on Engineering Management*, 56, 257–270

Gilovich, T, Griffin, D and Kahneman, D (2002) *Heuristics and Biases: The Psychology of Intuitive Judgement*. Cambridge University Press, Cambridge

Hartman, F (1997) Proactive risk management myth or reality? In: K Kahkonen and KA Artto (eds) *Managing Risks in Projects*. E & FN Spon, London

Hillson, D (2002) The risk breakdown structure (RBS) as an aid to effective risk management. In: *5th European Project Management Conference, PMI Europe*. Cannes France

Hillson, D (2003) *Effective Opportunity Management for Projects: Exploiting Positive Risk*. Marcel-Dekker, New York

HM Treasury (2003) *The Green Book: Appraisal and Evaluation in Central Government*. London

Hopkin, P (2010) *The Fundamentals of Risk Management*. The Institute of Risk Management, London

Hubbard, DW (2009) *The Failure of Risk Management: Why It Is Broken and How To Fix It*. Wiley, Hoboken, New Jersey

Huber, O (1997) Beyond gambles and lotteries, naturalistic risky decisions. In: R Ranyard, WR Crozier and O Svenson (eds), *Decision Making: Cognitive Models and Explanations*. Routledge, London

Institute of Chartered Accountants in England and Wales (1999) Internal Control: Guidance for Directors on the Combined Code. 'The Turnbull Report'. Institute of Chartered Accountants in England and Wales (ICAEW), London

International Standards Organization/IEC (2002) *Guide 73: Risk Management Vocabulary guidelines for use in standards*. London

International Standards Organization (2009a) *Guide 31000 Risk Management Principles and Guidelines*. Geneva

International Standards Organization/IEC (2009b) *Guide 31010: Risk management Risk Assessment Techniques*

IRM, AIRMIC and ALARM (2002) *A Risk Management Standard.* London

Jacob, WF and Kwak, YH (2003) In search of innovative techniques to evaluate pharmaceutical R&D projects. *Technovation*, April, 291–296

Janis, I (1972) *Victims of Groupthink: A Psychological Study of Foreign-Policy Decisions and Fiascos.* Houghton Mifflin, Oxford

Johnson, G, Scholes, K and Whittington, R (2006) *Exploring Corporate Strategy.* 7th edition. FT Prentice Hall, Harlow

Kahkonen, K (2006) Management of uncertainty. In: DJ Lowe (ed.) with R Leiringer, *Commercial Management of Projects: Defining the Discipline.* Blackwell Publishing, Oxford

Kahkonen, K and Artto, KA (eds) (1997) *Managing Risks in Projects.* E & F Spon, London

Kahneman, D and Tversky, A (1979) Prospect theory: an analysis of decision under risk. *Econometrica*, 47, 263–291

Kasperson RE and Stallen P (eds) (1991) *Communicating Risks to the Public: International Perspectives.* Reidel, Norwell, MA

Kasperson, JX, Kasperson, RE, Pidgeon, N and Slovic, P (2003) The social amplification of research: assessing fifteen years of research and theory. In: N Pidgeon, RE Kasperon and P Slovic (eds) *The Social Amplification of Risk.* Cambridge University Press, Cambridge

Keeney, RL and von Winterfeldt, D (1986) Improving risk communication. *Risk Analysis*, 6, 417–424

Lant, TK and Shapira, Z (2001) *Organizational Cognition: Computation and Interpretation.* Lawrence Erlbaum Associates, Mahwah, NJ

Lewin, C (2002) *RAMP: Risk Analysis and Management for Projects.* Thomas Telford, London

Loch, CH, DeMeyer, A and Pich, MT (2006) *Managing the Unknown: A New Approach yo Managing High Uncertainty and Risk in Projects.* John Wiley & Sons, New York

Lupton, D (1999) *Risk.* Routledge, Oxon

Lyneis, JM and Ford, DN (2007) Systems dynamics applied to project management: a survey, assessment and directions for future research. *Systems Dynamics Review*, 23, 157–189

March, JG and Shapira, Z (1987) Managerial perspectives on risk and risk taking. *Management Science*, 33, 1404–1419

March, JG and Simon, H (1958) *Organizations.* John Wiley & Sons, New York

Maule, JA (2008) Risk communication in organizations. In: GP Hodgkinson and WH Starbuck), *The Oxford Handbook of Organizational Decision Making.* Oxford University Press, Oxford

Maytorena, E, Winch, GM, Freeman, J and Kiely, T (2007) The influence of experience and information search styles on project risk identification performance. *IEEE Transactions on Engineering Management*, 50, 315–326

Miller, R and Lessard, DR (2000) *The Strategic Management of Large Engineering Projects.* MIT Press, Boston, MA

Morgan MG, Fischhoff B, Bostrom A and Atman C J (2002) *Risk Communication: A Mental Models Approach.* Cambridge University Press, Cambridge

Office of Government Commerce (OGC) (2007) *Management of Risk: Guidance for Practitioners.* 2nd edition. The Stationery Office, London

Office of Technology Assessment (1989) *Round Trip to Orbit: Human Spaceflight Alternatives Special Report.* US Government Printing Office, Washington D.C.

Pascal, B (1670) *Pensees.* G. Desprez, Paris

Patterson, FD and Nealey, K (2002) A risk register database system to aid the management of project risk. *International Journal of Project Management*, 20, 365–374

Project Management Institute (2008) *Guide to the Project Management Body of Knowledge (PMBoK).* 4th edition. Project Management Institute (PMI), Newton Square, PA, USA

Ramsey, FP (1931) *The Foundations of Mathematics and other Logical Essays.* Kegan Paul and Co. Ltd, London

Raz, T and Hillson, D (2005) A comparative review of risk management standards. *Risk Management*, 7, 53–66

Renn, O and Levine, D (1991) Credibility and trust in risk communication. In: RE Kasperson and PJM Stallen (eds) *Communicating Risks to the Public.* Kluwer, Dordrecht

Rosenhead, J and Mingers, J (2001) *Rational Analysis for a Problematic World Revisited: Problems Structuring Methods for Complexity, Uncertainty and Conflict.* John Wiley & Sons, Chichester

Salomo, S, Weise, J and Germunder, HG (2007) NPD planning activities and innovation performance: the mediating role of process management and the moderating effect of product innovativeness. *Product Innovation Management*, 24, 285–302

Saputelli, L, Hull, R and Alfonzo, A (2008) Front end loading provides foundation for smarter project execution. *Oil and Gas Financial Journal*, 5

Savage, LJ (1954) *Foundations of Statistics.* John Wiley and Sons, New York

Schoemaker, PJH (1982) The expected utility model: its variants, purposes, evidence and limitations. *Journal of Economic Literature*, XX, 529–563

Schwartz, P (1997) *The Art OF THE Long View: Planning for the Future in an Uncertain World*, 2nd edition. John Wiley & Sons Ltd, Chichester

Simon, H (1947) *Administrative Behavior*. Macmillan, New York

Snowden, DJ and Boone, M (2007) A leader's framework for decision making. *Harvard Business Review*, November, 69–76

Tversky, A and Kahneman, D (1974) Judgment under uncertainty: heuristics and biases. *Science*, 185 (4157), 1124–1131

van der Heijden, K (2004) *Scenarios: The Art of Strategic Conversation*. 2nd edition. John Wiley & Sons Ltd, Chichester

Vaughan, D (1987) *The Challenger Launch Disaster: Risky Technology, Culture and Deviance at NASA*. Chicago University Press, Chicago, IL

Voetsch, RJ, Cioffi, DF and Anbari, FT (2005) Association of reported project risk management practices and project success. *Project Perspectives*, XXVII, 4–7

Von Neumann, J and Morgenstern, O (1944) *Theory of Games and Economic Behavior*. Princeton University Press, Princeton, NJ

Vose, D (2000) *Risk Analysis, A Practical Guide*. John Wiley & Sons, New York

Weber, EU, Siebenmorgen, N and Weber, M (2005) Communicating asset risk: how name recognition and the format of historic volatility information affect risk perception and investment decisions. *Risk Analysis*, 25, 597–609

Weick, KE (1995) *Sensemaking in Organizations*. Sage, Thousand Oaks, CA

Weick, KE and Sutcliffe, KM (2001) *Managing the Unexpected: Assuring High Performance in an Age of Complexity*. Jossey-Bass, San Francisco

Williams, T (2002) *Modelling Complex Projects*. John Wiley & Sons Ltd, Chichester

Williams, TM (1996) The two-dimensionality of project risk. *International Journal of Project Management*, 14, 185–186

Williams, TM, Ackermann, F and Eden, C (1997) Project risk: systemicity, cause mapping and scenario approach. In: K Kahkonen, and KA Artto (eds) *Managing Risks in Projects*. E & FN Spon, London

Williams, TM, Samset, K and Sunnevag, KJ (2009) *Making Essential Choices with Scant Information, Front-End Decision Making in Major Projects*. Palgrave Macmillan, Basingstoke

Wilkinson, A, Elahi, S and Eidinow, E (2003) Risk world scenarios. *Journal of Risk Research*, 6, 297–334

Winch, G (2010) *Managing Construction Projects*. 2nd edition. Wiley-Blackwell, Oxford

Winch, G and Maytorena, E (2011) Managing risk and uncertainty on projects: a cognitive approach. In: PWG Morris, JK Pinto and J Soderlund (eds) *The Oxford Handbook of Project Management*. Oxford University Press, Oxford

Winter, M (2009) Using soft systems methodology to structure project definition. In: TM Williams, K Samset and KJ Sunnevag (eds), *Making Essential Choices with Scant Information, Front-End Decision Making in Major Projects*. Palgrave, Basingstoke

Chapter 5

Financial Decisions

Anne Stafford

Learning outcomes

When you have completed this chapter you should be able to:

- Understand the nature and purpose of financial information
- Be aware of the main user groups and their financial information needs
- Understand the content and function of the primary financial statements
- Explain the nature and use of some basic accounting control systems
- Be aware of how to use financial information to make short-term decisions
- Be aware of long-term investment appraisal techniques such as discounted cash flow, and understand how they are used in the creation of a business case

Introduction

Commercial managers are involved in both supporting and making financial decisions, and interact with the finance function to generate and report financial information. This chapter aims to provide a basic understanding of financial decision making, so that commercial managers are better equipped to understand not only the financial information provided both by their own organisations and by clients or customers, but also their role in contributing to this.

Traditionally, financial information developed as an aid to the stewardship function of management, to demonstrate how well business resources had been managed. Now financial information plays a far greater role as an aid to management in planning, decision making and control, with a primary objective being to provide useful and timely information to help managers make decisions.

Financial information is therefore useful to those on both the demand and the supply side of commercial management, and it is essential for managers to develop financial skills as part of their overall management skills package.

On the demand side, financial information is needed to assess whether a project will generate a sufficient return on an investment – for example, by a lender of finance investing in a construction and facilities management project, or whether a project is affordable in terms of the projected monthly/annual cost – for example, the outsourcing of the IT function.

On the supply side, the commercial manager needs to know whether the project costs are running at the amounts estimated, and the amount of and reason for any significant differences. The level of return that the project is making on the investment must also be monitored, together with the ability of the project to make debt repayments when due.

Commercial Management: theory and practice, First Edition. David Lowe.

This chapter covers the nature and purpose of financial information, the people who use financial information, and the financial environment within which companies operate. It then explains what the main financial statements are, and the accounting conventions behind their preparation. Next, it examines how managers can use financial information to help make financial decisions, through understanding cost behaviour and then applying this knowledge to techniques such as break-even analysis, performance measurement and cash-flow management. The chapter concludes by explaining investment appraisal methods such as discounted cash flow.

The nature and purpose of financial information

Financial information is provided by the accounting function in an enterprise or organisation. Here we can distinguish between the act of **recording financial information** and the act of **providing useful information**. Put very simply, **bookkeeping** is the recording of data (in a double-entry system) whereas **accounting** is the provision of financial information to users. The accounting process involves the classification and measurement of transactions, summarising them for a period of time (a week, a month or a year), and then communicating and analysing that financial information.

A major function of accounting is the preparation of financial statements. The **International Accounting Standards Board** (IASB), which regulates how published financial information should be prepared, states that financial statements have the following aim:

> 'The objective of general purpose financial reporting is to provide financial information about the reporting entity that is useful to present and potential equity investors, lenders and other creditors in making decisions in their capacity as capital providers. Information that is decision-useful to capital providers may also be useful to other users of financial reporting who are not capital providers.' (IASB, 2008, paragraph OB2)

The IASB discusses how such financial information can be provided, referring to the importance of the information provided by the **balance sheet**. This statement shows the financial position of a business, frozen at a point in time (the last day of the accounting period). The IASB also considers the need to assess financial performance, provided in the **income statement**, which gives a historical summary of performance over the past year. Furthermore, the IASB discusses how the **cash-flow statement** can help to identify financial flexibility, indicating how well the company can respond to changes in the financial environment in the future.

These three statements taken together help users to look backwards and assess how well the company has used the resources invested in it, as well as forwards in terms of making decisions about their involvement with the company in the future.

Activity 5.1 FirstTV

FirstTV plc is in business providing digital television and related services. It wishes to expand and has the opportunity to acquire one of two companies. The first, Localview Ltd, provides cable TV to homes in Greater Manchester. The second, Countrywide Ltd, offers an internet and TV email service.

- What accounting information would be relevant in choosing between the two companies?
- What other information would be relevant?

Who are the users of accounting information that the IASB refers to? Over the past 30 years national and international accounting regulators have produced reports identifying stakeholder groups interested in corporate information. Table 5.1 identifies representative user groups.

Table 5.1 User groups and their information needs.

User group	Information needs
Investors	How well is the business performing? They are interested in future prospects. They are also interested in the share price movement
Lenders	Are the funds loaned secure? Will the business be able to pay interest as it falls due and will loans eventually be repaid?
Suppliers and other creditors	Is the company high or low risk, in both the short and long term? They want to be sure that their bills will be paid when they fall due
Employees	Are their jobs secure? Are there career opportunities? They want to know how stable and profitable the company is
Customers	They want to know how stable is the company in the long run, particularly if they wish to build up a long-term relationship with the company or if warranties are involved
Government and their agencies	They are interested in what tax is due and in collecting national statistics. In some industries, such as utilities, there will be regulatory bodies setting financial and other criteria for companies to meet
The public	Obviously an extremely varied user group, which includes local community groups concerned about the impact of a large employer on their local environment, with respect to local employment, etc., as well as national and international pressure groups with interests ranging from human rights issues to environmental issues. The nature of the relationship between members of the public and companies is changing, with less formal groupings using social networking to get their point across, as has been evident with demonstrations in 2010 and 2011 against the perceived tax avoidance by large multinationals including Barclays, Arcadia and Vodafone
Management	They need timely and relevant financial information on a wider range of issues than those reported in the financial statements

Activity 5.2 Sue Peel

You meet Sue Peel at the gym. She has a small computer services business and her accountant has just finished preparing accounts for her first year of trading. She says that she cannot understand all the fuss made about balance sheets and accounts because 'it's only important to me, and I'm only interested in how much cash there is in the bank'.

- Who is likely to use Sue's accounts?
- Why does she need to have a balance sheet?
- What do you think of her final comment 'I'm only interested in how much cash there is in the bank'?

Desirable qualities of financial information

It is important that financial information is useful. If it isn't, then users will either come to the wrong conclusion or they won't bother to use financial information when they make decisions. Either way, companies and/or individuals are likely to be worse off. The UK Accounting Standards Board (ASB), in its *Statement of Principles* (1999) sets out desirable qualities that will help to make financial information useful as follows:

> 'Information provided by financial statements needs to be **relevant** and **reliable** and, if a choice exists between relevant and reliable approaches that are mutually exclusive, the approach chosen needs to be the one that results in the relevance of the information provided being maximised.'
>
> 'Information is **relevant** if it has the ability to influence the economic decisions of users and is provided in time to influence those decisions.'
>
> 'Information is **reliable** if:
>
> (a) it can be depended upon by users to represent faithfully what it either purports to represent or could reasonably be expected to represent;
> (b) it is free from deliberate or systematic bias (that is, it is **neutral**);
> (c) it is free from material error;
> (d) it is complete within the bounds of materiality; and
> (e) in its preparation under conditions of uncertainty a degree of caution (i.e. **prudence**) has been applied in exercising judgement and making the necessary estimates.'

In addition, information needs to be:

- **Comparable** (with similar information about the entity from different time periods and with similar information from other entities). Comparability also implies **consistency**, and also gives rise to the need for **adequate disclosure** to permit comparisons to be made
- **Understandable**, so that users can perceive its significance. This will depend on the ability of the various users, as well as the way in which the information is presented (that is, the degree of aggregation and classification)

Activity 5.3 Relevance and reliability

- Can you think of an example where there is a conflict between relevance and reliability?
- An item of information that managers would find useful is potentially available. It is relevant and reliable. Why would a company choose not to produce it?

Further reading

The nature and purpose of financial information

Abraham, A, Glynn, J, Murphy, M and Wilkinson, B (2008) The nature and purpose of accounting (Chapter 1). In: *Accounting for Managers*. 4th edition. Cengage Learning EMEA, UK

IASB conceptual framework. Information available at: http://www.ifrs.org/Current+Projects/IASB+Projects/Conceptual+Framework/Conceptual+Framework.htm (accessed 12 September 2012)

The distinction between financial accounting and management accounting information

The financial reports necessary for the smooth running of commercial activities can be divided into two types – those prepared for financial accounting purposes and those providing management accounting information – although there is overlap between the two.

Financial accounting reports are prepared for the purposes of supplying information to shareholders and other interested parties, and must be prepared according to statutory and other regulatory requirements in terms of content and format. They are needed if the project is set up as a limited company, as UK company law requires a copy to be filed at Companies House each year. They will also be used for tax purposes, although adjustments need to be made, and to enable companies to raise finance. Manager remuneration – for example, performance related pay – may also be partly dependent on financial accounting measurements. Statutory financial accounting relies almost wholly on historical financial information, as this can then be subject to assurance through the audit process.

Management accounts are concerned with the production of information in order to facilitate planning, evaluation and control of commercial activities. The content and format of the information is at the discretion of management, as part of the management role is not just to use and act on information provided, but also to decide what information is contained within management accounts and reports, how it is presented and who receives it.

Management accounts are used to make decisions on specific projects, to produce costing data to enable project costs and prices to be calculated, for monitoring and control of day-to-day performance and to feed into the organisation's strategic decision-making process, including investment decisions. Although management accounting uses some historical data, there is much more emphasis placed on forecasting information and non-financial measurements.

Activity 5.4 Financial and management accounting

Do you think the distinction made between financial and management accounting may be misleading?

The nature of the financial reporting environment

The nature and type of financial reporting will vary depending on the way in which a business has been formed. Small unincorporated businesses and partnerships are usually subject to less regulation and need only provide financial information for tax purposes or, for example, if there is a requirement for a bank loan. However, commercial managers typically work for larger incorporated enterprises, where the company has a legal identity separate from that of its owners. The company can enter into contracts, own property, sue and be sued in its own right. Owners' liability is limited to the nominal (face) value of the shares that they own. Such **limited liability companies** can be:

- **Private**, in which case the level of disclosure is reduced, but shares are not generally available for sale
- **Public**, in which case there are more onerous rules in order to protect the public. Shares are available to buy or sell on a stock exchange, and the market value may fluctuate considerably. Market value will be different from the nominal (face) value of the shares

Most public limited companies (plcs) trade as **groups** rather than as one company. A group consists of a holding company, which issues shares to the public, and various other companies over which the holding company has control or can exert influence on management decisions. It is more convenient for a group to be structured in this way because:

- It enables the group to easily acquire other companies without having to change its structure
- It is simpler for reporting purposes as subsidiary companies may be grouped together in divisions, usually according to business activity
- Multinational groups will often be required by law to have companies incorporated in each country of activity
- The investment in and earnings generated by companies over which the holding company can exert influence, such as associated undertakings, strategic partnerships and joint ventures, are shown in the group financial statements

An example of a group structure is shown in Box 5.1 using the Vinci organisation chart taken from its 2011 Annual Report. It shows how Vinci is structured primarily in terms of business divisions (concessions and contracting, the latter being further sub-divided into energy, transport and construction), with national subsidiaries being structured at sub-business division level.

The accepted financial objective for most firms is to **maximise wealth creation for shareholders** (Friedman, 1970). This is somewhat different from seeking to maximise profit in the short term, although many managers are criticised for going for short-term profit, perhaps in order to boost their annual bonuses, rather than long-term growth.

Activity 5.5 Wealth creation

Can you think of any reasons why making the maximum possible profit this year may not be in the best interests of overall wealth creation?

Published financial reporting by companies is governed by a number of national and international regulatory frameworks. These comprise:

- The legal framework – for example, in the UK this is the Companies Act 2006 plus European Union directives
- Stock exchange requirements for listed companies
- Financial reporting standards, both national and international. Most countries, including Europe, require companies to follow the International Financial Reporting Standards (IFRSs) published by the International Accounting Standards Board (IASB). The major exception is the US, which requires compliance with standards issued by its national standard setter, the Financial Accounting Standards Board (FASB)

The rules prescribed by these frameworks include the following:

- Content, formats and presentation of published financial statements, including extensive disclosures, permissible valuation methods for items such as buildings, inventory and financial instruments, and treatments of specific transactions (especially complex or problematic ones, for example leases, goodwill, research and development expenditure)

Box 5.1 Extract from Vinci Annual Report 2011

Relations between the parent company and subsidiaries

Organisation chart (*)

VINCI

- CONCESSIONS
  - VINCI Autoroutes
    - ASF 100%
    - Escota 99%
    - Cofiroute 83%
    - Arcour 100%
  - VINCI Concessions
    - VINCI Park
    - VINCI Airports
    - Rail infrastructure
    - Road infrastructure
    - Stadiums
- VINCI Immobilier
- CONTRACTING
  - VINCI Construction
    - VINCI Construction France
    - VINCI Construction UK
    - CFE (Belgium)
    - Sogea-Satom (Africa)
    - VINCI Construction Overseas France
    - Subsidiaries in Central Europe
    - Soletanche Freyssinet
    - Entrepose Contracting
    - VINCI Construction Grands Projects
    - VINCI Construction Terrassement
    - Dodin Campenon Bernard
  - VINCI Energies
    - VINCI Energies France
    - VINCI Energies International
    - Cegelec GSS
    - VINCI Facilities
  - Eurovia
    - French subsidiaries
    - ETF-Eurovia Travaux Ferroviaires
    - Specialised subsidiaries
    - Eurovia GmbH (Germany)
    - Eurovia CS (Czech Repub. and Slovakia)
    - Eurovia Polska SA
    - Eurovia Group Ltd (UK)
    - Hubbard Group (USA)
    - Eurovia Canada Inc.
    - Other foreign subsidiaries

(*) Simplified organisation chart of the Group at 31 December 2011 (main companies owned directly or indirectly and percentage of capital held)

VINCI's direct shareholdings in subsidiaries and affiliates are described on page 278. The list of the main consolidated companies (pages 254–260) gives an indication of the various subsidiaries that comprise the Group and of VINCI's equity interest (whether direct or indirect) in the various entities.

Source: Vinci Annual Report 2011

- Requirements for **directors**, which under UK legislation includes the following in relation to financial information:
  - the preparation of annual financial statements that show a true and fair view
  - the selection and consistent application of appropriate accounting policies
  - making judgements and estimates that are reasonable and prudent
  - stating whether applicable financial reporting standards have been followed
  - the preparation of financial statements on the going-concern basis unless this is inappropriate
  - ensuring that the company keeps proper accounting records
  - taking reasonable steps to safeguard the company's assets and prevent fraud

- Requirements in relation to **auditors**, as companies have a legal duty to appoint an auditor each year. The auditor is responsible for reporting their opinion to the shareholders as to whether the:
  - balance sheet and profit and loss account show a true and fair view of the company's financial state of affairs and results for the year
  - financial statements have been properly prepared in accordance with legal and financial reporting regulations
- Requirements in relation to **corporate governance**, which are necessary due to the separation of ownership (by shareholders) from control (by managers and directors) in the running of large companies. In order to ensure that directors are carrying out their legislative duties to run the company on behalf of the shareholders properly, corporate governance mechanisms have developed over time. Listed companies are required to comply with the relevant national code of practice, such as the Combined Code on Corporate Governance in the UK.

Companies are required to report on corporate governance issues including:
- reports from the various committees (executive, remunerations, pensions, audit)
- much more disclosure about the directors, their remuneration and their responsibilities, including the role of non-executive directors

An example of the corporate governance disclosure requirements is shown in Box 5.2 using an extract from the Vodafone plc Annual Report 2010, a group highly regarded in terms of its commitment to corporate governance.

Box 5.2 Extract from Vodafone plc Annual Report 2010

We are committed to high standards of corporate governance which we consider are critical to business integrity and to maintaining investors' trust in us. We expect all our directors, employees and suppliers to act with honesty, integrity and fairness. Our business principles set out the standards we set ourselves to ensure we operate lawfully, with integrity and with respect for the culture of every country in which we do business.

Compliance with the Combined Code

Our ordinary shares are listed in the UK on the London Stock Exchange. In accordance with the Listing Rules of the UK Listing Authority, we confirm that throughout the year ended 31 March 2010 and at the date of this document we were compliant with the provisions of, and applied the principles of, Section 1 of the 2008 FRC Combined Code on Corporate Governance (the 'Combined Code'). The Combined Code can be found on the FRC website (www.frc.org.uk). The following section, together with the 'Directors' remuneration' section on pages 57 to 67, provides detail of how we apply the principles and comply with the provisions of the Combined Code. We have been following the FRC consultation on further proposed changes to the Combined Code and intend to comply with such revisions should they be adopted.

Corporate Governance Statement

We comply with the corporate governance statement requirements pursuant to the FSA's Disclosure and Transparency Rules by virtue of the information included in this corporate governance section of the annual report together with information contained in the 'Shareholder information' section on pages 125 to 131.

Source: Vodafone plc (2010)

The financial reporting environment is dynamic and subject to constant change, as reporting issues have become more complex. Typically it has been reactive, rather than proactive, with changes being brought about as a result of corporate scandal or financial crisis rather than through the implementation of best practice. For example, there have been changes to the international regulatory regime that aim to reverse the undermining of public confidence following financial scandals such as Enron, Worldcom, Parmalat and the banking crisis. Regulations in relation to directors and auditors have increased as has disclosure in respect of complex financial transactions involving the use of financial instruments.

Further reading

Corporate governance

Abraham, A, Glynn, J, Murphy, M and Wilkinson, B (2008) Corporate governance and ethics (Chapter 2) and Regulation, audit and taxation (Chapter 14). In: *Accounting for Managers*. 4th edition. Cengage Learning EMEA, UK

OECD (2004) *OECD Principles of Corporate Governance*, available at: www.oecd.org/daf/corporate affairs/principles/text (accessed 10 May 2011)

Solomon, J (2010) *Corporate Governance and Accountability*. 3rd edition. John Wiley and Sons Ltd, Chichester

Presentation of financial statements

Financial reports routinely prepared by company accountants include the balance sheet, the income statement and the cash-flow statement, with *ad hoc* reports being prepared for special projects. In addition, financial projections, based on these formats, are an important part of the project procurement process.

The balance sheet

The balance sheet shows the way in which the company is funded, usually through a mixture of share capital and debt, and how the company has used the funds at its disposal. It represents the **statement of financial position** of the company – a 'snapshot' at one point in time. The balance sheet is made up of the three elements of the accounting equation:

Assets – Liabilities = Equity

In order to 'balance', the amount of resources that a company owns or controls (the assets) less the amounts that it owes to third parties, both in the short and long term (the liabilities) must equal the residual value that is due to the shareholders (equity). This is usually shown in the format of Net Assets (non-current plus current assets less liabilities) equals Equity.

Definitions of each of these elements are as follows:

- **Assets** are the rights or other access to future economic benefits controlled by a company as a result of past transactions or events
- **Liabilities** are the obligations of a company to transfer economic benefits as a result of past transactions or events
- **Equity** is the residual amount found by deducting all of the company's liabilities from all of its assets

Let's break each of these elements down into more detail.

Non-current assets (also known as fixed assets) are assets to be used on a long-term basis (that is, over the course of more than one year) to generate profits. They can be divided into further subcategories:

- **Tangible** non-current assets, which are made up of property, plant and equipment (PPE), such as land, buildings, machinery, motor vehicles, computers, other fixtures and fittings
- **Intangible** non-current assets, which are items with no physical substance such as goodwill, leases, development expenditure
- **Investments**, for example, the equity shares and loans in other companies

They are shown at their net value, which is usually their historic cost less **accumulated depreciation** (an amount deducted to represent the using up of their value over the length of time during which the company has the economic use of the asset).

Current assets are assets that generate cash within 12 months of the balance sheet date and include:

- **Inventory** (also known as stock) – for example, raw materials, work-in-progress, finished goods, goods for resale
- **Trade receivables** (also known as debtors), which are the amounts owed to the company by its customers
- **Prepayments**, which are payments the company has made before the balance sheet date, relating to a future period – for example, insurance or rent paid in advance
- **Cash and cash equivalents**, which is cash-in-hand and held at the bank in readily accessible accounts

Current assets are normally shown at the lower of cost and net realisable value. In most cases, therefore, items are shown at their purchase cost. An example where net realisable value would be lower than cost would be in the retail fashion industry, where last season's lines can be sold only at a price less than cost. They need to be shown at this lower amount so that profit is not overstated.

Current liabilities (also known as creditors due within one year) are amounts that the company must pay within 12 months of the balance sheet date and include:

- **Bank overdraft**
- **Borrowings**, but only those repayable within 12 months
- **Trade payables** (also known as trade creditors), which are the amounts owed by the company to its suppliers
- **Accruals**, which are amounts owed by the company but for which no invoice has been received at the balance sheet date – for example, the amount owed for electricity consumed by the company over the last month
- **Current tax liabilities**, relating to tax payable on profit earned during the year
- **Proposed dividends**, relating to dividends declared but not paid at the balance sheet date

Non-current liabilities (also known as creditors due after more than one year) are amounts payable more than 12 months after the balance sheet date, and can include:

- **Borrowings**, which are repayable after more than 12 months
- **Deferred tax liabilities**, which relates to tax which will potentially become payable at some point in the future
- **Pension scheme obligations**, relating to future payments which the company is due to make in respect of current employees' future retirement benefits
- **Provisions for liabilities and other charges**, which are amounts set aside to pay future liabilities in respect of events that have occurred by the balance sheet date but where the amounts involved cannot be determined with certainty. For example, legal claims, guarantees, reorganisation costs

Equity (also referred to as capital and reserves or shareholders' funds) is the long-term finance provided by the owners of the company. This represents amounts owed by the company to its owners (shareholders). It can be broken down into the following sub-categories:

Box 5.3 Balfour Beatty balance sheet

Statements of financial position at 31 December 2010

	Group			
	Notes	2010 £m	2009[1,2] £m	2008[1] £m
Non-current assets				
Intangible assets – goodwill	12	**1,196**	1,145	975
– other	13	**251**	298	223
Property, plant and equipment	14	**320**	315	296
Investments in joint ventures and associates	15	**488**	451	465
Investments	16	**95**	83	55
PPP financial assets	17	**327**	260	151
Deferred tax assets	24	**163**	191	132
Derivative financial instruments	20	–	1	3
Trade and other receivables	21	**70**	98	74
		2,910	2,284	2,374
Current assets				
Inventories	18	**89**	100	125
Due from customers for contract work	19	**591**	524	383
Derivative financial instruments	20	**4**	–	2
Trade and other receivables	21	**1,197**	1,329	1,193
Current tax assets		**4**	5	–
Cash and cash equivalents – PPP subsidiaries	23	**18**	10	2
– other	23	**566**	608	461
		2,469	2,576	2,166
Total assets		**5,379**	5,418	4,540
Current liabilities				
Trade and other payables	22	**(2,232)**	(2,412)	(2,168)
Due to customers for contract work	19	**(651)**	(607)	(540)
Derivative financial instruments	20	**(2)**	(1)	(66)
Current tax liabilities		**(29)**	(8)	(23)
Borrowings – PPP non-recourse loans	23	**(8)**	(19)	–
– other	23	**(37)**	(23)	(12)
		(2,959)	(3,070)	(2,809)
Non-current liabilities				
Trade and other payables	22	**(144)**	(163)	(152)
Derivative financial instruments	20	**(45)**	(24)	(40)
Borrowings – PP non-recourse loans	23	**(280)**	(239)	(145)
– other	23	**(11)**	(13)	(9)
Deferred tax liabilities	24	**(8)**	(9)	(10)
Liability component of preference shares	27	**(89)**	(88)	(87)
Retirement benefit obligations	25	**(441)**	(586)	(261)
Provisions	26	**(242)**	(227)	(166)
		(1,260)	(1,349)	(870)
Total liabilities		**(4,219)**	(4,419)	(3,679)
Net assets		**1,160**	999	861
Equity				
Called-up share capital	27	**343**	343	239
Share premium account	28	**59**	57	54
Equity component of preference shares	28	**16**	16	16
Special reserve	28	**30**	32	139
Share of joint ventures' and associates' reserves	28	**144**	157	226
Other reserves	28	**334**	288	79
Retained profits	28	**230**	102	104
Equity attributable to equity holders of the parent		**1,156**	995	857
Non-controlling interests	28	**4**	4	4
Total equity		**1,160**	999	861

[1] Restated for the adoption of IFRIC 12 (Notes 1.2 and 37)

[3] Restated for the amendments to the acquisition statement of financial position of Parsons Brinckerhoff Inc. (Notes 1.2 and 37)

Source: Balfour Beatty (2010)

- **Share capital**, which is the nominal value of the shares issued by the company. It may consist of different types (classes), the most common being:
  - **ordinary/equity shares**: they give the shareholders the right to vote at meetings. These shareholders are the risk takers and they are only entitled to a share of profits after all other claims have been met
  - **preference shares**: effectively another form of loan. They do not carry voting rights. Dividends are typically fixed and must be paid before ordinary dividends can be awarded
  - **Share premium** is the excess of the amount paid by shareholders over the nominal value of the shares when issued. Note that when the shares are subsequently traded, there is no impact on the company's share capital or share premium.
  - **Reserves** consist of the accumulated profits that have been retained by the company to fund future growth

The Balfour Beatty plc group statements of financial position (balance sheets) taken from the 2010 annual report are shown in Box 5.3. Note that often (but not always) accounting convention uses brackets to denote a negative number.

Many commercial managers may be involved with **contracts** that start within one accounting period and finish in another – for example, the construction of a stadium, a bridge or a hospital. This means that at the end of each year a valuation of the work to date must be carried out so that appropriate amounts are shown in the balance sheet. In addition, profits are usually recognised in proportion to the percentage of work completed, as although the profit may not yet have been realised in full, it is considered more appropriate to recognise the profit over the length of the contract rather than show all the profit at the end of the contract. Profits can be recognised in this way if a profitable outcome can be reliably estimated. If a loss is predicted then prudence dictates that this should be recognised in full.

Construction contracts must each be accounted for separately. Each contract will record costs incurred and revenue recognised. Customers will be billed as the contract progresses. However, usually there are differences between the amount of revenue recognised and the progress billings invoiced. Where the customer still owes the company for work done that has not yet been invoiced, a current asset is recorded. If, however, the customer has been invoiced for too much or the company is unable to recover all the costs incurred (for example, on a loss-making fixed-price contract), then a current liability will be recorded. Balfour Beatty shows amounts for both a current asset and a current liability in relation to outstanding contracts in its statements of financial position for the year.

It is important to remember that the balance sheet does **not** represent the current worth or value of the company. It includes only the measurable resources put into the company and what they are represented by. There are various resources that contribute value to a company, but which cannot be measured with sufficient objectivity to be included in a company balance sheet. Examples include intangible items such as business reputation, goodwill and the human resources used in the company. In contrast, market value is measured by the share price, which will fluctuate on a daily basis dependent on a number of factors. These include market expectation of future company performance, the market view of the value of intangibles excluded from the balance sheet, and other factors both internal (for example, a problem specific to the project such as an unexpected change in user demand) and external (for example, a global downturn in the stock market such as the financial crisis of 2008).

The income statement

The income statement, or **statement of financial performance**, gives a summary of income less expenditure for the year. Income is that earned from sales made during the year. Expenditure relates to the costs incurred to achieve those sales. For legal purposes this is an annual summary of historical performance over the past year. For management accounting purposes, it may be prepared on a monthly basis, and compared with budgeted, or expected, figures, and may be combined with a projected forecast of the remaining months' profits to give an estimated out-turn for the year.

The main headings of the income statement are as follows:

- **Revenue** is the total sales or operating revenue of a company. Sales can be either goods or services and exclude VAT
- **Cost of sales** represents the value of materials and services (including labour and overheads) bought during the year in order to make the goods or to provide the services sold
- **Distribution costs** are costs incurred after production of goods is completed and will typically include the following:
  - warehousing costs
  - promotion/marketing costs
  - selling costs
  - transport and logistics costs

  This section of the income statement thus records the costs relating to supply chain management, covered in Chapters 2 and 8.
- **Administrative expenses** are the additional costs incurred in running a company and will typically include the following:
  - administrative salaries, occupancy and other related costs
  - professional fees
  - any other central costs that cannot be allocated to production or distribution

  Distribution costs and administrative expenses are both **operating expenses** for the company.
- **Other operating income** relates to any income that is not recorded under turnover, for example, royalties, commissions, rents
- **Exceptional items** are items which are significant due to their size or incidence and need to be shown separately on the face of the income statement. They fall into three categories:
  - profits or losses on the sale or termination of a business
  - costs of a fundamental reorganisation or restructuring
  - profits or losses on the disposal of non-current assets
- **Operating profit** is the profit made from the main activities of the company. It represents how well management have carried out the task of running the company over the past year
- **Finance costs** include any payments or receipts relating to how the company is financed. Finance charges relating to leases and financial instruments are included here as well as bank and debenture interest
- **Taxation** is based on the profit reported for the year
- **Profit for the financial year** is the profit attributable to equity shareholders of the overall group
- **Dividends** are the proportion of profit for the financial year that is distributed to shareholders of the company/group
- **Retained profit for the financial year** is the profit added to the company's reserves for the year, to be reinvested into the business
- **Basic earnings per share** shows the amount of profit for the financial year earned by each ordinary share, and thus enables a comparison to be made to previous years and to other companies
- **Diluted earnings per share** shows the amount of profit for the financial year earned by each share if we assume that all future opportunities to increase the number of shares in issue have been taken up – for example, all share options granted to directors of the company have been exercised. This shows the effect on earnings per share of any potential increase in share capital, and therefore gives an indication of future risk to earnings and wealth creation

The Balfour Beatty plc Group income statement taken from the 2010 annual report is shown in Box 5.4 as an example. Note that it provides no breakdown of operating expenses, as regulations require very minimal disclosure in this area.

The income statement includes only expenditure that is regarded as revenue expenditure, that is, the costs incurred in purchasing, producing and selling the goods or services of the company, together with the day-to-day administrative costs. Capital expenditure is treated differently as it has a lasting impact on the company. It normally results in the acquisition or improvement of an asset that will be used by the company for the purpose of generating income over several years. The cost of the new capital asset is spread over the number of years during which the company will benefit from its use. This annual charge for the use of an asset is called depreciation.

Box 5.4 Balfour Beatty plc Group income statement

Group income statement for the year ended 31 December 2010

	Notes	2010 Before exceptional items* £m	2010 Exceptional items* (Note 8) £m	2010 Total £m	2009 Before exceptional items*[1] £m	2009 Exceptional items (Note 8) £m	2009 Total[1] £m
Revenue including share of joint ventures and associates		**10,541**	–	**10,541**	10,339	–	10,339
Share of revenue of joint ventures and associates	15.2	**(1,305)**	–	**(1,305)**	(1,385)	–	(1,385)
Group revenue	2	**9,236**	–	9,236	8,954	–	8,954
Cost of sales		**(8,132)**	–	**(8,132)**	(8,173)	–	(8,173)
Gross profit		**1,104**	–	1,104	781	–	781
Net operating expenses							
– amortisation of intangible assets	13	–	**(82)**	**(82)**	–	(48)	(48)
– other		**(851)**	**(23)**	**(874)**	(582)	63	(519)
Group operating profit/(loss)		**253**	**(105)**	**148**	199	15	214
Share of results of joint ventures and associates	15.2	**85**	**(27)**	**58**	81	–	81
Profit/(loss) from operations	4	**338**	**(132)**	**206**	280	15	295
Investment income	6	**46**	–	**46**	32	–	32
Finance costs	7	**(65)**	–	**(65)**	(47)	(15)	(62)
Profit/(loss) before taxation		**319**	**(132)**	**187**	265	–	265
Taxation	9	**(83)**	**39**	**(44)**	(69)	15	(54)
Profit/(loss) for the year attributable to equity holders		**236**	**(93)**	**143**	196	15	211

* and amortisation of intangible assets (Note 13)
[1] Restated for the adoption of IFRIC 12 (Notes 1.2 and 37)

	Notes	2010 Pence	2009[1,2] Pence
Basic earnings per ordinary share	10	**21.0**	37.1
Diluted earnings per ordinary share	10	**20.9**	37.0
Dividends per ordinary share proposed for the year	11	**12.7**	12.0

[1] Restated for the adoption of IFRIC 12 (Notes 1.2 and 37)
[2] Per share numbers have been restated for the bonus element of the 2009 rights issue (Note 28.2)
Source: Balfour Beatty (2010)

Box 5.5 Balfour Beatty plc Group statement of comprehensive income

Group statement of comprehensive income for the year ended 31 December 2010

	Notes	**2010 £m**	2008[1] £m
Profit for the year		**143**	211
Other comprehensive income/(expense) for the year			
Currency translation differences		**43**	(77)
Actuarial movements on retirement benefit obligations		**87**	(350)
Fair value revaluations – PPP financial assets		**61**	(81)
– PPP cash-flow hedges		**(67)**	5
– other cash-flow hedges		**(2)**	(2)
– available-for-sale investments in mutual funds	16.1	**4**	–
Changes in fair value of net investment hedges		–	18
Tax relating to components of other comprehensive income		**(25)**	120
Total other comprehensive income/(expense) for the year		**101**	(367)
Total comprehensive income/(expense) for the year attributable to equity holders	28.1	**244**	(156)

[1]Restated for the adoption of IFRIC 12 (Notes 1.2 and 37)
Source: Balfour Beatty (2010)

Current financial reporting practice dictates that certain gains and losses may not be included in the income statement because they are not yet realised in the form of cash. Instead they are shown in the **statement of comprehensive income**. This enables the user of the accounts to have a complete picture of all gains and losses for the company, thereby providing a better summary of performance for the year. Since the financial crisis of 2008 this has been particularly important, as many companies have been recording large losses in these areas as a result of falls in the global stock market.

The most notable examples of unrealised gains and losses are:

- Changes in the fair value of financial instruments
- Changes in the expected obligations in relation to retirement benefits
- Foreign exchange gains and losses

The Balfour Beatty plc Group statement of comprehensive income taken from the 2010 annual report is shown in Box 5.5 as an example.

The cash-flow statement

The cash-flow statement shows how cash has been managed during the period, giving details of the cash received and spent. It acts as a link between the balance sheet and the income statement, showing the capacity of a company to adapt to different circumstances, both internal and external. It shows whether a company has the ability to generate sufficient funds from operations to cover its investment needs, or whether further external funds are needed. Cash is the most vital asset of any company. It is also the asset least susceptible to manipulation. This statement enables a user of the accounts to make a better assessment of how cash has been generated and utilised by the company.

Assessments can be made of:

- How effective the company has been in turning profits into cash
- How much cash has been spent on investment in non-current assets and where the finance has come from – for example borrowings, the issue of new equity shares or profits
- The cash cost of financing the company in terms of interest and dividends paid

The elements of the cash-flow statement

The cash-flow statement is divided into three main sections, which help users to understand how well a company is managing its cash:

- **Net cash flow from operating activities**: this reconciles the operating profit figure from the income statement with operating cash flow through adjusting for non-cash items such as depreciation, changes in working capital and other items that relate to investing or financing activities
- **Cash flows from investing activities**: these include:
  - cash flows related to receipts resulting from ownership of investments or loans to other companies (that is, dividends and interest received)
  - cash flows relating to the sale or purchase of businesses, intangible or tangible non-current assets, and to the sale or purchase of debt instruments of other companies
- **Cash flows from financing activities**: these include:
  - payments to the providers of finance (dividends and interest paid)
  - long-term financing cash flows, such as issue and repurchase of equity shares, issue and redemption of borrowings, and repayment of finance leases

These are added together to show the overall net increase or decrease in cash for the financial period.

The Balfour Beatty plc cash-flow statements taken from the 2010 annual report are shown in Box 5.6 as an example.

The difference between profit and cash

In the income statement, income is included when it is earned, that is, when a sale is made, not when the cash is received. In some sectors – for example, retailing – the cash is received immediately, but in manufacturing and many service industries, payment may be several months later. Similarly, expenditure is included when it is incurred, not when the cash is paid. For example, a purchase of goods would be recorded when the goods are actually received. Payment may be made several months later.

In the same way, expenditure in the income statement is matched with the income it helps to generate. This is why the cost of a non-current asset such as a piece of production equipment is expensed by an annual amount being charged to the income statement that over the life of the asset will add up to the total purchase cost, rather than the full amount being expensed at the purchase date, as the equipment will be used to generate income over the life of the asset.

In addition, at the end of the accounting period, there are frequently bills outstanding that nevertheless relate to costs incurred during the period and which therefore need to be included in expenses for the year. A common example would be an electricity bill which is always received after the electricity has been consumed, but which needs to be included in expenditure for the period. In practice, an estimated amount is included as a cost for the period, which will be corrected when the actual bill is received.

So the income statement can be potentially misleading, as in any financial period there is a mixture of complete transactions (where cash has been exchanged) and incomplete ones (where cash has yet to be exchanged). Although estimates over the short term are unlikely to be far out (for example, an electricity bill can be estimated fairly accurately using meter readings and previous period comparisons), over the long term the figures are likely to become more distorted (for example, estimating the future life of a new piece of equipment is unlikely to be accurate).

This means that it is possible for a company to show very different figures for profit and cash at the end of a financial period. In particular, when companies are expanding, they can show healthy profit figures while experiencing cash shortages, due to the fact that they will be paying out cash to suppliers for the business expansion, while having to wait before they receive cash in from their customers.

Box 5.6 Balfour Beatty cash-flow statements

Statements of cash flows for the year ended 31 December 2010

	Group		
	Notes	2010£m	2009£m
Cash flows from operating activities			
Cash generated from/(used in) operations	35.1	**169**	294
Income taxes paid		**(21)**	(31)
Net cash from/(used in) operating activities		**148**	263
Cash flows from investing activities			
Dividends received from joint ventures and associates		**62**	75
Dividends received from subsidiaries		**–**	–
Interest received		**19**	17
Acquisition of businesses, net of cash and cash equivalents acquired	29.1	**(44)**	(300)
Purchase of intangible assets		**(14)**	(3)
Purchase of property, plant and equipment		**(85)**	(71)
Purchase of other investments		**(13)**	–
Investments in and loans made to joint ventures and associates		**(56)**	(50)
Investments in subsidiaries		**–**	–
Investments in PPP financial assets		**(22)**	(95)
Settlement of financial derivatives		**–**	(57)
Disposal of investments in joint ventures		**24**	–
Disposal of property, plant and equipment		**13**	19
Disposal of other investments		**7**	16
Net cash (used in)/from investing activities		**(109)**	(449)
Cash flows from financing activities			
Proceeds from issue of ordinary shares		**2**	356
Purchase of ordinary shares		**(3)**	(6)
Proceeds from new loans		**49**	121
Proceeds from new finance leases		**4**	–
Repayment of loans		**(30)**	(4)
Repayment of finance leases		**(5)**	(3)
Ordinary dividends paid		**(84)**	(63)
Interest paid		**(31)**	(19)
Preference dividends paid		**(11)**	(11)
Net cash (used in)/from financing activities		**(109)**	371
Net (decrease)/increase in cash and cash equivalents		**(70)**	185
Effects of exchange rate changes		**12**	(30)
Cash and cash equivalents at beginning of year		**608**	453
Cash and cash equivalents at end of year	35.2	**550**	608

Source: Balfour Beatty (2010)

The cash-flow statement, reporting the actual cash inflows and outflows of the company, is therefore seen to provide additional **relevant** and **reliable** information about the company. It is also less open to manipulation than the income statement.

When managers are planning new projects, therefore, it is essential that they consider the cash-flow projections as well as the estimated profits to be earned.

Further reading

Financial statements

Abraham, A, Glynn, J, Murphy, M and Wilkinson, B (2008) External financial reporting (Chapter 13). In: *Accounting for Managers*. 4th edition. Cengage Learning EMEA, UK

IASB and IFRS website: http://www.ifrs.org/Home.htm (accessed 13 September 2012)

Financial analysis

Once financial statements have been prepared, they can then be analysed and interpreted. Evaluation can take a number of forms. The financial statements themselves can be used to generate trends – for example, growth in sales, percentage change in profits, movement in earnings. Providers of finance will monitor debt covenants, whereby companies must remain within certain financial ratio parameters in relation to cash and net current assets, otherwise the provider will call in its lending. A return on the investment can be calculated and compared with the target return. Managers can add to the financial accounts to produce additional internal information comparing actual figures with those budgeted and planned for. However, it is important to acknowledge that financial accounting information can be incomplete or subject to creative accounting techniques such as off balance sheet financing, making it difficult to understand the true financial position of a company. The fact that not all company resources can be converted into financial numbers – for example, the value of a company's reputation can never be adequately reflected on a balance sheet – means that financial statements and their analysis can only ever be a starting point for further investigation.

Activity 5.6 **Annual report**

- Obtain a copy of a listed company's annual report – these are widely available on the Internet – try Googling the company name. Often the annual reports are available under the heading of 'Corporate' or 'Investors'. Alternatively, use PrecisionIR's free annual reports service at: https://www.orderannualreports.com/060/UI/GP/home.aspx?cp_2=P498 (accessed 4 December 2012)
- Look through the annual report, noting the different sections. Which do you find the most interesting, and why? The least interesting, and why? How does the company ensure it is meeting the needs of its users?
- What financial information does the company particularly want you to take note of?
- Would you trust the information provided in a company's annual report? Why/why not?
- Do you think sufficient information is given? What further information would you find helpful? Do you think too much information is given?

Further reading

Financial analysis

Abraham, A, Glynn, J, Murphy, M and Wilkinson, B (2008) Financial statement analysis (Chapter 15). In: *Accounting for Managers*. 4th edition. Cengage Learning EMEA, UK

Holmes, G, Sugden, A and Gee, P (2008) *Interpreting Company Reports*. 10th edition. Financial Times Press

Management accounting for planning, control and decision making

So far we have concentrated on **financial reporting**, where information tends to be highly standardised, and the same financial report must be used by users with many different needs and agendas. Now we move on to look at **management accounting**. Here the accounting information created can be individual to the needs of the specific users. There is much more emphasis on the **context** of the financial information, which in turn leads to more sophisticated analysis of **costs** and **cost behaviour**.

We have seen how financial reporting has become more complex in response to changes in the financial regulatory environment. In the same way, management accounting processes need to be considered within the wider business context:

> 'Accounting can now be seen as a set of practices that affects the type of world we live in, the type of social reality we inhabit, the way in which we understand the choices open to business undertakings and individuals, the way in which we manage and organise activities and processes of diverse types, and the way in which we administer the lives of others and ourselves.' (Hopwood and Miller, 1994, p. 1)

The first decade of the twenty-first century has seen changes in the underlying business context that have affected the way in which management accounting is used within organisations. Scapens *et al.* (2003) state that these include:

- An increased pace of change in the business world
- Shorter product life cycles and competitive advantages
- A requirement for more strategic action by management
- The emergence of new companies, new industries and new business models
- The outsourcing of non-value-added but necessary services
- Increased uncertainty and the explicit recognition of risk
- Novel forms of reward structures
- Increased regulatory activity and altered financial reporting requirements
- More complex business transactions
- Increased focus on customer satisfaction
- New ethics of enterprise governance
- The need to recognise intellectual capital
- Enhancing knowledge management processes

As a specific example of a response to these changes, Rolls-Royce has sought to transform its finance function to give more support to the overall business through:

- Establishing shared finance support operations that seek to simplify and standardise activities
- Continuing to operate corporate centres of excellence such as treasury and tax operations
- Developing business partnering to provide insight and decision support needed by the organisation
- Providing operational support through a business support centre (Perry, 2011)

Commercial managers need to use and contribute to providing management accounting information across a wide range of areas. These can include:

- Strategic planning
- Investment appraisal
- Budget/profit planning
- Financial management
- Communication of financial and operating information
- Financial control

Such a range of areas includes both short-term and long-term decisions. Short-term decisions usually involve decisions around how to make best use of existing resources in terms of controlling costs and generating the maximum profit, whereas long-term decisions will be concerned with new capital investment.

Further reading

Management accounting

Drury, C (2008) An introduction to management accounting (Chapter 1). In *Management and Cost Accounting*. 7th edition. Cengage Learning EMEA, UK

Cost determination and cost behaviour

It is essential for commercial managers to understand how costs are determined and how they behave because costing information feeds into both planning decisions and control procedures.

Planning decisions use estimates of future costs, usually based on experience of historic costs, and include pricing and outsourcing decisions.

Control procedures are in place to maintain and improve the efficiency with which resources are employed within a company by providing timely and accurate feedback on the efficiency and effectiveness of its operations. This involves comparing actual costs of current operations against planned costs. By highlighting discrepancies, current costs can be kept in line with planned costs and the quality (in terms of accuracy) of future planned costs can be enhanced.

A management accounting system for projects will accumulate costs in two forms:

1. In terms of their relationship to a person, item or location. This is known as responsibility accounting and aims to control costs by associating them with individuals in the management hierarchy and making those individuals responsible for their costs
2. In terms of their relationship to a product or service, through a process of direct and indirect allocation. **Direct costs** are those that can be directly related to one cost unit, that is, the individual product or service for which the business wishes to determine a cost. **Indirect costs** are those that cannot be identified directly and which are often referred to as overheads. Examples are rent and general administration costs

Understanding **cost behaviour** is important when planning commercial projects and transactions. Costs can be categorised as:

- **Variable costs**, which are those that vary (usually assumed to be in direct proportion) with changes in the level of activity of the cost centre to which they relate – for example, power-by-the-hour, or hourly labour charges associated with providing a help desk service
- **Fixed costs**, which accrue with the passage of time and are not affected by changes in the activity level – for example, rent. However, are costs ever really fixed? Obviously, fixed costs will only hold for reasonable limits. Outside these limits fixed costs will go up or down. The area inside these limits is called the relevant range, and will hold for the short term. In the long term all costs become variable, for example, more space can be rented or the lease can be surrendered
- **Semi-variable costs (mixed costs)**, which contain both a fixed and a variable element. When activity is nil, some cost is still incurred. The cost then increases like variable costs, as activity increases – for example, telephone charges: the fixed element is the quarterly rental and the variable element is the charge for the calls made. Most service and utility costs are semi-variable

- **Stepped costs** which are fixed over an activity range and then suddenly increase by a single amount. For example, one machine is hired to produce 200 items; if production exceeds 200, a second machine is hired. Most administrative staff costs are stepped costs and as such will affect project costing. Project models will build in the points at which additional staff would need to be employed

Once costs have been determined, they can be used in commercial management for pricing strategies. Usually, the key element is that the contract price must be more than cost (this may not be the case if the project investors wish to enter a new market and need to develop expertise). Since the sales revenue needs to cover all the project costs, this will usually require estimating both capital and revenue costs over a number of years and using financial modelling techniques to determine monthly charges under different scenarios.

One example of pricing policy is to calculate the total project cost and then add a percentage on top (cost-plus-profit pricing). The commercial manager can then check the resulting price to see that it is competitive. This topic is referred to in Chapter 8.

The allocation of indirect costs is all important here as:

- The less direct a cost is, the harder (more arbitrary) any allocation of that cost will be
- The more arbitrary the cost allocation, the more room there is for suboptimal (and therefore potentially incorrect) decision making

The example in Box 5.7 clearly demonstrates how important it is to get the best allocation possible, as this will lead to better decision making by management. Although this is a simple pricing example, the same principles hold for large commercial projects. Since many long-term contracts are awarded based on fixed prices, where the supplier must deliver a project within a certain price or where price increases are restricted to the rate of increase of indices, such as the Consumer Price Index or specific industry indices, it is crucial that cost relationships have been properly understood so that the initial cost build is as accurate as possible. Otherwise cost movement over time may lead to unexpected impacts on contract profits.

Box 5.7 Pricing and indirect cost allocation example

Technoproducts Ltd is a new company making two products: a mobile phone and a tablet computer. Management want to know how best to allocate costs between the two products. Their previous experience has been to allocate indirect costs on the basis of labour cost. However, the new company is using computer-controlled assembly equipment, which has increased indirect costs and reduced labour costs. Analysis shows that the number of product components may be a more appropriate cost driver for allocating indirect costs to the two products.

Relevant figures are:

	Mobile phone Per unit	Tablet computer Per unit
Materials	£20.00	£80.00
Labour	£10.00	£50.00
Number of components	3	30
Planned production	20000 units	4000 units

Indirect costs in total are £500,000. Technoproducts operates a cost-plus-pricing policy of 40% on full cost.

We will calculate prices based on each of the allocation methods and compare our results. Using **labour** as the basis of allocation, the calculation is:

$$\frac{\text{Total indirect costs}}{\text{Total labour cost}} = \text{Indirect cost per labour}$$
$$= \frac{£500000}{(£10 \times 20000) + (£50 \times 4000)}$$
$$= £1.25$$

Allocation is: Mobile phone: £1.25 × £10 = £12.50/phone
Tablet: £1.25 × £50 = £62.50/tablet

The pricing structure is therefore:

	Mobile phone £	Tablet computer £
Material	20.00	80.00
Labour	10.00	50.00
Variable cost	30.00	130.00
Indirect costs	12.50	62.50
Total cost	42.50	192.50
Price @140%	59.50	269.50

Using **number of product components** as the basis of allocation, the calculation is:

$$\frac{\text{Total indirect costs}}{\text{Total number of components}} = \text{Indirect cost per component}$$
$$= \frac{500000}{(3 \times 20000) + (30 \times 4000)}$$
$$= £2.78 / \text{component}$$

Allocation is: Mobile phone: £2.78 × 3 = £8.34/phone
Tablet: £2.78 × 30 = £83.40/tablet

The pricing structure is therefore:

	Mobile phone £	Tablet computer £
Material	20.00	80.00
Labour	10.00	50.00
Variable cost	30.00	130.00
Indirect costs	8.34	83.40
Total cost	38.34	213.40
Price @140%	53.68	298.76

The two different bases of allocation give rise to significantly different prices using the cost plus method.

Further reading

Cost determination and cost behaviour

Abraham, A, Glynn, J, Murphy, M and Wilkinson, B (2008) Cost concepts and measurement (Chapter 4). In: *Accounting for Managers*. 4th edition. Cengage Learning EMEA, UK
Drury C. (2008) A5n introduction to cost terms and concepts (Chapter 2). In: *Management and Cost Accounting*. 7th edition. Cengage Learning EMEA

Cost-volume-profit analysis

A useful way to predict costs at different levels of activity is to use **break-even analysis**. Any decision requires an assessment of the revenues and **relevant** costs resulting under such different alternatives. If we produce a graph showing total cost against volume of activity we can see that, even with a zero level of activity, fixed costs are being incurred. By definition, they are fixed, and will not change with the level of activity (Figure 5.1).

Therefore, in any decision that concerns varying the level of activity, **fixed costs** are **not** a **relevant** cost in the short term. In these circumstances, any product or service that can be sold for more than its variable cost is making a positive **contribution** to the fixed costs of the business. **Contribution** is the difference between the selling price and the **variable** cost of producing and selling that item. This is in contrast to **profit** per unit, which is the difference between the selling price and the cost of producing and selling that item. **Unit contribution** is therefore equal to **selling price** less **total variable cost per unit.**

Contribution analysis is an important tool in management decision making. It frees managers from having to consider fixed costs. They can then make decisions on discounting prices in order to increase contribution alone. For example, a contribution analysis approach is frequently used by the travel industry. Customers who are prepared to book at the last minute can often get a better deal because travel companies will discount from the full price. So long as they cover their variable costs, they are getting a contribution towards fixed costs or profit, and this is better than having empty airline seats or empty hotel rooms.

Break-even point (cost/volume/profit analysis)

The break-even point is the **volume** of sales at which **total contribution** is **equal** to **total fixed costs**. It is also the volume of sales at which neither a profit nor a loss is made (see Box 5.8).

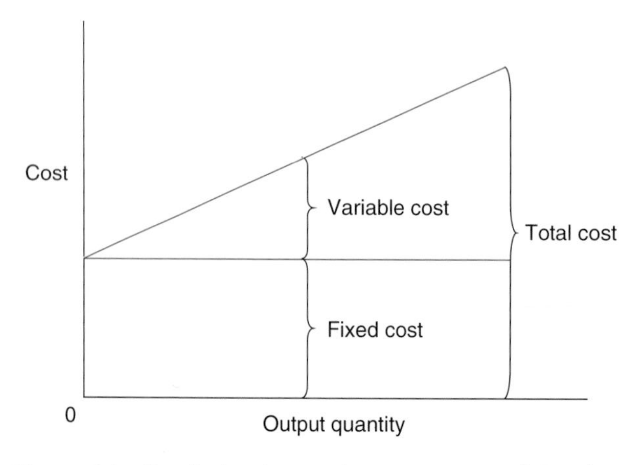

Figure 5.1 Graph showing total cost to output relationship.

Box 5.8 Illustration of cost-volume-profit analysis

Jamie, a restaurant owner, has expected sales of 15 000 meals at £20 each. His variable costs are £8 per meal and fixed costs are £120,000.

To calculate the **break-even point**, we need to calculate the point where **total contribution** is equal to **total fixed costs**.

We can calculate the unit contribution as follows:

	£
Selling price	20
Less variable costs per unit	8
	12

Dividing the fixed costs by the contribution per unit will give us the number of meals needed to cover the fixed costs and therefore break even.

$$\frac{\text{Fixed costs}}{\text{Contribution per unit}} = £120{,}000/12$$
$$= 10\,000$$

Therefore **break-even point** is at a volume of 10 000 meals, with a sales revenue of (10 000 meals × £20) = £200,000.

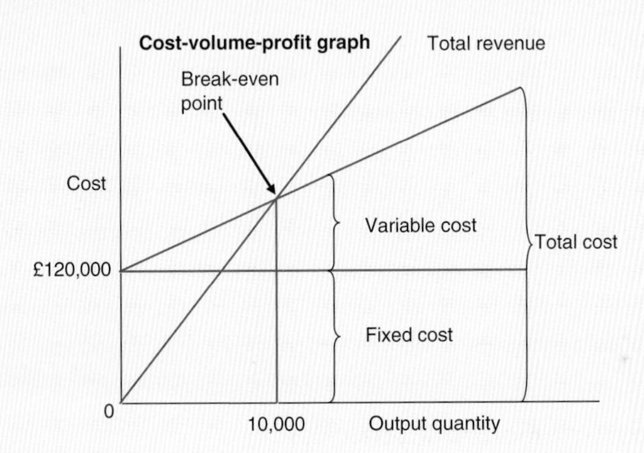

Jamie expects to make sales of 15 000 meals, which will give him a profit of:

	£
Contribution (15 000 meals × £12)	180,000
Less fixed costs	120,000
Profit	60,000

The **margin of safety** is the difference in volume between the break-even point and expected sales. Calculating the margin of safety makes it possible to compare the planned or expected level of activity with the break-even point and so make a judgement concerning the riskiness of the activity. Planning to operate only just above the level of activity necessary in order to break even may indicate that it is a risky venture, since only a small fall from the planned level of activity could lead to a loss.

For Jamie, the margin of safety is:

Actual number of units sold	15 000 meals
Less number of units at break even	10 000 meals
	5000 meals

The margin of safety is high at 50% above the break-even point.

Activity 5.7 Cost-volume-profit analysis

Following a visit by a Health and Safety officer, who requires a number of improvements to be carried out to bring the restaurant up to meet legal requirements, Jamie's costs alter as follows:

Fixed costs £140,000
Variable costs £9 per meal

What price should Jamie charge to make a profit of £55,000 on the sale of 15 000 meals?

Break-even analysis can be useful for short-term decision making. A graphical display can be provided and sensitivity analysis carried out. However, it can only be used in certain (relatively straightforward) situations and relies on a number of assumptions:

- Fixed costs remain fixed throughout the range considered
- Variable costs fluctuate proportionally with volume – that is, a linear relationship
- Efficiency and productivity do not change
- A single product or static mix of products is dealt with
- Volume is the only factor affecting cost
- Costs can be accurately and precisely predicted

Real-life scenarios are rarely this simple, as most companies will have more than one product or service line, and mix will therefore vary. Other limiting factors may come into play.

Cost prediction

In order to make decisions, management want to know how costs change in relation to activity. Actual knowledge is related to past experience of cost behaviour. Management will need to predict future behaviour, and can use a number of different techniques. These include:

- Engineering methods based on direct estimation of required production inputs
- Historical cost methods based on the information already in the accounting system, which is extrapolated to meet the new position
- Statistical methods such as linear regression analysis

Each method will yield different figures, and experience is needed to arrive at an appropriate conclusion. However, understanding cost behaviour will help management make better and more effective decisions. Cost estimating techniques are covered in Chapter 8.

Further reading

Cost-volume-profit analysis

Abraham, A, Glynn, J, Murphy, M and Wilkinson, B (2008) Cost analysis for management decisions (Chapter 5). In: *Accounting for Managers*. 4th edition. Cengage Learning EMEA, UK

Drury, C (2008) Cost-volume-profit analysis (Chapter 8). *Management and Cost Accounting*. 7th edition. Cengage Learning EMEA, UK

Performance measurement

Costs will also feed into the budgetary planning and control system, that is part of the performance measurement system for the business. Budgets help managers plan for the future in a controlled way. A budget can be described as a formal financial and/or quantitative statement, prepared prior to a defined period of time, of the policy to be pursued during that period for the purpose of attaining a given objective. It may include income, expenditure and the employment of capital.

The budget is the short-term projection of the long-term strategy for managing projects and project contracts, and will usually be set in terms of profit targets and a required return on capital. The revenue budget is usually planned first, taking into consideration external factors such as the general and sector business conditions, the trend of market size, probable changes in purchasing power and expected changes in the competitive situation, and internal factors such as changes in promotional policy, human resources and pricing strategies. The expenses, capital and cash budgets can then be prepared, bearing in mind that profit targets must be attainable.

Budgets are statements of expectations and should therefore be used to evaluate actual performance. This is achieved by incorporating them into a system of budgetary control, which uses the budget data to monitor and control progress. Once a budget is operational, actual figures should be collected and compared against the budget to show any variances, with significant amounts, whether positive or negative, being investigated and explained so that timely remedial action can be taken when appropriate. Thus the budgetary process provides a continuous measure of business performance. Although the level of detail in terms of budgets varies from business to business, with some managers only having to report to investors on the key performance indicators of earnings and return on investment, it will still be important to use a system of budgetary control to maintain careful management of project revenues and costs.

Any difference between a budget estimate and an actual result is called a **variance**. It is good practice to investigate and explain any variances that are significant:

- This provides a check on the accuracy of budgets
- It identifies problems at an early stage and so should enable timely remedial action to be taken
- It provides a continuous measure of company, department or individual performance

Investigation and explanation of variances forms part of the process of **management control**. This is a process for motivating and inspiring people to perform organisational activities that will further the organisation's goal. It is also a process for detecting and correcting unintentional performance errors and intentional irregularities.

With fixed budgets, the actual figures arising are compared with the original (budget) estimates. However, the original budget was based on a particular level of activity (for example, sales or output). If the actual activity level differs from that predicted, then some costs will also differ for no other reason than the variance in the expected level of activity.

Thus, any individual cost difference may be made up of two parts:

1. Difference arising purely because of activity change
2. Difference for other reason(s)

We can use a **flexed budget** where the original budget is adjusted to reflect **actual** levels of activity, thereby identifying separately differences arising from changes in activity and differences arising for other reasons (Box 5.9). This is much more useful to management.

Advantages and disadvantages of budgeting

Effective budgeting improves efficiency in that it demands the following **requirements**:

- Careful planning and the provision of information/data for management
- The participation of both managers and other employees

Box 5.9 Flexible budget example

PCS Ltd is a call centre processing insurance claims for large insurance companies. Here is some information about the planned and actual performance for one day, 1 August.

Budget for 1 August

Activity level – process 2000 transactions.
Budgeted fee to the client is £1.00 per transaction
It takes one hour to process 20 transactions

Expenses	£	£
Wages – 100 hours @£6/hour		600
Overheads:		
Telephone	100	
Other	800	
		900
Total cost		1,500

Actual results for 1 August

The number of transactions processed was 1800.
Total revenue earned was £1,950.

Expenses	£	£
Wages – 108 hours		702
Overheads:		
Telephone	96	
Other	690	
		786
Total cost		1,488

Variance analysis on the above figures, comparing actual results to the original budget, gives us the following results. Note that variances are shown as **adverse** (A) if they are **less** than budgeted **income**, or **greater** than budgeted **costs**. They are shown as **favourable** (F) if they are **greater** than budgeted **income**, or **less** than budgeted **costs**.

	Original budget £	Actual results £	Variance £
Sales	2,000	1,950	50A
Wages	(600)	(702)	102A
Telephone	(100)	(96)	4F
Other	(800)	(690)	110F
Profit/(loss)	500	462	38A

If we calculate the variances between the budgeted and the actual figures we don't get very useful information, as some of the expenses are likely to vary with the volume of transactions processed, and the sales revenue and actual volume are quite different to the budget. So it is more useful to draw up a flexible budget for comparison.

The flexible budget takes the **actual volume** of transactions, and applies the **budgeted rates**. Thus we can calculate variances against the result we would have expected if we had originally planned the actual volume of transactions.

	Original budget	Flexed budget	Actual results	Variance
Sales	2000	1800	1950	150 F
Wages	(600)	90 h × £6 (540)	(702)	162A
Telephone	(100)	(100)	(96)	4 F
Other	(800)	(800)	(690)	110 F
Profit/(loss)	500	360	462	102 F

Obviously fixed costs do not change with a flexed budget – they remain the same as the original budget. It is assumed here that expenses for telephone and other are fixed costs.

Now it appears that, although the volume of transactions was down against budget, the actual sales revenue is better than budgeted for that level of transactions. Can this level of revenue be maintained? We can also see that there is a problem on wages that needs to be investigated. Why is this so much higher than the flexed budget?

We can drill down further to calculate both a volume and a price variance for wages.

We know that it should have taken 1800/20 = 90 hours but it actually took 108 hours
So the volume variance is (108 – 90) = 18 hours × £6 = £108 A

It should have cost 108 × £6 = £648, but it actually cost £702.
So the price variance is 702 – 648 = £54A
(we can check these add back to the overall wage variance – 108A + 54A = £162A)

So there are problems on both variances. The volume variance may be because there were abnormal technical problems with some of these transactions. The price variance may be because there was a wage increase that hadn't been budgeted for, or because higher grade staff had to be used than budgeted.

- Coordination and cooperation
- A sound accounting system

Effective budgeting can then deliver the following benefits:

- New trends and inefficiencies are detected at an early stage of the planning and control process
- The delegation of duties/authority
- Control by responsibility
- Management by exception
- A sound evaluation system for comparing/reporting on budgeted and actual results
- The motivation of employees
- Good clear communications
- Corrective action by management to remedy adverse conditions

These benefits can happen only if all those who are involved understand what budgeting is trying to do and are able to express their views during the budget preparation process, and if the budget is flexible enough to take account of changes in circumstances.

However, most companies will experience some difficulties in implementation of the budget process. These include problems related to behavioural issues, as employees may regard the techniques as pressure instruments and be unwilling to cooperate, or conversely may be satisfied with a lower level of achievement that meets the budget target when better performance could have been reasonably attained. Usually the checking of estimates is difficult, and conditions may change once budgets have been agreed. This means that management must take care over the way in which the budgetary control system is instigated and operated to avoid suboptimal problems such as the stifling of opportunities to innovate and be creative, or the wasting of surplus funds at the end of the year so as not to jeopardise the budget requirement for the following period.

Further reading

Budgeting

Abraham, A, Glynn, J, Murphy, M and Wilkinson, B (2008) Budgetary planning and control (Chapter 7). In: *Accounting for Managers*. 4th edition. Cengage Learning EMEA, UK

Drury, C (2008) The budgeting process (Chapter 15). *Management and Cost Accounting*. 7th edition. Cengage Learning EMEA, UK

Cash-flow management

An important element of planning and control is cash-flow management. More businesses go bankrupt through lack of cash than for any other reason, and it is possible to be highly profitable yet still have cash-flow problems, particularly at the start of a long-term project. The cash-flow forecast brings together the elements of capital and revenue income and expenditure to help forecast the cash needed to keep the business solvent. With careful planning, cash borrowing can be forecast and arranged in advance, or else avoided through adopting alternative sales and payment strategies.

Cash-flow management is closely related to working capital control – that is, the management of net current assets – as the amount of cash tied up in trade receivables (money owed to the business by its customers) will affect when cash can be paid out to trade payables (suppliers). This can be a particular problem for a start-up business when there are many calls for cash to be paid out to suppliers for both capital and day-to-day expenditure before any cash is due to come in from customers for products and/or services provided.

Cash-flow management is particularly important for implementing commercial project pricing strategies based on determining monthly charges to customers. Here the way in which the underlying asset is financed will affect the cash requirements of the business. If there is a large cash outflow to pay for capital expenditure at the start of the project, then the cash flow will start from a negative position, and high finance and interest charges will be incurred that will be reflected in a higher monthly price to the customer. Such a price may then not be competitive. Some businesses seek to receive a large cash payment at the start of the project to cover the capital and start-up costs. The ideal strategy will involve a large cash payment at the start of the project to cover the capital and start-up costs, followed by monthly rental to cover ongoing personnel and maintenance costs. Any unrecovered up-front charges will be built into finance charges. This will deliver a positive cash flow, as shown in Box 5.10.

However, with many types of contract – such as PFI contracts where in order to provide the contracted service, the operator has to design, build, finance and operate the project – up-front charges will not be covered directly. In order to cover the capital costs, the business must either borrow the necessary funding, or, more likely, enter into a strategic partnership with a leasing house. Here the leasing house effectively has legal ownership of the assets, with the business paying a rental, including financing charges, over the life of the lease. This in turn means that the business must increase the monthly charge to the customer in order to ensure that investment targets are hit. The way in which prices are agreed will also affect cash flow, as a fixed-price contract must take into account expected changes in costs, whereas a contract incorporating

Box 5.10 Cash management example 1

A new contract has a three-year life. Initial capital costs are £500,000 and ongoing monthly costs (to cover wages, maintenance, utilities and administration) are £30,000 in year 1, £31,000 in year 2 and £32,000 in year 3. These are all payable in the month in which they are incurred. Depreciation charges amount to £14,000 per month (But depreciation is not a cash figure – it is merely the allocation of the initial capital expenditure over the life of the contract – and so it is not included in the cash budget.)

Payment is received one month in arrears. The first payment is £600,000. Payments for the remaining 11 months of year 1 are £35,000, for year 2 £36,000 per month and for year 3 £37,000 per month.

The opening cash balance is £40,000 and the bank is prepared to allow an overdraft up to any limit. Interest is charged at 6% p.a. on any overdraft. (We will show this as a simple charge of 0.5% on the overdraft outstanding at the end of the month, merely to make the point. Obviously, banks will work out compound interest on a daily basis.)

We will prepare a cash budget for the first six months of the contract.

	Month 1	Month 2	Month 3	Month 4	Month 5	Month 6
Inflows						
Cash from sales	–	600,000	35,000	35,000	35,000	35,000
Outflows						
Capital costs	500,000	–	–	–	–	–
Expenses	30,000	30,000	30,000	30,000	30,000	30,000
Total outflows	530,000	30,000	30,000	30,000	30,000	30,000
Net cash inflow in month	(530,000)	570,000	5,000	5,000	5,000	5,000
Cash balance at start of month	40,000	(492,450)	77,550	82,550	87,550	92,550
Cash balance at end of month	(490,000)	77,550	82,550	87,550	92,550	97,550
Interest on overdraft	(2,450)	–	–	–	–	–
Revised balance	(492,450)	77,550	82,550	87,550	92,550	97,550

The cash budget shows a very positive position, with plenty of surplus cash being generated. (Again, to keep it simple, no interest is being given on positive cash balances. In reality, the treasury department must work out the best place to keep surplus cash in order to maximise returns.) This cash will be used to finance other projects.

price changes allows costs and revenues to be matched more closely. Indeed, many companies lease assets solely in order to improve their monthly cash flow position, as demonstrated in Box 5.11. The drawbacks are from a strategic position, in that a leasing company has now become involved in the project, and will take a share of the overall profits.

The choice of exit strategies (i.e. how the assets are treated when the contract ends) will also affect cash flow. For example, the treatment of assets is significant, because of the sums involved. One alternative is to require the customer to purchase the assets at an agreed value, such as net book value at the end of the contract, although the example in Box 5.12 shows that this results in significant cash outflows at the start of the contract. (Net book value is the capital cost of the asset on purchase, less any amounts charged for depreciation. As depreciation is an accounting adjustment, it has no cash effect.) An alternative is to fully

Box 5.11 Cash management example 2

We use the same figures as in the previous example, except that this time the large up-front payment is not available. Instead, the assets are leased on a finance lease and a payment of £18,000 per month is made to cover this. (The accounting treatment of finance leases is complex and outside the scope of this chapter. We consider only the cash flow implications here.) The monthly price goes up to £54,000 in year 1, £55,000 in year 2 and £56,000 in year 3.

Again, we prepare a cash budget for the first six months of the contract.

	Month 1	Month 2	Month 3	Month 4	Month 5	Month 6
Inflows						
Cash from sales	–	54,000	54,000	54,000	54,000	54,000
Outflows						
Finance lease	18,000	18,000	18,000	18,000	18,000	18,000
Expenses	30,000	30,000	30,000	30,000	30,000	30,000
Total outflows	48,000	48,000	48,000	48,000	48,000	48,000
Net cash inflow in month	(48,000)	6,000	6,000	6,000	6,000	6,000
Cash balance at start of month	40,000	(8,040)	(2,050)	3,950	9,950	15,950
Cash balance at end of month	(8,000)	(2,040)	3,950	9,950	15,950	21,950
Interest on overdraft	(40)	(10)	–	–	–	–
Revised balance	(8,040)	(2,050)	3,950	9,950	15,950	21,950

The cash budget here shows that the initial shortfall is quickly eliminated and a steady positive inflow follows.

depreciate the assets over the life of the contract, so that their net book value at the end of the contract is nil. A higher price would be charged to reflect the full use of the assets.

Additional complications arise on large projects where there are a number of services being provided over differing time periods. Ragged-end roll-outs are more difficult to account for than co-terminus roll-outs, where all services will cease at the same time.

Finally, it is important to schedule project start dates carefully so that cash outflows are staggered. Ideally cash flows should complement each other so that cash outflows on a new project are matched by cash inflows from a mature project, thus levelling out cash reserves.

Further reading

Cash flow management

Abraham, A, Glynn, J, Murphy, M and Wilkinson, B (2008) Working capital management (Chapter 6). In: *Accounting for Managers*. 4th edition. Cengage Learning EMEA, UK

Box 5.12 Cash management example 3

We use the figures as before, except this time the customer is required to purchase the assets at the end of the three-year period. It is assumed that the assets have a five-year life and their net book value at the end of the contract is £200,000. The monthly sales figure is £35,000 in year 1, £36,000 in year 2 and £37,000 in year 3. In addition there is an up-front charge of £20,000 to cover start-up costs.

The first six months of the contract show the following:

	Month 1	Month 2	Month 3	Month 4	Month 5	Month 6
Inflows						
Cash from sales	20,000	35,000	35,000	35,000	35,000	35,000
Outflows						
Capital costs	500,000	–	–	–	–	–
Expenses	30,000	30,000	30,000	30,000	30,000	30,000
Total outflows	530,000	30,000	30,000	30,000	30,000	30,000
Net cash inflow in month	(510,000)	5,000	5,000	5,000	5,000	5,000
Cash balance at start of month	40,000	(472,350)	(469,687)	(467,010)	(464,320)	(461,617)
Cash balance at end of month	(470,000)	(467,350)	(464,687)	(462,010)	(459,320)	(456,617)
Interest on overdraft	(2350)	(2,337)	(2,323)	(2310)	(2,297)	(2,283)
Revised balance	(472,350)	(469,687)	(467,010)	(464,320)	(461,617)	(458,900)

And the last six months show the following:

	Month 31	Month 32	Month 33	Month 34	Month 35	Month 36
Inflows						
Cash from sales	37,000	37,000	37,000	37,000	37,000	237,000
Outflows						
Expenses	32,000	32,000	32,000	32,000	32,000	32,000
Net cash inflow in month	5,000	5,000	5,000	5,000	5,000	205,000
Cash balance at start of month	(400,094)*	(397,069)	(394,030)	(390,975)	(387,905)	(384,819)
Cash balance at end of month	(395,094)	(392,069)	(389,030)	(385,975)	(382,905)	(179,819)
Interest on overdraft	(1,975)	(1,961)	(1,945)	(1,930)	(1,914)	(899)
Revised balance	(397,069)	(394,030)	(390,975)	(387,905)	(384,819)	(180,718)

* Following through in the same way from month 6 to month 30 we arrive at an opening balance as shown.

It is clear that this strategy, where the business buys the assets at the start of the contract, selling them to the consumer at the end of the contract, has significant cash-flow consequences and should be avoided. At the very least, it would need to be used in conjunction with leasing.

Capital investment decisions

Commercial managers need to understand how cash-flow forecasting is used in the making of long-term investment decisions. The issue here is how to make a decision in relation to whether or not to invest in a project that:

- Will involve a large outlay of resources
- Will take place over a long period of time
- May involve significant risk for the business if, for example, the conditions on which the cash flows were based change substantially

Making the right decision is therefore important as in many cases the size and nature of these projects mean that the outcome will have an impact on shareholder value. So businesses need to make sure that they are using an appropriate method of investment appraisal.

There are four main methods of investment appraisal:

1. **Accounting rate of return (ARR)**, which calculates the percentage value of the average annual operating profit divided by the average investment over the life of the project, and compares it with a target percentage rate. If the ARR is greater than the target rate, the project should be accepted
2. **Payback period**, which calculates the amount of time it takes to recover, or pay back, the initial investment. If the project pays back within the maximum payback period required by the business, it should be accepted
3. **Net present value (NPV)**, which compares the present value of cash inflows (benefits) with the present value of cash outflows (costs). If a project generates an NPV greater than zero, it should be accepted
4. **Internal rate of return (IRR)**, which calculates the interest (discount) rate whereby NPV equals zero. If a project generates a discount rate higher than the minimum discount rate set by the business, then it should be accepted

Although all of these methods are commonly used in business, there is a major disadvantage with ARR and payback period in that they both ignore the **time value of money**. This is an important concept in financial appraisal because managers are investing at the current time in projects that will generate cash flows at various points in the future. How can we value these future cash flows? The value in a year's time of a pound today is obviously worth more than a pound, as we can invest our pound and earn interest on it.

For example, a company invests £100,000 today at an interest rate of 5% ($r = 0.05$) over the next year. At the end of one year the company receives the initial investment of £100,000 plus the interest of 5% on it, which is £5,000. So it receives:

$$\text{FV} = £100{,}000(1+r) = £105{,}000$$

This is the **future value (FV)** of £100,000. In this example, the future value of £100,000 is £105,000.

Alternatively, we can adopt the opposite approach. For example, if a company wants to earn £105,000 from its investment in a year's time, it needs to know how much money it needs to invest today. This time we are trying to determine today's value – that is the **present value (PV)** – of an amount of money received in the future:

$$\text{PV} = £105{,}000\,/\,(1+0.05)^1 = £100{,}000$$

where the superscript 1 = 1 year's time.

Using this discounted cash-flow technique, which takes into account the time value of money, means that we can compare project cash flows that occur in different time periods. We discount all the cash flows to **present value**. If the present value for all the project cash flows is greater than zero, then the project is worthwhile undertaking (Box 5.13).

Box 5.13 Net present value calculation example

Buenavista plc is considering a project with the following estimated cash flows:

Year	Cash flow £
0	(200,000)
1	40,000
2	60,000
3	80,000
4	80,000

The company's cost of capital (discount rate) is 7%.
We can calculate the NPV of the proposed project as follows:

Year	Cash flow £	Discount factor at 7%		Present value £
0	(200,000)	1/1	1.000	(200,000)
1	40,000	$1/1.07^1$	0.946	37,440
2	60,000	$1/1.07^2$	0.873	52,380
3	80,000	$1/1.07^3$	0.816	65,280
4	80,000	$1/1.07^4$	0.763	61,040
NPV				16,140

The NPV is positive at £16,140 and so the project should be undertaken.

Activity 5.8 Infraco plc

Infraco plc has the opportunity to invest in three different projects, A, B and C, which have the following estimated cash flows:

Year	Project A	Project B	Project C
0	(200,000)	(200,000)	(200,000)
1	70,000	40,000	40,000
2	70,000	50,000	40,000
3	70,000	70,000	40,000
4	70,000	80,000	200,000

Infraco has a cost of capital (discount rate) of 8%.
Infraco plc can invest in only one project. Which should it choose and why?

Calculating the project's **internal rate of return (IRR)**, that is, the discount rate that gives an NPV of zero, will give us the project yield. This needs to be deduced by trial and error, which can be done quickly using a computer model, otherwise it is time-consuming and laborious (see Box 5.14). However, some companies may use IRR to set target rates of returns from investment.

NPV and IRR are both techniques that take the time value of money into account. However, IRR demonstrates a number of pitfalls:

- It ignores the scale of the investment, instead focusing on the percentage return. This means that a project with a small return but a high IRR will always rank above a project with a larger absolute return in PV terms, but a lower IRR
- The IRR rule may disagree with the NPV rule if the project profile shows cash inflows first, followed by cash outflows at a later date
- It is possible for a project to show multiple IRRs if, for example, project cash flows change signs over time

Box 5.14 IRR example

We know from the NPV example (Box 5.13) that Buenavista plc had a positive NPV of £16,140 at a discount rate of 7%.

To calculate an approximation of IRR, we need to do a further calculation of NPV at a higher discount rate, ideally one where NPV is negative. Using a discount rate of 12% gives us a negative NPV as follows:

Year	Cash flow £	Discount factor at 11%		Present value £
0	(200,000)	1/1	1.000	(200,000)
1	40,000	$1/1.12^1$	0.901	36,040
2	60,000	$1/1.12^2$	0.812	48,720
3	80,000	$1/1.12^3$	0.731	58,480
4	80,000	$1/1.12^4$	0.659	52,720
NPV				(4,040)

We can then interpolate between the two answers to give us an approximate IRR. We know that the IRR is between 7% and 12%.

Using the formula

$$\text{IRR} = \text{Lower \% rate} + \left(\frac{\text{NPV at lower rate}}{\text{Range between NPVs}}\right) \times (\text{Higher \% rate} - \text{Lower \% rate})$$

We can calculate IRR for the project as being

$$\begin{aligned}\text{IRR} &= 7\% + \left(\frac{16{,}140}{16{,}140 + 4{,}040}\right) \times (11\% - 7\%)\\ &= 7\% + (0.8 \times 4\%)\\ &= 7\% + 3.2\%\\ &= 10.2\%\end{aligned}$$

Consequently, NPV is the only investment appraisal measure that ensures the **maximisation of shareholder value**. It is important to recognise that the absolute level of costs associated with the project is not necessarily relevant since shareholder wealth will be increased, providing any costs can be passed onto willing customers, thus ensuring a positive cash flow. Irrespective of the absolute level of cost, shareholder interest is best served by projects that maximise the gap between costs to the company and price charged to customers.

In the context of commercial management, NPV is the technique usually used for the preparation of business cases. So, for example, in the case of PFI projects, there has been a focus on the use of NPV since the early 1990s (Shaoul *et al.*, 2005). However, although the theory demonstrates the superiority of NPV, in practice many companies continue to use the other three methods (Drury and Dugdale, 1996), partly because other suboptimal behaviour is taking place (Berkovitch, 2004).

A further consideration for companies undertaking capital investment is how to finance that investment. The two usual sources of finance for capital investment are equity capital (including reserves) and long-term debt. Companies need to consider the cost of each of these types of capital. The cost of equity capital takes into account the company's dividend policy together with the return required by investors. This may mean that debt finance is cheaper. However, although there are advantages for a company in having some of its activities financed by debt capital, an excessive level of debt can be seen as risky and the cost of taking out additional debt will increase.

Further reading

Capital investment decisions

Abraham, A, Glynn, J, Murphy, M and Wilkinson, B (2008) Capital budgeting (Chapter 9). Capital structure and dividend decisions (Chapter 11). *Accounting for Managers*. 4th edition. Cengage Learning EMEA, UK

Drury C. (2008) Capital investment decisions: appraisal methods (Chapter 13). *Management and Cost Accounting*. 7th edition. Cengage Learning EMEA, UK

Conclusion

Commercial managers need to have a working understanding of financial reporting in order to ensure that projects are being properly run. The combination of financial and management accounting reports described above enable day-to-day financial monitoring and a more long-term financial review of projects to be carried out.

Furthermore, management accounting information can help commercial managers to plan and control costs, and make informed business decisions. Unlike financial accounting, which collects historical accounting data in order to prepare official reports, management accounting information looks both to the future and to the past. Income and costs can be estimated and put together to produce budgets, whether revenue or capital, as part of an overall strategy for achieving business targets, frequently expressed in financial terms such as a percentage return on investment or a percentage profit on sale. Cash-flow forecasts can be modelled to assist with estimating borrowing requirements, setting pricing strategies, or for using in NPV projections. Performance evaluation can be carried out by collecting actual costs and comparing them to budgeted figures to determine areas of good and poor performance, and to identify where and when management action should be taken.

Good quality financial information, and some understanding of how it is produced, enables commercial managers to work in conjunction with other business functions to make effective decisions.

Box 5.15 Suggested solutions for activities

Activity 5.1: First TV plc

- Accounting information that would be relevant:
  - published accounts information (including the balance sheet, income account and cash-flow statement) about the two companies, and financial analysis of this information (these companies are fictitious, but data about companies would be obtainable from electronic databases such as Fame (UK companies), Orbis or Thomson One Banker
  - if they are listed companies, information about the share prices and recent movement of the share price
  - information about any borrowings that the companies may have, including details of how non-current assets are financed
  - any financial statistics available about the industry
- Other information that would be relevant (this list is not exhaustive and you may have thought of additional items):
  - sales and marketing information about the companies and the regions in which they operate
  - technical information about the systems and networks that each company has
  - human resources information

Activity 5.2: Sue Peel

- Those likely to use Sue Peel's accounts include:
  - the 'taxman' to assess her liability to income tax
  - banks to assess whether Sue's business is a good credit risk
  - other existing and prospective creditors (for example, leasing companies) wishing to know how good their debts are and whether to make further advances
  - Sue herself should find the accounts of use in determining how well she is running her business and how to improve her efficiency
  - any further investors (or partners) in Sue's business
- A balance sheet is normally prepared to show the balances on the accounts in the books at the year end. However, the only good reason for preparing a balance sheet (apart from the legal requirement if she is trading as a limited company!) is that it will be used. We will therefore consider the users of the accounts as identified above:
  - the **Inland Revenue** will calculate income and other taxes based on the self-assessment form that Sue or her accountant will complete
  - the **bank** will have a working knowledge of how Sue manages her bank account but if it is considering lending to Sue it will want to know:
    - what proportion of the business finance is provided by the bank and what proportion is provided by Sue?
    - whether she is getting a good return on the capital she employs – in particular is it enough to cover interest on a further loan?
    - how liquid is Sue's business? Sue may have large creditor balances that could suddenly fall to be payable: if her assets are long term (non-current assets and other long-term investments) she could find herself in difficulties

    - how well Sue manages the financial side of her business. Does she collect her debts swiftly and in full? Is she making the best of the credit available to her from suppliers? Is the level of any inventory (stock) uneconomically high?
- **Prospective creditors** will be concerned that the business is sound and will use the balance sheet in the same way as the bank.
  - by analysing the information in the balance sheet **Sue** should be able to detect weaknesses in her financial control and endeavour to correct them. In particular she should consider:
    - average credit given and taken
    - stock levels
    - return on capital employed
- **Future investors and partners** will be concerned, as other creditors, with the financial health of the business and may use the balance sheet to evaluate that health as noted above. They will also be concerned with the amount invested by Sue in the business
- **'I'm only interested in how much cash there is in the bank'**
  Whilst cash availability is important for a business, there is more to cash management than just having a positive cash balance. Consider the following:
    - for a business to expand, a principal requirement is the availability of cash. Money in the bank is one source of cash, but sometimes taking out a loan can be sensible as well (For example, in our personal lives most of us will take out a mortgage as it is the only way in which we can finance the purchase of a house)
    - If Sue is generating large cash surpluses she can look for other opportunities that will deliver a higher return than simply leaving the money in a bank account
    - If there is insufficient money in the bank to pay creditors when debts fall due, then it is not a good indicator of financial health. Cash flows need to be managed to ensure that cash is available at the right time

Activity 5.3: Relevance and reliability

One example would be where a manager needs to decide whether or not to accept an offer for specialist equipment owned by the business. The manager would need to check whether the price being offered is reasonable, so would need to work out what the equipment is worth in its current used state. It would therefore be highly relevant to the decision to know what the current value of the equipment is, but any such value established is not likely to be very reliable, as the equipment is specialist and therefore it would be difficult to find information about second-hand market values.

Activity 5.4: Financial and management accounting

A company may decide not to produce the information because it considers the cost of doing so outweighs the potential benefit of having the information.

The distinction between financial and management accounting suggests that there are differences between the information needs of external users and internal users (managers). Although differences undoubtedly exist, there is also a great deal of overlap between the needs of external users and managers. For example, managers will at times be interested in receiving an historic overview of business operations. Equally, external users would be interested in receiving information relating to the future, such as the forecast level of profits, and non-financial information, such as the state of the order book, product innovations, marketing data, and so on. Commercial sensitivity obviously restricts what information can be made publicly available, and forecast information cannot be sensibly audited. These factors limit the usefulness of published information.

Activity 5.5: Wealth creation

Some reasons are:

- Risk. A company can be particularly profitable one year by taking large risks in relation to new business – for example, choosing not to have expensive quality control mechanisms, or entering into contracts where there is a risk that the customer may default on payment. This may make the business profitable, but it could lead to disaster sooner or later.
- Short-termism. Concentrating on the short term and ignoring the long term brings short-term results at the expense of the long-term performance and reputation of the company. For example, cutting out spending on things that are likely to have pay-off in the long term, such as research and development and training, will have immediate profitability benefits, at the cost of longer term ones. The UK stock market is often criticised for its short-term approach.
- Increased size. Expanding the business, through increased investment, could lead to higher profit in the long term, but there will be a time lag in achieving this.

Activity 5.6: Annual report

What you find interesting is dependent both on the company chosen and what you as an individual are interested in. However, research shows that most people don't read much of the annual report, and the financial information (not surprisingly!) is least popular. How well companies take account of their users' needs is very variable. Some companies pride themselves on using easy to understand language and visuals, others will just supply the minimum information required to comply with the regulations.

Usually there is a section of financial highlights at the start of the report, which we might assume have been chosen particularly to place the company in the best light possible.

A company's annual report is made up of three main sections: a report on the business, a governance report and the financial statements. Parts of the governance report and the whole of the financial statements are audited by an independent auditor, and the remainder of the annual report is checked to see that it is consistent with the financial statements. The auditor has to follow professional standards in carrying out their work, and therefore the financial statements should be reliable (although you only need to look at the number of professional investigations against auditors to have some doubts!). Some parts of the review material at the front of an annual report may be biased towards the positive, and, as it will frequently be commenting on the future, it cannot yet be checked for accuracy.

The extent to which you trust the information depends on your level of cynicism – but a healthy 'pinch of salt' is always useful!

It's a difficult balance to provide the right amount of detail to allow a proper understanding without giving information overload. The complexity of many of the issues requires a lot of explanation and technical detail. One problem is that financial reporting is moving towards international convergence, requiring standardisation with the US position, which is largely rules-based and very lengthy. Much more explanation of the business position is now being given, which put the financial statements into more of a context, although the notes to the accounts remain pretty impenetrable to most people. Issues such as accounting for financial instruments are very complex and require reams of disclosure.

In terms of further information that might be helpful, the annual report provides no information on share price or share price movement. Industry sector statistics might be useful as well.

Activity 5.7: Cost–volume–profit analysis

We know the profit required, so we need to work backwards from the profit figure as follows:

Profit required = £55,000
Therefore contribution required = profit + fixed costs
= £55,000 + £140,000
= £195,000
Contribution per meal = total contribution/number of meals
= £195,000/15,000
= £13
Price per meal = contribution + variable costs
= £13 + 9
= £22

Activity 5.8: Infraco plc

We can calculate the NPV of each project as follows:

Year	Discount factor 8%		Project A £	Project B £	Project C £
0	1/1	1.000	(200,000)	(200,000)	(200,000)
1	$1/1.08^1$	0.926	64,820	37,040	37,040
2	$1/1.08^2$	0.857	59,990	42,850	34,280
3	$1/1.08^3$	0.794	55,580	55,580	31,760
4	$1/1.08^4$	0.735	51,450	58,800	147,000
NPV			**31,840**	**(5,730)**	**50,080**

Project B has a negative NPV so should not be chosen. Project C has a higher NPV than Project A so should be the first choice for Infraco plc. However, the cash inflow profiles between the two projects are very different. Project A is offering a steady cash inflow each year, whereas for Project C there is a large cash inflow in the final year of the project. Managers would need to consider carefully how accurate these projected cash flows are likely to be, and also whether the company's cost of capital is likely to change, as any increase would make Project C less attractive.

References

Abraham, A, Glynn, J, Murphy, M and Wilkinson, B (2008) *Accounting for Managers*. 4th edition. Cengage Learning EMEA, UK

Accounting Standards Board (1999) *Statement of Principles for Financial Reporting*. Financial Reporting Council Ltd, London

Balfour Beatty (2010) Balfour Beatty Report and Accounts 2010, Balfour Beatty, London. http://annualreport10.balfourbeatty.com/ (accessed 1 June 2012)

Berkovitch, E (2004) Why the NPV criterion does not maximise NPV. *The Review of Financial Studies*, 117, 239–255

Burns, JE, Ezzamel, M, Scapens, RW and Baldvinsdottir, G (2003) *The Future Direction of UK Management Accounting Practice*. Elsevier/CIMA, London

Drury, C (2008) *Management and Cost Accounting*. 7th edition. Cengage Learning EMEA, UK

Drury, C and Dugdale, D (1996) Surveys of management accounting practice. In: C Drury (ed.) *Management Accounting Handbook*. 2nd edition. Butterworth-Heinemann Ltd, London

Friedman, M (1970) The Social Responsibility of Business is to Increase its Profits. *The New York Times Magazine*, September 13

Holmes, G, Sugden, A and Gee, P (2008) *Interpreting Company Reports*. 10th edition. Financial Times Press

Hopwood, AG and Miller, P (1994) *Accounting as Social and Institutional Practice*. Cambridge University Press, Cambridge

International Accounting Standards Board (2008) *Exposure Draft of an Improved Conceptual Framework for Financial Reporting: Chapter 1: The Objective of Financial Reporting, Chapter 2: Qualitative Characteristics and Constraints of Decision-useful Financial Reporting Information*. IASCF, London

Perry, M (2011) Future focus. *Accounting and Business UK*, 1, 28–29

Shaoul, J, Stafford, A and Stapleton, P (2005) *Capital Appraisal and Evaluation in Trunk Roads and Motorways*. Report to British Accounting Association Public Sector Special Interest Group

VINCI (2011) *2011 Annual Report*. VINCI, Paris, France. http://www.vinci.com/vinci.nsf/en/finance-documentation-annual-reports/pages/2011.htm#top (accessed 1 June 2012)

Vodafone plc (2010) *Annual Report for the Year Ending 31 March 2010*. Vodafone plc, Newbury, Berkshire. http://www.vodafone.com/content/annualreport/annual_report10/governance/index.html (accessed 1 June 2012)

Chapter 6

Legal Issues in Contracting[1]

David Lowe and Edward Davies

Learning outcomes

When you have completed this chapter you should be aware of the main legal and regulatory issues that influence commercial and contract management practice. Specifically, you should be able to:

- Identify elements of a valid contract
- Differentiate between alternative contract strategies and types
- Describe the key components of business-to-business contracts for the supply of goods and services
- Evaluate alternative dispute resolution frameworks
- Identify relevant procurement legislation and competition (antitrust) laws
- Demonstrate an awareness of the regulations that govern international trade and protect intellectual property

Introduction

Commercial practitioners become involved in a wide range of legal issues, for example:

- Briefing, liaising with and managing internal and external legal advisers
- Drafting, negotiating and administering contracts, licensing and confidentiality agreements, etc.
- Ensuring compliance with organisational legal requirements
- Interpreting legislation, regulatory frameworks and rights and obligations under contracts and agreements
- Appraising contracting frameworks

These activities require the development of legal awareness, including an understanding of the principles of contract law, regulatory frameworks and applicable legislation.

The aim of this chapter is to develop an appreciation of some of the key legal issues that influence commercial practice. The intention is to provide a selective overview of contract provisions and procedures, together with an international perspective on the legalisation and regulations that affect business-to-business (b2b) exchanges within a project and programmes environment. The chapter presents an outline of commercial contracts for the supply of goods and/or services in terms of generic contract provisions: the key components of contracts. To illustrate these principles, reference is made, where appropriate, to a number of industry standard form contracts.[2] It also provides an introduction to procurement/acquisition regulations, competition and antitrust legislation, international trade law, intellectual property and freedom of information, insolvency and employment rights on the transfer of an undertaking (TUPE).

Commercial Management: theory and practice, First Edition. David Lowe.

Contractual issues

Winch (2006) asks the question: how are project coalitions governed, or put another way, how are all the multiple transactions that occur over the life of a project coordinated and managed in order to satisfy the client's needs? Adopting a transactional cost economics (TCE) approach, Winch views project governance in terms of a 'nexus of treaties': a connected group or succession of formal agreements. Within his framework, commercial management is seen as the function responsible for selecting, formulating and maintaining appropriate governance mechanisms, necessitating the management of contracts.

Contract management has been defined as:

> '...the process which ensures that both parties to a contract fully meet their respective obligations as efficiently and effectively as possible, in order to deliver the business and operational objectives required from the contract and in particular to provide value for money.' (HM Treasury, 2005)

This process commences with the identification of the purchaser's needs and concludes with the completion of the contract. Further, the process has two dominant characteristics:

- **Risk** identification, apportionment and management – related to contract performance
- **Relationship** management – between the purchaser and supplier

The procedural aspects of contract management are described in more detail in Chapter 7.

A prudent commercial manager and contract management team will need a thorough understanding of:

- Procurement process and post-tender (bid) negotiation (and any legislation that governs these activities)
- Assumptions made by the purchaser and the supplier
- The parties' expectations of the relationship
- Contract terms and conditions; for example:
  - purchaser's duties and responsibilities under the contract
  - supplier's obligations under the contract
  - main cost determinants, how they relate to the outputs and quality standards, and how they will be measured
  - certification and payment mechanism
  - purchaser's and supplier's rights if things go wrong
- Legal implications of the contract for which they are responsible (CUP, 1997)

In addition, with the increasing globalisation of business relationships, commercial practitioners should be aware of the regulations that govern international trade and protect intellectual property.

There is a view that, to facilitate successful projects, contracts should be 'left in the drawer'. Latham (1994) is sympathetic to this view. He considers that the function of a contact is to serve the contract process, not vice versa. However, Hughes and Maeda (2003) contend that this view is incorrect; planning for future events in the contract process could be very problematical without knowledge of the contract. For example, there may be a requirement for one party to notify the existence of particular problems within a fixed time frame, or (if no notice is given) that party may lose the right to claim extra payment for dealing with the problem. Also, from a project management perspective, it is often better to anticipate what is to be done in the event of a problem occurring by setting out a procedure in writing, especially where the problem is a common one.

As discussed in Chapter 1, Macneil (2000a) has adopted the term 'essential contract theory' which is composed of:

> '... patterns and norms that supply a checklist for isolating all elements of the enveloping relations that might affect any transaction significantly. They supply a framework both for understanding those relations and for analyzing them.' (Macneil, 2000b)

Macneil has identified ten common contract behavioural patterns and norms, which he asserts exist in all contracts:

1. Role integrity (requiring consistency, involving internal conflict and being inherently complex)
2. Reciprocity (the principle of receiving something in return for something given)
3. Implementation of planning
4. Effectuation of consent
5. Flexibility
6. Contractual solidarity
7. The restitution, reliance, and expectation interests (which he considers to be the 'linking norms')
8. Creation and restraint of power (the 'power norm')
9. Propriety of means (the mechanisms through which relationships are maintained; these may include both formal and informal processes, and customary behaviour)
10. Harmonization with the social matrix (Macneil, 2000a; Ivens and Blois, 2004)

The power norm is considered to have the most influence on exchange relationships.

Definitions

Goods and services necessary for the completion of projects are procured through the use of contracts with suppliers. Those suppliers use separate contracts to procure goods and services from subcontractors. Although subcontracts are contracts in their own right, they are generally referred to as subcontracts (denoting a relationship to the principal contract with the purchaser).

Contracts

A **contract** is an agreement between two parties under which one party promises to do something for the other in return for a consideration, usually a payment. This places obligations on both parties to fulfil their part of the agreement. It is also the foundation for the relationship between the parties.

A valid contract usually needs the following ingredients:

- An agreement between the parties that is intended to be legally binding (this can be implied in certain circumstances)
- Both parties must have the capacity to enter into the contract
- The contract must be for a lawful purpose; in some legal systems the purpose of the contract has to be properly defined to demonstrate this purpose

The parties to a contract are:

- **The purchaser**: the party that acquires or obtains goods or services by payment or at some cost. Alternatively referred to as the buyer, client, customer, employer, owner, proposer, sponsor, user, etc.
- **The supplier**: the provider of goods and services. Also referred to as the contractor, main supplier, main contractor, prime contractor, prime supplier, seller, vendor, etc.

This text will refer to the purchaser and supplier unless reference is made to a specific form of contract. Although most b2b contracts are formed between two entities, some standard forms permit a multiparty approach; see for example, the Association of Consulting Architects Ltd (ACA) Project Partnering Contracts PPC 2000 and SPC 2000 (ACA, 2003a, 2003b).

Although not a party to the contract, contracts may also refer to:

- **A project manager**: the person who leads the purchaser's contract management team. This person may be an employee of the purchaser or an external professional organisation. The terms architect, contract

Box 6.1 Understanding a contract

It is possible to let a construction contract in a very simple way. It used to be said that all that was required was scope, time and price. As long as these three elements were agreed and there was an intention between the parties to be contractually bound then a construction contract would come into force. Any matters not specifically dealt with could either be identified using 'implied terms' or by the application of the general law.

The position now is even simpler under English Law. If a start and finish time are not agreed, then (under the general law) 'reasonable' times will be implied. If no price is agreed, then contracts subject to the Housing Grants, Construction and Regeneration Act 1996 will import a statutory scheme (Scheme for Construction Contracts (England and Wales) Regulations 1998), which provides a full set of payment terms. It could, therefore, be argued that for the majority of projects the parties only need to agree the scope of work and the law will do the rest.

It is therefore possible (as a matter of law) to use a short written contract between the parties which sets out a scope of work, a price (perhaps with payment milestones) and some outline dates. The rest of the arrangements would then be left to discussions between the parties as the project proceeds. In a commercial environment it is possible that these discussions would be guided by the 'wisdom' which has been received from the use of standard form contracts with which both parties are familiar.

There are many construction projects (particularly, for example, between small companies and jobbing builders) that are already implemented in this way. There are several reasons why this approach may be preferred:

- There is an existing relationship between the parties
- There is a possibility of future work if all goes well
- The amounts involved are not large enough to justify the effort of compiling (and/or agreeing) a comprehensive contract

When these contracts are the subject of major disputes, the task of resolving these disputes can be more uncertain and more complicated than if a comprehensive contract had been entered into. This is because the first task will be to understand the arrangements entered into, not just in terms of what has been expressly agreed – the task of 'filling in the gaps' can itself create further disputes as to how the rights and liabilities of the parties should be distributed.

On the other hand, many suppliers will say that they have worked perfectly well without any formal contracts for years and would regard the introduction of such a document as showing a lack of trust on the part of the purchaser. The JCT House Owners Contract goes some way to addressing these concerns by being short, easy to read and written, to the extent that is possible, in 'plain English'.

On larger projects, however, the lack of a comprehensive contract will create an immediate need for a way of organising, on a day-to-day basis, the arrangements between the parties. For example: What are the invoicing and payment arrangements? Who is authorised to order variations (changes)? What is the role of the purchaser's project manager? Again the parties may look to a standard form contract which they are accustomed to using as a reference point for their procedures. This is particularly the case in relation to the UK building industry where JCT contracts (and increasingly NEC contracts) are regarded by many as having almost a statutory basis. However, for complex contracts this approach is not recommended. The prime reason for this is not any legal reason as such. The law can always (eventually) fill in the gaps assuming there is certainty of scope. However, the lack of pre-agreed management procedures and powers can be a significant disadvantage, as their use is of primary importance in delivering a successful project.

manager, or engineer may be used, while the NEC3 Supply Contract refers to a supply manager. This text will refer to the project manager

- **Subcontractors**: suppliers to the main supplier, main contractor, prime contractor, prime supplier, etc.

Caution should always be exercised where the descriptions above (or similar ones) are used. There are no 'official' definitions (beyond those set out in individual contracts) and there is a range of possible meanings associated with each.

Although contracts can take the form of a single document, generally commercial contracts comprise several documents (or distinct elements within documents). For example:

- **The contract agreement**: this itemises the documents that comprise the contract. It includes the identities of the parties and identifies other documents that set out the scope of work, the contract price and the schedule for its execution
- **General specification and scope of work, product or service**: these describe the scope of work to be undertaken (or product/service to be supplied) and the technical standards required
- **General conditions of contract**: often a recognised standard form of contract. This will detail the obligations to produce and to pay, allocate risks, describe the consequences of failure to pay or produce and include relevant commercial arrangements, relating for example to insurances, bonds, safety procedures, industrial relations, defects and disputes, etc. It will also set out and the administrative procedures to manage the implementation of the project
- **Special conditions of contract**: these cover additions and amendments to the general conditions as required by the purchaser or the specific circumstances of the project

In practice, however, these documents will be interlinked, with some having greater importance than others. Due to the potential for conflicting provisions within these documents an order of priority needs to be established.

Letters of intent

Occasionally, letters of intent are used as an interim arrangement to permit a successful bidder to start work in advance of signing a formal contract. This may be necessary because there are items of work that can be progressed without too much financial exposure on the part of the purchaser. Alternatively, it may be that some products or components (for example, steel) need to be ordered a long time before they are needed. A letter of intent of this type should have the properties of a contract and it should be clear as to which elements of the project have been agreed and which have not. Ideally, a letter of intent of this type should state that the purchaser:

- Intends to place a contract with the supplier
- Wishes the supplier to begin work in advance of the contract being signed
- Authorises the supplier to begin work in advance of the contract being signed (setting out any limits on that authority)
- Agrees to pay the supplier for the work carried out in the event that no contract is entered into (with a procedure setting out how and when such payment will be calculated and made)
- Will subsume the work and any payment made into the formal contract, once signed

A risk for a purchaser going down this route is the potential loss of 'bargaining position' in relation to any terms not yet agreed. It is much harder to negotiate with an incumbent supplier than with a bidder. An alternative solution would be to use a **'start-up' contract**, incorporating appropriate controls, limits and safeguards. The JCT has published a specific agreement for 'pre-contract services'(JCT, 2011).

A letter of intent may not be legally enforceable and could be revoked by the issuer without any redress if it is merely an announcement of one party's present intentions.

Labels and descriptions

Care must be taken not to assume that a document with a particular title (for example, 'contract', 'order' or 'letter of intent') has a particular effect associated with that title. There are very few labels of this type that have an official legal meaning, albeit they may have a particular meaning for the parties. The courts are more inclined to read the document to find out what it says. For example, there is no legal distinction between the terms 'contract' and 'order'. Both refer to legally binding agreements for the supply of goods and/or services in return for some form of remuneration. Commercially, however, the term 'contract' is usually used in relation to an agreement involving a longer time period and a greater outlay than an order.

Subcontracts

In parallel to the contract between the purchaser and the supplier (the main contract), the supplier employs subcontractors and suppliers of materials, plant, equipment and services, etc. The supplier is generally held responsible for any subcontracted work and the purchaser sometimes retains the right to vet subcontractors and to limit the extent to which the work is subcontracted. The supplier may be free to choose the terms of subcontracts or alternatively the terms may (to a greater or lesser extent) be required to correspond, or to be '**back-to-back**', with those of the main contract.

Purchasers may also reserve the right to nominate subcontractors, requiring the supplier to enter into a subcontract with a subcontractor chosen by the purchaser or the project/commercial manager, usually to carry out specialist work. Contracts need to deal with the particular risks introduced by nomination – for example, in relation to the default or non-availability of the nominated subcontractor – and to ensure that the supplier pays the nominated subcontractor.

Standard and model forms

Many industry sectors use standard or model conditions of contract as opposed to bespoke contracts. Wright (1994) contends that standard forms are used because they:

- Provide a recognised and predictable contractual basis
- Save time, both in writing and in negotiating the contract
- Are familiar to the project/contract management teams, resulting in smoother running projects, or at least in the avoidance of some misunderstandings that could disrupt progress

Some organisations have their own standard conditions of contract. Invariably, these contracts are (at least to some extent) biased in favour of the party that composed them. This bias might manifest itself both in terms of risk allocation and procedural convenience. Alternatively, model conditions are used. Model conditions tend to be drawn up by an industry association including representatives of all parts of an industry and are, therefore, generally held, according to Wright (amongst others) to represent a reasonable basis upon which organisations within that industry might be prepared to do business with each other – as risks are (in theory at least) equitably distributed between the parties. However, they do have some disadvantages – for example, a tendency to be complex, inflexible, unclear and unwieldy. The 'committee' approach results in compromises. These disadvantages are being addressed and many improvements have been made in the last few years to most standard forms.

Legal interpretation of contracts

Contracts operate within and are subject to the framework of the applicable law. Legal systems vary throughout the world but many have common 'roots' such as:

- The English 'common law'
- Code-based systems originally emanating from mainland Europe
- Religious laws such as the Shariah

As a broad assertion, however, the general principles of commercial contract law are much the same the world over, but the detail will vary. Specific legal advice should be sought on the implications of the interpretation of contract clauses due to a particular legal system. For an overview of English, Finnish, French and German contract law, and Islamic and Roman law see Jaeger and Hök (2010).

International contracts need to state which country's law or jurisdiction will apply. The choice of law can have a significant effect. The law determines how (and whether) a contract is formed and provides guidance on how it should be interpreted in the case of ambiguity or gaps. It also provides regulations within which the parties must function – for example, licensing of certain activities, safety, tax, etc. These can vary considerably. For example, in some jurisdictions, the levying of interest on late payments is forbidden and interest provisions in the contract will be ineffective. In other jurisdictions, interest is compulsory, and will be implied by operation of law if not provided for in the contract. The area of the law that exhibits the most and widest variety relates to where and how disputes are resolved. Where a choice of legal systems exists, it is important to contemplate the consequences of that choice before entering into the contract.

Where the contract is produced in more than one language, the contract needs to specify the ruling language and the language for communication purposes. This is important because, in the event of a dispute occurring, the exact words used in the contract will be analysed in order to determine the terms of the agreement made (or deemed to have been made). Language is also important for day-to-day operations and more than one language may be used. For example, an international contract written in (and being managed in) English may provide that certain communications (such as safety instructions) should also use the languages spoken by the people working on (or visiting) the site.

Exercise 6.1 Produce a contract in less than 100 words

Task: produce a contract for the repair of a fence in less than 100 words

The background proposition is that you are building a large facility, the operation of which is conditional on having a good-quality fence because equipment in the facility could be dangerous to unaccompanied visitors.

Solution

A rather short solution might say something like:

'Restore the fence to its original standard commencing on 1 September and completing by 30 November of this year for £20,000 payable on completion'.

The contract is then tested through a number of scenarios. For example, what happens when:

- The supplier does not attend site for the first six weeks
- The supplier does attend site but cannot start work because working areas are blocked by equipment belonging to the purchaser
- The fence is destroyed by a storm when partially built.

It turns out that the most regularly useful parts of construction contracts are actually the management powers and procedures for such situations. People do need rules and boundaries (try playing tennis without them) so that they can, both in the literal and metaphorical sense, 'play the game'.

Contract terms

The terms of a contract are all the rights and obligations (commitments) agreed between the parties, together with any terms implied by law.

Express and implied terms, incorporation by reference

- **Express terms**: are those terms that are stated (written) within a contract
- **Implied terms**: are, for example within English law, those terms that form part of a contract but are not expressed. Implied terms can include:
  - conformity with statutes
  - supplier's responsibility for subcontractors
  - duty to use reasonable skill and care in design work
  - supplier's obligation to execute the work at a reasonable rate of progress having regard to the completion date
- **Incorporation by reference**: for example, where the contract makes reference to terms contained within other documents

Commercial contracts

Complexity of contracts

Commercial contracts are generally complex documents. For example, construction projects generally require contracts to set out the legal, financial and technical arrangements for the project. As a result, one potential source of risk is the contract document itself. The contract conditions, according to Bubshait and Almohawis (1994), need to be assessed for clarity, conciseness, completeness, internal and external consistency, practicality, fairness, and effect on project performance – that is on quality, cost, schedule and safety. They present a simple and systematic instrument to evaluate these attributes.

Box 6.2 Pathclearer

Scottish & Newcastle, when faced with an overwhelming legal workload, decided to undertake a fundamental review of their contracting strategy. They concluded that, in relation to ongoing long-term commercial relationships, it was often better to rely on free market forces and behaviours than to attempt to set out comprehensive contractual obligations in a contract. They determined that the general law would usually provide a fair remedy in most circumstances. As an alternative to the traditional comprehensive contracts previously used by the organisation, they introduced the 'Pathclearer' contract. Based on lean contracting principles, Pathclearer avoids detailed contract terms unless there is an overriding need for their inclusion. The agreement between Scottish & Newcastle and their suppliers comprised a short letter stating the key provisions of the 'deal' (the goods/services to be procured, the price agreed, etc.) and the applicable general law that would cover any additional matters. The document also included an indemnity for Scottish & Newcastle against third party claims arising from their use of the supplier's goods or services. Further, as the approach was implemented globally, it stated that English law applies. Dependent on the scope, additional clauses could be included if necessary.

Whilst acknowledging that the approach is not applicable in all cases – for instance, that it is more appropriate for 'continuing relationships', rather than 'one-off transactions' – Pathclearer has been used in a variety of situations. The approach was found to be most applicable, resulting in significant savings in both time and money, to procurement and sales and distribution contracts. However, it has also been used for manufacturing, supply chain, information systems, marketing, outsourcing and even corporate development contracts.

Source: Weatherley (2005)

Further reading

Contractual issues

Chapter 1 – Contract Formation. In: Griffiths, M and Griffiths I (2002) *Law for Purchasing and Supply*. 3rd edition. Prentice Hall Financial Times, Harlow. pp. 3–23

Cordero-Moss, G (2011) *Boilerplate Clauses, International Commercial Contracts and the Applicable Law*. Cambridge University Press, Cambridge

McKendrick, E (2010) *Contract Law: Text, Cases, and Materials*, 4th edition. Oxford University Press, Oxford

Murdoch, J and Hughes, W (2008) *Construction Contracts: Law and Management*, 4th edition. Taylor & Francis, Abingdon, Oxon

Poole, J (2010) *Textbook on Contract Law*, 10th edition. Oxford University Press, Oxford

Contract strategy and type

The provisions of a contract should (Smith and Wearne, 1993; Wearne, 1999):

- **Define the responsibilities of the parties**: for example, set out the projects objectives and priorities; financing arrangements, innovation, development, design, quality, standards, procurement, scheduling, implementation, installation; project management, safety, inspection, testing, commissioning , operating and maintenance
- **Allocate risk**: for example, financial investment in the project, errors in design, failure to achieve performance specification, subcontractor default, site productivity problems, delays, mistakes and insured events
- **Determine effective payment terms**: for example, for development, design, demolition, construction, fabrication, implementation, management and others services

Contracts are usually classified in terms of strategy (procurement methodology or organisational choice) – for example, traditional, design and build, turnkey and management contracts – or by type (allocation of risk and payment terms) – for example, lump sum, re-measurement and target cost contracts. Contract strategy and type should be planned concurrently.

Contract strategy

For a discourse on the various procurement/contract strategies, see Chapter 7. The following are examples of contracts classified in terms of strategy.

Design combined with production

- **Design-build contracts**: examples include: FIDIC Conditions of Contract for Design, Build and Operate Projects (2008); FIDIC Conditions of Contract for Plant and Design–Build (1999); JCT 05 – Design and Build Contract (DB) – Main Contract – (Revision 2: 2009); ConsensusDOCS 410 – Standard Design–Build Agreement and General Conditions Between Owner and Design-Builder (Cost Plus Fee with GMP) (2011); ConsensusDOCS 415 – Standard Design–Build Agreement and General Conditions Between Owner and Design-Builder (Lump Sum Based on the Owner's Program Including Schematic Design Documents) (2007)
- **Turnkey contracts**: example: FIDIC Conditions of Contract for EPC[3] (Engineering, Procurement and Construction)/Turnkey Projects (1999)

Design separate from production

Two alternative organisational structures exist where design is separate from production:

- **Sequential contracts**: conventional or traditional contracts, examples include: FIDIC Conditions of Contract for Construction (FIDIC,1999); JCT 05 – Standard Building Contracts (Revision 2: 2009)
- **Parallel contracts**:
  - *management contracting*: examples include: JCT 05 – Management Building Contract (MC) – Main Contract (2008); NEC Engineering and Construction Contract: Management Contract (2005)
  - *construction management*: examples include: JCT 05 – Construction Management Appointment (CM/A) (2008); AIA (American Institute of Architects) A132™ – 2009 Standard Form of Agreement Between Owner and Contractor, Construction Manager as Adviser Edition/A232™ – 2009 General Conditions of Contract for Construction, Construction Manager as Adviser Edition/B132™ – 2009, Standard Form of Agreement Between Owner and Architect, Construction Manager as Adviser Edition/ C132™ – 2009, Standard Form of Agreement Between Owner and Construction Manager as Adviser; ConsensusDOCS 500 – Standard Agreement and General Conditions Between Owner and Construction Manager (CM is at Risk) (2011); ConsensusDOCS 510 – Standard Agreement and General Conditions Between Owner and Construction Manager (Cost of Work with Option for Preconstruction Services) (2007); ConsensusDOCS 801 – Standard Owner and Construction Manager Agreement (Where the Construction Manager is the Owner's Agent and Owner Enters Into Trade Contracts) (Revised 2009)

Parallel contracts are advantageous where: an early project completion date is crucial; the design requirements are uncertain at the outset; supplier involvement in the design process is advisable; there is a requirement to maintain the operation of existing installations, the segmented work is of a specialist nature and/or suppliers have a limited capability.

Alternative organisational arrangements

- **Term contracting**: refers to a particular type of work to be executed over a given time period. It is commonly used for the provision of a service – for example, repair and maintenance work – where the general nature of the work is known but the number and exact nature of each item of work are not. Each individual order issued under the term contract becomes a discrete contract governed by the umbrella terms. Example: JCT 05 – Measured Term Contract (MTC) – Main Contract (Revision 2: 2009)
- **Framework agreements**: similar to term contracting but generally are used where individual call-off orders are more substantial and can be individually priced. Examples include: JCT Framework Agreement (FA) – Main Contract (2007); NEC3 Framework Contract (FC) (June 2005)
- **Programme management**: example: ConsensusDOCS 800 – Standard Program Management Agreement and General Conditions Between Owner and Program Manager (2007)

Cooperative arrangements

- **Joint ventures**: this usually refers to the situation where the supplier is a 'stand-alone' grouping of more than one commercial entity. According to Smith *et al.* (1995), there are two major areas of operational difficulty in joint ventures, which have implications for both bid preparation and project implementation: conflict and culture. Likewise, Walker and Johannes (2001) found that equalisation of power is crucial within joint venture partnerships, and the need to understand organisational cultural diversity was also seen to be pivotal. Specific contract conditions are required, therefore, to ensure the establishment of a suitable organisational structure that will encourage the successful completion of the project and that will safeguard the purchaser – for example, in the event of the default or liquidation of one of the joint venture members. Example: AIA C101™ - 1993 Joint Venture Agreement for Professional Services
- **Partnering arrangements**: these were introduced to facilitate collaborative and integrated working arrangements between multiple parties. Examples include: JCT – Constructing Excellence Contract (CE) (Revision 1: 2009); NEC Partnering Option X12 (2001); ACA Standard Form of Contract for Project

Partnering PPC2000 (Amended 2008); ACA PPC International – ACA Standard Form of Contract for Project Partnering (2008); ACA Standard Form of Contract for Term Partnering TPC2005 (Amended 2008); ConsensusDOCS 300 – Standard Form of Tri-Party Agreement for Integrated Project Delivery (2007)

Contract type

Essentially, there are two categories of project contract payment terms: price based and cost based.

Price based

- **Fixed price**: this is where the supplier is paid a fixed price or lump sum (a single tendered price), for the entire project. The terms 'fixed', 'firm' and 'lump' sum have no 'official' meaning; only the payment terms actually used within the contract are determinative. A fixed or firm price may vary if, for example, extra work is ordered, if certain problems are encountered or there might be a fluctuations clause that operates to increase the price of the cost of materials increases. Examples include: AIA A101™ – 2007 Standard Form of Agreement Between Owner and Contractor where the basis of payment is a Stipulated Sum/ A201™ – 2007 General Conditions of the Contract for Construction; ConsensusDOCS 200 – Standard Agreement and General Conditions Between Owner and Constructor (Lump Sum) (2011); IChemE Form of Contract: Lump-sum Contract (The Red Book), 4th edition (2001)
- **Measurement**: this is where a provisional list of the items and quantities of the work to be executed under the contract (bill of quantities) is incorporated into the bid/contract documentation: the purchaser pays a standard unit price based on agreed productivity rates and unit rates. Examples include: JCT 05 – Standard Building Contract with Quantities (SBC/Q) – Main Contract (Revision 2: 2009); NEC Engineering and Construction Contract: Priced contract with bill of quantities (2005)
- **Re-measurement**: this is where the actual work carried out by the supplier is measured on completion, as implemented, based upon either:
  - *an approximate bill of quantities*: similar to a measurement contract but with more flexibility. Example: JCT 05 – Standard Building Contract with Approximate Quantities (SBC/AQ) – Main Contract (Revision 2: 2009)
  - *a schedule of rates*: a list of potential items to be executed under the contract; where the purchaser reimburses the supplier for the work actually done using agreed unit rates
  - *a bill of materials*: a list of the materials expected to be used together with a unit of measurement; where the purchaser pays a standard rate based on a pre-agreed composite unit of measure
- **Target price**: This is where the supplier is paid its actual cost including overheads and profit. However, if the cost is greater than a pre-agreed target, the supplier will only receive part of the excess. If the cost is lower than the target, the supplier will be paid part of the saving. There is usually provision for the target cost to be altered in the event of scope changes and other risks borne by the purchaser

Price-based contracts incentivise the supplier; by working efficiently, cost can be controlled and profit maximised. However, the supplier is also incentivised only to supply goods and services that meet the absolute minimum required by the specification. With regard to risk, price-based contracts require the supplier to bear a comparatively high level of risk: they are required to perform all the necessary work to meet the specification within a specified timescale. From the purchaser's perspective, the major limitation of a price-based contract is that it establishes a relatively inflexible contract structure.

Cost based

- **Cost plus**: this is where the supplier is reimbursed all its entitled expenditure plus an agreed overhead and profit margin, which can be either a percentage of the final cost (cost plus percentage fee), or a fixed amount (cost plus fixed fee). Examples include: NEC Engineering and Construction Contract: Cost reimbursable contract (2005); AIA A102™ – 2007 Standard Form of Agreement Between Owner and Contractor where the basis of payment is the Cost of the Work Plus a Fee with a Guaranteed Maximum Price; AIA A103™ – 2007 Standard Form of Agreement Between Owner and Contractor where the basis

Box 6.3 Proposition

Fixed-price contracts lead to adversarial behaviour

A fixed-price contract is rarely actually 'fixed' because there is usually an allocation of the financial consequences of various risks. It is usually more helpful to talk about risk rather than a default because it is very rare that there is a deliberate action on the part of either of the parties to prejudice the project.

Typical risks that would be borne by the purchaser are denial of access to the site and change in the scope of the work, but sometimes there are less obvious 'purchaser risk' items such as force majeure or events which relate to insurable matters where the purchaser has arranged the insurance. Some risks will be shared risks. For example, there is an extension of time but no extra money is paid to the supplier. Weather (and particularly adverse weather) is a risk which is typically placed in that category. The supplier's risks are most obviously the work costing more than anticipated due to incorrect estimating, but also inefficiency and rising materials/labour costs.

It is often the case that construction contracts are delayed (or the cost of implementing them increases) because of the concurrent occurrence of a number of risk events, some of which are to be borne by the purchaser and some by the supplier. There is a body of law and practice dealing with how concurrent delays and cost increases are to be dealt with. However, there is usually at least some scope for a dispute as to which of the causes was the most significant. Typically the purchaser will say the cause which is at the supplier's risk caused the problem and the supplier will take the opposite position.

It is generally considered that a target-price contract is less likely to lead to disputes because of the greater sharing of all risks between the purchaser and supplier. Put simply, the financial effects of the occurrences of these problems are mitigated because they are shared and the amounts of money at stake – the difference between winning and losing a dispute – will be much less than if there was an all or nothing approach.

of payment is the Cost of the Work Plus a Fee without a Guaranteed Maximum Price; ConsensusDOCS 410 – Standard Design-Build Agreement and General Conditions Between Owner and Design-Builder (Cost Plus Fee with GMP) (2011); IChemE Form of Contract: Reimbursable Contract (The Green Book), 3rd edition (2001); IChemE Form of Contract for use in the process industries: Target Cost Contract (The Burgundy Book), 1st edition (2003)

Cost-based contracts have the benefit of being more collaborative but they impose a much lower degree of control on the supplier, requiring more managerial effort by the purchaser. Compared with price-based contracts, the level of risk borne by the supplier will reduce, while that of the purchaser will rise. However, the contract will more easily contend with high levels of change.

Incentives and contract type

Incentive provisions can be incorporated within fixed-price and cost-reimbursable contracts. Herten and Peeters (1986) describe and illustrate three specific types: cost incentives, schedule incentives and performance incentives. They also refer to multiple-incentive contracts, where two or more of these incentives are combined, either dependently or independently, in the same contract. Bubshait (2003) puts forward a fourth type: safety incentives, although he found only limited support for its value.

Incentive contracts are still not as extensively used as they might be. According to Ward and Chapman (1994) this is perhaps due to a lack of appreciation of the limitations of conventional fixed-price contracts and/or of the ability of incentive contracts to motivate suppliers. However, within industrial projects,

Bubshait (2003) highlights the variation in the perception of purchasers and suppliers concerning incentive/disincentive (I/D) contracting. Although he found a general agreement on the effectiveness of I/D contracting in encouraging supplier performance, few organisations incorporate I/D principles into their contracts. Moreover, penalty systems were used, rather than incentive systems, to penalise the supplier for late completion. Over the first decade of the twenty-first century, as indicated by Volkman and Henebry (2010), there has been increased interest in and adoption of incentive contracts by corporations. However, there is a tendency to resort to fixed-price contracts where budgets are limited due to economic conditions.

Examples of incentive contracts include: NEC Engineering and Construction Contract: Target contract with bill of quantities (2005); NEC Engineering and Construction Contract: Target contract with activity schedule (2005); ConsensusDOCS 410 – Standard Design-Build Agreement and General Conditions Between Owner and Design-Builder (Cost Plus Fee with GMP) (2011).

Choice of contract

The choice of contract type is one of the most significant strategic decisions, since it determines how the supplier is paid and how risk is allocated between the parties. This interrelationship between risk allocation and payment terms is illustrated in Figure 6.1, shown as a dual continuum based on the distribution of risk to the parties derived from the perceived level of risk at contract formation and an arm's-length (market) and collaborative (hybrid) approach to contractual relationships. As a general principle, contract type should aim to give the maximum likelihood of attaining the objectives of a project (Wang *et al.*, 1996); it should be regarded as a 'means to an end'.

Griffiths (1989) summarises the advantages, problems and resource requirements of the major contract type alternatives.

Based upon 93 R&D defence projects, Sadeh *et al.* (2000) found that contract type has a considerable impact on project success. Under increasing technological uncertainty, both parties to the contract benefit

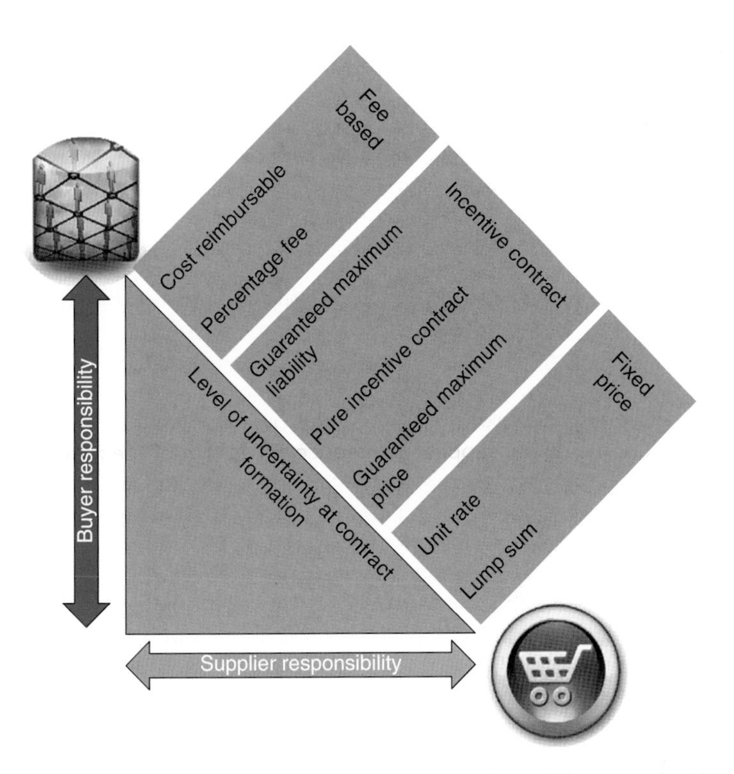

Figure 6.1 Risk allocation via contracts. Source, Adapted from Winch, 2010.

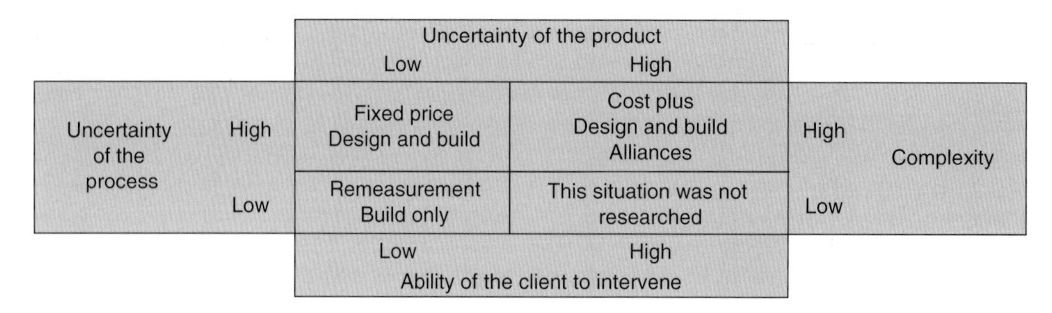

Figure 6.2 Selection of contract types. Source: Turner and Cochrane (1993).

from cost-plus contracts, whereas fixed-price contracts generate more benefits when uncertainty is lower. They recommend two-stage projects. At the first stage – the preliminary design and feasibility study stage, where technological uncertainty is very high – they recommend the use of cost-plus contracts. At the second stage – the full-scale design and development stage – a fixed-price contract is preferable.

Likewise, Turner and Simister (2001) have demonstrated that, when using transaction cost analysis to indicate when alternative contract pricing terms should be adopted, it is uncertainty of the final product and not risk *per se* that determines the most appropriate type of contract. Further, they suggest that if the purpose of a contract is to create a project organisation, based on a system of cooperation not conflict, then the requirement for goal alignment is more significant. This, they consider, requires that all parties to a contract should be properly incentivised, and is accomplished by incorporating contract pricing terms as illustrated in Figure 6.2.

Turner and Simister (2001) conclude that the main criterion for selecting contract pricing terms is goal alignment. However, transaction costs are minimised *en passant*.

In addition to complexity, uncertainty (of the product and process) and flexibility (the ability of the client to intervene), the choice of contract type (solution) is influenced by the:

- **Context** of the transaction
- **Governance** structure adopted to regulate the exchange

The **context**, referred to as **atmosphere** in TCE, is the environment within which the exchange takes place, and includes factors that are both internal and external to the organisation:

- **Internal factors**: internally, the choice of contract type may be influenced by the availability of suitable resources, existing agreements (alliances, framework agreements, etc.), previous decisions made at the programme or portfolio level, the predisposition of senior managers and internal politics, and *institutional* dynamics
- **External factors**: these include the following:
  - *institutional aspects of the environment*: such as the potential impact of both home and host government systems and intervention; cultural values; legal systems and frameworks – for example, the implications of civil codes and precedent based legal models, procurement legislation/directives, competition/antitrust laws, corporate governance legislation and regulatory obligations; organisation of labour; and financial system – for example, the impact of the World Bank and other funding agencies
  - *prevailing and forecasted economic conditions*: the choice of contract type and the means of its negotiation may vary depending upon the current and projected economic conditions; the power balance between the parties may change relative to variations in and the cyclic nature of the economic cycle. The commercial function in conjunction with other analysts, therefore, needs to monitor market trends with the aim of identifying market turning points
  - *configuration of the supply network*: the decision also involves an assessment of the nature of the suppliers' business and its underlying supply network. For example, consideration of the degree of

competition within a specific market segment: are there an adequate number of appropriately qualified suppliers, or does it exhibit features of either an oligopolistic or monopolistic market? Additional factors include: the maturity and composition of the supply chain (web/network) – for example, is it highly fragmented or integrated?; the need to appoint local or global suppliers; and the financial, technological and collaborative capability of potential suppliers relative to the proposed transaction. Kaufman *et al.* (2000) have developed a strategic supplier typology that classifies prospective partners as commodity supplier, technology specialist, contractor, or problem-solver

Governance, relates to the form of transaction, which is derived from the decision by a purchaser to either 'make or buy' – that is, utilise either the market or hierarchy as its governance structure. Alternatively, the purchaser could adopt a hybrid (collaborative) approach. The principal driver of the choice is the purchaser's ability to 'safeguard' post-award (ex post) opportunistic behaviour by the supplier (as discussed in Chapter 1).

A key feature of both the **context** of the exchange and its **governance** structure is the dynamic of the **power** balance between the parties (see, for example, Cox *et al.*, 2000, 2001, 2002; Cox and Ireland, 2006).

Frequently, a single contract type is adopted across the project life cycle; however, on major projects it may be appropriate to determine distinct contract types for individual work packages at the concept, assessment, demonstration, manufacture, in-service and disposal stages. In doing so, decision makers would need to reconcile the potentially conflicting issues of complexity, uncertainty, flexibility, context and governance in order to create a rational contract strategy.

From the 1990s onwards, there has been a marked change in the way that construction projects have been procured; while the traditional approach of separate design and construction remains an important feature of the sector, Design-build is now an established approach. Moreover, design-build is one of only four procurement routes sanctioned by the Cabinet Office; the others are PPP/PFI (public–private partnerships/Private Finance Initiative), prime contracting and framework arrangements (Cabinet Office, 2012). Also, the most recent RICS (Royal Institution of Chartered Surveyors) Survey Building Contracts in Use during 2007 (Davis Langdon, 2010) reveals the emergence of partnering agreements. Changes over time in the types of construction contracts used in the UK, linked to procurement approach, are shown in Table 6.1.

Globally, there has been a significant shift towards guaranteed maximum price (GMP) forms of contract (KPMG International, 2007). The KPMG global construction survey, found that although fixed price remains the most popular approach with 35% of the respondents stating that this was the only form of contracting they used, GMP accounted for 29%, unit price 14%, cost plus fixed fee 11%, time and materials 4% and other 7%. A more recent survey (KPMG International, 2010) found that although PPPs are an increasing aspect of the international construction environment, most suppliers appear to be reluctant to assume the levels of risk they had previously accepted. This position may of course have changed materially since 2007.

Table 6.1 Contract types (procurement approach) in use by project value.

Procurement method	1984 %	1985 %	1987 %	1989 %	1991 %	1993 %	1995 %	1998 %	2001 %	2004 %	2007 %
Lump sum	72	70	70	62	55	50	56	38	40	34	31
Measure and value	7	5	3	4	3	4	2	2	3	3	2
Cost plus	4	3	5	1	0	0	1	<1	<1	<1	<1
Design-build	5	8	12	11	15	36	30	41	43	43	33
Management type	12	14	10	22	27	10	11	18	12	2	11
Target contracts	–	–	–	–	–	–	–	–	–	11	7
Partnering	–	–	–	–	–	–	–	–	–	7	16
Totals	100	100	100	100	100	100	100	100	100	100	100

Source: compiled and adapted from various RICS Contracts in Use surveys.

Further Reading

Contract type and choice

Hughes W (2006) Contract management (Chapter 15). In: DJ Lowe with R Leiringer (eds) *Commercial Management of Projects: Defining the Discipline*. Blackwell Publishing, Oxford. pp. 344–355

Exercise 6.2 Contract type and choice

For a recent project that you have been involved in:

- Evaluate the selection criteria for the contract adopted (preferably a standard form)
- Review the amendments made to this form and assess their impact on project delivery and relationships

Review the contracting strategy adopted on the three football stadia. What do you consider the prime factors that determined the approach taken?

Compare and contrast the contracting strategy adopted by Wembley National Stadium Ltd with the one implement by BAA on Terminal 5.

Roles, relationships and responsibilities

Roles
A contract defines the roles of the parties: the purchaser and the supplier. Additionally, the contract apportions roles: for example, project management, design execution and integrity, production supervision and dispute determination.

The role of the parties to the contract
Generally, the purchaser will be involved in:

- Defining exactly what services are to be provided
- Setting service levels
- Providing relevant and timely information to the project manager and supplier
- Informing the supplier of dissatisfaction or underperformance

The degree of empowerment of a supplier is dependent upon the procurement approach adopted and the terms of the contract. Generally, the supplier will be responsible for:

- Deciding how to provide the asset, product or service
- Delivering the asset, product or service to specification
- Deciding priorities to realise the asset, product or service
- Meeting the purchaser's requirements within the contract terms and budget
- Monitoring the asset, product or service delivery performance
- Developing and implementing agreed procedures
- Providing information as required by the contract (CUP, 1997)

The contract may also define the role of other parties (agents of the purchaser), such as project manager, engineer, architect, landscape architect, interior designer, commercial manager/quantity surveyor, superintending officer, or clerk of works and the relationships with other suppliers such as subcontractors, nominated subcontractors and nominated suppliers.

Relationships

Rules and procedures

The establishment of transparent procedures, from the outset, will enhance contract management, while reducing the disruption that problems may generate. Whereas some will relate to the routine contract management activities, others will operate only when needed.

Procedures will be required for:

- Performance/service management
- Risk assessment
- Contingency planning
- Payment submission, processing and certification
- Budget review and control
- Change management – instigated by either the supplier or the purchaser
- Price adjustments
- Interrelationship management and control
- Security
- Problem management
- Disputes resolution
- Compliance monitoring
- Termination requirements

Effective implementation of a project will rely upon the relationships between the parties, not necessarily on the contract or role definition. Further, it is crucial that these relationships be established at the outset, continually reviewed and actively managed. Relationships need to be balanced between flexibility and openness, on the one hand, and professionalism and businesslike behaviour on the other.

Risk and responsibilities

Contracts set down the rights and obligations of the parties to the contract and describe the responsibilities and procedural roles of those named within the contract. Loosemore and McCarthy (2008) assert that a common awareness of the distribution of risk within a contract is a key feature of '*harmonious, effective and efficient*' projects. Unfortunately, their study on the perception of risk allocation in construction supply chains found '*worryingly high*' differences of opinion between contracting parties, especially between those organisations lower down the supply chain. They advocate '*communication, consultation and involvement in contractual decision-making*' in order to align the perception of risk allocation between the parties.

For any project, achievement of its objectives is the principal risk; this is borne by the purchaser. Likewise, the purchaser bears the key risks of any project – for example, in deciding to instigate the project, defining the project's scope and specification, selecting a contract strategy, and choosing a supplier. Other risks relate to the design, implementation and delivery of the project; contracts seek to allocate these risks to the parties. However, both parties may be at risk irrespective of the contract – for example, where forces outside their control frustrate the work.

Ideally, the allocation of risk between the parties should be based upon:

- **Managerial principles**: satisfactory project completion is more likely to be achieved through effective planning, management and supervision rather than requiring guarantees and imposing rights to damages for default

- **Commercial principles**: risk should be borne by the party best able, economically, to control, manage or insure against its consequences
- **Legal principles**: unfair contract terms, penalties (designed to punish or deter rather than to compensate) may not be enforceable (Wearne, 1999)

Ultimately, it is not in the purchaser's financial interest to ask a supplier to absorb all risks. The purchaser's objectives are more likely to be attained through the use of contract terms that motivate the supplier to perform on time, economically, etc., and where the risks transferred are not so great as to be detrimental to either party in the short or long term (Barnes, 1983). This principle was appreciated by BAA (see Case Study B). When formulating their contracting approach they concluded that they were unable to effectively transfer the risk of the project onto their supply chain. Instead they decided to retain and manage the risk.

The obligations and entitlement of the purchaser

The purchaser has three main obligations: to enable the supplier to complete the works/product/service; to pay the agreed price; and to accept the works/product/service on completion. Contracts typically also include entitlements, such as the right to appoint a project manager or engineer and the right to employ and pay others to complete work or services if the supplier fails to perform in accordance with the contract.

The purchaser discharges its contractual responsibility by paying the supplier the accepted contract amount, amended where required under the contract. For a discussion of the role and responsibility of the purchaser under the FIDIC suite of contracts see Van Houtte (1999).

The obligations of the supplier

Contracts generally contain numerous clauses that require the supplier to comply with an instruction or do something; the FIDIC Construction Contract includes over 80. For example, the supplier is to: design, execute and complete the Works in accordance with the contract; comply with instructions given by the Engineer; remedy any defects in the Works; and institute a quality assurance system.

The supplier discharges its contractual responsibility by fulfilling its obligations under the contract. For example:

> 'The Contractor shall complete the whole of the Works... including achieving the passing of the Tests on Completion, and completing all work which is stated in the Contract as being required for the Works or section to be considered to be completed for the purposes of taking over...' (FIDIC Construction Contract: Sub-clause 8.2)
>
> 'The Supplier Provides the Goods and Services in accordance with the Goods Information.' (NEC Supply Contract: Clause 20.1)

Management and supervision

Purchasers often delegate the functions of contract administration – which Wearne (1992) refers to as a '*concierge de contrat*' – and project supervision to third parties via contracts. These functions are often combined: for example, in contract strategies where design is separate from production, the initial designer, architect or engineer usually undertakes both roles. More uncommonly, contracts separate these functions: for example, in the NEC suite of contracts the project manager is responsible for contract administration and the supervisor for ensuring the works are implemented in accordance with the contract (NEC, 2006a).

The supervision of a supplier, in terms of what, how and when to supervise, is dependent upon the risks inherent in the project, the contract terms and the inclusion of incentives to encourage satisfactory performance. Normally, the supervisor has no (independent) authority to amend the contract or to relieve either party of any duties, obligations or responsibilities under the contract.

Time, payment and change provisions

This section addresses the key areas of time issues, payment provisions and change mechanisms within contracts.

Time issues

Commencement

Contracts include a date for commencement of the project, usually determined by the purchaser or by negotiation between the parties. The commencement date should be set so that it enables the supplier to mobilise resources. Further, contracts should determine what happens in the event of the purchaser failing to provide access to a site or make available plant, service or any other resources required under the contract, as such failure could frustrate the contract.

Schedule

Generally, contracts include statements regarding the progress of the project. For example, both the NEC3 Supply Contract and the FIDIC Construction Contract require the supplier to submit a detailed programme (schedule) within a stipulated time frame and to submit revised schedules whenever the previous schedule is inconsistent with actual progress or with the supplier's obligations.

Suspension

Contracts can include provisions that enable the purchaser to suspend the project. The consequences of suspension of the project, the entitlement to payment for plant and materials in event of suspension, the resumption of the project, and the possible termination of the contract as a result of prolonged suspension need also to be addressed within the contract.

Completion

Usually purchasers specify a completion date or alternatively the number of calendar or working days authorised for executing the work. Failure to complete a project within this stipulated time limit can be grounds for a significant dispute between the parties. Numerous contracts include two completion targets: substantial completion and final acceptance. Generally, the contract will stipulate the procedures to be used to determine substantial completion – that is, where the supplier has achieved substantial performance. Substantial completion usually means that the project is finished to the point that it can be used and tested, subject only to minor outstanding items that are capable of rectification without impeding use.

Payment provisions

The contract states how, what and when the supplier will be paid – for example, stage payments based on work completed at monthly intervals or milestone based. Further, it will determine whether payment is incentivised – by, for example, the use of a key performance indicator (KPI) system – and the level of retention (monies retained by the purchaser until the supplier has remedied defaults). When planning a contract, a purchaser should consider what payment terms are most likely to motivate the supplier to achieve the purchaser's objectives for the project.

Fixed-price terms of payment

Fixed-price terms of payment are appropriate for projects that are fully specified prior to inviting potential suppliers to bid and where the completion date of the project is more important to the purchaser than the flexibility to make changes to the specification or any contract terms.

Advance payments

Advance payments, alternatively referred to as down-payments or payment for preliminaries, are principally used to reduce the supplier's financing charges where there is a considerable site establishment set-up cost or the need to prepay for materials before work can start. They can also be used where the supplier considers the purchaser to be a poor credit risk. A potential problem, however, is that the purchaser could lose the value of an advance payment if the supplier subsequently defaults. In order to avoid this, the supplier can be required to provide a bond before receiving payment.

Milestone and planned progress payments

The supplier can receive payments **on account** in a series of payments for achieving defined stages of progress. Two examples are: **milestone** payments, where payment is made based upon progress in completing defined segments of the project; and **planned payment systems**, where payment is activated upon achieving defined percentages of a supplier's schedule.

Stage payments provide the supplier with an incentive to complete the work promptly. Incorporating additional **bonus** payments for attaining a milestone ahead of schedule can increase this incentive. However, it has the disadvantages that the contract and its management are more complex, and disputes may arise if the milestones or equivalents have not been adequately defined or their achievement proved. Additionally, the contract should state what happens when a stage is achieved ahead of schedule, and what payment is due if a stage is missed but the subsequent one is attained. There is also the question of what should happen if a milestone is achieved late due to purchaser delay, or if the project is terminated between milestones.

Box 6.4 Influencing supplier behaviour: The use of different payment systems

Payment on completion

One of the most important features of a contract which will influence contract behaviour is the payment system. A common example of a payment system is payment on completion. It is arguable that payment on completion will generally incentivise a supplier to finish early because that will reduce the amount of his unrecovered outlay.

There are, however, some other effects. The first is that if there is a significant outlay on the part of the supplier and the supplier has to finance that outlay then this will be added into the construction cost. Although interest rates have been relatively low in recent years compared with, say, the 1970s, it is not always clear that the supplier will have obtained good financing arrangements and actually it might be very expensive to obtain credit. Furthermore, suppliers sometimes have a total facility available to them across all of their projects and if one contract is going to utilise that facility significantly then this can be disproportionately disadvantageous for suppliers who might want to recover a higher margin to offset this disbenefit or alternatively not bid for the work at all.

Another effect is that by withholding large amounts of money from a supplier until completion, there is a possibility that the supplier may 'rush' the work. This may mean that there is a reduction in quality because of the need to be paid quickly.

Finally there is potential for disputes. For example, a supplier might claim interest on a deferred payment where the delay is caused by the purchaser.

Bonus for early completion

Another payment mechanism is to award a bonus for early completion. Although this would not necessarily have the same effect as payment on completion it does have similar qualities. First, the supplier may assume that this bonus is not going to be paid because early completion is not likely to be achieved and therefore there will be no cost saving.

Alternatively there may be the same risk, as in payment on completion, of the supplier's work being rushed in order to achieve this bonus particularly if it is substantial. There is scope for dispute as to whether or not the bonus should be paid if (for example) the supplier says that there has been delay by the purchaser. If the purchaser has issued numerous variations to the contract the supplier might argue that had those variations not been issued then early completion would have been achieved, and the supplier would have received the bonus, and it is only because of the number of variations that this has not occurred.

Liquidated damages

The most common method of incentivising suppliers to finish on time is the use of liquidated damages, this being a predetermined level of damages (for example, per day or per week) for which the suppliers will be liable in the event of late completion.

The use of liquidated damages may have some of the same negative effects (and positive effects) as described in relation to payment on completion, including the potential for rushed work and also risk pricing if the supplier believes that there is a strong likelihood of having to pay those liquidated damages, or just simply an increase in the construction price. Again, there is also the possibility of disputes similar to that of payment on completion where there is a dispute in relation to the reason for late completion.

Key performance indicators

Another payment regime which can influence supplier behaviour is to use KPIs. This might operate as part of what is usually termed a painshare/gainshare system. The system usually works by having a contract with a target price. The supplier is paid the actual cost of the work. If the actual cost exceeds the target price the supplier will only be paid a percentage of the excess (pain). If the actual cost is less than the target price the supplier will receive a percentage of the saving (gain).

One of the ways of using a KPI regime to influence the supplier's behaviour is to change the percentage levels for the supplier in accordance with different categories of KPIs. These KPIs might be quality, safety or other topics which are important for that particular project, but can also be used to monitor compliance with interim milestones. For example, if interim time milestones are missed then this would have a negative effect on the KPI score leading to a reduction in the gain available to a supplier or an increase in the pain.

Other KPIs which are used in long-term framework contracts are collaboration KPIs where the supplier is expected to come up with ideas and assistance (even for other suppliers on the purchaser's panel), and to the extent that they do this they will be scored against that KPI.

Other ideas for KPIs are innovation, overall quality and even a KPI for a tidy site. Where these KPIs attract significant amounts of money or reduce the supplier's entitlement then the supplier will be incentivised to perform better in these areas than might otherwise be the case.

However, KPI systems can be too complicated or be too remote from the supplier's day-to-day endeavours and this can result in them having little effect in practice.

Probably the best way of influencing a supplier's behaviour is the offer of repeat business. However, this can be a somewhat imprecise approach if the supplier does not fully understand the purchaser's concerns. Purchasers should generally try to tell suppliers what they think is important and thereby what they think is less important. Even if the KPI regime is relatively light in the amount of money it attracts it is a good way of communicating the owner's wishes to the supplier especially where there is likely to be repeat business.

Payment based upon agreed rates

This is where payment for work executed is based upon rates (unit prices) provided by the supplier when bidding for the contract, with the anticipated quantities of each item of work listed in a **bill of quantities** or **schedule of measured work**. Unit rate terms of payment provide a basis for paying a supplier relative to the extent of work completed. The final contract price is calculated using fixed (pre-agreed) rates but is adjusted if the quantities change.

Alternatively, some contracts incorporate a **schedule of rates**, where the rates are provided on the basis of indications of possible total quantities of each item of work in a defined period or within a limit of variation in quantity. Schedules of rates have the potential advantage of creating a basis for payment when the type of work is known but not the exact quantities. However, there is the potential disadvantage that suppliers will incorporate uneconomical rates which will skew the value for money or the profitability of the project if a particular element of work increases unexpectedly.

Contract price adjustment/price fluctuations/variation of price

Contracts can include terms for reimbursing suppliers for escalation of their costs as a result of inflation – for example, a clause that allows a supplier to raise their prices in line with a pre-agreed index. Such terms have the potential advantage that suppliers are relieved of the task of attempting to forecast inflation rates, which may result in the submission of higher prices; however, a disadvantage for purchasers is that they generate uncertainty over the final project cost.

Where such terms are not incorporated in contracts, supplier's prices are prone to be higher in periods of inflation. However, the final contract sum will be independent of inflation. Also, in periods of inflation the suppliers will be incentivised to complete their work more quickly to reduce the impact of increases in their costs.

Cost reimbursable terms of payment

Cost-based terms of payment can be referred to as cost reimbursable, prime cost, dayworks, time and materials, etc. There are two variants:

- **Cost plus fixed fee**: where the purchaser reimburses all the supplier's reasonable costs: employees on the contract, materials, equipment and payments to subcontractors, plus usually a fixed sum for financing, overheads and profit
- **Cost plus percentage fee**: as above, but the fee is added as a fixed percentage

Cost-plus contracts can be let competitively with the supplier providing their rates per hour or per day for categories of personnel, equipment and other services. Although this is also a unit rate system, it varies from those mentioned above as payment is for cost not performance.

Under cost-plus contracts the supplier's risks are limited at the expense of potential profit. For the purchaser, cost-plus contracts are appropriate where all the potential categories of resources can be predicted, even though the exact extent of the work initially remains uncertain.

Target-incentive contracts

Target-incentive contracts are a development of the reimbursable type of contract, where the purchaser and supplier agree at the beginning a probable cost for an as yet undefined project. However, they also agree to share any savings in cost relative to the target. However, if the target is exceeded the supplier will be reimbursed at less than cost plus.

Cost-plus incentive fee is where the fee may fluctuate either up or down within set limits and in accordance with a formula linked to permissible actual costs. Veld and Peeters (1989) consider that the most important aspect in cost incentives is the sharing factor. They note that the sharing formula can be non-linear, vary between overrun and underrun, and sometimes incorporate a neutral zone.

Al-Subhi Al-Harbi (1998) explains how purchasers and supplier's select the supplier's sharing fraction based upon a risk-averse, risk-taking or risk-neutral perspective. Veld and Peeters (1989), Ward and

Chapman, (1994), Al-Subhi Al-Harbi (1998) and Berends (2000) provide examples of incentive fees expressed as equations.

Berends (2000) believes that to be effective, an incentive scheme must be aligned with the owner's overall project objectives, not just cost. Further, it must provide a positive relationship between the supplier's performance and the supplier's profit margin. Subject to both parties having the ability to prepare realistic cost estimates, the contract negotiation process affords a means to deliver an effective incentive scheme.

Convertible terms of payment

A convertible contract incorporates an agreement that after any significant uncertainties have been decided it will be changed into a fixed-price or unit-rate-based contract. Potentially, such an agreement has the advantage of limiting the contract price once it is converted. However, there is little or no opportunity for competitive bidding.

Periodic payments

Contracts may incorporate clauses that entitle the supplier to **interim payments**: payments on account. These payments are usually on a monthly basis, based on the estimated value of the project executed or undertaken by the supplier in the preceding month, and include any amount to be added or deducted under the terms of the contract – for example, retention.

Retention

Many large contracts often incorporate a clause whereby a percentage of any payment due to the supplier (for example, as fixed price, milestone achievement or value of work completed) is retained by the purchaser for a specified period (generally up to one year). The retained amount (retention) is paid (released) once the supplier has satisfactorily fulfilled its obligations, for example, the rectification of any defective work.

Such a clause has the potential advantage, for the purchaser, of motivating the supplier to complete the project appropriately the first time, thereby activating the release of the retention at the earliest opportunity. A potential disadvantage is the possibility that suppliers may increase their prices to account for the delay in receiving full payment.

Incorporating change

A purchaser's needs may alter during the period of a contract – for example, in quality, quantity, and even character. Where a supplier provides an extra service or additional work they are entitled to request extra payment. Contracts should therefore incorporate terms that facilitate the effective **management of change**. Where, in the course of a contract, major **variations** are expected, suppliers should be granted greater empowerment to initiate change, possibly in the form of **value management**. This can be of significant advantage to the purchaser where there is the potential to incorporate advances in new technologies or new techniques.

Contracts should also provide a mechanism for pricing any changes in the project deliverables, , although, with regard to the purchaser, these mechanisms are usually less competitive than those in the original contract. Any changes need to be managed by both the purchaser and the supplier, and inevitably this adds to the cost of the contract

Change management requires the following stages: identification of a potential change requirement; contemplation of the full impact of the change; production of a formal change order; and notification to the parties of the agreed change. Additionally, each stage of the process should be documented and the decision maker should possess the appropriate authority to agree the change.

Latham (1994) identified variations as one of the main problems confronting the UK construction industry, regarding them as probably being a significant cause of disruption, disputes and claims.

For example, under the FIDIC Construction Contract the engineer can instigate variations at any time before issuing the taking-over certificate for the works, either by issuing an instruction or by requesting that the supplier submit a proposal. The supplier is required to comply with a variation, although the supplier can object to a variation where the goods required for the variation cannot readily be obtained. Further, the contract prescribes how a variation is to be measured and evaluated. Generally, appropriate rates or prices specified in the contract will be used or form the basis to derive new rates or prices to value the work. Where no appropriate rates or prices exist, the work is to be valued based on the reasonable cost of carrying out the work, plus reasonable profit.

Remedies for breach of contract

Contracts need to incorporate provisions to manage the consequences of **default** by either party. This section introduces the concepts of **performance indicators, liquidated damages** and **termination**.

Performance indicators

The contract should contain appropriate and achievable performance indicators (PIs). Although it is essential to include the general principles of PIs within the bid documentation, the actual indicators are often determined during the contract negotiation stage with the preferred supplier – that is, before the contract price is agreed and the contract signed.

Determining an effective performance measurement system requires the identification of:

- Assets and or services to be provided
- Critical success factors for a particular contract
- KPIs, targets and measures
- Components to be measured or assessed, both in terms of the outputs and outcomes of the asset and or service(CUP, 1997)

Liquidated damages

Where it is possible to derive a pre-estimate of the loss to be suffered under certain circumstances, it is generally prudent to incorporate liquidated damages into the contract. Liquidated damages, however, should be a bona fide pre-estimate of the loss in the given situation.

'Liquidated damages' places a limited liability on the supplier to pay a specified sum for a defined breach in performance – for example, late delivery of a product or late completion of a project. The aim is to encourage suppliers to meet their contract obligations. However, their effectiveness may be limited – for example, where the cost of performing their obligations is more than the liquidated damages.

In addition to liquidated damages, contract terms can be used to motivate the supplier, for instance by offering additional payments where the supplier completes on time or recovers time after a delay in meeting their contract obligations.

Rebates

Minor defaults by a supplier can be dealt with using rebates, alternatively termed service credits. Generally included in service contracts, their aim is to:

- Promote satisfactory service
- Signal falling standards of service
- Caution the supplier over the provision of poor service

To be effective, a rebate mechanism should be straightforward and proportional to the specific aspects of service and various levels of default.

Termination

Under certain circumstances a contract may need to be terminated. As this action results in a 'lose–lose' situation it is generally only used as a last resort, due, usually, to the unacceptable performance of the supplier. However, other exceptional circumstances may arise necessitating the termination of the contract – for example, a change in government systems, a change of policy, or a change in user needs.

Termination of a contract requires contingency plans to ensure continuity of supply, construction or completion of a service or product. This is essential when the service is critical to the business and the supplier has the ability to terminate the contract. These contingency plans may need to be incorporated into the contract – for example, to effect transition from one supplier to another the existing supplier could be required to cooperate with the new supplier(s). Also, special contract provisions may be needed to deal with intellectual property rights (IPR) arising from the contract. IPR is covered in more detail later in this chapter.

Under the FIDIC Construction Contract, for example, the employer (purchaser) is entitled to terminate the contract if the supplier: abandons the works, subcontracts the whole of the works or assigns the contract without the required agreement, becomes bankrupt or insolvent or gives or offers to give a bribe (this list is indicative not exhaustive). Likewise, the supplier is entitled to terminate the contract if: the employer fails to provide reasonable evidence of their financial arrangements, or an amount due under an interim certificate; the engineer fails to issue a payment certificate; or the employer substantially fails to perform his obligations under the contract. (Again, this list is indicative not exhaustive and most clauses contain a time frame for compliance.)

Similarly, the NEC3 Supply Contract contains provisions by which the parties may terminate the contract (Clause 91); further, it specifies both the procedure to be followed and the amounts due on termination (Clauses 92 and 93).

Bonds, guarantees and insurances

The principles of bonds and guarantees

Financial guarantees are issued by banks, insurance companies, surety companies or a parent company so that funds will be available should the purchaser have a legitimate claim against the supplier. Chaney (1987) classifies the following types of guarantee:

- **Tender (bid) guarantee (bond)**: typically required by purchasers to ensure that once a bid has been submitted the supplier can be held to it, and, so far as possible, to exclude suppliers who lack the necessary financial resources to complete the contract. It will virtually always be conditional and have a time limit
- **Repayment guarantee**: generally incorporated where an advance payment is involved. It can either be 'on-demand' or 'conditional'. An on-demand bond permits the purchaser to invoke the guarantee without having first to establish default, whereas under a conditional guarantee the purchaser can only invoke the guarantee once the supplier has admitted a breach of contract, or upon a ruling of a court or an arbitrator that the supplier is in default
- **Performance guarantee**: seeks to ensure that a purchaser can recover damages in the event of a supplier's failure to perform. Again, it may be classified as on-demand or conditional
- **Retention guarantee**: an alternative to a retention clause, discussed earlier. Whereas a purchaser may consent to this arrangement from the start of the contract, retention guarantees are more commonly used during the maintenance period – that is, upon attaining practical completion
- **Surety bond**: the usual form of guarantee in North America. It varies significantly from the principle of indemnity: rather than merely paying the amount of the bond following a default, the emphasis is on the surety arranging for the completion of project. A performance and payments bond is usually required on all publicly bid construction projects and may be required on some private projects (Carty, 1995)

The liability of the guarantor is for costs incurred by the purchaser (limited to the amount of the guarantee), whereas a surety has the added responsibility of arranging for the completion of the work by a third party. The guarantor does not insure the supplier, merely provides a guarantee. Therefore, as a condition of issuing the guarantee, the supplier will be required to indemnify the guarantor to the extent that, if the guarantee is called in, the guarantor will only suffer financially if the supplier is or becomes insolvent. The guarantor will generally require the supplier to provide details of its financial position and capability and resources to undertake the contract. Depending upon who issues the guarantee, premiums can amount to 2% of the sum guaranteed per annum, a cost either directly or indirectly borne by the purchaser.

The NEC3 Supply Contract provides the option of including a parent company guarantee (Option Clause X40) and performance bond (Option Clause X13) within the agreement.

Again, caution should be used with the 'label' attached to any document under consideration. The terms set out above are not 'terms of art' and there are documents in circulation that 'mix and match' some of the features described.

Insurance

Contracts typically state what types of insurance are required, determine who is to be responsible for obtaining the insurance and specify the particular terms of the policies and limits of coverage.

Claims

Smith and Wearne (1993) define a claim as a demand or request, usually for extra payment and/or time. Usually, a 'claim' is an assertion of a right contained in a contract; however, in others it is the converse, a submission outside the terms of a contract. In lieu of the more common term 'claim', the NEC3 suite of contracts uses 'compensation events'.

Suppliers

Contracts include provisions for the submission of claims for time and money where the supplier's work is likely to be disrupted or delayed by events that are at the purchaser's risk. In such circumstances the supplier is generally required to give notice to the project manager of such disruption or delay. Further, the supplier will be required to give specific details of what caused the delay, its nature and to quantify the likely delay or disruption.

In the first instance, contracts usually empower the project manager (the engineer under the FIDIC Construction Contract) to agree or determine any matter, by consulting with each party so as to reach an agreement. If there is a failure to reach an agreement then the project manager is usually authorised to make a fair determination. If the project manager concludes that completion is or will be delayed (by a cause that is not at the supplier's risk), the supplier will be entitled to an extension of time, an extension of time and payment of any cost incurred, an extension of time and payment of any cost incurred plus reasonable profit, or payment of any cost incurred plus reasonable profit.

Extension of time

Events that might typically lead to the supplier having an entitlement to an extension of time and/or extra payment include:

- A substantial change (ordered by the purchaser) in an item of work included in the contract
- The discovery of unforeseeable ground conditions and the steps necessary to deal with them
- Delay, impediment or prevention caused by or attributable to the purchaser, the purchaser's personnel, or the purchaser's other suppliers on the site
- The discovery of antiquities and the like found on the site
- Suspension of the works

- Rectifying loss or damage to the works, goods or supplier's documents due to risk that the purchaser takes
- Failure of the purchaser to issue any necessary drawing or instruction within an agreed (or reasonable) time
- Failure by the purchaser to give proper access to, or possession of, the site
- The purchaser taking over and/or using a part of the works, other than such use as is specified in the contract or agreed by the supplier

Words such as reasonable and unforeseen can be further defined – for example 'not reasonably foreseeable by an experienced supplier by the date for submission of the tender'.

Purchaser's claims

Likewise, the purchaser can claim against the supplier – that is, include an amount as a deduction in the contract price or payment certificate. Typical claims against the supplier, according to Bubshait and Manzanera (1990), relate to:

- Use of non-specified materials
- Defective work
- Damage to property
- Late completion by the supplier

Purchaser's claims can be broadly categorised as: liquidated damages, claims explicitly provided for by the contract and claims for damages for breach of contract by the supplier (Corbett, 2000).

For a discussion of force majeure (Van Dunne, 2002) and claims under the FIDIC Construction Contract see Seppala (2000) and Corbett (2000).

Formal procedures for submitting a claim

Generally, contracts require the project manager (supported by commercial managers) to assess and award extensions of time and/or expenses. However, most contracts place the onus for substantiation entirely on the claimant. Further, they may state specific time frames for submission of details and assessment, and specify the course of action available if the claimant disagrees with the decision.

Kumarasawamy and Yogeswaran (2003) review the techniques and approaches available to substantiate claims for an extension of time and give recommendations on their use. These include: global impact, net impact, time impact, snapshot, adjusted as-built critical path method and isolated delay type techniques. Chapter 7 addresses the practical implications of submitting claims.

Box 6.5 Proposition

Conditions precedent on extension of time loss and expense claims force suppliers into making claims

A common tactic in relation to extension of time and other claims is to say that the supplier has to give notice of the circumstances that he says entitles him to claim at the earliest possible opportunity.

On the face of it, this is to enable the purchaser to take steps to mitigate the problem. What also sometimes happens is that the purchaser is quite clear that the supplier must give this notice to allow an opportunity to mitigate the problem and will put into a contract an obligation on the supplier to give such a notice immediately.

Some contracts will follow this requirement for giving a notice by including a sanction that if the supplier does not put in the notice then the right to claim is lost. In this situation suppliers are forced into making claims in relation to any event, regardless of how minor, because even if they do not know whether a situation ultimately will be something which is at the purchaser's risk a protective notice must be put in to avoid losing the claim. Sometimes parties will see claims as being adversarial and the mere sending of a claim notice to the purchaser may therefore cause a deterioration in the relationship between the parties.

Although there is no doubt that the earlier a problem is identified the earlier it will be resolved, and it will be of benefit to the project for problems which could give rise to claims to be identified early in the process and resolved, it might be better to remove the pejorative nature of a claims notice by writing a management procedure where such things are not regarded as aggressive but simply part of the project management obligations of the parties.

Whether or not a failure to give such notices will result in loss of remedy can be a difficult point. Some contracts provide that the supplier will only be paid what he would have been paid had the purchaser been told of the problem at the appropriate time and therefore had the ability to solve the problem. It is arguable that this is not particularly satisfactory because in hindsight it will be very difficult to prove what the purchaser would have done and what the financial consequences would have been.

On balance, from a purchaser's point of view, it will probably be better to have the notice requirement. After all, strict compliance can always be waived if this would lead to genuine injustice towards the supplier.

Compensation events

The NEC3 Supply Contract includes a list of compensable circumstances (Compensation events Clause 60); like the FIDIC Construction Contract, it includes clauses outlining the process for notifying, assessing and implementing compensation events (Clauses 61, 63 and 65 respectively). However, the form favours a collaborative approach to resolving the compensation events, initially requiring discussion between the parties in order to find a solution; it also includes provisions for the submission of quotations by the supplier to deal with the issue(s) encountered (Clause 62) and assessment by the Supply Manager (Clause 64).

Dispute resolution

Adversarial and non-adversarial dispute resolution

Problems can arise when implementing contracts regardless of the relationship between the parties and the type of project. In addition to those mentioned earlier, problems can arise as a result of one or both of the parties having conflicting objectives, failing to anticipate significant risks or variations, failing to consult, making erroneous assumptions, and – at a basic level – making a mistake.

A major role of the project manager and commercial manager is in predicting potential difficulties in implementing the project and in intervening to prevent problems turning into disputes. Commercial managers should endeavour to reduce the effects of problems by instituting transparent procedures that are adhered to, by promoting collaboration and by engendering a shared aspiration to resolve difficulties. The effect of problems can be reduced by:

- Establishing approved procedures
- Establishing and adhering to boundaries of delegated authority

Conflict management

Non-binding

Dispute review advisors
Dispute review boards
Negotiation
Quality matters:
  Quality assurance
  Total quality management
Procurement systems

Dispute resolution

Non-binding	*Binding*
Conciliation	Adjudication
Executive tribunal	Arbitration
Mediation	Expert determination
	Litigation
	Negotiation

Figure 6.3 Conflict management/dispute resolution taxonomy. Adapted from Fenn *et al.* (1997).

- Making contingency plans
- Setting up regular reviews with both the purchaser and the supplier
- Instigating timely recognition and corrective action
- Implementing appropriate contract changes
- Escalating, where appropriate, the problem to senior management (CUP, 1997)

Traditionally, both in the US and the UK, the preferred method of dispute resolution arising out of contracts was litigation: a slow and costly process. Contracts, therefore, generally incorporate other methods of dispute resolution to avoid disrupting the implementation of the project and to resolve any dispute in a fair and timely manner. Examples include: mediation, conciliation, dispute review boards, adjudication, and arbitration.

Fenn *et al.* (1997) distinguish between **conflict** and **disputes**. Conflict they assert, although an unavoidable fact of organisational life, has positive aspects to do with commercial risk-taking; alternatively, disputes afflict industry. They propose a conflict management/dispute resolution taxonomy (Figure 6.3).

Elsewhere, Cheung (1999) presents a dispute resolution 'stair-step' chart comprised of the following steps:

- **Prevention**: includes risk allocation, inceptive for cooperation, and partnering
- **Negotiation**: includes direct and step negotiation
- **Standing neutral**: includes dispute review board, and dispute resolution adviser
- **Non-binding resolution**: includes mediation, mini-trial, and adjudication
- **Binding resolution**: arbitration
- **Litigation**: judgment by the courts

Antagonism and cost increase as one goes down the list (that is, up the stairs). Moreover, resorting to arbitration or litigation is unlikely to improve the implementation of the project and should, therefore, only be used as a final measure. Furthermore, if they are used, the contract has effectively failed.

In the Hong Kong construction industry, Cheung found that when deciding upon a method of dispute resolution, the parties are mainly interested in the benefits, be they tangible or perceived, that may be gained. These benefits include prompt resolution, low cost and preservation of the relationship between the parties.

Box 6.6 Case study: Runway resurfacing project

Consider a project where a runway was being resurfaced. The technique being used was that strips of the runway would be removed and the underlying strata examined. Any problems would be resolved and the relevant portion of the runway reinstated and concreted over. This project was to take place during the night with the airport being closed for the purpose. However, there was always a flight due very early the next morning and it was important to get the runway functional again so that it could be used. One of the issues that the parties foresaw as being a likely cause of a dispute was whether or not the concrete had been compacted adequately.

This was not a question of whether or not further compaction should take place; it was a question of whether the supplier should be paid extra (and possibly given an extension of time) for such compaction in the event that it was unnecessary. If the supplier had to spend a significant amount of time re-compacting concrete he might not have the time to complete the next strip during the night. It was agreed between the parties that if the owner's engineer declared that the runway needed more compacting before aircraft could safely land on it then this would be done. However, they realised that it would be almost impossible after the event to decide whether it was strictly necessary because the effect of excessive compacting on the runway surface was negligible.

It was therefore decided that a structural engineer who lived near the airport would be asked to stay 'on standby' during the night and that if there was a dispute as to whether or not the concrete was adequately compacted he would be asked to attend the airport – he lived about ten minutes away – and he would adjudicate on the question of whether or not the supplier would be paid extra or given an extension of time. This decision would not be capable of challenge except in the case of fraud or obvious error.

The effect of this instant dispute resolution system was to deter the parties from making spurious claims because it was recognised by both of them that somebody with very accurate measuring equipment could immediately tell the parties whether or not one of them was being unreasonable. Over a contract period of nearly a year the engineer was never called out.

Generally, governments are reluctant to legislate in regard to commercial contracts. However, in the case of construction contracts there are exceptions:

- In 1981 the US state of California established mandatory arbitration for state agencies' construction contracts (Carty, 1995)
- In the UK, the Housing Grants, Construction and Regeneration Act 1996 ('the Construction Act'), subsequently amended by the Local Democracy, Economic Development and Construction Act 2009, established the right for a party to a construction contract to refer a dispute arising under the contract for adjudication, as outlined in the Scheme for Construction Contracts (England and Wales) Regulations 1998. The decision of the adjudicator will be binding until the dispute is finally decided by legal proceedings, by arbitration or by agreement. For a detailed review of the implications of this act see Paterson and Britton (2000)

Dispute resolution under the FIDIC Construction Contract and NEC3 Supply Contract

Initially, under the FIDIC Construction Contract the engineer will proceed to agree or determine any matter. In doing so:

> '... the engineer shall consult each Party in an endeavour to reach agreement. If agreement is not achieved, the engineer shall make a fair determination, in accordance with the Contract, taking due regard of all relevant circumstances.' (Sub-clause 3.5)

The contract also makes provision for the establishment of a **dispute adjudication board** (DAB) (Sub-clause 20.2), comprising either one or three suitably qualified persons, to adjudicate disputes that may arise among the parties. Where a dispute has been referred to the DAB and a decision, if any, has not become final and binding, the contract states that it shall ultimately be settled by international adjudication

For a discussion on the resolution of construction disputes by disputes review boards see Shadbolt (1999) and specifically under the FIDIC Construction Contract see Seppala (2000).

The NEC3 Supply Contract (Clause 94) requires a dispute arising under or in relation with the contract to be notified and referred to and decided by an adjudicator, who is appointed at the start of the contract by the parties. The contract includes an adjudication table that sets out what, when and by whom a dispute may be referred to the adjudicator. Following a decision by the adjudicator, if a party is dissatisfied with the outcome, the dispute can be referred to a tribunal, which may be arbitration if stated in the contract.

Again, Chapter 8 covers the practical implications of conflict management and dispute resolution.

Box 6.7 Proposition

Final certificate regimes force suppliers into disputes

There is often a long period of time during which the parties agree a final account. Some contracts provide that at the end of this period the purchaser will issue a final certificate stating how much is to be paid to the supplier

The effect of the final certificate is sometimes that (unless the supplier challenges the certificate by starting a dispute resolution procedure) the certificate will become binding on the supplier. In other words if the supplier has a claim and the purchaser rejects it then the supplier has to make a decision whether or not to pursue this claim. If the supplier decides to continue with the claim, this can only be done by issuing proceedings.

It may be that if such a provision was not in the contract then the supplier might either decide not to pursue the claim once it has been fully investigated or there might have been time for negotiation.

Further reading

Conflict management and dispute resolution

Fenn P (2006) Conflict management and dispute resolution (Chapter 11). In: DJ Lowe with R Leiringer (eds) *Commercial Management of Projects: Defining the Discipline*. Blackwell Publishing, Oxford. pp. 234–26

Fenn, P (2011) *Commercial Conflict Management and Dispute Resolution*. Taylor & Francis, Abingdon, Oxon

Exercise 6.3 Conflict management and dispute resolution

For a recent project that you have been involved in where a dispute or conflict occurred:

- Describe how the issue(s) were resolved
- Evaluate how the contract aided or hindered the resolution process

Flexibility, clarity and simplicity

Effective contracts require flexibility, clarity and simplicity, and should stimulate good management. Standard forms have been criticised for failing to provide these attributes. For example, standard forms of contract within the UK construction industry have been criticised for their complexity and lack of clarity – criticisms that could be levelled at standard forms of contract in many industries and countries. Possible reasons for this, according to Broome and Hayes (1997), relate to their origin, being derived from very old precedents: age, the language and phrasing being derived from English contracts of the late nineteenth century; development by committee; partisanship; lack of direction; and amendment, where specific users heavily amend and supplement the standard form.

Likewise, a review of procurement and contractual arrangements in the UK construction industry (Latham, 1994) criticised the existing standard forms of contract. Contracts, it suggests, should be fair, comprise simple phrasing, set transparent management procedures and encourage teamwork. As a result, several forms including the ACA PPC2000 Standard Form of Contract for Project Partnering, the NEC3 Engineering and Construction Contract, and the JCT – Constructing Excellence Contract have been drafted to address these concerns.

The New Engineering and Construction Contract (NEC, 2005), for example, which was designed to be used internationally, includes virtually all Latham's recommendations. The NEC suite of contracts has been written to form a manual of project management procedures rather than an agenda for litigation. However, a survey by Hughes and Maeda (2003) found their respondents to be ambivalent about the concept of a spirit of mutual trust; moreover, they held that authoritative contract management would improve performance. This finding is divergent from the principles behind current steps towards innovative working.

Collaboration

This section examines three construction contracts considered by Sir Michael Latham, author of the influential report on the UK construction sector *Constructing the Team* (Latham, 1994), to encapsulate his recommendations and principles for a modern construction contract. The contracts are:

- ACA PPC2000 Standard Form of Contract for Project Partnering (ACA, 2003) (PPC2000)
- NEC3 Engineering and Construction Contract (NEC, 2006a) (NEC3)
- JCT – Constructing Excellence Contract (CE) Revision 1: 2009 (JCT, 2006a)/JCT – Constructing Excellence Contract Project Team Agreement (CE/P) Revision 1: 2009 (2006b)

While PPC2000 is a multiparty partnering contract, both NEC3 and JCT-CE are bilateral contracts, which embrace collaborative methodologies. NEC3 incorporates a partnering option (Option X12) for use when more than two parties are involved on a project, while the JCT contract has an associated Project Team Agreement (JCT-CE/P), a multiparty arrangement that facilitates and reinforces cooperation.

Table 6.2 presents a comparison of the main features (terms and conditions of these three standard forms of contract) that are considered to promote collaborative working.

Besides Latham's support, all three contracts have received industry-wide backing. Currently, however, only the NEC form is endorsed by the Office of Government Commerce (OGC) as being fully compliant with the Achieving Excellence in Construction (AEC) principles. In response to queries concerning whether or not OGC should extend its endorsement to PPC2000 and JCT-CE, Arup Project Management was appointed by OGC to review the three contracts. Evaluation criteria were based upon the encouragement of collaborative working arrangements as outlined in the AEC initiative. Specifically, the review considered the features of the contract that:

- Encouraged:
  - collaborative working
  - project processes necessary for successful projects
  - the achievement of value
  - supply chain management

Table 6.2 Comparison of three recent construction contracts that promote collaborative working.

	PPC2000	NEC3 Engineering and Construction Contract	JCT 05 – Constructing Excellence Contract (CE)
General features			
Multiparty contract	Yes	No – the basic NEC form is a partnering contract between two parties	No – the approach is a series of bilateral contracts (within a common framework), which embraces a collaborative methodology
Partnering option	N/A	Yes – Option X12 (Partnering) employed for partnering between more than two parties involved on a project (or programme)	Yes – associated CE/P JCT – Constructing Excellence Contract Project Team Agreement - a multi-party Project Team Agreement which facilitates/reinforces a cooperative approach Under this agreement Project Team members do not owe each other a duty of care; further no Project Team member has any liability to any other member for any act or omission of the Project Team or its members. However, this does not apply in respect of payments provisions under the agreement
Single/multiple projects	Single	Single (although the NEC3 Framework Contract is available)	Single (although the form can be used in conjunction with the JCT Framework Agreement for a series of projects)
Intended parties	Client and first tier suppliers – related Standard forms for specialist contracts SPC2000 and SPC International available The form has been utilised by the public, voluntary and private sectors to deliver offices, residential, educational, healthcare, leisure and public buildings, plus road and rail infrastructure The contract can be used on any form of partnered project in any jurisdiction	Client and first tier suppliers – related NEC3 Subcontract (ECS), Short Subcontract, and Professional Services Contract (PSC) available Applicable to public and private sector clients both in the UK and internationally	Client, contractors, consultants, or subcontractors at all tiers. Parties are referred to as the 'Purchaser' and 'Supplier' While applicable to both the private and public sectors, the contract has been developed to meet the particular needs of local authorities and other public sector clients

(Continued)

Table 6.2 (*cont'd*).

	PPC 2000	NEC3 Engineering and Construction Contract	JCT 05 – Constructing Excellence Contract (CE)
Contract philosophy	Collaborative working is central to the contract, which incorporates processes (a base and route map) to encourage effective project delivery as part of a Partnering Team	Designed to address all facets of the management of engineering and construction projects, the contract sets project management procedures within a legal framework Stated benefits include: stimulus to good management, flexibility and simplicity, which can be applied to any project, large or small	The elimination of waste and the successful delivery of a project is achieved by the identification, collaboration between and management of the entire supply chain Project members work together for the benefit of the project, which in turn creates value for the individual participants
Payment terms	A mechanism is provided to enable the parties to: • Develop and agree prices for all elements of the project • Establish a Maximum Price within a Budget included in a Price Framework – at the lowest price that delivers best value (Clause 12.3) A fixed rate is included for the Constructor's profit, site and central office overheads (Clause 12.4) The contract also provides for the payment of: • Services carried out prior to the date of the Commencement Agreement (Clause 12.1) • Activities under any Pre-Possession Agreement (Clause 12.2)	Contract variants: • Priced contract with activity schedule (Option A) • Priced contract with bill of quantities (Option B) • Target contract with activity schedule (Option C) • Target contract with bill of quantities (Option D) • Cost reimbursable contract (Option E) • Management contract (Option F) Option selected prior to appointment	Section 7 Payment: • Target Cost option based on a target cost and guaranteed maximum cost • Contract Sum option by reference to a contract sum Option selected prior to appointment Under the optional risk and reward sharing arrangements (Section 3 of the Project Team Agreement) provision is made for the establishment of a Project Target Cost by the members of the Project Team
Features that promote collaborative working			
Collaboration	*The Partnering Team members shall work together and individually, in accordance with the Partnering Documents, to achieve transparent and cooperative exchange of information in all matters relating to the Project and to organise and integrate their activities as a collaborative team* (Clause 3.1)	*The Employer, the Contractor, the Project manager and the Supervisor shall act as stated in this contract and in the spirit of mutual trust and co-operation* (Actions 10.1 and Option X12)	*The Overriding Principle guiding the Purchaser and the Supplier in the operation of this Contract is that of collaboration. It is their intention to work together with each other and with all other Project Participants in a co-operative and collaborative manner in good faith and in the spirit of mutual trust and respect. To this end the Purchaser*

			and the Supplier agree they shall each give to, and welcome from, the other, and the other Project Participants, feedback on performance and shall draw each other's attention to any difficulties and shall share information openly, at the earliest practicable time. They shall support collaborative behaviour and address behaviour that does not comply with the Overriding Principle (Clause 2.1)
Core group/project team	A Core Group is established by the members of the Partnering Team (Clause 3.3). Its role is to: • To review and stimulate progress of the Project • Implement the Partnering Contract Core Group decision making is by consensus (i.e. all members present at a particular meeting) – members of the Partnering Team are to comply with any decision made within specified terms of reference (Clause 3.6).	Option X12: Partnering establishes a 'Core Group' from the partners listed in a Schedule of Core Group Members. The Core Group, for example: • Acts and takes decisions on behalf of the Partners on specified issues (Clause X12.2.3) • Decides how the Core Group operates (Clause X12.2.4) • Issues and revises the Schedule of Core Group Members and the Schedule of Partners (Clause X12.2.5) Unless otherwise agreed, the Client's representative heads the Core Group	Each Project Team member is represented on the Project Team, an advisory body, whose function is to guide the successful delivery of the Project through its design and construction. [Section 2 Working Together (Contract) and Section 2 Working together as the Project Team (Project Team Agreement)] Tasks undertaken by the Project Team may include: • Reviewing progress of the project • Reviewing and revising the project Risk Register • Considering risk avoidance and/or mitigation measures • Reviewing the results of any project planning, risk or value engineering workshops • Monitoring the performance of a contributor to the project • Considering any Project Team Relief Event or dispute • Reviewing the performance of Project Team members against their KPIs • Considering opportunities to deliver improved value • Where applicable, considering the proposed Final Cost of the project under its risk sharing arrangements

(*Continued*)

Table 6.2 (*cont'd*).

	PPC 2000	NEC3 Engineering and Construction Contract	JCT 05 – Constructing Excellence Contract (CE)
Partnering objectives	Partnering relationships are to be established, developed and implemented by members of the Partnering Team The aim is to deliver: (i) *trust, fairness, mutual co-operation, dedication to agreed common goals and an understanding of each other's expectations and values* (ii) *finalisation of the required designs, timetables, prices and supply chain for the Project* (iii) *innovation, improved efficiency, cost-effectiveness, lean production and reduction or elimination of waste* (iv) *completion of the Project within the agreed time and price and to the agreed quality* (v) *measurable continuous improvement…* (vi) *commitment to people…* plus any further objectives included in the Partnering Documents for the benefit of the Project and the common benefit of the Partnering Team (Clause 4.1)	That Partners work together as required by the Partnering Information and '… *in a spirit of mutual trust and co-operation*' (X12.3: Working together) A Partner: • May request information from another Partner, if required, to complete the work under its own contract (the Partner shall comply with the request) • Is required to forewarn its Partners of any issue that may impact on the realisation of their objectives • Implements Core Group decisions under its own contracts • Uses common information systems as specified in the Partnering Information • Gives (full, open and objective) advice, information and opinion to the Core Group and its Partners when requested The Core Group: • May issue instructions to the Partners to amend the Partnering Information • Prepares and maintains the project timetable • Issues revised copies of the timetable to the Partners	*If not already prepared, the Parties shall draw up and adopt a project protocol setting out the aims and objectives of the Project Team with regard to the delivery of the Project and the development of their working relationships* (Clause 2.8 – Project Team Agreement)
Partnering facilitator	A Partnering Adviser is appointed to advise and support the Partnering Team either individually or collectively (Clause 5.6).	N/A	N/A

Intellectual property rights (IPR)	IPR is retained by the individual Partnering Team members, who grants the Client and the other Partnering Team members '... *an irrevocable, non-exclusive, royalty-free licence to copy and use all such designs and documents for any purpose relating to completion of the Project...*' and in respect of the Client '... *the Operation of the Project*' (Clause 9.2)	Not specifically mentioned, however, under Clause 22.1: *The Employer may use and copy the Contractor's design for any purpose connected with construction, use, alteration or demolition of the works unless otherwise stated in the Works Information and for other purposes as stated in the Works Information*	IPR (Copyright) is retained by the individual Supplier (or the relevant member of its Supply Chain), who grants (or ensures that the member of its Supply Chain grants) to the Purchaser '... *an irrevocable, assign-able, royalty free licence to use, copy and reproduce all designs and related documents prepared in connection with the Services for any purpose relating to the Project including, without limitation, the construction, maintenance, letting, sale, promotion, advertisement, reinstatement, refurbishment and repair of the Project...*'. (Clause 4.10)
Integration of the supply chain	Clause 10 (Supply Chain) defines the procedures for the integration of the individual partnering team member's supply chains. The principles include the incorporation/provision of: • Open-book arrangements • Terms and conditions to reflect the requirements of the Client, the interests of the Partnering team members and the project • The 'best available' warranties and support • Maximum in novation Plus the establishment and demonstration of best value for the client and, wherever practical, establishment of partnering relationships consistent with the Partnering Contract	Limited references to the extended supply chain; the contract includes clauses relating to the appointment and position of subcontractors (Clause 26)	*The Supplier shall endeavour to work together with, and fully involve, his Supply Chain (if any) in the delivery of the Services, and shall organise or take part in project planning, risk and value engineering workshops involving all or relevant members of his Supply Chain and other Project Participants as necessary or appropriate to the stage of the Project* (Clause 4.16) *The Supplier shall use reasonable endeavours to engage all members of his Supply Chain using the JCT – Constructing Excellence Contract (CE) or otherwise on terms that fully reflect the terms of this Contract... The Supplier acknowledges that terms imposing more onerous obligations on members of the Supply Chain are to be avoided...* (Clause 4.16).

(*Continued*)

Table 6.2 *(cont'd)*.

	PPC 2000	NEC3 Engineering and Construction Contract	JCT 05 – Constructing Excellence Contract (CE)
Incentivisation Gain/pain share	Incentives (shared savings arrangements and/or added value inducements) may be included in the Partnering Documents to influence the behaviour of Partnering Team members to '*maximise their efforts*' for the benefit of the Project. Also, the Core Group is required to consider and seek to agree additional appropriate incentives Clause 13.1), which are then recommended to the Client for approval (Clause 13.2). All incentive arrangements are to be implemented by the members of the Partnering Team (Clause 13.2).	The following incentives are available under Option C: Target contract with activity schedule and Option D: Target contract with bill of quantities: The responsibility for assessing the Contractor's share of the difference between the sum of the Prices and the price for 'Work Done to Date' lies with the Project Manager in accordance with the methodology for its calculation provided in Clauses 53.1 and 53.5 Where the price for 'Work Done to Date' is lower than the sum of the Prices, the Contractor receives a predetermined proportion of the saving. Conversely, where the price for 'Work Done to Date' is more than the sum of the Prices, the Contractor pays the Client a predetermined proportion of the over spend (Clauses 53.2 and 53.6)	The contract incorporates both gain share and pain share mechanisms: • Gain share: where the Actual Cost of delivering the Services is lower than the agreed Target Cost, a proportion of the difference is shared between the Parties as specified in the Contract Particulars (Clause 7.11) • Pain share: where the Actual Cost of delivering the Services is greater than an agreed *Guaranteed Maximum Cost*, a proportion of the difference is borne by the Supplier as specified in the Contract Particulars (Clause 7.13) Optional risk and reward sharing arrangements between members of the Project Team are also provided in Section 3 of the Project Team Agreement: • Surplus (when a project Final Cost has been established): where the Final Cost of delivering the project is lower than the agreed Project Target Cost, a proportion of the difference is paid by the Client to each Party, as agreed, up to the their individual Maximum Benefit (Clause 3.10) • Deficit: where the Final Cost of delivering the project is greater than the agreed Project Target Cost, a proportion of the difference is paid to the Client by each Party, as agreed, up to the their individual Maximum Liability (Clause 3.11)

Bonus for early completion	Incentive payments can be linked to the project completion date or any other target established as a KPI (Clause 13.5).	Option X6 enables bonus payments to be made at a predetermined rate/day	Provisions can be made for a bonus payment or payments to be made where the Service or a predetermined section of the Service is completed ahead of schedule (Clause 7.28)
Risk management:	Asserts that the Partnering Team members recognise the risks (and their associated costs) involved in the design, supply and construction of the Project. Both at the Partnering Team level and supplier level, Risk Management exercises are to be undertaken to analyse and manage these risks effectively. The aim being to: • Identify risks and their potential cost impact • Eliminate or reduce risks and their potential cost impact • Insure risks when relevant and cost effective • Distribute/allocate risks to the Partnering Team member most able to manage a particular risk (Clause 18.1) Generally the supplier (Constructor) is responsible for managing all risks connected with the project and its site, from the date of the Commencement Agreement until the Completion Date. Any exceptions to this are to be stated in the Partnering Terms or in the risk sharing arrangement(s) specified in the Commencement Agreement (Clause 18.2).	Clause 80.1 (Employer's risks) itemises the general risks borne by the client and allows for additional Employer's risks to be included in the Contract Data Clause 81.1 (The Contractor's risks) states that all risks not borne by the Employer are to be carried by the Contractor A Risk Register is incorporated as part of the Contract Data, which is revised by the Project Manager to include early warning events identified by the Project Manager or Contractor (plus the Core Group if X12 used). Also, the contract allows for the convening of collaborative risk reduction meetings (Clauses 11.2.14 and 16)	Active project risk identification and management is promoted by the inclusion of a mandatory Risk Register (Clause 5.1) A Risk Allocation Schedule, usually derived from the Risk Register, forms part of the Agreement, enabling the identified risks to be described, valued (i.e. a statement of any amount contained within the Target Cost/Contract Sum to meet the risk) and a time period attributed (for which the Supplier is responsible). The schedule also enables the cost and time consequences of the risk to be apportioned between the Purchaser and Supplier. Two Risk Allocation Schedule options are available, where the method of adjustment is either based on: • Each risk (Schedule A) or • The total amount/total period of all the risks (Schedule B) Responsibility for preparing, updating and amending the Risk Register can be assigned to the Supplier (Clause 5.1 and 5.2) The cost/time consequences of risks are to be allocated on a fair/practical basis. The contract also deals fairly with the occurrence of risks: 'Relief Events' (Clauses 5.7 to 5.16)

(*Continued*)

Table 6.2 (*cont'd*).

	PPC 2000	NEC3 Engineering and Construction Contract	JCT 05 – Constructing Excellence Contract (CE)
Performance measurement – key performance indicators (KPIs)	Includes: • Regular performance reviews (by reference to KPIs) of each Partnering Team member by the Core Group • Provision of information, on an open-book basis, by each Partnering Team member to indicate progress against its KPIs • Review of continuous improvement proposals by the Core Group (Clause 23)	If Option X12 adopted and KPI's agreed	If Part 6 is implemented and KPIs established, KPIs are used by the Supplier to monitored the performance of the Purchaser and likewise by the Purchaser to monitor the performance of the Supplier (Clause 6.1) Additionally, regular formal reviews of both Purchaser and Supplier performance (against their KPIs) are required, together with a discussion of ways to improve their performance (Clause 6.2) Under the Project Team Agreement, monitoring may be undertaken jointly by the Project Team (Clause 6.3)
Transparency (open book approach)	Transparency in the establishment of the Price Framework	• If 'target cost' option selected transparency of cost and adjustment of Target • If 'Lump Sum' option selected – transparency of Fee and Defined Cost of Compensation Events	• If 'target cost' option selected – definition of Actual Cost and records • If 'contract sum' option selected – no transparency of pricing and no clear basis of valuing Relief Events (Clauses 7.2 to 7.4)
Insurance	Project insurance (including the site, any structures on it or any identified risk) is obtained in the joint names of the parties by the Partnering Team member designated in the Commencement Agreement (Clause 19.1)	Insurance is provided by the Contractor as required by the 'Insurance Table', plus any further insurance as itemised in the Contract Data. Similarly, any insurance provided by the Employer is also listed in the Contract Data (Clause 84.1)	The Purchaser and the Supplier are both required to maintain insurance cover as designated in the Contract Particulars (Clause 8.1) Where available, comprehensive project insurance should be considered (Footnote 14)
Dispute resolution	A staged problem solving and dispute avoidance/resolution process is included:	NEC3 contains two dispute resolution procedures, one (Option W2) to be used in the United Kingdom when the Housing Grants, Construction and Regeneration Act 1996 applies, the second (Option W1) where it does not	The aim is for disputes to be resolved by the Project Team using a collaborative approach. Failing this the contract allows for resolution by mediation, adjudication or litigation. Adjudication is to be conducted in accordance with

	• Initially differences or disputes are referred to a Problem-Solving Hierarchy, which seeks to achieve an agreed solution • If the Problem-Solving Hierarchy fails to resolve the issue or find a solution the issue is referred to the Core Group for review • Following these two stages, if the difference or dispute is still unresolved, then the issue may be referred to conciliation, mediation or any other form of alternative dispute resolution • Any party to the difference or dispute has the right to refer the issue to adjudication • If adjudication is unsuccessful any issue can either be referred to the courts (as set out in the Project Partnering Agreement) or, if permitted, to an arbitrator (Clause 27: Problem solving and dispute avoidance or resolution)	Both procedures enable a dispute arising under or in connection with the contract to be referred to and decided by an Adjudicator. Following adjudication, a dissatisfied party may refer a dispute to a tribunal. If the tribunal is arbitration, the arbitration procedure is to be as stated in the Contract Data	The Scheme for Construction Contracts. There is no provision for arbitration Corresponding provisions are provided under section 4: Dispute Resolution in the Project Team Agreement
Additional information			
Introduced/launched	September 2000 – The first standard form of Project Partnering Contract – (amended in 2003 and October 2008)	NEC3 June 2005 (amended June 2006) NEC2 January 1995 NEC March 1993	1 March 2007 (attestation update February 2008)
Impetus	Developed in response to the recommendations of the *Rethinking Construction* ('the Egan') report and contains principles proposed in the Construction Industry Council's *Guide to Project Team Partnering* Plus, a recognition that the construction and engineering sectors required a process document that encompassed all the functions	To create a contract that met both the present and future requirements of the engineering, building and construction industries; the objective being to improve upon existing forms of contract in terms of: • Flexibility • Clarity and simplicity • Stimulus of good management	To provide a contract that supported collaborative and integrated team working within the supply chain, with the aim of: • Encouraging collaborative behaviour • Promoting and requiring the implementation of risk management during the pre-tender stage (to aid the delivery of successful projects)

(*Continued*)

Table 6.2 (*cont'd*).

	PPC 2000	NEC3 Engineering and Construction Contract	JCT 05 – Constructing Excellence Contract (CE)
	involved in the design and delivery of a project during its pre-construction and construction stages	Edition 2 of the contract was revised so as to be compliant with the principles for a modern contract as recommended by the 'Constructing the Team' (The Latham) report	• Providing flexibility in use • Being applied throughout the supply chain

Endorsements	• Endorsed by The Construction Industry Council
	• Recommended by Constructing Excellence.
	• Supported by The Housing Corporation
	• Sir Michael Latham, Chairman of ConstructionSkills confirmed that the contract incorporates all his recommendations and principles for a modern construction contract
		• OGC (Office of Government Commerce):
		  ◦ Recommends the use of NEC3 by public sector construction procurers
		  ◦ Considers NEC3 to be fully compliant with the Achieving Excellence in Construction (AEC) principles
		• Sir Michael Latham, Chairman of CITB-ConstructionSkills
			• Constructing Excellence in the built environment
			• The Local Government Association (LGA)
			• Sir Michael Latham, Chairman of CITB-ConstructionSkills

Associated forms	PPC(S)2000 – the Scottish supplement, PPC International SPC2000* SPC International* * = Standard forms for specialist contracts	NEC3 Engineering and Construction Subcontract (ECS) NEC3 Engineering and Construction Short Subcontract NEC3 Professional Services Contract (PSC) NEC3 Engineering and Construction Short Contract (ECSC) NEC3 Adjudicator's Contract (AC) NEC3 Term Service Contract NEC3 Framework Contract (FC)	CE/P JCT – Constructing Excellence Contract Project Team Agreement Attestation Update (February 2008)
Publishers/authors	Association of Consultant Architects, 98 Hayes Road, Bromley, Kent BR2 9AB Authors: Trowers & Hamlins, Sceptre Court, 40 Tower Hill, London EC3N 4DX	NEC, 1 Heron Quay, London, E14 4JD Published by Thomas Telford Ltd	The Joint Contracts Tribunal (JCT), 4th Floor, 28 Ely Place, London EC1N 6TD/Constructing Excellence Published by Sweet and Maxwell Principal authors, Giles Dixon and Martin Howe, solicitors

- dispute prevention
- early dispute resolution
- risk management
- client and supply chain involvement in design development
- Provided:
  - processes for dealing with variation control and pricing
  - performance management
  - risk allocation
  - clear terms regarding variation pricing and impact of variations on programme
- Incentivised supply chain performance

plus the user-friendliness of the documentation (Arup, 2008).

Arup established that all three contracts fulfilled OGC's evaluation criteria – that is, used as intended, each contract would enable the contracting parties to realise OGC's AEC principles. Further, they concluded that:

> 'No single contract is superior to the other two in all respects – each has its own strengths and weaknesses and each is highly adaptable. The difference in the way that each contract is applied by users will be at least as significant as the differences in the processes or terms and conditions provided within the contract.' (Arup, 2008)

Procurement/acquisition regulations

The procurement of assets and services by government bodies and agencies is generally regulated by statutes. For example, in Europe public procurement is governed via a series of the European Union Directives and Regulations, which are subsequently incorporated into the national legislation of the member counties. Similarly, in the US the 'Federal Acquisition Regulation' (FAR) underpins a 'Federal Acquisition Regulations System', which codifies and issues standardised acquisition policies and procedures for use by all executive agencies. A comprehensive knowledge of the appropriate regulations is required by commercial practitioners working for both public sector customers and suppliers seeking to contract with government entities. As contracting operations become increasingly global, knowledge of country-specific procurement and acquisition regulations is crucial.

EU Procurement Directives and the corresponding UK Regulations

Procurement procedures for the award of public works, supply and service contracts are regulated by the Public Contracts Directive 2004/18/EC within the European Union (EU).[4] This legislation is implemented within the UK under the Public Contracts Regulations 2006,[5] which regulate the tendering procedures for supplies, services and works contracts offered by public authorities, such as central government, local and regional authorities, health authorities and similar bodies. Works in the utilities sector (water, energy, transport and postal services sector) are covered by the Utilities Contracts Regulations 2006,[6] which implements the EU legislation: Utility Contracts Directive 2004/17/EC.[7] The applicable EU Directives and their associated UK Regulations are provided in Table 6.3.

Although certain contracts are currently excluded from the regulations – for example, works that are declared to be secret by a member state, works covered by special security measures, and contracts entered into in accordance with an international agreement – measures have been taken to restrict these exemptions wherever possible. For instance the European Defence and Security Directive (directive 2009/81/EC), seeks increased competition within the sector by opening up the European defence market. The new Directive, however, still permits member states, in exceptional circumstances, to exempt specific contracts from competition.

The current threshold values for supplies, services and works covered by the EU procurement rules are given in Box 6.8.

Chapter 8 outlines the principle implications of the European Procurement Regulations on procurement practice.

Table 6.3 EU Procurement Directives and the corresponding UK Regulations.

European legislation	UK implementation
Public Contracts Directive 2004/18/EC: Procedures for the award of public works contracts, public supply contracts and public service contracts	*Statutory Instrument 2006 No. 5*: The Public Contracts Regulations 2006 (see amendments below) *Scottish Statutory Instrument 2012 No. 88*: The Public Contracts (Scotland) Regulations 2012
Utility Contracts Directive 2004/17/EC: Procedures of entities operating in the water, energy, transport and postal services sectors	*Statutory Instrument 2006 No. 6*: The Utilities Contracts Regulations 2006 (see amendments below) *Scottish Statutory Instrument 2012 No. 89*: The Utilities Contracts (Scotland) Regulations 2012
Defence & Security Directive 2009/81/EC: Procedures of entities operating in the fields of defence and security	*Statutory Instrument 2011 No. 1848*: The Defence and Security Public Contracts Regulations 2011
Directive 2007/66/EC Remedies Directive: Amends the existing Remedies Directives (89/665/EEC and 92/13/EEC) to improve the effectiveness of review procedures concerning the award of public contracts	*The Public Contracts (Amendment) Regulations 2009* and *The Utilities Contracts (Amendment) Regulations 2009* Incorporated into The Public Contracts (Scotland) Regulations 2012 and The Utilities Contracts (Scotland) Regulations 2012 as above
Standard Forms – Regulation 842/2011: Standard forms for the publication of notices in the framework of public procurement procedures pursuant to Directives 2004/17/EC and 2004/18/EC	
Threshold amendments – Regulation 1251/2011: Amending Directives 2004/17/EC, 2004/18/EC and 2009/81/EC of the European Parliament and of the Council in respect of their application thresholds for the procedures for the awards of contract	
Communication 2011/C 353/01: The corresponding values in the national currencies other than the Euro of the thresholds as amended by Commission Regulation (EC) No 1251/2011	
	The Public Contracts Regulations 2006 and The Utilities Contracts Regulations 2006 have been amended as below: *The Money Laundering Regulations 2007* *The Public Contracts and Utilities Contracts (Amendment) Regulations 2007* *The Public Contracts and Utilities Contracts (Postal Services Amendments) Regulations 2008* *The Public Contracts and Utilities Contracts (CPV Code Amendments) Regulations 2008* *The Bribery Act 2010 (Consequential Amendments) Order 2011* *The Public Procurement (Miscellaneous Amendments) Regulations 2011*

Source: Reprinted from European & UK Procurement Regulations, © Millstream Associates Ltd, 2012 (http://www.mytenders.org/sitehelp/help_legislation.aspx). Reproduced by permission of Millstream Associates Ltd.

Box 6.8 EU procurement thresholds

The European Public Contracts Directive (2004/18/EC) applies to public authorities including, amongst others, government departments, local authorities and NHS Authorities and Trusts. The European Utilities Contracts Directive (2004/17/EC) applies to certain utility companies operating in the Energy, Water and Transport sectors.

The directives set out detailed procedures for the award of contracts whose value equals or exceeds specific thresholds. Details of the thresholds, applying from 1 January 2012 are given below. Thresholds are net of VAT.

Public Contracts Regulations 2006 – from 1 January 2012

	Supplies	Services	Works
Entities listed in Schedule 1[1]	£113,057 (€130,000)	£113,057[2] (€130,000)	£4,348,350[3] (€5,000,000)
Other public sector contracting authorities	£173,934 (€200,000)	£173,934 (€200,000)	£4,348,350[3] (€5,000,000)
Indicative Notices	£652,253 (€750,000)	£652,253 (€750,000)	£4,348,350 (€5,000,000)
Small lots	£69,574 (€80,000)	£69,574 (€80,000)	£869,670 (€1,000,000)

[1] Schedule 1 of the Public Contracts Regulations 2006 lists central government bodies subject to the World Trade Organisation's (WTO) Government Procurement Agreement (GPA). These thresholds will also apply to any successor bodies

[2] With the exception of the following services, which have a threshold of £173,934 (€200,000):

- Part B (residual) services
- Research & Development Services (Category 8)
- The following Telecommunications services in Category 5
  - CPC 7524 – Television and Radio Broadcast services
  - CPC 7525 – Interconnection services
  - CPC 7526 – Integrated telecommunications services
- Subsidised services contracts under regulation 34

[3] Including subsidised works contracts under regulation 34

Utilities Contracts Regulations 2006 – from 1 January 2012

	Supplies	Services	Works
All sectors	£347,868 (€400,000)	£347,868 (€400,000)	£4,348,350 (€5,000,000)
Indicative Notices	£652,253 (€750,000)	£652,253 (€750,000)	£4,348,350 (€5,000,000)
Small lots	£69,574 (€80,000)	£69,574 (€80,000)	£869,670 (€1,000,000)

Source: http://www.ojec.com/Threshholds.aspx

US Federal Acquisition Regulations

In the US, all public executive agencies abide by the 'Federal Acquisition Regulations System' (FARS) policies and procedures, which incorporate the Federal Acquisition Regulation (FAR) and additional agency acquisition regulations. FARS seeks to:

> '... deliver on a timely basis the best value product or service to the customer, while maintaining the public's trust and fulfilling public policy objectives. Participants in the acquisition process should work together as a team and should be empowered to make decisions within their area of responsibility.'

Additionally, the Defense Acquisition Regulations System (DARS), which integrates FAR and the Defense Federal Acquisition Regulation Supplement (DFARS), facilitates the acquisition of goods and services for the Department of Defense (DoD) by developing and maintaining acquisition policy and support. Its policies stress '... *flexibility, responsiveness, innovation, discipline, and streamlined and effective management*', while the Department is committed to '... *providing maximum practical opportunities for competition, both for the initial contract award and for orders placed against multiple award contracts*'.[8]

Competition and antitrust legislation

It is generally held that dynamic, open competition and fair exchanges within markets are beneficial to consumers. Following on from this principle, many countries have introduced competition or antitrust legislation that seeks to protect consumers from biased commercial practices by:

- *Encouraging competition and efficiency in business exchanges*
- *Restraining monopolies and preventing restrictive practices*

An awareness of the implications of competition and antitrust legislation, both within home markets and globally, is of major importance to commercial practitioners. The following overview is provided as an illustration of the areas covered by such legislation and the implications of breaching their provisions.

UK laws on anti-competitive behaviour

Anti-competitive behaviour in the UK is prohibited under Chapters I and II of the Competition Act 1998. Similarly, it may be proscribed under Articles 81 and 82 of the EC Treaty. Principally, the legislation bans:

- **Anti-competitive agreements between businesses** (and deals made by trade associations): for example, those that '... *prevent, restrict or distort competition or are intended to do so and which affect trade in the UK and/or EU*', under Chapter I of the Act and Article 81 of the EC Treaty. Such agreements may include: price fixing of goods and services; apportioning contract opportunities within a market, restricting production; and unfairly discriminating between customers. Cartels – arrangements between entities (suppliers or manufacturers) not to compete with each other, thereby reducing competition and generating artificially high prices – are an extreme example of anti-competitive agreements
- **The abuse of a dominant market position**: defined as the '... *ability to behave independently of competitive pressures, such as other competitors, in that market*', under Chapter II of the Act and Article 82 of the EC Treaty

The Office of Fair Trading (OFT) is responsible for enforcing competition legislation in the UK, although the regulators of specific industries/utility services (for example, air traffic control, communications, electricity, gas, railway and water/waste water) have been granted 'concurrent powers' to implement and enforce the legislation within their remit. The consequences of breaching UK competition legislation can include fines of up to 10% of an organisation's global turnover. Additionally, any third party affected by a breach (for example, a customer, competitor or consumer group) can instigate claims for damages. In the case of cartels, individuals implicated in such activity can face prison terms of up to 5 years and/or receive fines, and directors of undertakings found to be engaged in such activities can be disqualified from holding directorships for up to 15 years.

Australian laws on anti-competitive behaviour

Australia's principal competition legislation is established by Part IV of the Competition and Consumer Act 2010 (CCA2010), although further provisions in respect of anti-competitive behaviour by entities in the telecommunications industry have been made. The Act prohibits the following activities:

- **Cartel conduct** (under Division 1 of Part IV), which is both a criminal and a civil offence; although a limited defence is permissible in the case of joint venture contracts. Individuals convicted in the criminal courts of a cartel offence can receive a prison term of up to 10 years or fines of up to $220,000 per offence
- **Anti-competitive agreements** – both vertical and horizontal (under Section 45 of CCA2010), although the Act provides an exemption for joint ventures, and further legislation allows collective bargaining arrangements in some instances
- **Exclusionary provisions** or primary boycotts (also under Section 45); although, once again, a limited defence is permissible in the case of joint venture contracts
- **Misuse of market power** (under Section 46(1) of CCA2010) – predatory pricing is specifically addressed and prohibited under a recently added clause 46(1AA)
- **Exclusive dealing** – anti-competitive vertical transactions (under Section 47 of CCA2010), which include: *the conditional supply (or purchase) of goods and/or services, and the refusal to supply on specific grounds*. The Act acknowledges the potential benefits that could arise from exclusive dealing and, as a result, a system of notification and authorisation is permissible if considered to be of public benefit
- **Mergers** where it can be shown that such action is likely to or will lead to reduced competition in a specific market. However, clearance or authorisation for a proposed merger can be granted

The Australian Competition and Consumer Commission is the key enforcer of Part IV of CCA2010, with powers to instigate proceedings and obtain evidence where it deems a contravention to have occurred. It also has the authority to grant 'authorisation', when beneficial to the public interest, of activities that may contravene the Act, the exception being in the case of mergers where it has an advisory role. Appeals against decisions, concerning authorisation and notification, made by the Australian Competition and Consumer Commission are considered by the Australian Competition Tribunal (ACT).

In Australia, jurisdiction on all matter relating to competition legislation resides with the Federal Court, which also holds sole jurisdictions under both state and territory Competition Codes. Contraventions of Part IV of the Competition and Consumer Act 2010 can result in the enforcement of both civil remedies and criminal penalties. Civil remedies, for example, include pecuniary penalties, damages, injunctions, divestiture (in the case of mergers), non-punitive orders, punitive orders (for example, adverse publicity orders), and the disqualification of directors. Infringement of the cartel provisions of the Act can result in the imposition of fines (criminal) and imprisonment (with a maximum term of 10 years).[9]

Further examples of competition and antitrust legislation

Administered and enforced by the Competition Bureau, the Canadian Competition Act (R.S.C., 1985, c. C-34 - Act amended last on 12 March 2010) covers: mergers; agreements between Federal financial institutions, vertical agreements between purchasers and suppliers; bid-rigging; and abuse of dominance.[10]

While most states in the US have their own competition legislation (referred to as antitrust laws), the Federal government enforces three key Federal antitrust laws:

- The Sherman Antitrust Act 15 U.S.C. §§ 1-7
- The Clayton Act 15 U.S.C. §§ 12-27
- The Federal Trade Commission Act (15 U.S.C. §§ 41-58, as amended)[11]

In addition to three Acts, further applicable statutes include:

- The Wilson Tariff Act, 15 U.S.C. §§ 8-11
- The Antitrust Civil Process Act, 15 U.S.C. §§ 1311-14
- The International Antitrust Enforcement Assistance Act of 1994, 15 U.S.C. §§ 6201-12

Taken together, these laws seek to prevent activities and business practices that undermine competition and cause inflated prices for goods and services.

Details of the most frequent antitrust violations and information on the signs that may point to the occurrence of anti-competitive collusion can be found at: http://www.justice.gov/atr/public/guidelines/211578.htm (accessed 14 September 2012)

Further reading

Competition and antitrust law

Australian Competition and Consumer Commission (2010) *Competition and Consumer Law: An overview for Small Business.* Australian Competition and Consumer Commission, Canberra, Australia http://www.accc.gov.au/content/item.phtml?itemId=960959&nodeId=831070cac0344078514ad9e60084f469&fn=Competition%20and%20Consumer%20Law.pdf (accessed 27 June 2011)

Competition Bureau (2009) *Competitor Collaboration Guidelines.* Competition Bureau, Gatineau, Quebec, Canada http://www.competitionbureau.gc.ca/eic/site/cb-bc.nsf/vwapj/Competitor-Collaboration-Guidelines-e-2009-12-22.pdf/$FILE/Competitor-Collaboration-Guidelines-e-2009-12-22.pdf (accessed 27 June 2011)

Department of Justice (2009) *Antitrust Division Manual.* 4th edition. Department of Justice, Washington, DC http://www.justice.gov/atr/public/divisionmanual/atrdivman.pdf (accessed 27 June 2011)

Dabbah, MM (2010) *International and Comparative Competition Law (Antitrust and Competition Law).* Cambridge University Press, Cambridge

Jones, A and Sufrin, B (2010) *EU Competition Law: Text, Cases and Materials*, 4th edition. Oxford University Press, Oxford

Office of Fair Trading (OFT) (2005) *Competing Fairly: An Introduction to the Laws on Anti-Competitive Behaviour.* The Office of Fair Trading, London http://www.oft.gov.uk/shared_oft/business_leaflets/ca98_mini_guides/oft447.pdf (accessed 27 June 2011)

International trade law

With the globalisation of trading relationships, commercial practitioners should be mindful of the numerous regulations governing international trade. The following, for example, are the principal international conventions (and the UK legislation that implements them) and additional UK legislation, relating to the transportation of goods by:

- **Air**: the Warsaw and Montreal Conventions (implemented by the Carriage by Air Act 1961 and the Carriage by Air and Road Act 1979)
- **Rail**: the Convention concerning International Carriage by Rail (implemented by the International Transport Conventions Act 1983)
- **Road**: the Convention on the Contract for the International Carriage of Goods by Road (CMR Convention) (implemented via the Carriage of Goods by Road (Parties to Convention) Order 1967). Additional UK legislation includes the Carriage by Air and Road Act 1979 and the Carriage of Dangerous Goods and Use of Transportable Pressure Equipment Regulations 2009 (which incorporates the European Agreement concerning the International Carriage of Dangerous Goods by Road, generally referred to as ADR)
- **Sea**: the Hague–Visby rules (implemented by the Carriage of Goods by Sea Act 1971). Additional legislation includes the Carriage of Goods by Sea Act 1992[12]

Moreover, an appreciation of the standard terms and conditions used in international contracts is essential – for example, those promoted by the International Chamber of Commerce (ICC): Incoterms® 2010.

International commercial terms (Incoterms®)

Originally published in 1936, the Incoterms® rules devised by the ICC are a globally acknowledged set of definitions, standards and conventions used for interpreting commercial terms included in both international and domestic sale of goods contracts. The current version, Incoterms® 2010, became operable on 1 January 2011 and is recognised by the United Nations Commission on International Trade Law (UNCITRAL). Their main purpose is to delineate the costs, risks and responsibilities involved in the transfer (delivery) of goods between vendors and purchasers, thereby reducing the incidence of misinterpretation between the trading parties. An appreciation of the Incoterms® rules is vital for commercial managers involved in international trade and exchanges between EU members.[13]

The Incoterms® 2010 comprise 11 rules classified into two groups (see Table 6.4).

The first seven rules may be used where goods are conveyed by either a single method or multiple means of transport. In the latter case, part of the carriage may be ship. However, the second set of rules requires both the location to which the goods are transported to the purchaser and the point of delivery to be ports.[14]

The Incoterms® 2010 rules also cover the status of electronic communication, assign duties to provide information regarding insurance, allocate obligations to acquire or to assist in acquiring security-related clearances, and address the issue of terminal handling charges.

Letters of credit

A letter of credit, or 'documentary credit', is a guarantee provided by a bank (at a cost) that assures a seller receives payment (a fixed sum) from a buyer within a stipulated time frame. Predominantly used where goods are bought and sold internationally, they provide collateral: transferring the risk of default to the bank. Generally, such agreements will impose stringent terms and require proof of compliance – for example, confirmation of transportation. Payment is conditional, requiring strict conformity with the terms of the letter of credit. Although normally of benefit to both seller and buyer, they can in some instances prolong the transaction and in others result in administrative inconveniences (Business Link, 2011).

Table 6.4 Classification of the Incoterms® 2010 rules.

Rules for any mode or modes of transport	
EXW	Ex Works
FCA	Free Carrier
CPT	Carriage Paid To
CIP	Carriage and Insurance Paid To
DAT	Delivered At Terminal
DAP	Delivered At Place
DDP	Delivered Duty Paid
Rules for sea and inland waterway transport	
FAS	Free Alongside Ship
FOB	Free On Board
CFR	Cost and Freight
CIF	Cost Insurance and Freight

Source: http://www.iccwbo.org/Incoterms/index.html?id=40772 (accessed 14 September 2012)

Further reading

International trade law

See Chapter 20 – International trade. In: Griffiths, M and Griffiths I (2002) *Law for Purchasing and Supply*. 3rd edition. Prentice Hall Financial Times, Harlow. pp. 307–316
Carr, I (2009) *International Trade Law*, 4th edition. Routledge, Abingdon, Oxford
Chuah, J (2009) *Law of International Trade*, 4th edition, Sweet & Maxwell, London
SITPRO (Simplifying International Trade) (2007) *SITPRO Financial Guide: Letters of Credit - Best Practice*. SITPRO Ltd, London http://webarchive.nationalarchives.gov.uk/20100918113753/http://www.sitpro.org.uk//trade/lettcredbest.pdf (accessed 20 July 2011)

Intellectual property and freedom of information

Responding to the opportunities and threats arising from globalisation and rapid technological change, commercial managers frequently become involved in safeguarding an organisation's competitive advantage – for example, by protecting the knowledge and expertise developed and held by the business. The areas of law particularly applicable to this aspect of the function's activities are: intellectual property and freedom of information.

Intellectual property

Intellectual property (IP) is any innovative (original) work that can be traded, whereas IP rights have been defined as:

> '... a bundle of rights that protects applications of ideas and information that have commercial value.' (Cornish and Llewelyn, 2003)

According to Gowers (2006) IP serves three key purposes, to:

1. Incentivise knowledge (and as a result wealth) creation
2. Accrue knowledge in a culture
3. Protect a distinctive identity

However, in realising these objectives a balance needs to be struck between an individual's rights and those of society. For example, as Gowers (2006) states:

> 'Unlike physical property, knowledge, ideas and creations are partial "public goods". Knowledge is inherently non-rivalrous. That means one person's possession, use and enjoyment of the good is not diminished by another's possession, use and enjoyment of the good.'

Notably, IP is considered a source of competitive advantage and revenue (Rivette and Kline, 2000) for individuals, organisations and nations, a point supported by Gowers:

> 'The UK's competitive advantage in the changing global economy is increasingly likely to come through high value added, knowledge intensive goods and services. The Intellectual Property (IP) system provides an essential framework both to promote and protect the innovation and creativity of industry and artists.' (Gowers, 2006)

An alternative perspective is provided by Doctorow (2008) who asserts that the term 'intellectual property' merely relates to '... *knowledge – ideas, words, tunes, blueprints, identifiers, secrets, databases*'. He argues that the term is ideologically loaded and an unsafe euphemism that can result in a variety of flawed interpretations of knowledge. At best '*troublesome*', these flawed concepts are damaging to any nation attempting to transform itself into a 'knowledge economy'. Similarly, Boldrin and Levine (2002) present a case against IP, arguing that it has led to misconceptions and abuses. They consider that current IP legislation

> '... confuses the protection of property rights on objects in which ideas are embodied with the attribution of monopoly power on the idea itself and, furthermore, with restrictions on the usage of such goods on the part of the buyers.' (Boldrin and Levine, 2002)

A variety of IP rights have been developed to protect the diverse applications of knowledge. There are various means of protecting an individual's or organisation's IP, but the principal commercial mechanisms are:

- Confidentiality
- Copyright
- Designs
- Patents
- Trademarks

The Copyright, Designs and Patents Act 1988 (as amended) is the principal legislation that deals with IP rights in the UK. It establishes the right of the originators of 'works' to control how their innovation may be used and generally entitles them to be acknowledged as the creator (author) of the work. The act includes the following categories:

- Literary
- Dramatic
- Musical
- Artistic
- Typographical arrangement of published editions
- Sound recordings
- Films

Commercial documents, company manuals, and computer programs, for example, fall under the literary classification, whereas architectural designs, technical illustrations, photography, logos, etc. are deemed to be artistic works.

Copyright applies to works that are considered to be original, and which display elements of aptitude, effort or judgement. It emerges when an entity creates the work but is restricted to the 'output' as opposed to the concept underpinning the work. As a rule, the rights are exclusively acquired by the individual or group that created the work, although where it has been generated by an employee, as part of their contract of employment, the rights normally belong to the employer. However, where work has been commissioned, for example as part of a supply contract, IP rights will normally be retained by the creator (supplier), subject to any alternative arrangements. For example, the NEC3 Supply Contract, while not explicitly stated, implies that IP in a supplier's design is retained by the supplier but grants the purchaser the right to:

> '... use and copy the Supplier's design and use the services for any purpose connected with use or alteration of the goods and services unless otherwise stated in the Goods Information and for other purposes as stated in the Goods Information.' (Clause 22.1)

Other commercial contracts include more complex arrangements, defining pre-existing (background) intellectual property rights (IPR) and emerging (foreground) IPR as those developed in connection with the agreement.[15] Whereas suppliers generally retain rights to the former, customers seek either to obtain the rights to the latter or acquire a non-exclusive royalty free perpetual and irrevocable licence to the latter. Further, if the purchaser has the dominant position, they may impose terms that transfer all IP rights. Negotiation of IP terms is a key feature of the commercial role during the deal creation phase.

Actions for redress, when an infringement of an IP right has occurred or is believed to have occurred, can only be brought by the owner of the right or an exclusive licensee.

These rights are administered through a network of UK, European and global organisations and bodies. In the UK, the Intellectual Property Office (IPO: the operating name of the Patent Office: an Executive Agency of the Department for Business Innovation and Skills (BIS)),[16] is accountable for all the key IP mechanisms. Specifically; it:

- Grants UK patents (plus registers UK trademarks and designs)
- Raises awareness with business and consumers of IP-associated matters
- Advises ministers on IP policy
- Provides a tribunal function concerning disputes over patents, trademarks and both registered and unregistered design rights
- Advances UK enforcement strategy

Internationally, The Berne Convention for the Protection of Literary and Artistic Works (The Berne Convention), initially introduced in 1886, establishes a 'common framework and agreement between nations' regarding IP rights. Administered by the World Intellectual Property Organization (WIPO), the latest form of the convention is the 1971 Paris Act. Alternatively, the Universal Copyright Convention (UCC) was introduced in 1952 to afford international protection to those in countries unwilling to sign up to the Berne Convention, the most important being the USA. The relevance of the UCC is somewhat diminished, as most countries now abide by the Berne Convention, including the US. Other bodies include:

- The European Patent Organisation
- World Trade Organization (WTO) through the Trade-Related Aspects of Intellectual Property Rights Agreement (TRIPS)

In addition to IP rights, there are other methods of promoting innovation, for example:

- **Confidentiality agreements (confidentiality-disclosure agreements [CDAs] or non-disclosure agreements [NDAs])**: these are used, for example, where an inventor needs to reveal details of a development to a third party before acquiring a patent, or where a supplier has to disclose its IP to a prospective purchaser during tender negotiations. These agreements utilise the law of confidentiality to protect the IP owner's rights. Frequently, contracts of employment will include confidentiality terms. The IPO booklet *Non-Disclosure Agreements*, part of its IP Healthcheck Series, provides an overview of NDAs along with examples of a One-Way Non-Disclosure Agreement and a Mutual Non-Disclosure Agreement (see http://www.ipo.gov.uk/cda.pdf)
- **Secrecy (trade secrets)**: although many organisations have their 'trade secrets', they are notoriously difficult to keep and tend to be adopted as a protection mechanism where the innovation is difficult to replicate, where IP protection is unsuitable or where protection is sought for longer than the period granted under a patent. Importantly, they offer no protection where a third party autonomously develops an identical invention or procedure. When used in combination with NDAs, trade secrets are protected under the law of confidentiality
- **Open source**: seen either as a practical approach or a philosophical position, this allows third parties 'open access' to a product's underlying principles and data. Open source practices gained prominence in the development of computer software but have since been applied to other areas of product development

Freedom of information

The UK Freedom of Information Act 2000, which came into effect in January 2005, confers the legal right for any individual to request and be given information held by a public body. In the UK, the Ministry of Justice is responsible for freedom of information policy[17]; developing:

- *The framework for the Act to be enacted properly by public authorities, providing them with guidance and best practice*
- *Guidance for requesters to ensure that they are able to easily use and understand the Act*

The Act includes 23 exemptions, whereby a request for information can be declined. These grounds include: commercial interests, defence, health and safety, information provided in confidence, law enforcement and national security.[18] Under Section 43: commercial interests, for example, an exemption is granted where revelation would be liable to damage the commercial interests of an individual; additionally, a specific exemption is provided for trade secrets. Guidance is available for all 23 exemptions.[19]

The Act places a requirement on all public bodies to establish and maintain a 'publication scheme'. The scheme should detail the categories of information published (or planned to be published), the method of publication and whether or not a charge is to be made to access the information. The Ministry of Defence, for instance, has developed guidelines of the implementation of the Act (Commercial Policy Group Guideline No 10 – The Freedom of Information Act 2000 and Environmental Information Regulations 2004).[20] Commercial contracts between public bodies and the private sector contain 'Freedom of Information' clauses; see for example the OGC Model ICT Services Agreement.[21]

Exercise 6.4 Intellectual property rights and freedom of information

- For a recent contract entered into by your organisation, examine the IP clauses. Analyse whether the treatment of the contracting parties' IP was dealt with equitably
- For a recent project (subject to the provisions of the freedom of information legislation), examine the freedom of information clauses contained within a contract and consider whether they adequately protect your organisation's interests

Further reading

Intellectual property rights and freedom of information

Brazell, L (Ed.) (2008) *Intellectual Property Law Handbook*. The Law Society, London

Gowers, A (2006) *Gowers Review of Intellectual Property*. HM Treasury/HMSO. http://www.hm-treasury.gov.uk/media/6/E/pbr06_gowers_report_755.pdf (accessed 27th July 2011)

Insolvency

Insolvency is generally held to be the inability of a company or individual to honour debts that become payable and/or where liabilities are greater than the value of assets held. Additionally, it refers to the numerous legal measures applicable when a company, partnership or individual is deemed to be **insolvent**.

The UK legal framework under which insolvency is administered includes, among others, the following legislation: the Insolvency Acts 1986 and 2000; the Companies Acts 1985 and 2006; the Company Directors

Disqualifications Act 1986; the Employment Rights Act 1996; and the Enterprise Act 2002. Plus a raft of related secondary legislation, such as the Insolvency (Amendment) Rules 2011; the Insolvent Companies (Disqualification of Unfit Directors) Proceedings (Amendment) Rules 2007; the Limited Liability Partnerships (Amendment) Regulations 2005; and the Enterprise Act 2002 (Disqualification from Office: General) Order 2006. Additionally, the EC Regulation on Insolvency Proceedings[22] came into force in the UK in 2002 (in common with all EU member states apart from Denmark), and the Cross-Border Insolvency Regulations 2006 introduced further amendments and establish the United Nations Commission on International Trade Law (UNCITRAL) Model Law within the UK.

The aim of the legislation is to provide an objective process for managing the assets of an insolvent entity, settling creditors' claims and determining the fate of the company, partnership or individual at the end of the process. The Insolvency Service – an Executive Agency of the Department for Business Innovation and Skills (BIS)[23] – is responsible for the framework and mechanisms used to manage financial failure and for dealing with any related misconduct in the UK.

Insolvency procedures

The procedures listed below are available in respect of companies, individuals and partnerships (England, Wales and Northern Ireland).[24]

Companies

- **Administration**: overseen by an administrator (an authorised insolvency practitioner) assigned by the court, the company or its directors, or holders of a floating charge, administration defers action by creditors. Potentially it enables the entity to sell assets in order to meet its debts (particularly those of secured or preferential creditors), to achieve a higher return than would be possible if the concern was immediately wound up and to allow, where possible, the business to survive
- **Administrative receivership** (frequently referred to as receivership): this is where an administrative receiver is appointed by a creditor or creditors with a floating charge security (debenture holders: predominantly lending institutions – for example, banks, credit unions, or finance companies that have provided finance secured on the whole or a significant portion of an organisation's assets) following a default or breach of security terms. Although a company's directors may request the appointment of a receiver, only a lender can legally make the appointment. Acting as an agent of the company, administrative receivers have similar powers to administrators (see above), although they are not authorised to deal with the claims of unsecured creditors
- **Company voluntary arrangement (CVA)**: a formal arrangement where an organisation, assisted by a court-authorised insolvency practitioner, attempts to reach a mutually satisfactory agreement with all its creditors. If a settlement is reached, the insolvency practitioner would oversee the agreement and reimburse the creditors accordingly. The process requires the organisation's directors to make an application to the courts
- **Liquidation**: a legal process whereby a liquidator (sometimes the official receiver) is appointed to 'wind up' a limited company. Its purpose is to make sure the organisation's affairs have been appropriately resolved. Once in liquidation the directors of the business relinquish control to the liquidator. The duties of the liquidator/official receiver include: establishing why the business failed; collecting any due payments; terminating all business and trading contracts; resolving outstanding legal disputes; selling all the company's assets; dispensing monies raised to creditors; and (where there is a surplus) returning monies to shareholders. The liability of shareholders is limited to the value of any unpaid shares they may hold. The process concludes with the dissolution of the company. There are three forms of liquidation:
  - *compulsory liquidation*: following a petition by an appropriate person, for example, a creditor or director of a company, a 'winding-up order' in respect of the company is made by the courts
  - *creditors' voluntary liquidation*: where a company's shareholders elect to place the business into liquidation
  - *members' voluntary liquidation (MVL)*: regulated under the Insolvency Act 1986, MVL only applies where a company is solvent and its shareholders decide to place it into liquidation

Individuals (personal insolvency), including an individual member of a partnership

- **Bankruptcy**: where a court makes a bankruptcy order, in respect of an individual judged to be insolvent, following the presentation of a bankruptcy petition by either one or more creditors (creditor's petition) owed an unsecured amount greater than £750 or an individual (debtor's petition). The process involves the disposal of an individual's assets and the fair distribution of any funds raised to the creditors. It also gives the individual, subject to some limitations, an opportunity start again
- **Voluntary arrangements**:
  - *individual voluntary arrangements (IVAs)*: a formal arrangement where an individual, aided by a court-authorised insolvency practitioner, attempts to reach a mutually satisfactory agreement with all his or her creditors. The process requires the individual to make an application to the courts
  - *fast-track voluntary arrangements*
- **Debt relief order (DROs)**: introduced in April 2009, a DRO is a simpler, less costly and expeditious alternative to bankruptcy for individuals with unsecured debts of less than £15,000, who are on a low income and have negligible assets

Partnerships

- **Administration**
- **Bankruptcy** (of individual members)
- **Compulsory liquidation**
- **Voluntary arrangements**
  - *company voluntary arrangements*
  - *individual voluntary arrangements* (in respect of individual members)

As an alternative to liquidation, both companies and individuals can seek to enter into an **informal arrangement** with their creditors in order to try and reach a mutually satisfactory agreement.

Insolvency provisions in commercial contracts

Commercial contracts invariably include insolvency provisions. Under the FIDIC Construction Contract, for example, the employer (Clause 15.1(e)) and contractor (Clause 16.2(g)) have identical entitlements to terminate the contract if either party, for instance, becomes bankrupt or insolvent, enters into liquidation, or has an administration or receiving order made against them. Similarly, under the NEC3 Supply Contract both the purchaser and supplier can terminate the contract if the other has, in the case of a company or partnership:

- *Had a winding-up order made against it*
- *Had a provisional liquidator appointed to i,*
- *Passed a resolution for winding-up (other than in order to amalgamate or reconstruct)*
- *Had an administration order made against it*
- *Had a receiver, receiver and manager, or administrative receiver appointed over the whole or a substantial part of its undertakings or assets, or*
- *Made an arrangement with its creditors* (Clause 91.1)

or in the case of an individual:

- *Presented his petition for bankruptc,*
- *Had a bankruptcy order made against him*
- *Had a receiver appointed over his assets*
- *Made an arrangement with his creditors* (Clause 91.1)

Further reading

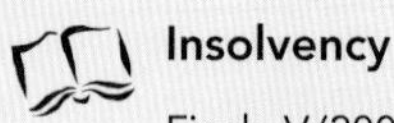

Insolvency

Finch, V (2009) *Corporate Insolvency Law: Perspectives and Principles*, 2nd edition. Cambridge University Press, Cambridge

Goode, R (2011) *Principles of Corporate Insolvency Law*, 4th revised edition. Sweet & Maxwell, London

The Insolvency Service (2009) *A Guide for Creditors*. Department for Business Innovation and Skills (BIS), London. http://www.bis.gov.uk/assets/bispartners/insolvency/docs/publication-pdfs/guideforcreditors.pdf (accessed 27 July 2011)

The Insolvency Service (2010) *Guide to Bankruptcy: When – Where – How – What*. Department for Business Innovation and Skills (BIS), London. http://www.bis.gov.uk/assets/bispartners/insolvency/docs/publication-pdfs/guidetobankruptcy.pdf (accessed 27 July 2011)

The Insolvency Service (2011) *A Guide for Directors: When – Where – How – What*. Department for Business Innovation and Skills (BIS), London. http://www.bis.gov.uk/assets/bispartners/insolvency/docs/publication-pdfs/11-1016-guide-for-directors.pdf (accessed 27 July 2011)

In such circumstances the contract stipulates both the procedure to be followed and the amounts due on termination (Clauses 92 and 93).

However, due to the major impact termination invariably has on achieving the purchaser's objectives and depending on the stage of the project or contract, a purchaser may, working with the receiver, allow the supplier to complete the contract.

Employment rights on the transfer of an undertaking

Effective since April 2006, the revised Transfer of Undertakings (Protection of Employment) (TUPE) Regulations 2006, generally referred to as the 'TUPE Regulations', put into operation in the UK the EC Acquired Rights Directive (77/187/EEC, as amended by Directive 98/50/EC and consolidated in 2001/23/EC), in the UK.[25] The Regulations apply equally to private, public or third sector private sector organisations (undertakings), irrespective of their size. The Regulations are relevant to commercial exchanges that involve the transfer of an 'economic entity' (for example, a discrete unit of a business or organisation) or an 'organised grouping of employees' as a result of outsourcing, insourcing or the compulsory competitive tendering of a public service.

Essentially, the TUPE Regulations safeguard an employee's continuity of and terms and conditions of employment in event of this being assigned to a new employer following a 'relevant transfer'. Conditional on specific prerequisites, such as the requirement to be located in the UK immediately prior to the transfer or change, the Regulations are applicable in the following circumstances:

- **Business transfer**: occurs when '... *a business, undertaking or part of one is transferred from one employer to another as a going concern... This can include cases where two companies cease to exist and combine to form a third*' (BIS, 2009)
- **Service provision change**: Occurs when '... *a client engages a contractor to do work on its behalf, or reassigns such a contract (for example, where a contractor takes on a contract to provide a service for a client from another contractor) — including bringing the work "in-house"* ' (BIS, 2009)

The legislation ensures that employees of the 'transferor' (the initial or preceding employer) retain the same terms and conditions of employment, with the exception of some occupational pension rights, following a transfer. The transferred employees automatically become the employees of the 'transferee' (the new employer) and their employment contract is as though it had originally been entered into with the transferee employer. Although the Regulations afford some degree of flexibility in specific circumstances, where agreed with the transferring workforce, for the transferee or transferor to alter the terms and conditions of employment.

The main provisions of the legislation include:

- **Safeguards to protect employees from dismissal**: specific provisions are included that prevent employers dismissing staff before or after a relevant transfer. However, dismissal is permitted where changes to the workforce are made on the grounds of an 'economic, technical or organisational' (ETO) reason – for example, redundancy
- **A consultation process**: representatives of the affected workforce must be consulted over:
  - a proposed transfer
  - any proposed action by either the transferor or transferee employer that would impact on the affected employees
- **Provision of data**: the transferor employer is required to provide the new employer, before the transfer takes place, with information on the transferring employees
- **Insolvency provisions**: where the transferor employer is insolvent, the Regulations allow the parties to alter employment contracts, if by doing so it will preserve jobs by enabling a relevant transfer to proceed
- **Remedies for non-compliance**: employees and their representative have the right to seek redress through employment tribunals and the civil courts, while transferees can submit a complaint to an employment tribunal when a transferor fails to observe the requirement to furnish 'employee liability information'

Further reading

Employment rights on the transfer of an undertaking (TUPE)

Department for Business, Innovation and Skills (BIS) (2009) *Employment Rights on the Transfer of an Undertaking: A Guide to the 2006 TUPE Regulations for Employees, Employers and Representatives.* Department for Business, Innovation and Skills, London. http://www.berr.gov.uk/files/file20761.pdf (accessed 27 June 2011)

Derbyshire, W and Hardy, S (2009) *TUPE: Law and Practice: Transfer of Undertakings (protection of Employment) Regulations 2006*, 2nd edition. Spiramus Press, London

Fairclough, M (2006) *Managing Business Transfers: TUPE and Takeovers, Mergers and Outsourcing*, 2nd edition. Tottel Publishing, Haywards Heath

Conclusion

A contract, according to the Association of Project Management, Specific Interest Group on Contracts and Procurement (SIGCP, 1998), should be designed to be the basis for successful project management: being right in principle (contract strategy) and right in detail (contract terms). This chapter has discussed the factors that influence commercial contract practice, providing the commercial manager with an overview of generic contract provisions and procedures.

To summarise, a good contract stipulates what, where and when something is to be provided; identifies the supplier and purchaser; defines various roles, relationships and responsibilities; sets standards; deals with issues of time, payment and change; provides remedies for breach of contract; covers the issues of bonds, guarantees and insurances; and contains mechanisms for the submission of claims and the resolution of disputes. Further, effective contracts require flexibility, clarity and simplicity, and should stimulate good management. The following best practice, in terms of planning the contract, is provided by the Association of Project Management, Specific Interest Group on Contracts and Procurement (SIGCP, 1998):

- Understand how a contract is formed and how it can be discharged
- Choose the terms of a contract logically, taking into account the nature of the work, its certainty, its urgency, and the competence, objectives and motivation of all parties
- Consider how the contract will impact on other contracts and projects, and plan the coordination of the work carried out under the contract in relation to existing facilities and systems
- Plan, before starting, how the contract will terminate
- Envisage what can go wrong, utilise risk management and allocate risks to best motivate their control
- Define the obligations and rights of each party
- Specify only what can be tested
- Agree criteria for the assessment of satisfactory performance
- Determine and incorporate effective payment terms
- Say what you mean and be clear about what you want
- Finally, realise that a contract should be a means to an end

Finally, in relation to contracts, although the fundamental legal ingredients are an important platform it is only in rare cases where they will be relied on. Leaving aside the three most important points, which are: what is being procured, who is it being procured from and how much is being paid, the management powers and those provisions and procedures that influence behaviour are the most important in the contract.

The chapter has also introduced the areas of procurement and acquisition regulations, competition and antitrust legislation, international trade law, intellectual property and freedom of information, insolvency and employment rights on the transfer of an undertaking (TUPE) applicable to commercial practitioners, and has provided links and suggestions for further reading on the topics.

Box 6.9 Case study: Port refuelling project

This case study is an illustration of techniques used to influence supplier behaviour. The facts have been changed to protect the identity of the project.

The project concerns a site which is a major undertaking to link fuel suppliers with customers. For the purpose of this example assume that it is a large port at which ships are refuelled. One of the most important things about shipping is quick turnaround time. The prospect of there being no fuel so that a ship could not leave is considered 'unthinkable'. For this reason a very large amount of fuel is stored on the site having been pumped from various refineries. There are a number of different oil companies who supply fuel to ships at this port; they pool the fuel in a shared storage area and then it is pumped out by a jointly owned company. The project was to provide additional storage because of the growth of the port.

The risk in the project was disproportionately high compared with the cost. In particular, as mentioned, the risk of disrupting supplies was very important. There was a possibility that if the new storage arrangements were not connected up properly the operation of the entire system could be prejudiced. Also for other reasons – because fuel was being dealt with and because the site was in the middle of a busy port – there was a significant safety risk.

It was decided that the priorities were (in descending order) safety, quality, time and cost. It was somewhat unusual to have cost at the bottom of this list but the project was relatively straightforward and the cost relatively small compared with the risk of running out of fuel. Although the owner wanted value for money, the actual cost was not of itself of great importance.

Safety

One tenderer was rejected at pre-qualification stage as their safety record was not good enough and the other tenderers were told about this. This action (as well as being a legitimate action) was also designed to 'send a message' to the tenderers explaining how important safety was going to be. The first action that was put into effect was a very thorough discussion and presentation by each tenderer as to what steps they would take to achieve health and safety, and there was then an inspection regime (described by the suppliers as bordering on the 'obsessive') on site including the monitoring of whether people were wearing personal protective equipment. This included a walk around site every day by the owner's project manager at a different time each day.

There was an owner-funded bonus system for safety compliance which in cash terms represented something slightly under 1% of the project value. The benefit to the supplier was partly subjective based on overall impression and partly objective based on KPIs. Half of the benefit for compliance was given directly to site operatives – not in cash but as merchandise, mainly high-quality branded goods.

Quality

The payment system was cost plus with a target and a painshare/gainshare system. However, there was no gainshare on supervision. In other words if the cost was less than the target in relation to supervision – supervision was separately identified in the target and separately accounted for – there would be no gainshare. So the supplier was not incentivised to reduce his cost. There was also reduced painshare on supervision.

There were named supervisors and quality personnel appointed at the beginning of the project and they were paid enhanced hourly rates. If they were changed for any reason the replacements were to be paid at a lower hourly rate than the original personnel – and the target would be reduced accordingly.

The work was divided up into a number of packages which were let sequentially. Although the supplier had in theory been awarded all the packages in the event of poor performance on quality, any of the packages could be removed and put out to competitive tender.

A final strategy was that the end user participated in all quality inspections which took place on a regular basis and certainly more than twice a week.

Time

The project was started very early and well before the extra storage was actually needed. There were linear design and implementation phases for packages. In other words the design was done first and taken as far as possible before construction was started.

The supplier was responsible for his own management with an imposed large management team. The contract stated how many people were to be employed on project management tasks.

There was a full time project manager employed by the owner and outside consultants were also used.

There was an open book on communications with subcontractors with frequent inspections. All the correspondence in and out between the supplier and subcontractors was capable of review by the purchaser's project manager and this right was frequently exercised.

Cost

The project was a fully open book at all levels including subcontract tendering. For each package the target cost (if too high or too low) could be rejected with the owner's right to directly let any parts of the work. The project manager shadowed all procurement. Finally and possibly most importantly the supplier was given a healthy profit margin to incentivise him to get the packages and to devote the appropriate resources to the job.

Endnotes

1 This chapter is based upon an extended and updated version of material first published in D Lowe (2004) Contract management. In: PWG Morris and JK Pinto (eds) *The Wiley Guide to Managing Projects*. John Wiley and Sons, Hoboken, New Jersey.

2 In particular the Fédération Internationale des Ingénieurs-Conseils (FIDIC) *Conditions of Contract for Construction*, 1st Edition 1999 (FIDIC Construction Contract) (FIDIC, 1999) and the NEC3 Supply Contract (NEC, 2009). The FIDIC contract is primarily intended for building and engineering works designed by the employer; however, it does allow for the inclusion of some elements of supplier-designed civil, electrical, mechanical and/or construction works. Members of FIDIC come from over 60 countries worldwide and the FIDIC Construction Contract is an internationally recognised standard form. Similarly, the NEC3 Supply Contract, part of the internationally used NEC3 suite of contracts and designed to be used for the procurement of high-value goods and associated services (including design aspects), is intended for both local and global procurement. For a detailed review of the FIDIC suite of contracts see Bunni (2008), Jaeger and Hök (2010) and Robinson (2011); Trebes and Mitchell (2005), Eggleston (2006) and Rowlinson (2011) provide overviews of the NEC3 suite of contracts.

3 Engineering, Procurement and Construction.

4 See: http://eur-lex.europa.eu/LexUriServ/site/en/oj/2004/l_134/l_13420040430en01140240.pdf (accessed 20 July 2011)

5 See: http://opsi.gov.uk/si/si2006/20060005.htm (accessed 14 September 2012)

6 See: http://opsi.gov.uk/si/si2006/20060006.htm (accessed 14 September 2012)

7 See: http://eur-lex.europa.eu/LexUriServ/site/en/oj/2004/l_134/l_13420040430en00010113.pdf (accessed 14 September 2012)

8 See: https://www.acquisition.gov/far/index.html (accessed 14 September 2012)
http://www.acq.osd.mil/dpap/dars/index.html (accessed 14 September 2012)

9 See: http://www.australiancompetitionlaw.org/overview.html (accessed 14 September 2012)

10 See: http://www.competitionbureau.gc.ca/eic/site/cb-bc.nsf/eng/home (accessed 14 September 2012)

11 http://www.justice.gov/atr/about/antitrust-laws.html (accessed 14 September 2012)

12 See: http://www.businesslink.gov.uk/bdotg/action/detail?itemId=1082147334&type=RESOURCES (accessed 14 September 2012)

13 See: http://www.iccwbo.org/incoterms/ (accessed 14 September 2012)

14 For a glossary of Incoterms® 2010 rules see http://www.ism.ws/files/Products/UpdatedGlossaryIncoterms.pdf (accessed 20 December 2012)

15 See, for example: Section G - Intellectual Property, Data and Confidentiality of the OGC Model ICT Services Agreement: http://www.ogc.gov.uk/policy_and_standards_framework_model_ict_contract.asp (accessed 27 July 2011)

16 See: http://www.ipo.gov.uk/search.htm (accessed 27 July 2011)

17 See: http://www.justice.gov.uk/about/moj/what-we-do/our-responsibilities.htm (accessed 14 September 2012)

18 See: http://www.justice.gov.uk/information-access-rights/foi-guidance-for-practitioners/exemptions-guidance (accessed 27 September 2012)

19 http://www.justice.gov.uk/downloads/information-access-rights/foi/foi-exemption-s23.pdf (accessed 20 December 2012)

20 See: http://www.mod.uk/NR/rdonlyres/2189959A-0014-49A0-A5CB-7FAD3B8E637C/0/cpg_guideline_foi_sv.pdf (accessed 14 September 2012)

21 See: http://webarchive.nationalarchives.gov.uk/20110405225302/http:/www.ogc.gov.uk/policy_and_standards_framework_model_ict_contract.asp (accessed 27 September 2012)

22 32000R1346: Council regulation (EC) No 1346/2000 of 29 May 2000 on insolvency proceedings. http://eur-lex.europa.eu/LexUriServ/LexUriServ.do?uri=CELEX:32000R1346:EN:HTML (accessed 14 September 2012)

23 See: http://www.bis.gov.uk/insolvency/about-us (accessed 14 September 2012)

24 See: Association of Business Recover Professionals (2008) *Understanding Insolvency*. Association of Business Recovery Professionals, London. http://www.r3.org.uk/media/documents/publications/public/Understanding_insolvency_-_October_2008.pdf (accessed 14 September 2012)

25 The legislation supersedes the Transfer of Undertakings (Protection of Employment) Regulations 1981, as amended.

References

ACA (2003a) *Guide to the ACA Project Partnering Contracts PPC 2000 and SPC 2000*. Association of Consultant Architects, Bromley

ACA (2003b) *PPC2000 ACA Standard Form of Contract for Project Partnering*. Association of Consultant Architects, Bromley

Al-Subhi Al-Harbi, KM (1998) Sharing fractions in cost-plus-incentive-fee contracts. *International Journal of Project Management*, 5, 231–236

Arup Project Management/Office of Government Commerce (OGC) (2008) *Partnering Contract Review*. Ove Arup & Partners Ltd, London

Barnes, M. (1983) How to allocate risks in construction contracts. *International Journal of Project Management*, 1, 24–28

Berends TC (2000) Cost plus incentive fee contracting – experience and structuring. *International Journal of Project Management*, 18, 165–171

BIS (Department for Business, Innovation and Skills) (2009) *Employment Rights on the Transfer of an Undertaking: A Guide to the 2006 TUPE Regulations for Employees, Employers and Representatives*. Department for Business, Innovation and Skills, London. http://www.berr.gov.uk/files/file20761.pdf (accessed 29 July 2011)

Boldrin, M and Levine, DK (2002) *The Case against Intellectual Property*. Department of Economics; Centre for Economic Policy Research (CEPR) Discussion Paper No. 3273, Universidad Carlos III de Madrid. http://papers.ssrn.com/sol3/papers.cfm?abstract_id=307859 (accessed 27 July 2011)

Broome, JC and Hayes, RW (1997) A comparison of the clarity of traditional construction contracts and the New Engineering Contract. *International Journal of Project Management*, 15, 255–261

Bubshait, AA (2003) Incentive/disincentive contracts and its effects on industrial projects, *International Journal of Project Management*, 21, 63–70

Bubshait, A and Almohawis, SA (1994) Evaluating the general conditions of a construction contract, *International Journal of Project Management*, 12, 133–136

Bubshait, K and Manzanera, I (1990) Claim management. *International Journal of Project Management*, 8, 222–228

Bunni, NG (2008) *The FIDIC Forms of Contract: The Fourth Edition of the Red Book, 1992, The 1996 Supplement, The 1999 Red Book, The 1999 Yellow Book, The 1999 Silver Book*. 3rd edition. Blackwell Publishing, Oxford

Business Link (2011) *Information Support Compliance: Letters of Credit*. Business Link, Department for Business, Innovation and Skills (BIS), London http://www.businesslink.gov.uk/bdotg/action/layer?topicId=1084535824 (accessed 27 July 2011)

Cabinet Office (2012) *Government Construction: Common Minimum Standards for Procurement of the Built Environments in the Public Sector*. Cabinet Office, London. https://update.cabinetoffice.gov.uk/sites/default/files/resources/CMS-for-publication-v1-2.pdf (accessed 27 September 2012)

Carty, GJ (1995) Construction. *Journal of Construction Engineering and Management*, 121, 319–328

Central Unit on Procurement (CUP) (1997) *Contract Management, CUP Guidance No. 61*. HM Treasury, London

Chaney, AR (1987) Financial guarantees. *International Journal of Project Management*, 5, 231–236

Cheung, S (1999) Critical factors affecting the use of alternative dispute resolution processes in construction. *International Journal of Project Management*, 17, 189–194

Corbett, E (2000) FIDIC's New Rainbow 1st Edition – an advance. *The International Construction Law Review*, 17, 253–275

Cornish, W and Llewelyn, D (2003) *Intellectual Property: Patents, Copyright, Trademarks and Allied Rights*. Sweet and Maxwell Limited, London

Cox, A and Ireland, P (2006) Strategic purchasing and supply chain management in the project environment – theory and practice. In: DJ Lowe with R Leiringer (eds) *Commercial Management of Projects: Defining the Discipline*. Blackwell Publishing, Oxford. pp. 390–416

Cox, A, Ireland, P, Lonsdale, C, Sanderson, J and Watson, G (2002) *Supply Chains, Markets and Power: Mapping Buyer and Supplier Power Regimes*. Routledge, London

Cox, A, Sanderson, J and Watson, G (2000) *Power Regimes: Mapping the DNA of Business and Supply Chain Relationships*. Earlsgate Press, Boston, UK

Cox, A, Sanderson, J and Watson, G (2001) Supply chains and power regimes: toward an analytical framework for managing extended networks of buyer and supplier relationships. *Journal of Supply Chain Management*, 37, 28–35

Davis Langdon (2010) *Contracts in Use: A Survey of Building Contracts in Use during 2007*. RICS (Royal Institution of Chartered Surveyors), London http://www.rics.org/site/scripts/download_info.aspx?downloadID=4748&fileID=5853 (accessed 25 July 2011)

Doctorow, C (2008) *'Intellectual property' is a silly euphemism*, *Guardian*, 21 February. http://www.guardian.co.uk/technology/2008/feb/21/intellectual.property (accessed 27 July 2011)

Eggleston, B (2006) *NEC 3 Engineering and Construction Contract: A Commentary*, 2nd edition. Blackwell Publishing, Oxford

Fenn, P, Lowe, D and Speck, C (1997) Conflict and dispute in construction. *Construction Management and Economics*, 15, 513–518

FIDIC (1999) *FIDIC Conditions of Contract for Construction.* Thomas Telford Ltd, London

Gowers, A (2006) *Gowers Review of Intellectual Property.* HM Treasury/HMSO. http://www.hm-treasury.gov.uk/media/6/E/pbr06_gowers_report_755.pdf (accessed 27 July 2011)

Griffiths, F (1989) Project contract strategy for 1992 and beyond. *International Journal of Project Management*, 7, 69–83

Herten, HJ and Peeters, WAR (1986) Incentive contracting as a project management tool. *International Journal of Project Management*, 4, 34–39

HM Treasury (2005) *Government Accounting 2000: A Guide on Accounting and Financial Procedures for the Use of Government Departments: Amendment 4/05.* The Stationery Office, London. http://www.government-accounting.gov.uk/current/frames.htm (accessed 1 May 2007)

Hughes, W and Maeda, Y (2003) *Construction Contract Policy: Do We Mean What We Say?* FiBRE – Findings in Built and Rural Environments, RICS Foundation. The Royal Institution of Chartered Surveyors, London. www.rics-foundation.org/publish/documents.aspx

Institution of Chemical Engineers (IChemE) (2003) *IChemE Forms of Contract.* www.icheme.org. IChemE, Rugby

Ivens, BS and Blois, KJ (2004) Relational exchange norms in marketing: a critical review of Macneil's contribution. *Marketing Theory*, 4, 239–263

Jaeger, AV and Hök, GS (2010) *FIDIC - A Guide for Practitioners.* Springer-Verlag, Berlin/Heidelberg

JCT (2006a) *JCT – Constructing Excellence Contract.* Sweet and Maxwell Ltd, London

JCT (2006b) *JCT – Constructing Excellence Contract: Project Team Agreement.* Sweet and Maxwell Ltd, London

JCT (2011) *JCT Pre Construction Services Agreement: General Contractor 2011: (PCSA).* Sweet and Maxwell Ltd, London

Kaufman, A, Wood, CH and Theyel, G (2000) Collaboration and technology linkages: a strategic supplier typology. *Strategic Management Journal*, 21, 649–663

KPMG International (2007) *Construction Procurement for the 21st Century – Global Construction Survey 2007.* KPMG International Cooperative, Switzerland http://www.kpmg.com/global/en/issuesandinsights/articlespublications/pages/global-construction-survey-2007.aspx (accessed 25 July 2011)

KPMG International (2010) *Adapting to an Uncertain Environment – Global Construction Survey 2010.* KPMG International Cooperative, Switzerland http://www.kpmg.com/CN/en/IssuesAndInsights/ArticlesPublications/Documents/Global-construction-survey-O-201012.pdf (accessed 25th July 2011)

Kumaraswamy, MM and Yogeswaran, K (2003) Substantiation and assessment of claims for extensions of time. *International Journal of Project Management*, 21, 27–38

Latham, M (1994) *Constructing the Team: Final Report of The Government/Industry Review of Procurement and Contractual Arrangements in the UK Construction Industry.* HMSO, London

Loosemore, M and McCarthy, CS (2008) Perceptions of contractual risk allocation in construction supply chains. *Journal of Professional Issues in Engineering Education and Practice*, 134, 95–105

Macneil, IR (2000a) Contracting worlds and essential contract theory. *Social and Legal Studies*, 9, 431–438

Macneil, IR (2000b) Relational contract theory: challenges and queries. *Northwestern University Law Review*, 94, 877–907

NEC (2006a) *NEC3 Engineering and Construction Contract.* Thomas Telford Ltd, London

NEC (2006b) *NEC3 Engineering and Construction Contract Guidance Notes ECC.* Thomas Telford Ltd, London

NEC (2009) *NEC3 Supply Contract.* Thomas Telford Ltd, London

OGC (Office of Government Commerce) (2009) *Managing Contracts and Service Performance – Document Library, Briefings, Delivery Lifecycle.* Office of Government Commerce, London. http://www.ogc.gov.uk/delivery_lifecycle_managing_contracts_and_service_performance.asp (accessed 29 July 2011)

Paterson, FA and Britton, P (Eds) (2000) *The Construction Act: Time for Review.* Centre of Construction Law and Management, King's College, London

Rivette, KG and Kline, D (2000) Discovering new value in intellectual property. *Harvard Business Review*, 78, 54–66

Robinson, M (2011) *A Contractor's Guide to the FIDIC Conditions of Contract.* Wiley-Blackwell, Oxford

Rowlinson, M (2011) *Practical Guide to the NEC3 Engineering and Construction Contract*, 3rd edition. Wiley-Blackwell, Oxford

Sadeh, A, Dvir, D and Shenhar, A (2000) The role of contract type in the success of r&d defense projects under increasing uncertainty. *Project Management Journal*, 31, 14–22

Seppala, CR (2000) FIDIC's New Standard Forms of Contract – force majeure, claims, disputes and other clauses. *The International Construction Law Review*, 17, 125–252

Shadbolt, RA (1999) Resolution of construction disputes by dispute review boards. *The International Construction Law Review*, 16, 101–111

Smith, C, Topping, D and Benjamin, C (1995) Joint ventures. In: J.R. Turner (ed.), *The Commercial Project Manager*. McGraw-Hill Book Company, Maidenhead

Smith, NJ and Wearne, SH (1993) *Construction Contract Arrangements in EU Countries*. European Construction Institute, Loughborough

Specific Interest Group on Contracts and Procurement (SIGCP) (1998) *Contract Strategy for Successful Project Management*. The Association for Project Management, Norwich

Trebes, B and Mitchell, B (2005) *NEC Managing Reality: Procuring an Engineering And Construction Contract*. Thomas Telford Publishing, London

Turner, JR and Cochrane, RA (1993) The goals and methods matrix: coping with projects with ill-defined goals and/or methods of achieving them. *International Journal of Project Management*, 11, 93–102

Turner, JR and Simister, SJ (2001) Project contract management and a theory of organisation. *International Journal of Project Management*, 19, 457–464

Van Dunne, J (2002) The changing of the guard: force majeure and frustration in construction contracts: the foreseeability requirement replaced by normative risk allocation. *The International Construction Law Review*, 19, 162–186

Van Houtte, V (1999) The role and responsibility of the owner. *The International Construction Law Review*, 16, 59–79

Veld, J in't and Peeters, WA (1989) Keeping large projects under control: the importance of contract type selection. *International Journal of Project Management*, 7, 155–162

Volkman, DA and Henebry, K (2010) The use of incentive contracting and firm reputation. *Corporate Reputation Review*, 13, 3–19

Walker, DHT and Johannes, DS (2001) Construction industry joint venture behaviour in Hong Kong – designed for collaborative results? *International Journal of Project Management*, 21, 39–49

Wang, W, Hawwash, KIM and Perry, JG (1996) Contract type selector (CTS): a KBS for training young engineers. *International Journal of Project Management*, 14, 95–102

Ward, S and Chapman, C (1994) Choosing contractor payment terms. *International Journal of Project Management*, 12, 216–221

Wearne, SH (1992) Contract administration and project risks. *International Journal of Project Management*, 10, 39–41

Wearne, SH (1999) Contracts for goods and services. In: G Lawson, S Wearne and P Iles-Smith, *Project Management for the Process Industries*. Institution of Chemical Engineers, Rugby

Weatherley, S (2005) Pathclearer: a more commercial approach to drafting commercial contracts. PLC*Law Department Quarterly*, October–December, 39–46. www.practicallaw.com/lawdepartment

Winch, GM (2006) The governance of project coalitions – towards a research agenda. In: DJ Lowe with R Leiringer (eds) *Commercial Management of Projects: Defining the Discipline*. Blackwell Publishing, Oxford. pp. 324–343

Wright, D (1994) A 'fair' set of model conditions of contract – tautology or impossibility? *International Construction Law Review*, 11, 549–555.

Part 3

Approaches to Commercial Practice

Chapter 7	Best-Practice Management	239
Chapter 8	Commercial Strategies and Tactics	288
	Part A: Intent	296
	Part B: Deal Creation	332
	Part C: Execution	359

Chapter 7

Best-Practice Management

Learning outcomes

When you have completed this chapter you should be able to:

- Explain the common causes of project failure
- Describe the components of internal control
- Identify and evaluate best practice management processes, procedures and techniques relevant to commercial practice

Introduction

Major product (asset) and service (capability) acquisition projects impose significant demands on both the purchaser and supplier; however, project failure in many cases can be attributed to inadequate demand-side management. This chapter identifies and outlines 'best practice' in managing projects and programmes and in particular those aspects pertinent to commercial practice, as identified in Chapter 1. This includes governance issues associated with the Turnbull recommendations and the OGC Gateway Process, the use of process protocols, such as, PRINCE2, MSP, MoP, and MoV, plus guidance on best practice in procurement and contract management.

The perspective taken is predominantly that of the demand side, although the underlying principles are applicable to both sides of an exchange transaction. This chapter explores how purchasers can articulate their requirements for new assets and services effectively, and then manage the process of asset/service definition and delivery successfully. From the demand side, this chapter provides insight into the effective management of acquisition projects; from the supply side it seeks to improve understanding of how to create greater purchaser value. Further, as the bid process itself can be viewed as a 'project', this chapter provides guidance for the management of the bid/proposal development process.

Project success or failure

As stated above, the failure of numerous major acquisition projects can be attributed to poor management on the part of the purchaser (client). A survey of significant risks by Deloitte & Touche (Jones and Sutherland, 1999), for example, established that the risk of most concern to organisations was **failure to manage major projects**. Other risks included: failure of strategy, failure to innovate, poor reputation/brand management, and lack of employee motivations and poor performance. The survey suggested that the most significant risks were generally operational or strategic in nature, and the major projects that were identified commonly had a technology component.

Commercial Management: theory and practice, First Edition. David Lowe.

Common causes of project failure

Within the UK, a review of all the National Audit Office (NAO, 2004) Value for Money reports established that the principal single source of risks on the projects they considered were related to governance issues, while most of the remaining issues were linked to the client's organisation. The NAO and the Office of Government Commerce[1] (OGC – now part of the Efficiency and Reform Group of the Cabinet Office) have compiled a list of the common causes of project failure. These comprise:

1. *Lack of clear links between the project and the organisation's key strategic priorities, including agreed measures of success*
2. *Lack of clear senior management and Ministerial[2] ownership and leadership*
3. *Lack of effective engagement with stakeholders*
4. *Lack of skills and proven approach to project management and risk management*
5. *Too little attention to breaking development and implementation into manageable steps*
6. *Evaluation of proposals driven by initial price rather than long-term value for money (especially securing delivery of business benefits)*
7. *Lack of understanding of, and contact with, the supply industry at senior levels in the organisation*
8. *Lack of effective project team integration between clients, the supplier team and the supply chain* (OGC, 2005a)

In response, the OGC has published a best-practice guide: *OGC Best Practice – Common Causes of Project Failure* (OGC, 2005a) which presents a series of questions for managers to consider under each cause of project failure. While the guide is aimed primarily at those managing or similarly involved in the delivery of projects across government, the content is generic and equally applicable to private sector organisations. These questions include, for example:

- *Do we know how the priority of this project compares and aligns with our other delivery and operational activities?*
- *If the project traverses organisational boundaries, are there clear governance arrangements to ensure sustainable alignment with the business objectives of all organisations involved?*
- *Have we secured a common understanding and agreement of stakeholder requirements?*
- *Do we have adequate approaches for estimating, monitoring and controlling the total expenditure on projects?*
- *Has sufficient time been built-in to allow for planning applications in Property & Construction projects for example?*
- *Do we have a proposed evaluation approach that allows us to balance financial factors against quality and security of delivery?*
- *Have we checked that the project will attract sufficient competitive interest?*
- *Has a market evaluation been undertaken to test market responsiveness to the requirements being sought?* (OGC, 2005a)

Key performance indicators

In order to measure the effectiveness of UK industry central government has instigated the collection and publication of specific key performance indicators (KPIs). For construction, Constructing Excellence (CE) in partnership with the Department for Business, Innovation and Skills (BIS) generate a series of KPIs. First published in 1999 and annually ever since, the KPIs are intended to be used by organisations within the sector to measure and compare their individual performance. Table 7.1 presents some of these KPIs.

Generally, these KPIs reveal a high level of satisfaction by clients of the construction industry in terms of the product and service they received. However, based on an average of the last five years, they show that 48% of projects were not delivered within budget, while 53% were not delivered on time. To illustrate this, the recently completed Wembley Stadium (9 March 2007) was originally to have opened in 2003 and cost

Table 7.1 Summary of construction industry KPIs (1999–2011).

Headline KPI	1999	2000	2001	2002	2003	2004	2005	2006	2007	2008	2009	2010	2011
Client satisfaction													
Product[1]	72%	73%	72%	73%	78%	80%	83%	84%	82%	83%	86%	87%	87%
Service[1]	58%	63%	63%	65%	71%	74%	77%	79%	75%	77%	84%	82%	80%
Value for money[1]			67%	69%	73%	74%	79%	80%	75%	75%	82%	77%	81%
Contractor satisfaction													
Performance overall[1]					64%	65%	63%	62%	62%	62%	64%	69%	69%
Provision of information overall[1]					57%	59%	58%	56%	56%	56%	59%	63%	64%
Payment overall[1]					67%	66%	65%	65%	63%	63%	67%	71%	77%
Predictability cost													
Design[2]	65%	64%	63%	63%	65%	62%	63%	66%	64%	65%	61%	67%	79%
Construction[2]	37%	45%	48%	50%	52%	49%	48%	44%	49%	48%	46%	47%	59%
Project[2]		50%	46%	48%	52%	50%	48%	45%	46%	49%	48%	52%	63%
Predictability time													
Design[2]	27%	37%	41%	46%	53%	55%	52%	57%	58%	58%	53%	69%	51%
Construction[2]	34%	62%	59%	61%	59%	60%	62%	60%	65%	58%	59%	57%	60%
Project[2]		28%	36%	42%	44%	44%	46%	44%	58%	45%	45%	43%	45%

Measures: 1 = % scoring 8/10 or better; 2 = % on cost/target or better
Source: adapted from Glenigan (2011)

Table 7.2 Project cost overruns.

Project	Estimate	Actual
Holyrood Parliament	£40 million	£414.4 million
Sydney Opera House	$2.5 million	$87 million
Thames Barrier	£23 million	£400 million
Barbican	£8 million	£187 million
Chunnel	£2 billion	£8 billion

£326.5 million compared with the final figure of £798 million. Similarly, the British Library exceeded its budget by almost £450 million and was completed 15 years behind schedule, and the Sydney Opera House was delivered 10 years late. Table 7.2 illustrates some notable project cost overruns.

Cost and time overruns, however, are not a recent phenomenon. The construction of the UK Houses of Parliament caused controversy in the 1870s when it overran its budget by 200%. Nor are they restricted to construction and infrastructure projects. There have been numerous IT/IS, defence and engineering projects that have suffered similar fates; for example, Concorde, the joint UK/French supersonic airplane cost 12 times its predicted figure. It could be argued that these examples are all high-profile, complex projects involving a high level of risk and uncertainty. The CE/BIS figures, however, suggest that the problem of time and cost overrun is not limited solely to such projects.

The KPIs from the construction sector have been included as they are now well established and more comprehensive than those from other industry sectors (for example, ICT/communications and aerospace/defence). KPIs for manufacturing have been produced; the last was published by the Department for Business, Enterprise and Regulatory Reform (BERR), now BIS, in 2008[3]. Additionally, KPIs for the telecommunications sector are provided by the Office of the Telecommunications Adjudicator, and BT publishes its own KPIs[4].

Project success

Project success, as with project failure, is a nebulous concept. Its perception may vary between stakeholders and over time. Identifying drivers of project success is, therefore, somewhat difficult. Having said this it is apparent that some organisations are clearly more successful than others.

Baker *et al.* (1988) suggest that it is driven by a combination of: *goal commitment of project team, accurate initial cost estimates, adequate project team capability, adequate funding to completion, adequate planning and control techniques, minimal start-up difficulties, task (vs social) orientation, absence of bureaucracy, on-site project manager, and clearly established success criteria.* Alternatively, Pinto and Slevin (1988) identified 10 '*critical success factors*': *project mission, top management support, project schedule/plan, client consultation, personnel recruitment, technical tasks, client acceptance, monitoring and feedback, communication and feedback, and trouble-shooting.* More recently, a KPMG Global IT Project Management Survey (KPMG International, 2005) established that organisations with lower project failure rates:

- Had a project management office (PMO) that actively managed all projects (22% of respondents)
- Reported to the board regularly on major projects (30% of respondents)
- Had a very formal benefits process (18% of respondents)
- Had formally qualified project managers (24% of respondents claim this to be the case in their organisation)
- Always undertook rigorous risk analysis during the initial planning stage (29% of respondents)

Project governance

Winch (2003), advocates that the issue of project success or failure is primarily one of project management: specifically, a problem in how the interactions between the stakeholders in the process are governed. This is supported by the KPMG survey which suggested that having an appropriate governance framework in place was a key feature of successful projects.

However, there is also an essential commercial management dimension to the issue and its resolution. Commercial practitioners formulate, negotiate and agree the contracts that underlie project governance and are responsible for the management of the contract once awarded. Moreover, several recent NAO and OGC reports on public sector procurement capability have indicated deficiencies in the commercial capability of UK central government departments. For example, the NAO report: *Commercial Skills for Complex Government Projects* found '*generally weak*' commercial capabilities across all 16 departments, with the largest skill gaps being in '*... contract management, the commissioning and management of advisers, risk identification and management, and business acumen*' (NAO, 2009a). The resulting commercial transformation programme resulted in:

> '... commercial issues and risks featuring more at the top tables across government and more senior commercial directors being appointed. There has also been significant process efficiency.' (Smith, 2010)

However, despite this, the subsequent *Efficiency Review* undertaken by Sir Philip Green (Green, 2010) reported that the UK Government '*does not leverage its buying power, nor does it follow best practice*'. It concluded, once again, that one of reasons for this was '*... inconsistent commercial skills across departments*'.

The sections that follow identify and describe best practice in developing the purchaser's capacity to manage projects and programmes, and highlight some of initiatives that have driven change within the UK. However, the existence of this acknowledged 'best practice' does not in itself ensure the application of good practices within organisations. The majority of the best management practices described later in this chapter were developed by UK government agencies.

Governance issues

Corporate governance is now undeniably established as an essential component of the organisational milieu. Starting with the 1992 Cadbury Committee recommendations that directors should report on the effectiveness of internal control, further reports followed: Rutteman, Greenbury, Hampel, and Turnbull. The recommendations were incorporated into the Combined Code of the London Stock Exchange (the Combined Code) now termed the UK Corporate Governance Code. The document outlines standards of good practice pertaining to company boards. Sections cover: leadership, effectiveness, remuneration, accountability and relations with shareholders[5].

Risk management and internal control requirements of the UK Corporate Governance Code

The current code states that:

> 'The board is responsible for determining the nature and extent of the significant risks it is willing to take in achieving its strategic objectives. The board should maintain sound risk management and internal control systems.' (Principle C.2)

While Provision C.2.1 of the code states that:

> 'The board should, at least annually, conduct a review of the effectiveness of the company's risk management and internal control systems and should report to shareholders that they have done so. The review should cover all material controls, including financial, operational and compliance controls.'

Further, for those organisations with a premium listing of equity shares in the UK, their annual report and accounts must contain the following items:

- *A statement of how the listed company has applied the Main Principles set out in the UK Corporate Governance Code, in a manner that would enable shareholders to evaluate how the principles have been applied*

- *A statement as to whether the listed company has:*
  - *complied throughout the accounting period with all relevant provisions set out in the UK Corporate Governance Code; or*
  - *not complied throughout the accounting period with all relevant provisions set out in the UK Corporate Governance Code, and if so, setting out:*
    1. *those provisions, if any, it has not complied with;*
    2. *in the case of provisions whose requirements are of a continuing nature, the period within which, if any, it did not comply with some or all of those provisions; and*
    3. *the company's reasons for non-compliance*

First issued in 1999, the Internal Control Guidance for Directors on the Combined Code (the Turnbull guidance) constructed a framework that adopts a flexible, principles/risk-based approach, for establishing a system of internal control and for reviewing its effectiveness. The document is considered to be a significant factor in improving the overall standard of risk management and internal control within UK businesses and for adding value.

The Financial Reporting Council (FRC) established the Turnbull Review Group in 2004, which published its *Revised Guidance for Directors on the Combined Code in October 2005: The Revised Turnbull Guidance*[6] (FRC, 2005).

Likewise, the Institute of Chartered Accountants in England and Wales has published a briefing document *Implementing Turnbull: A Boardroom Briefing*[7] (Jones and Sutherland, 1999), which lists a series of questions that company directors could be posing and several useful steps that they might take in order to fulfil the recommendations. It also provides several case studies.

The purpose of the revised Turnbull guidance is to:

- Reflect sound business practice whereby internal control is embedded in the business processes by which a company pursues its objectives
- Remain relevant over time in the continually evolving business environment
- Enable each company to apply it in a manner that takes account of its particular circumstances (FRC, 2005)

Directors, according to the guidance, are required to apply their judgement in evaluating how the company has put into practice the requirements of the code on internal control and the reporting of such to their shareholders.

Internal controls, according to the FRC guidance document comprise '*... all types of controls including those of an operational and compliance nature, as well as internal financial controls*'.

The guidance document comprises four sections: an introduction, maintaining a sound system of internal control, reviewing the effectiveness of internal control, and the board's statement on internal control. An appendix – assessing the effectiveness of the company's risk and control processes – is also provided, which contains a series of questions that may be considered by the board and reviewed with its management when undertaking its annual assessment and periodically reviewing reports on internal control. These questions relate to risk assessment, control environment and control activities, information and communication, and monitoring.

According to the guidance, internal control and risk management are important to a company as:

1. *A company's system of internal control has a key role in the management of risks that are significant to the fulfilment of its business objectives. A sound system of internal control contributes to safeguarding the shareholders' investment and the company's assets*
2. *Internal control facilitates the effectiveness and efficiency of operations, helps ensure the reliability of internal and external reporting and assists compliance with laws and regulations*
3. *Effective financial controls, including the maintenance of proper accounting records, are an important element of internal control. They help ensure that the company is not unnecessarily exposed to avoidable financial risks and that financial information used within the business and for publication is reliable. They also contribute to the safeguarding of assets, including the prevention and detection of fraud*

4. *A company's objectives, its internal organisation and the environment in which it operates are continually evolving and, as a result, the risks it faces are continually changing. A sound system of internal control therefore depends on a thorough and regular evaluation of the nature and extent of the risks to which the company is exposed. Since profits are, in part, the reward for successful risk-taking in business, the purpose of internal control is to help manage and control risk appropriately rather than to eliminate it* (FRC, 2005)

Jones and Sutherland (1999) suggest that effective risk management and internal control should result in:

- Competitive advantage
- Less unexpected and unwelcome surprises
- Early mover advantage
- Improved likelihood of achieving business goals
- Higher share prices over the longer term
- Less management time spent 'firefighting'
- Improved likelihood of change initiatives being achieved
- More internal focus on doing the right things properly
- Lower cost of capital
- Better basis for agreeing strategy

Risk and control, therefore, are clearly associated with attaining business objectives. Moreover, Jones and Sutherland stress that managers should not concentrate solely on the negative aspects of risk but should consider the opportunities to be gained from focusing on risk and control, rather than merely focusing on controls.

Maintaining a sound system of internal control

This section outlines the responsibilities of the board and its management, and describes the elements of a sound system of internal control.

The responsibility for an organisation's system of internal control resides with the board of directors, whereas its management has the task of executing the board's procedures on risk and control. The guidance states that the board should establish suitable procedures on internal control and undertake appropriate oversight to assess the effectiveness of the system. Further, it must ensure the effectiveness of the procedures in managing risk. Likewise, management is responsible for identifying and evaluating the risks faced by the company for consideration by the board. They are also responsible for devising, administering and monitoring an appropriate system of internal control that applies the policies implemented by the board. In developing these policies the board should take into account the following factors:

- The degree and type of the risks facing the company
- The degree and categories of risk considered to be appropriate for the company to accept
- The probability of these risks occurring
- The company's ability to minimise the incidence and effect of the risks that occur on the business
- The benefits to be gained from managing these risks compared with the cost of operating a particular system of control (FRC, 2005)

A company's internal control system comprises the policies, procedures, activities, actions and the like, and includes:

- Control activities
- Information and communications processes
- Procedures for overseeing its ongoing effectiveness (FRC, 2005)

To be effective, the system should:

- Be ingrained in the company's working practices and culture
- Be able to react promptly to developing business risks
- Contain appropriate reporting procedures when significant failings occur (FRC, 2005)

Reviewing the effectiveness of internal control

A key responsibility of the board is to review the effectiveness of the company's internal control system (the board is required to define its review process), whereas management is answerable to them for monitoring the system and for verifying that this has been performed.

Continuous, effective monitoring is a vital element of a good internal control system; therefore, the board should obtain and evaluate reports on the company's internal control on a regular basis. Also, they are required to perform an annual assessment as a prerequisite to their public statement on internal control.

The board's statement on internal control

In order to aid shareholders' appreciation of the major features of its risk management procedures and internal control system, the company's annual report and accounts should contain helpful, high-level information on its application of Code Principle C.2 and Code Provision C.2.1 as considered appropriate by the board.

Sarbanes-Oxely Act 2002

The Turnbull guidance is acknowledged by the US Securities and Exchange Commission (SEC) as an appropriate framework for fulfilling the requirements to report on internal controls over financial reporting within the US, as stipulated in Section 404 of the Sarbanes-Oxeley Act 2002 and associated SEC rules. A guide for UK and Irish companies registered with the SEC has been published by the FRC[8] (FRC, 2004), which describes how companies can use the Turnbull guidance to meet these requirements.

Internal control and the public sector

The UK government has incorporated the principles of the Turnbull guidance into their accounting and financial procedures (HM Treasury, 2005).

Exercise 7.1 Governance issues related to Turnbull

Evaluate your organisation's compliance with the Turnbull guidance

Best management practice and process improvement

Commonly, most complex projects assemble (both in-house and from third parties) aspects of design, innovation, expertise, resources, and technology to realise, within the private sector, corporate goals and to produce a commercial advantage for the commissioning organisation, and in the public sector, generally, to deliver social benefits and value for money. The efficient use of these elements through the application of effective project management techniques facilitates the realisation of these goals within the allocated budget, on time and to the quality specified. Further, an essential part of the project management function is the identification and management of potential risks.

Various 'best practice' approaches, frameworks, methodologies, principles, processes, procedures and guidance have been developed. The following interrelated management practices, predominantly developed by the UK Office of Government Commerce (OGC[9]), but incorporating feedback from both the public and private sectors, seek to improve the delivery and control of projects:

- **PRINCE2®** (Projects in Controlled Environments): a project management methodology or approach to the management of projects
- **MSP®** (Managing Successful Programmes): an approach to the management of programmes

- **MoP**™ (Management of Portfolios): guidance on portfolio management
- **P3M3**® (Portfolio, Programme and Project Management Maturity Model): a framework that enables organisations to evaluate their existing portfolio, programme and project management capability
- **M_o_R**® (Management of Risk): a framework and guidance on risk management at the strategic, programme, project or operational level
- **MoV**® (Management of Value): guidance on value management related to portfolios, programmes and projects[10]

The following section provides an overview of these products (collectively referred to as the Best Management Practice Portfolio), plus an overview of best-practice procurement guidance, particularly OGC's guidance aimed at improving the procurement of construction projects (**Achieving Excellence in Construction Procurement Guides**), the National Audit Office (NAO) and 4Ps guidance on contract management, the IACCM Contract and Commercial Management Operational Guide and the UK MOD's Commercial Skills Framework, and an outline of the **OGC Gateway**™ **Process**[11], a programme or project stage review process.

Although not included in this text, the related Best Management Practice methodology **ITIL**® (IT Infrastructure Library[12]) – an approach to IT service management, and **P3O**® (Portfolio, Programme and Project Offices[13]) – guidance on instituting, developing and maintaining structures to support the management of portfolios, programmes and projects are also available.

PRINCE2®

Introduction

PRINCE (**PR**ojects **IN C**ontrolled **E**nvironments) was initially developed in 1989 for the UK government by the Central Computer and Telecommunications Agency (CCTA), subsequently renamed the OGC (the Office of Government Commerce), as its standard approach to IT project management. Extensively used by the public and private sectors, both in the UK and internationally, it has effectively become the industry benchmark – the project management methodology of choice for both IT and non-IT projects. Its application is mandatory on all projects commissioned by the UK government.

The PRINCE methodology is acknowledged as a 'best practice' product that integrates the requirements, knowledge and know-how of its users, and promotes the principles of good project management. Its structure is configured to include the broad array of disciplines and activities involved in a project. The current 'refreshed' edition of the method, PRINCE2, was launched in 2009, and is a straightforward, generic, process-based and structured approach for effective project management, which is both modifiable and scaleable. The method, which conforms to ISO9001, concentrates on organisation, management and control, and presumes that projects which employ a standard project methodology are more successful than those that do not.

There are two guides to PRINCE2 – '*Managing Successful Projects with PRINCE2*®' (OGC, 2009e) and '*Directing Successful Projects with PRINCE2*®' (OGC, 2009f) – which provide non-proprietary best-practice project management support. Although the subject matter is unquestionably generic common sense that anyone can use, the approach copyright is retained by the Crown.*

PRINCE2 operates as a common language for the project participants: customer, users and suppliers. It provides a framework that enables costs, timescales, quality, scope, risk and benefits to be planned, delegated, monitored and controlled. The method is **product-based**, with the **business case** (the rationale and justification for the project) driving all the project management processes on a project, from **starting up a project** to **closing a project**.

Although PRINCE2 was developed to be applicable for a range of customer/supplier arrangements, the manual assumes that the project will be run on behalf of the customer employing the services of a single

* The following overview of PRINCE2® is derived from the following sources: APMG Group (2012a); Best Management Practice (2012a); APMG UK (2012a); OGC (2006a, 2009e, 2009f, 2009g); Wideman (2002); Siegelaub (2004); PRINCE2.com (2012); 12manage (2008)

(prime) supplier from start to completion within the framework of a contract. The main characteristics of PRINCE2 include:

- A central business case
- A clear defined life cycle
- A process-based approach
- An emphasis on disaggregating projects into manageable and controllable stages
- An organisation structure, with defined responsibilities, to manage the project
- Defined and quantifiable business products, with an associated set of activities to realise them
- Defined resources (OGC, 2009g; PRINCE2.com, 2012)

Management levels, roles and responsibilities

PRINCE2 acknowledges four parallel levels of management: **corporate or programme management**, **directing a project** (the project board), **managing a project** (the project manager's level) and **managing product delivery** (team-level implementation management).

Dependent upon the scale of a project, PRINCE2 defines various roles and functions, which may be assigned, subdivided or amalgamated, which are described below.

The project board

A key aspect of PRINCE2 is the notion of an accountable **project board**, which provides clearly defined oversight and support. The board's composition includes representatives of the customer, the eventual user and the supplier or specialist input: those most able to make decisions concerning project viability. These representatives are termed the **customer**, **senior user** and **senior supplier** respectively. The process model calls for the board to be established at an early stage, during **starting up a project**.

The duties of the project board include:

- **Executive**: the customer – chairs the project board, owns the business case and ultimately accepts responsibility for the project. The function has the responsibility for ensuring that the project or programme retains its business focus, that it has unambiguous authority, and that the product (including risks) is actively managed, achieves its objectives and delivers the anticipated benefits
- **Senior User**: provides support to the executive and is responsible for specifying the requirements (in terms of quality, functionality and usability) of the end users of the product(s), liaising with the project team on behalf of the end users and monitoring the final solution so that it meets these requirements within the limits of the business case
- **Senior supplier**: provides support to the executive and represents the interests of those designing, developing, facilitating, procuring, implementing and possibly operating and maintaining the project products. The senior supplier is answerable for the quality of products provided by the supplier(s) and must have the power to commit or acquire resources as necessary

The project board is not intended to micro-manage, rather it manages by exception, monitors progress by means of reports, and controls via a series of decision points.

The project board grants authority to the project manager by explicitly assigning resources as the project evolves. The board supports the project manager in making key decisions, in supplying the project manager with the required decisions to resolve problems and for the project to progress, and by giving the project manager access to and authority in sections of the organisation as required.

The project manager

PRINCE2 defines the project manager as:

> 'The person given the authority and responsibility to manage the project on a day-to-day basis to deliver the required products within the constraints agreed with the Project Board' (APM Group, 2009)

These constraints, labelled **tolerances**, prescribe the limitations in terms of time, cost scope, benefit and risk within which the project manager is required to perform. Any deviation outside these limits develops into an **issue** and must be drawn to the attention of the project board.
The responsibilities of the project manager include:

- Preparing the project plans – defining what is to be done and when it is expected to be completed
- Selecting appropriate people or suppliers to undertake tasks on the project
- Ensuring the work is completed on time and to the appropriate standards

Further, the project manager frequently reports to the project board on the progress of the project emphasising any anticipated problem.

In addition to the roles of project board and project manager, PRINCE2 introduces a number of additional functions to facilitate its methodology. These include:

- Change authority
- Project assurance
- Project support
- Team manager

A glossary of PRINCE2 terms is currently available in 19 languages.[14]

The structure of PRINCE2

PRINCE2 is based upon four interrelated elements: **principles**, **themes**, **processes** and the **project environment**, which may be customised depending upon the type and size of specific projects. Crucially, the method assumes that an experienced project manager is involved, who is able to draw upon their existing knowledge and experience, and from project management bodies of knowledge.

Principles

In order for a project to be classified as a PRINCE2 project the following **principles** – *guiding obligations for good practice* – have to be adhered to:

- **Business justification**: a continuous focus on the business justification of the project
- **Learn from experience**: an acknowledgement that project teams learn from experiential learning
- **Roles and responsibilities**: clear and established roles and responsibilities allocated within an appropriate organisational structure
- **Manage by stages**: planning, delegating, monitoring and controlling projects sequentially
- **Manage by exception**: defined tolerances for each project objective enabling authority to be delegated within agreed limits
- **Product focus**: a clear focus on the '*definition and delivery*' of products
- **Tailor**: customising the approach to meet specific project characteristics

Themes

PRINCE2 comprises the following seven **themes**:

1. **Business Case** (*Why?*): this describes the justification, rationale and commitment for the project, based on anticipated costs, risks and benefits, and forms the principal control provision of a PRINCE2 project. It is reviewed and verified by the project board prior to the commencement of a project and subsequently at each significant decision point during the project's lifespan to evaluate the project's continuing desirability, viability and achievability. If the business case ceases to be viable then the project should be terminated
2. **Organisation** (*Who?*): this defines and establishes the project's governance structure to facilitate effective decision making and project authority. Project oversight is provided through the project board,

to ensure the commitment of resources to the project and that decisions concerning the viability of the project are taken by those with a direct interest in the outcome of the project

3. **Quality** (*What?*): this defines and describes the means by which the project's products are created (the scope) and verified (the quality), and how the project is managed via the adoption of a quality system (quality management) – the purpose being to deliver products that are 'fit for purpose'. The theme also covers implementing of continuous improvement and 'capturing and acting on' lessons learnt. Under PRINCE2 the quality requirements of the project's deliverables are derived from **product descriptions**
4. **Plans** (*How? How much? When?*): this provides a framework that underpins the design, development and maintenance of project plans (**project plan**, **stage plans** and **team plans**) that enables in order to facilitate 'communication and control'
5. **Risk** (*What if?*): uncertainty (risk) management is a core aspect of project management. This theme addresses how to identify, assess and control uncertainty through the application of risk management techniques with the aim of supporting informed decision making and improving the likelihood of project success. A prerequisite of continued business justification, risk management is undertaken throughout the project life cycle
6. **Change** (*What's the impact?*): this provides a '*systematic and common*' approach to the identification, assessment and control of change, and any resulting issues, which are inevitable during the life cycle of a project. It enables the likely effect on the business case of prospective changes in the scope of a project (in terms of their importance and impact on the cost and time aspects of the project) to be assessed, thereby ensuing informed decision making by the project board. **Configuration management**, a core component of a quality system, provides a mechanism for monitoring and managing project deliverables. Similarly, it provides a procedure to track project issues
7. **Progress** (*Where are we now? Where are we going? Should we carry on?*): this covers the procedures for monitoring and controlling the project that enable its ongoing viability to be determined. The advice is applicable to all management levels, both within and outside the project management team, supporting decision making so that the project:
   - Keeps to its programme and conforms to its plans
   - Continues to be viable in accordance with its business case
   - Results in the delivery of the required product(s) in accordance with previously agreed **acceptance criteria**

Essentially, PRINCE2 simply emphasises these **themes** as being fundamental to project success, while the methodology systematises them into a process model.

Process

The method divides projects into convenient **stages**, to provide suitable decision gates at appropriate points in the project and to promote the efficient management of resources and frequent monitoring of progress. Although the use of these stages is obligatory, dependent upon the complexity of the project their number is variable. Further, PRINCE2 distinguishes between technical stages and management stages.

PRINCE2 is constructed around the project life cycle. Six of the seven **processes** progress from **starting up a project** to **closing a project**, while the remaining **directing a project** run throughout the project supporting the other six. A definition is provided for each **process** together with their main inputs and outputs – the precise objectives to be attained and the activities to be undertaken.

- **Starting up a project**: this is a process that occurs once in the project life cycle and facilitates a controlled project start. The process is activated following the production of a **project mandate** – a high level definition of the rationale for the project and its required product. Start up establishes the foundations for oversight, management and appraisal of the project. Further, it ensures that the prerequisites for commencing the project are in position. The aim of the process is to approve projects that are viable and to reject those that aren't

- **Directing a project**: running throughout the project life cycle, the process defines the duties of the project board and provides the procedures (based on the business case) for project authorisation, continuity approval at the completion of each stage, and project closure. It is the structure for interacting with the project manager – a means by which the project board is accountable for the project's success
- **Initiating a project**: the vision of how the entire project is to be run is encapsulated in the **project initiation document** (PID), '*a logical set of documents that brings together the key information needed to start the project on a sound basis and that conveys the information to all concerned with the project*'.[15] The function of the process is to develop a common understanding of the crucial aspects of the project and the work to be undertaken to deliver the project's products prior to investing significant expenditure. The process takes place once in the project life cycle
- **Controlling a stage**: an iterative and repeatable project stage, which provides routine direction for the project manager on the management of the project. It includes: work authorisation and receipt of work; risk management; issue management and change control; status collection, analysis and reporting; viability consideration; remedial action; and escalation of concerns to the project board. PRINCE2 differentiates between **tolerance**, **contingency** and **change control**. Tolerance is a variation in works that the project manager is permitted to sanction without having first to seek approval from the project board. Contingency is a plan (incorporating time and funds to implement the plan) that will only be activated in the event of an anticipated risk transpiring. Change control is a management practice devised to control the processing of all project issues, including their submission, analysis and decision making
- **Managing product delivery**: this is the process that controls the relationship between the project manager and the team manager(s) by creating formal mechanisms for accepting, executing and delivering work packages
- **Managing stage boundaries**: this is a process undertaken at the end of (or near to the end of) each management stage. It addresses the handover from one completed work stage to the start of the next. The process incorporates strategic decision points that enable the project board to determine whether or not to continue with the project. It also verifies that work within a particular stage has been completed as specified, reviews the updated project plan, and confirms the continued business justification for the project and the acceptability of its risks. The process requires the project manager to provide the project board with sufficient information to carry out these tasks
- **Closing a project**: this is the process which provides closure to the project either by early termination or through completion of the work (project product) by achieving the objectives established in the original project initiation documentation or in any subsequent approved changes to these objectives

Techniques

In addition to risk management, change control and configuration management mentioned above, PRINCE2 provides an explanation of several **techniques**. For example:

- **Product-based planning**: the **product breakdown structure** identifies the components of the project deliverables, clarified by means of **product descriptions** for each product/deliverable. Product descriptions identify why the product/deliverable is being produced, the resources required to produce it, its form/construction, approval criteria and the measures to ensure conformity with these criteria
- **Quality reviews**: this is a technique for implementing quality control. The technique utilises product descriptions as the basis for appraisal, and establishes the process and resources required to evaluate conformity
- **Project assurance**: on a PRINCE2 project, assurance comprises three components corresponding with the perspective of each project board member. Assurance involves the confirmation that the project continues to be viable in relation to costs and benefits (business assurance), to meet the users' requirements (user assurance), and to provide an appropriate solution (supplier assurance). Responsibility for project assurance is held jointly by the executive, senior user and senior supplier, and cannot be delegated to the project manager. However, on certain projects, the task can be delegated to a separate group referred to as the **project assurance team**

Management products

PRINCE2 identifies and utilises 26 **management products**, which are subdivided into three categories:

1. **Baseline management products**: these define aspects of the project and are covered by change control once approved:
   - Benefits review plan
   - Business case
   - Communication management strategy
   - Configuration management strategy
   - Plan
   - Product descriptions
   - Project brief
   - Project initiation documentation
   - Project product description
   - Quality management strategy
   - Risk management strategy
   - Work package
2. **Records**: these are dynamic management products that record information on the progress of a project:
   - Configuration item records
   - Daily log
   - Issue register
   - Lessons log
   - Quality register
   - Risk register
3. **Reports**: these are management products that provide a record at a point in time of the status of particular aspects of a project:
   - Checkpoint report
   - End project report
   - End stage report
   - Exception report
   - Highlight report
   - Issue report
   - Lessons report
   - Product status account

Tailoring

A principle of the PRINCE2 approach is that the methodology should be tailored (adapted) to match the '*scale, complexity, geography or culture*' of the project, taking into account whether or not it is run on an individual basis as part of a programme: the method, therefore, should be appropriate to the context of the project.

Benefits of using PRINCE2

PRINCE2 provides benefits to the managers and directors or executives (senior responsible owners) of a project and to an organisation, through the greater control of resources, and the capability to manage business and project risk more effectively. Its strength lies in its structured method which offers a standard and common-sense methodology for the management of projects, incorporating established and recognised best practice in project management. It is widely acknowledged and understood, providing a common language for all participants in a project. Further, PRINCE promotes formal acknowledgment of responsibilities within a project and concentrates on what a project is to deliver, why, when, and for whom (Siegelaub, 2004; OGC, 2006a; 12manage, 2008; OGC 2009g; PRINCE2.com, 2012).

PRINCE provides projects with:

- A familiar, consistent methodology
- A controlled and structured start, middle and end
- Established terms of reference prior to starting a project
- Frequent evaluation of progress against plan
- Manageable stages for more accurate planning
- Assurance that the project continues to meet the business case
- Adaptable decision points
- Management commitment to provide appropriate resources as part of any approval to proceed
- Management control of any divergence from the plan
- Appropriate management and stakeholders involvement
- A defined structure for delegation, authority and communication
- Suitable communication channels between the project, project management, and stakeholders
- Frequent but brief management reports
- A process for encapsulating and imparting lessons learned
- A means to enhance the organisation's project management skills and competences. (OGC, 2006a; OGC 2009g; PRINCE2.com, 2012)

Limitations of PRINCE2

Not all aspects of project management (for instance, leadership and people management skills, conflict and stakeholder management, or a comprehensive explanation of the various project management tools and techniques) are addressed by PRINCE2. Likewise, contract management is not dealt with in detail; PRINCE2 merely provides the essential controls and limitations necessary for all the project participants to function within the constraints of whichever appropriate contract is adopted. There are, however, numerous texts that provide this information, together with several bodies of knowledge – for example, those produced by PMI (in the US) and APM (in the UK).

Summary

PRINCE2, therefore, is a robust, clear method for managing projects, which enables organisations to undertake '... *the right projects, at the right time, for the right reasons*', by ensuring that projects are initiated and continue based on a valid business case. It provides a framework for selecting project participants and allocating responsibilities. Further, it presents a set of processes to perform and summarises the type of information that should be assembled during the project life cycle.

Exercise 7.2 **Prince2**

Compare and contrast the project management maturity of your organisation against the PRINCE2 framework

Managing Successful Programmes (MSP®)

According to the Cabinet Office's Managing Successful Programmes 2011 Glossary of Terms and Definitions, programme management may be defined as:

> 'The coordinated organization, direction and implementation of a dossier of projects and transformation activities (i.e. the programme) to achieve outcomes and realize benefits of strategic importance.'[16]

Similarly, Thiry (2004) is of the opinion that programme management may effectively connect strategic decision making with its realisation via projects. However, he asserts that programme and project management rely on different paradigms. Project management falls within a performance paradigm, with an established track record in achieving immediate tactical-level deliverables; however, it has not demonstrated its capability for delivering improvement projects or strategic changes. In contrast, experience has shown that programmes need to consider a learning paradigm derived from strategic management and value.

Additionally, he states that an appropriate programme management methodology (focused on stakeholder expectations, quantifiable anticipated benefits related to project output, and allowing for organisational culture) will enhance the efficiency of organisational processes, support planned change, encapsulate innovation and continuous improvement, and facilitate the testing of evolving strategies (Thiry, 2004).

A programme comprises a series of coordinated or integrated projects that collectively will generally deliver a specified strategic level goal or goals – enabling the organisation to realise operational business benefits and improvements. Programme management, therefore, is a methodology for establishing and administering a programme, providing a framework for the management of risk and complexity essential to the resolution of interdependences and divergent priorities inherent in significant change and interrelated projects. As with project management, programme management is a management device that coordinates information, activities and individuals to accomplish something (OGC 2009d; APMG Group, 2012b; Best Management Practice, 2012b).

MSP®

Managing Successful Programmes (MSP®), like PRINCE®, was initially developed by the CCTA and provides best-practice guidance describing a structured approach to the effective management of programmes and change, based on experience within government and across the private sector. The current guide to MSP® *Managing Successful Programmes*, 4th edition (Sowden and the Cabinet Office, 2011) is published by the Stationery Office.*

Globally acknowledged and adopted, MSP® is a flexible, adaptable and non-prescriptive framework, providing a précis of programme management and describing the various principles and processes used to manage programmes. Its purpose is to assist both public and private organisations, manage and control the activities involved in managing a programme and to achieve genuine business advantage through a prescribed process of identifying, measuring, managing and realising benefits. MSP® supplements the approaches and methodologies recommended by PRINCE2.

MSP® explains how to identify the programme vision (the **vision statement**) – for example, to attain competitive advantage or to deliver a new service. How the vision statement will be implemented by the organisation is described in the programme's **blueprint**: a comprehensive explanation of the business processes, people, data, information systems and facilities required to create the outcome described in the vision statement.

MSP® comprises seven **principles**, nine **governance themes** and six **processes** linked to the programme life cycle (**the transformational flow**).

MSP® principles

MSP® is underpinned by the following principles of successful programme management:

1. **Remaining aligned with corporate strategy**: to respond to the emergent nature of strategy, and to deliver corporate performance targets
2. **Leading change**: by providing unambiguous leadership, involving stakeholders, promoting trust, actively engaging stakeholders, accepting a degree of uncertainty and assigning appropriate personal at the appropriate time
3. **Envisioning and communicating a better future**: in order to realise transformational change, programme leaders must propose and consistently communicate a clear vision of the proposed future

* The following overview of MSP® is derived from the following sources: OGC (2005b, 2009d); Sowden (2011a, 2011b); Sowden and the Cabinet Office (2011); Best Management Practice (2012b); APMG Group (2012b)

4. **Focusing on the benefits and threats to benefit realisation**: by establishing the programme's boundaries and constituent parts, and managing risk effectively
5. **Adding value**: by periodically evaluating the value derived from utilising a programme structure. If a programme ceases to add value, it is preferable to disband it, enabling individual projects to continue independently
6. **Designing and delivering a coherent capability**: to ensure the final capability or business structure delivers the maximum benefit and becomes operational in line with the delivery schedule and with minimum overall negative impact
7. **Learning from experience**: by reviewing and introducing performance improvement based on experiential learning

MSP® processes

MSP® identifies and provides guidance on the following **governance themes**; continuous throughout the programme life cycle, these concepts and associated activities facilitate programme control:

1. **Programme organisation**: this addresses how to apply a governance mechanism. The theme covers establishing unambiguous roles, responsibilities and communication routes. For example, instituting a sponsoring group and appointing a **senior responsible owner** (SRO) with accountability for ensuring the programme meets its objectives and delivers its anticipated benefits, a **programme manager** with responsibility for setting-up, managing and delivering the programme, and a **business change manager**
2. **Vision**: this creates the programme vision and compiling its **vision statement** (an explanation of the programme's required benefits and outcomes) – the foundations for the programme. The vision statement is a communication document that encapsulates the proposed outcomes and benefits of the programme, which is used to encourage and align stakeholder buy-in to the programme
3. **Leadership and stakeholder engagement**: stressing the need for leadership in transformational change, and ensuring the participation of all stakeholders in the programme, the theme covers the proactive analysis, understanding and management of, and communication and engagement with, stakeholders. It identifies appropriate analytical tools and introduces the concept of the stakeholder register
4. **Benefits management**: this theme identifies, optimises and tracks anticipated benefits to be derived from the programme to make sure that they are realised. The theme explicitly addresses the relationship between change and benefit realisation, and focuses on the need to exploit project outputs and delivered capability (opportunity exploitation)
5. **Blueprint design and delivery**: this theme expands and develops the programme's vision statement into a blueprint: a definition of the capabilities (processes, organisational structures, technology and knowledge) to be delivered by the programme. The theme explains how the blueprint (the basis for the programme) is used to measure and benchmark capability delivery, and how tranches create incremental transformation
6. **Planning and control**: this theme stresses the link between planning and control and programme success, and acknowledges the differences between the two concepts. It describes how programme plans are developed and internal control maintained, including sections on change control processes, monitoring and control strategy and transition planning
7. **The business case**: assessing the viability of the programme, a business case is a requirement of all programmes. The theme addresses the continuing viability of the programme, seeking answers to the question 'Is the investment in this programme still valuable?'
8. **Risk and issue management**: this theme implements a strategy for identifying, managing and resolving existing and potential issues and risks, and responding to change. The theme is based on the Management of Risk (M_o_R®) methodology and provides guidance on the management of threats and opportunities, risk aggregation, the use of early warning indicators and issue management. It covers risk at the strategic, programme, operation and project level

9. **Quality and assurance management**: this theme ensures quality and that the outcomes of the programme are fit for purpose. In respect of quality management the theme centres on achieving the principles and on changes to the scope of quality, while assurance management focuses on integrated assurance and outlining a strategy to deliver this. Additionally, the theme advocates the need for people development

The transformational flow

MSP® describes six **processes** and associated activities linked to the programme life cycle (transformational flow). The processes are:

1. **Identifying a programme**: this is the first process, which outlines a series of activities that transforms the programme from 'an idea to a concept' and in particular the creation of a vision that the organisation is able to sponsor. This involves needs and stakeholder analysis, plus engaging with the market
2. **Defining a programme**: during this process, the initial programme vision is confirmed and refined into the vision statement and a blueprint is prepared. This requires option analysis. Concurrently, procedures and strategies are generated for managing individuals, costs, benefits, risks, issues, quality, progress and communication. The programme is, therefore, defined and the sponsoring group will determine whether or not to commit to the programme
3. **Managing the tranches**: this provides guidance on the function of a tranche ('a group of projects structured around distinct step changes in capability and benefit delivery'). It describes the requisite management activities, and stresses the importance of tranche reviews
4. **Delivering the capability**: centred on the Blueprint, this process provides advice on the configuration, interface, management and control of the projects, plus their associated activities, and a programme
5. **Realising the benefits**: this includes activities aligned to the benefits management theme, plus guidance on performance measurement and transition planning
6. **Closing a programme**: this comprises the activities that ensure the programme meets its ends, and that its achievements are recognised. It provides guidance on how to consolidate and embed change, close down the programme, and disengage from stakeholders

Benefits

Benefits are derived from MSP® through the application of 'best-practice' programme management guidance. In particular by:

- Focusing on the objectives underlying the business change
- Providing a framework for the control of the change process
- Encouraging a more economical use of resources by prioritising and integrating projects
- Managing risks, opportunities and benefits more effectively
- Improving quality, standards and cost control
- Managing the business case more effectively
- Providing more efficient control of a complex range of activities
- Establishing clearly defined roles and responsibilities
- Delivering efficient and effective business transformation (OGC 2009d; APMG, 2012b)

Portfolio management (MOP®)

The Cabinet Office[17] define **portfolio** as:

> 'The totality of an organization's investment (or segment thereof) in the changes required to achieve its strategic objectives'

and **portfolio management** as:

> '... a co-ordinated collection of strategic processes and decisions that together enable the most effective balance of organisational change and business as usual'.

Portfolio management, therefore, involves the strategic management of projects and programmes, with the objective of delivering an organisation's entire corporate change portfolio. Additionally, it enables organisations to make informed decisions about revising '*business as usual*', via mechanisms that enable senior managers to assess whether the organisation is undertaking appropriate activities, in the best way, and whether the change process adopted is delivering maximum benefit via the best use of resources. These transformations are delivered via projects and programmes.

MoP®

Introduced in 2011 by the Cabinet Office, the Management of Portfolios (MoP®) guidance is published in two formats*:

- **Management of Portfolios**: this offers applied generic guidance to managers of portfolios
- **An Executive Guide to Portfolio Management**: this adopts a strategic perspective of the topic and provides senior executives with an appreciation of how portfolio management can help them address both individual and organisational challenges

MoP® is based on five flexible portfolio management **principles**:

1. Senior management commitment
2. Governance alignment
3. Strategy alignment
4. Portfolio office
5. Energised change culture

It also incorporates two **portfolio management cycles**: the **portfolio definition cycle** and the **portfolio delivery cycle**. Practices within the portfolio definition cycle include **understand**, **categorise**, **prioritise**, **balance** and **plan**. Practices within the portfolio delivery cycle include management control, **benefits management**, **financial management**, **risk management**, **stakeholder engagement**, **organisational governance** and **resource management**.

Portfolio, Programme and Project Management Maturity Model (P3M3®)

The Cabinet Office released its latest version of the Portfolio, Programme and Project Management Maturity Model (P3M3®) (2.1) in February 2010 (Sowden and OGC, 2010). Its roots can be traced through the Project Management Maturity Model to the Software Engineering Institute's (SEI) Capability Maturity Model (CMM). The update reflects the growth, in terms of maturity and recognition, of programme management as a discipline and an activity.

P3M3® provides a framework that enables organisations to evaluate their existing portfolio, programme and project management function, processes and procedures, and level of performance. Importantly, it provides a mechanism for the instigation of improvement plans. The approach utilises three models:

1. Portfolio management (PfM3)
2. Programme management (PgM3)
3. Project management (PjM3)

* The following overview of MoP® is derived from the following sources: Jenner *et al.* (2011); APMG Group (2012d); Best Management Practice (2012d)

These descriptive reference models facilitate independent assessment of the three levels of management activity, enabling the development of separate process improvement programmes for each (see Sowden and OGC, 2010; APMG Group, 2012h).

Exercise 7.3 PRINCE2, MSP®, MoP™ and P3M3®

By adopting a systemised approach to project and programme management, as advocated above, there is a danger that managers will be over burdened with bureaucracy, resulting in the various principles, processes and reviews becoming merely paper exercises.

Either:

(a) Appraise the benefits and disadvantages of introducing any or all of the above on a project within your own organisation.

Or

(b) If you have used PRINCE2®, MSP®, MoP™ and P3M3® evaluate their efficiency and effectiveness.

Management of Risk (M_O_R®)

> 'The term "risk management" incorporates all the activities required to identify and control the exposure to risk which may have an impact on the achievement of an organisations business objectives.' (Murray-Webster and OGC, 2010)

Introduction

Increasingly, modern business, and especially projects, are undertaken in dynamic, complex and turbulent circumstances (Kähkönen, 2006), particularly where joint working and partnering arrangements are adopted, with virtually every decision an organisation takes having positive and negative features, involving some level of risk. While a degree of risk taking is to be expected in order for an organisation to realise its objectives, it must, however, achieve an appropriate balance between these opportunities and threats. Risk management (or the management of uncertainty – see Chapman and Ward, 2002) can provide an appropriate framework to support consistent, transparent and replicable decision making.

As discussed earlier, the Turnbull guidance places an obligation on the board of companies to introduce, maintain, oversee and report on an effective control and risk management system. Also, recently the UK government has issued a number of publications addressing particular aspects of risk management, these include: *Supporting Innovation: Managing Risk in Government Departments* (NAO, 2000), *Successful IT: Modernising Government in Action* (Cabinet Office, 2000), and the HM Treasury's (2004) *Orange Book*.

M_o_R®

Management of Risk (M_o_R)® was developed by OGC after the publication of the Turnbull guidance in 1999 (described earlier) and provides best-practice guidance, describing a structured approach to the effective management of risk, again based on experience within government and across the private sector. The current guide to M_o_R®, *Management of Risk: Guidance for Practitioners* (Murray-Webster and OGC, 2010) addresses the implication and requirements of subsequent corporate governance and internal control legislation.*

* The following overview of M_o_R® is derived from the following sources: OGC (2009c); Murray-Webster and OGC (2010); APMG Group (2012c); Best Management Practice (2012c)

Adopted and acknowledged globally, MSP® is a generic, structured, yet flexible and adaptable framework that enables organisations to manage risk at the strategic, programme, project and operational level by applying a process comprising a sequence of clear-cut stages. Its purpose is to assist both public and private organisations achieve their objectives by making well-informed decisions, ensuring that crucial risks are identified and assessed, and that the right responses to these inherent threats and opportunities are taken. Equally applicable to large and small organisations in both the public and private sectors, it includes all the activities necessary to identify and control uncertainty.

While the M_o_R® approach is designed to complement OGC's (now the Cabinet Office's) best -practice guidance on programme, project and service management (MSP®, PRINCE2® and ITIL®), it recognises terminologies, roles and responsibilities used in other domains. A glossary is available, which covers the common language used across M_o_R®, PRINCE2® and MSP® and that contained in BSI's evolving Risk Standard.

A principles-based approach applicable to any organisation, M_o_R® presents an ordered context for proactive decision making: a **route map** for risk management. It combines **principles**, a recommended **approach**, interconnected **processes, checklists**, and **links** to extensive resources and guidance on risk management tools, techniques and specialisms, within an organisational **framework**, enabling risks to be identified, analysed and assessed, and a risk response to be planned and implemented. It also explains how the guidance should be embedded, appraised and applied.

The M_o_R® framework consists of four key concepts: principles, approach, process, and embed and review.

M_o_R® principles

The framework is founded on eight principles derived from corporate governance standards and ISO 31000: 2009 – the international standard for risk management: the prerequisites of mature risk management practice. These comprise:

1. Aligns with objectives
2. Fits the context
3. Engages stakeholders
4. Provides clear guidance
5. Informs decision making
6. Facilitates continual improvement
7. Creates a supportive culture
8. Achieves measurable value

Although the application of these principles is organisation-specific, they underpin the development of risk management practices.

M_o_R® approach

The M_o_R® approach describes how these practices might be applied within an organisation. It covers the modification, approval and delineation of these principles to meet its specific requirements. The approach revolves around the generation of a **risk management policy**, **risk management process guide** and **risk management plans**. Additionally, it advocates the use of **records** (**risk register** and **issue register**), **plans** (**risk improvement plan**, **risk communications plan** and **risk response plan**) and a **risk progress report**, and describes the interrelationship of these documents.

M_o_R® processes

The M_o_R® process describes, in terms of their **goals**, **inputs**, **outputs**, **techniques** and **tasks**, the following four iterative and often recurring steps: **identify**, **assess**, **plan** and **implement**.

Embedding and reviewing M_o_R®

Embedding and reviewing M_o_R® explains how an organisation can effectively introduce these principles, approaches and processes, and ensure that they are consistently employed throughout the organisation within the context of continuous improvement. Incorporating the principles of culture change, it seeks to address common barriers to the successful implementation of a risk management framework.

Additionally, *Management of Risk: Guidance for Practitioners* (Murray-Webster and OGC, 2010) contains a chapter on **perspectives**, which considers the diverse facets of risk management within an organisation, including strategic, programme, project and operational perspectives, within the context of successful business change, plus the following appendices: management of risk document outlines, common techniques, management of risk health check and management of risk maturity model.

Benefits

It is suggested that the benefits derived by organisations from applying the M_o_R® framework include:

- Improved corporate decision making derived from its transparent approach to risk exposure at all levels throughout the organisation
- Compliance with corporate governance requirements (for example, the Turnbull guidance)
- Effective monitoring and proactive management of risk due to appointing individuals who own and are accountable for risk and its management
- More certainty in the realisation of organisational and stakeholders goals due to the timely recognition and proactive management of threats and opportunities
- Improved financial management of projects/programmes and an enhanced capability to achieve 'value for money'
- Reduced risk of individual project failure as the management of project risk is undertaken within a broader programme context
- Consistency of approach to the management of risk due to high-level monitoring and control
- Informed, conscious acceptance of business risks within a support environment and culture of continuous improvement
- Improved contingency and business continuity plans (OGC 2009c)

Further reading

Management of risk

Kähkönen, K (2006) Management of Uncertainty. In: DJ Lowe (ed) with R Leiringer. *Commercial Management of Projects: Defining the Discipline*. Blackwell Publishing, Oxford. pp. 211–233

Murray-Webster, R and OGC (2010) *Management of Risk: Guidance for Practitioners*, 3rd edition. The Stationery Office, London

OGC (2007) *Achieving Excellence in Construction Procurement Guide 4–Risk and Value Management*. OGC, London

Management of Value (MOV®)

Kelly *et al.* (2004) view value management as the process by which:

> '… the functional benefits of a project are made explicit and appraised consistent with a value system determined by the client.'

The clients' value system, when determined, can be used to audit:

- The end user's operation of a product or facility compared with the organisation's corporate strategy
- The project brief

- The emerging design
- The method of construction/implementation

Maximum value is realised from a required degree of quality at minimum cost, the maximum degree of quality for a specified cost, or from optimising between the two. Value management, then, is the management of a process to maximise value on a continuum established by the client.

Kelly and Male (1993) consider the factors that distinguish value management from an accounting vision of an audit are:

- A positive and pro-active methodology to create alternative solutions in addition to that which has been proposed, by utilising a creative multidisciplinary team approach
- The application of a structured systems methodology
- The association of function with value

Further, they believe that value management should not be viewed as a cost-cutting exercise, a standardisation task or a conflict-oriented design review.

Kelly and Male provide the following definition of value management:

> 'A service which maximizes the functional value of a project by managing its development from concept to completion and commissioning through the audit (examination) of all decisions against a value system determined by the client.' (Kelly and Male, 1993)

An alternative definition is provided by the OGC:

> 'A systematic method to define what value means for organizations, and to communicate it clearly to maximize value across portfolios, programmes, projects and operations'[18]

There are currently three British Standards relating to value management: BS EN 1325-1:1997: *Value Management, Value Analysis, Functional Analysis Vocabulary – Value Analysis and Functional Analysis*; BS EN 12973:2000: Value Management; and BS EN 1325-2:2004: *Value Management, Value Analysis, Functional Analysis Vocabulary – Value Management.*

MoV®

Launched in November 2010 by the OGC, the Management of Value (MoV)* guidance is published in two formats:

- **Management of Value**, which offers applied generic guidance to those responsible for directing, managing, supporting or delivering portfolios, programmes and projects – for example, programme managers, project managers, change managers and risk managers.
- **An Executive Guide to Value Management**, which provides an overview of the methodology

A functional approach applicable to any organisation, MoV® can be applied at the portfolio, programme, project or operational level and is designed to complement and augment the Cabinet Office's best-practice guidance on project, programme, and portfolio management (PRINCE2®, MSP®, and MoP®) and the management of risk MoR®.

The purpose of MoV® is value maximisation: enhancing the value of the product, capability or change delivered and promoting the efficient, effective and economic use of resources, while achieving programme

* The following overview of MoV® is derived from the following sources: Dallas *et al.* (2010); APMG Group (2012e); APMG UK (2012b); Best Management Practice (2012e)

and project objectives and fulfilling stakeholder requirements. The methodology has much in common with **lean** principles.

The guidance incorporates four integrated concepts: **principles**, **processes and techniques**, **approach** and **environment**. It also explains how the guidance can be embedded within an organisation.

MoV® principles

The approach is founded on seven principles: the prerequisites of established value management practice. These comprise:

1. Aligning with organisational objectives
2. Focusing on functions and required outcomes
3. Balancing the variables to maximise value
4. Applying MoV throughout the investment decision
5. Tailoring MoV activity to suit the subject
6. Learning from experience and improving
7. Assigning clear roles and responsibilities and building a supportive culture

MoV® processes and techniques

The MoV® process is based upon the following activities:

- Framing the programme or project
- Gathering information
- Analysing information
- Processing information
- Evaluating and selecting
- Developing value-improving proposals
- Implementing and sharing outputs

Regarding value management techniques, the guidance describes **function analysis**, **function cost analysis** and **value engineering/analysis** and outlines other frequently used techniques.

MoV® approach to implementation

The MoV® approach describes how these practices might be applied within an organisation. It covers planning the MoV activities plus understanding and articulating, prioritising, improving, quantifying and monitoring improvements in value. It also advocates the need to record lessons learnt.

MoV® environment

This section of the guidance deals with the need to respond to both external and internal environmental influences. Additionally it addresses portfolio, programme, project and operational considerations.

Embedding MoV® into an organisation

MoV® explains how an organisation can effectively introduce these principles, processes and techniques, and approach, and ensures that they are consistently employed throughout the organisation. Additionally, *Management of value* (Dallas *et al.*, 2010) contains appendices addressing MoV document outlines, common techniques, health check and maturity model.

Benefits of using MoV

As with the previously outlined approaches, frameworks and methodologies, benefits are derived from MoV® through the application of 'best practice' value management guidance. It is claimed that MoV® enables managers to sustain or enhance benefits, as the approach:

- Promotes the efficient and effective use of resources
- Reduces expenditure by optimising both investment and whole-of-life costs

- Encourages innovative solutions
- Enables difficult decisions to be made within an auditable framework
- Expedites delivery (APMG UK, 2012b; APMG, 2012e)

Further reading

Value management

Dallas, MF (2006) *Value and Risk Management: A Guide to Best Practice.* Blackwell Publishing, Oxford
Dallas, M, OGC and Clackworthy, S (2010) *Management of Value.* The Stationery Office, Norwich
Kelly, J (2006) Value management of complex projects. In: DJ Lowe (ed) with R Leiringer, *Commercial Management of Projects: Defining the Discipline.* Blackwell Publishing, Oxford. pp. 211–233
Murray-Webster, R and OGC (2010) *Management of Risk: Guidance for Practitioners.* 3rd edition. The Stationery Office, London

Exercise 7.4 **Value management**

Within the construction sector there is a move to combine risk management and value management into a single activity (e.g. Dallas, 2006).

Consider the advantages and disadvantages of doing this within an industry sector of your choice.

Support for PRINCE2®, MSP® , MOP®, M_O_R®, etc.

A variety of resources are available to support both the implementation of and personal development in the Cabinet Office's best-practice methodologies, and two qualifications exist at foundation and practitioner levels for each product. The responsibility for maintaining standards in training and certification lies with Cabinet Office's partner, the APM Group Ltd (APMG).

Procurement best practice

Good practice procurement guidance is pervasive, published, for example, by central and local government departments and agencies, financial institutions such as the World Bank and the International Monetary Fund, and professional organisations representing the procurement and supply chain management function.[19] Within the UK, the OGC has been instrumental in disseminating procurement best practice.

Established in April 2000, OGC's function is to: promote best-value procurement through the provision of guidance and circulation best practice throughout central government departments; develop the government's market place; and to formulate policy standards. Specifically, it:

> '... promotes and fosters collaborative procurement across the public sector to deliver better value for money and better public services; and it provides innovative ways to develop the Government's commercial and procurement capability, including leadership of the Government Procurement Service.' (Cabinet Office, 2012)

OGC's six key goals are to:

1. *Deliver value for money from third party spend*
2. *Deliver projects to time, quality and cost, realising benefits*
3. *Get the best from government estate*
4. *Improve the sustainability of the government estate and operations*
5. *Help achieve delivery of further government policy goals*
6. *Drive forward the improvement of central government capability in procurement, project and programme management, and estates management through the development of people skills, processes and tools* (Cabinet Office, 2012)

Key contributions made by OGC in the area of procurement, in addition to those already reviewed, include:

- Gateway reviews
- The Successful Delivery Toolkit
- The promotion of Centres of Excellence

Following the UK general election in 2010, OGC became part of the new Efficiency and Reform Group (ERG) within the Cabinet Office; its role is currently under review.

OGC publishes *The Government Procurement Code of Good Practice: For Customers and Suppliers* (OGC, 2010a) which sets out the core values and behaviour for the UK government supply chain. The current code encourages open and collaborative working and the establishment of relationship to deliver reciprocal benefit. Part of CGC's guidance portfolio, the code is consistent with EC Rules and Procurement Policy Guidelines.[20]

The following overview of **Achieving Excellence in Construction** is provided as an example of the work undertaken by OGC in developing best-practice procurement guidance for a project-orientated sector.

Achieving Excellence in Construction

Initially introduced in 1999 as a three-year strategy but subsequently extended, the Achieving Excellence in Construction (AEC) initiative sought to improve the performance of the UK government as a construction industry client. Via AEC, central government departments and public sector bodies commit to: '... *maximise, by continuous improvement, the efficiency, effectiveness and value for money of their procurement of new works, maintenance and refurbishment.*'

The impetus for the initiative was a series of significant failures by the UK government as a construction industry client. Two studies emphasised these failures. The first by Bath University (Graves *et al.*, 1998) highlighted failings in the following areas:

- Poor management
- A risk-averse culture
- A lack of integration
- Poor project flow
- Focus on low-cost rather than value for money
- Short-term relationships

The second (Graves and Rowe, 1999), a benchmarking study, indicated that 73% of UK government construction contracts were not being delivered within budget, while 70% were delivered late.

To realise an appropriate level of improvement, a step change in culture within government construction procurement practice was called for. To deliver this step change, the main features of AEC include:

- Partnering
- Creating long-term relationships
- Reducing the number of financial and decision-making approval chains
- Empowering and improving the expertise of project participates

- Adopting performance measurement indicators
- Applying value and risk management and whole-life costing tools

By 2002, clear improvements in the delivery of public sector construction projects in terms of time and budget were apparent (OGC, 2003). The initiative was, therefore, extended for a further two years, concentrating on two strategic targets (AE STs):

- Improving the successful delivery of construction projects
- Challenging departments to accelerate the project progress at the critical procurement stage

Against these strategic targets (AE STs), by 2005 significant improvements had been made. For example, the results show (when compared with the 1999 statistics) that:

- 65% of projects were being delivered on time, compared with 34%
- 61% of projects were being delivered to budget, compared with 25%

Moreover, it was established that by implementing AEC best-practice principles an £800 million overspend on construction projects had been averted. It was also estimated that if these principles were applied throughout the public sector, further annual savings on construction expenditure of approximately £2.6 billion were possible (NAO, 2005).

Project data was subsequently collected by OGC on a six-monthly cycle, to enable the identification of strengths and weaknesses in construction procurement within individual departments.

The overall assumption that underpins the AEC best-practice guidelines is that:

> '... successful delivery requires an integrated process in which design, construction, operation and maintenance are considered as a whole – together with an understanding of how the project will affect business efficiency and service delivery over the lifetime of the project. It also requires effective use of project management techniques such as risk and value management.' (Source: OGC, 2007i; OGC, 2009a)

The AEC initiative has resulted in the publication of a series of:

- Guides and brochures
- Construction case studies

AEC procurement guidance

AEC procurement guidance is contained within two high-level brochures, three core documents and eight supporting guides. The guidance, based on current experience, supports future strategy and is consistent with the OGC Gateway™ process.

High-level brochures

- **A Manager's Checklist**: this is a companion document to *The OGC Gateway™ Process: A Manager's Checklist* described earlier. It summarises the additional key questions that a senior responsible owner (SRO) should ask (within the Gateway™ process) specific to construction projects. It also describes the main principles and considerations addressed by AEC (OGC, 2007g)
- **Achieving Excellence Construction Projects Pocketbook**: this presents an outline of construction procurement. It describes the main aspects and issues pertinent to construction projects and summarises (within the Gateway™ process) the principal project stages (OGC, 2007h)

Core documents

- **Achieving Excellence Guide 1 – Initiative into Action**: this presents a synopsis of the aims and objectives of Achieving Excellence, its key proposals and achievements to date. It also outlines the contents of the other procurement guides and puts forward key pointers for senior management (OGC, 2007i)

- **Achieving Excellence Guide 2 – Project Organisation**: this lists the critical factors for success, and clarifies the principal roles and responsibilities of the stakeholders in construction procurement projects: the investment decision maker, senior responsible owner, project sponsor, project manager and independent client advisers, etc. The integrated supply team approach is compared with the traditional procurement method. Also, desirable and essential attributes of the senior responsible owner, project sponsor and project management are given, in terms of management and technical abilities. Overall, it suggests an adaptable project organisational framework (OGC, 2007j)
- **Achieving Excellence Guide 3 – Project Procurement Lifecycle**: this describes the principles and processes of delivering a construction project within an integrated process. The principles include: why project management is important, the role of timely planning, and integrating the process. The process section presents a framework for construction procurement set within the OGC Gateway™ review process and provides a step-by-step description of each stage in the process identifying the responsible role, the relevant AEC procurement guide and referencing sources of further information (OGC, 2007k)

Supporting guides

- **Achieving Excellence Guide 4 – Risk and Value Management**: this describes the principles and practice of managing risk and value. Moreover, AEC advocates the use of risk and value management techniques as being fundamental for the successful delivery of construction projects. The principles section advocates that risk and value management should be carried out in parallel as they are interrelated activities, and provides an overview of and key messages about the two processes. The practice section provides a framework for risk and value management within the OGC Gateway™ review process and covers the key risk and value management activities during the project life cycle (OGC, 2007l)
- **Achieving Excellence Guide 5 – The Integrated Project Team**: this explains how the project participants can work together as an integrated project team. The AEC guidance advocates that:

  'client and suppliers working together as a team can enhance whole-life value while reducing total cost, improve quality, innovate and deliver a project far more effectively than in a traditional fragmented relationship that is often adversarial. Collaborative working should be a core requirement for each element of every project.'

The guide describes the principles and processes of team working and partnering. The principles section defines supply chain, integrated supply teams (IST), integrated project team (IPT), team working, and partnering; discusses why partnering is worth doing, when to adopt a partnering approach and lists critical success factors. The process section explains how to assemble and run IPTs. Topics include: preparation, assembling the IPT, the need for specialist advisers (linked to the OGC Gateway™ review process), team working and partnering workshops, partnering charters, dispute resolution, practical considerations and good practice: *Building Down Barriers*. Illustrative case studies are provided (OGC, 2007m)

- **Achieving Excellence Guide 6 – Procurement and Contract Strategies**: this describes how to establish appropriate procurement routes that will deliver best value for money. The guide describes the principles and processes of procurement and contract strategy. The principles section defines value for money, procurement strategy, procurement route and contract strategy. It also defines the three preferred integrated procurement routes: PFI, prime contracting and Design-Build. The process section considers the practicalities of the procurement process. It links the procurement process to the stages of the OGC Gateway™ review process. It also addresses the following issues: the level of risk transfer and funding arrangements, determining the contract strategy, framework agreements, deciding on the form of contract, and selection and award criteria, etc. (OGC, 2007n)
- **Achieving Excellence Guide 7 – Whole-Life costing**: this guide describes how to manage costs over a facility's life cycle and in particular whole-life costs (which are defined as: *the cost of design and construction, the long-term operational and maintenance costs and the costs associated with disposal*).

An overview of whole-life cost management principles is presented, along with a 'best-practice' process comprising: a cost management framework; baseline cost determination – the anticipated likely costs of operating the asset; whole-life cost estimation, including all potential costs and associated risk from project inception to disposal; and cost reporting and management (OGC, 2007p)

- **Achieving Excellence Guide 8 – Improving Performance**: AEC guidance asserts that performance measurement is crucial for enabling planned improvements in the quality, cost and time aspects of construction projects to be realised. The guide describes the principles, processes and techniques of performance evaluation. The principles section defines: performance measurement, key performance indicators (KPIs), project evaluation, post-project review, post-implementation review, and benchmarking. It also describes what clients have been able to achieve by adopting performance measurement, and good practice. The process section covers project evaluation, providing a framework for formal reviews at key decision points based on the OGC Gateway™ review process, ongoing design and construction phase, performance review, and feedback. The techniques section provides guidance on a performance management framework, KPIs, design quality indicators, benchmarking and a review process. The guide also provides case studies as examples, key questions for clients to consider in quantifying improvements in construction performance, and establishing performance evaluation and benchmarking systems (OGC, 2007o)
- **Achieving Excellence Guide 9 – Design Quality**: design is perceived as being fundamental to AEC and to the realisation of value for money. This guide describes the features of good design and how the procurement process can influence design quality (OGC, 2007q)
- **Achieving Excellence Guide 10 – Through Health and Safety**: health, safety and welfare matters are considered to be fundamental to the whole project process and the facility's life span, not solely limited to its construction stage. This guide identifies how health and safety issues are influenced by the client's actions and choices, and likewise how they can have a positive influence on the implementation of the contract and on realising value for money (OGC, 2007r)
- **Achieving Excellence Guide 11 – Sustainability**: this guide emphasises the value of and supports the government's commitment to sustainable development. Although still focusing on optimising whole-life value for money, it outlines the processes by which public sector clients can acquire and deliver construction projects that encourage sustainable development best practice. It focuses on the issues of sustainable development that should be addressed at each stage in a project's life cycle (based on the OGC Gateway™ review process). These issues are considered to apply to all construction projects irrespective of the procurement route adopted. The guide is intended to assist clients determine appropriate standards for their projects. Each section contains an introduction and a table, that outlines the key issues to be addressed at each stage and directs the client to additional sources of information (OGC, 2007s)

Supplementary guidance

- **Common Minimum Standards**: this presents the current public sector standards for the procurement of construction projects (OGC, 2006b)
- **Making competition work for you**: this presents practical, best-practice guidance on the application of competition. The guide addresses the inherent risks in competition – for example the likelihood of anti-competitive behaviour – and presents some useful measures that can be taken to mitigate these risks and illustrates the assistance that can be obtained from the Office of Fair Trading (OFT) (OGC, 2006c)
- **Getting Value for Money from Construction Projects through Design – How Auditors Can Help**: guidelines to assist auditors consider good design in public sector built environment projects (NAO, 2004)

Construction (achieving excellence) case studies

A series of best-practice construction case studies have been compiled that illustrate how the AEC principles have improved construction procurement throughout the public sector.

Exercise 7.5 Achieving excellence in construction

While suitable in some circumstances, the collaborative/integrated approach promoted by the AEC guidance may not be appropriate for all projects. (Cox and Townsend, 1998)

Review the AEC Guides, in particular the core guides and case studies, and consider under what circumstances the collaborative/integrated approach would be inappropriate.

Best-practice contract management

> 'Contract management is the process of managing and administrating the... contract from the time it has been agreed at contract award, through to the end of the service period; referred to as the "operational period" in this guide.' (4ps, 2007)

The vast majority of definitions of contract management, as with the one reproduced above, tend to focus primarily on the post-award nature of the function/activity. However, effective contract management has been shown to be delivered when the function has been involved both pre- and post-award, and the following are present: clear contract administration procedures, which receive executive support and are enabled by **contract lifecycle management** (CLM) software, and the adoption of standard contract language and risk-assessment (Aberdeen Group, 2004). CLM has been defined as:

> '... the process of systematically and efficiently managing contract creation, execution and analysis for maximising operational and financial performance and minimising risk'. (Aberdeen Group, 2004)

As mentioned in Chapter 1, CLM and **enterprise contract management** (ECM) are frequently used interchangeably in relation to the provision of organisation-wide contractual services based on current best practice and IT enabled processes, which efficiently and effectively support the management of a business' contracts and agreements (for example, see Saxena, 2008). Research undertaken by the Aberdeen Group (2004) found that enterprises derive the most value when their contract management function exhibited 10 common practices (see Table 7.3).

Table 7.3 Ten common contract management practices.

1. *Audit internal contract management processes, systems, and controls before investing in a contract management solution*
2. *Create a compelling business case with both benefit and crisis*
3. *Ensure proper executive and stakeholder support for both contract management initiative and automation investment*
4. *Define detailed functional requirements for a contract management solution – and stick to them*
5. *Dedicate and empower a contract management programme champion*
6. *Establish a contract management governance council to ensure support from functional and business unit leaders*
7. *Clearly define and communicate procedures and protocols for the complete contracting and contract administration*
8. *Where possible, use templates to streamline contracting cycles, minimise risk, and maximise compliance*
9. *Measure programme performance and market results*
10. *Identify areas for continuous improvement*

Source: Aberdeen Group (2004)

The main purpose of contract management from the demand and supply perspective is to ensure that effective contracts are drafted, negotiated, awarded and administered, and that:

- The signed contract reflects the 'deal' agreed between the parties and incorporates terms and conditions that are most likely to deliver the required objectives of the exchange
- The agreed contractual position is safeguarded post-award
- Risks are appropriately allocated and subsequently borne in accordance with the contract
- The asset(s)/services is/are delivered in line with the contract
- Value for money (best value) or alternatively the anticipated profit margin is realised
- Performance measurement is undertaken so that in the event a party's failure to execute/perform appropriate action can be taken
- Payments are made in accordance with the terms of the contract (for example, in the case of a service contract, conditional on the quality of the delivery of the service)
- Changes to the contract (variations) are made and valued in accordance with the contract
- Continuous improvement is delivered as required under the terms of the contract
- Relationships are maintained

Aligned with the commercial management framework and the procurement process, contract management guidance, therefore, should address the following stages:

- **Intent**: including, for example, developing of the commercial strategy for the project, drafting preferred contractual terms and conditions, defining the product, work or services to be procured; contributing to the generation of the requirement specification, solution selection, and the ITT or ITP (demand side) or contributing to proposal identification (supply side)
- **Deal creation**: including, for example, participating in the asset/service procurement/bid process, amending the contract documents or contributing to the development of the supplier's proposition, negotiating commercial terms, and participating in the contract award decision
- **Execution**: including, for example, managing risk and relationships, measuring performance and administering the contract

Elsey (2007) presents the following list of pre- and post-award contract management activities (see Table 7.4), although, again, it adopts a demand-side perspective.

Alternatively, the key activities undertaken by the supplier's contract manager are presented in Table 7.5.

A vast quantity of best-practice contract management guidance is available, again predominantly derived from the public sector,[21] although most is generic and equally applicable to the private sector. Similarly, the perspective taken, for the most part, is that of the demand side and there are slight differences in emphasis in respect of the contract management of asset delivery projects and service contracts such as PFI/PPP. In the latter, project management (pre-award) and contract management (post-award) are often viewed as sequential activities, taking responsibility for the project pre-award and post-award respectively. On major asset delivery projects, project management and contract management are concurrent processes.

Contract management guidance is generally based around the following area: establishing the contract management team, risk management, relationship management, performance measurement and contract administration (see Figure 7.1).[22] For example, the guidance produced by Capital Ambition (2011) includes the following steps to implementing good contract and relationship management (see Table 7.6).

The OGC document *Contract Management in Long Term or Complex Projects* (OGC, 2010b) advises taking into account the following factors:

- **Getting the requirement right**: defining the project's requirements based on output specifications and outcomes that permit the development of innovative proposals
- **Contract duration**: establishing appropriate contract periods consistent with the degree of inherent uncertainty in the project, the need for innovation, the potential for change (for example, as a result of technological change), and the level of investment required of the supplier

Table 7.4 Pre- and post-award contract management activities.

Upstream or pre-award activities

These include:

- Preparing the business case and securing management approval
- Assembling the project team
- Developing contract strategy
- Risk assessment
- Developing contract exit strategy
- Developing a contract management plan
- Drafting specifications and requirements
- Establishing the form of contract
- Establishing the pre-qualification, qualification and tendering procedures
- Appraising suppliers
  - Why appraise?
  - When to appraise?
  - What should be appraised?
  - Who should appraise?
  - How to appraise?
- Drafting ITT documents
- Evaluating tenders
- Negotiation
- Awarding the contract

Downstream or post-award activities

These include:

- Changes within the contract
- Service delivery management
- Relationship management
- Contract administration
- Assessment of risk
- Purchasing organisation's performance and effectiveness review
- Contract closure

Source: Elsey (2007)

Table 7.5 Responsibilities of the supply-side contract manager.

- *Track the interpretation of the business requirement into contractual provisions*
- *Monitor contract performance and report at service/business outcome level as appropriate*
- *Monitor subordinate performance metrics*
- *Identify and manage exceptions*
- *Represent the provider's interests to the customer*
- *Respond to changing customer needs*
- *Marshal and apply the provider's resources*
- *Determine and take remedial actions by agreement with the customer*
- *Negotiate remedies with the customer*
- *Escalate problems as necessary*
- *Operate the contract to specification*
- *Operate subordinate services/contracts*
- *Maintain/develop service components*
- *Set/maintain/develop infrastructure strategy according to the contractually allocated responsibilities*
- *Maintain/develop supporting infrastructure according to the contractually allocated responsibilities.*

Source: OGC (2002)

Figure 7.1 Aspects of contract management.

Table 7.6 Six steps to implementing good contract and relationship management.

1	Define your requirements
2	Develop your performance management framework
3	Develop your relationship management framework
4	Develop your risk management framework
5	Enable continuous improvement
6	Decide what resources you need to manage the contract

Source: Capital Ambition (2011)

- **Driving appropriate behaviours and taking a balanced approach**: constructing the commercial arrangements in order to influence appropriate supply-side (and demand-side) behaviour and deliver value for money (or from the supplier's perspective, profit). These are considered to be delivered through flexible and impartial contracts that promote and maintain good working relationships
- **Proactive contract management**: involving both parties in ongoing performance measurement and the review of contractual risks
- **Incentivisation**: adopting a balanced approach appropriate to the project's objectives and emphasising the successful delivery of the project/service
- **Value for money**: ensuring the underlying principles of best value are applied and adhered to (OGC, 2010b)

The following good practice contract management framework published by the OGC and the National Audit Office (NAO) (NAO/OGC, 2008) is provided as an example.

Good practice contract management framework

The contract management framework is described as a '... *good practice guide for managing a broad range of contracts*' and in particular contracts that involve the delivery of services over periods in excess of five years (NAO/OGC, 2008). Although the framework is primarily aimed at post-award contract management, it acknowledges the impact that decisions made during the pre-award stage may have

Table 7.7 The good practice contract management framework.

1. **Structure and resources** Area 1: Planning and governance Area 2: People Area 3: Administration	3. **Development** Area 8: Contract development Area 9: Supplier development
2. **Delivery** Area 4: Managing relationships Area 5: Managing performance Area 6: Payment and incentives Area 7: Risk	4. **Strategy** Area 10: Supplier relationship management Area 11: Market management

Source: adapted from NAO/OGC (2008)

on the execution (operational) phase contract management. It therefore recommends, for example, consideration of the following issues:

- The early involvement of the contract management function during the pre-award phase
- The effect the manner in which the tendering process is conducted may have on the resulting contract relationships
- The 'cultural fit' between customer and supplier
- The equity of the proposed contract terms
- The impact of KPIs and service-level agreements (SLAs) on the design and effectiveness of contract management

It regards the pre-award (tendering/contract award) phase and post-award contract management as a continuum.

In addition to the pre-award activities, it also excludes certain issues specifically related to complex PFI projects, such as refinancing, or contract termination – for example, the disposal of assets.

The framework document comprises three sections:

- **The good practice contract management framework**: an overview of the activities to be considered when planning for and managing contracts
- **Assessing the appropriate level of contract management**: an explanation of 'how to' assess risk and value opportunities contained in contracts
- **Linking the good practice framework with the risk and value opportunity assessment**: incorporates an example of how to combine the activities and methodology from the first two sections to generate contract management tactics and priorities

The NAO/OGC framework comprises four **blocks** and 11 **areas** under which key contract management activities are collated (see Table 7.7).

The framework

Structure and resources

- **Area 1: Planning and governance**: preparing for contract management and providing oversight

  Activities addressed under Area 1 include: planning the handover (transition) from the pre-award (tendering/contract award) phase to the post-award (contract management) phase; establishing unambiguous ownership of the contract and defining the roles of the key participants; defining appropriate contract management processes and plans; appointing a contract management senior responsible owner; aligning contract processes and reporting mechanisms with organisational governance structures and processes; ensuring regular assessment and evaluation is undertaken; embedding knowledge management and cascading contract management guidance

- **Area 2: People**: ensuring the right people are in place to carry out the contract management activities

  Activities addressed under Area 2 are considered under two headings:
  - *the contract manager (or contract management team)*: has/have, for example, appropriate delegated authority; continuity between pre- and post-award phases; a detailed understanding of the contract; appropriate skills; accurate job descriptions, appropriate remuneration, supervision and career path
  - *wider staff issues*: for example, ensuring the contract management team is adequately resourced; has an appropriate range of skills to meet specific requirements of the contract over its life cycle; and establishment of an organisation-wide contract management 'community of interest'

- **Area 3: Administration**: managing the physical contract and the timetable for making key decisions

  Activities addressed under Area 3 include: establishing a repository for 'hard copy' contracts and the production of a contract operations guide; utilising contract management software; instigating mechanisms to identify key contract 'trigger points', reporting regular and ad hoc contract management information; administrating contract closure/termination; and considering the cost impact of contract management regime on the supplier

Delivery

- **Area 4: Managing relationships**: developing strong internal and external relationships that facilitate delivery

  Activities addressed under Area 4 are considered under two headings:
  - *roles and responsibilities*: for example, ensuring that individual contract managers understand their role; supplier-side roles and responsibilities are explicit and well-structured; and the responsibilities of both the contract manager and the supplier are clear and defined
  - *continuity and communications*: for example, maintaining continuity of key supplier staff or instigating an appropriate handover process; instigating and utilising formal and informal communication channels between the contract manager and supplier; co-locating customer and supplier staff if appropriate; clearly communicating the customer's expectations, understanding of the contract and services/performance levels; ensuring effective communication between the stakeholders; defining and applying appropriate dispute resolution processes

- **Area 5: Managing performance**: ensuring the service is provided in line with the contract

  Activities addressed under Area 5 are considered under two headings:
  - *service delivery*: for example, ensuring that service management is well structured; the provision of timely information by the customer organisation; that a performance management framework and SLAs are in place (the latter should be derived from the business needs and monitored); that the performance of suppliers is assessed using '*clear, objective and meaningful*' metrics; that a focused, '*by exception*' reporting methodology is adopted; and that a transparent and timely issues resolution process is instigated
  - *feedback and communications*: for example, providing feedback to suppliers on their performance; establishing clear points of contact between the service users and both the supplier and contract manager; capturing and considering changes in user requirements; and instigating formal performance reviews

- **Area 6: Payment and incentives**: ensuring payments are made to the supplier in line with the contract and that appropriate incentive mechanisms are in place and well managed

  Activities addressed under Area 4 are considered under two headings:
  - *payment and budgets*: for example, ensuring that clear payment mechanisms are recorded and well understood by the parties; that well-defined and efficient payment processes are initiated; costs are

appropriately mapped against budgets; and that post-award payment changes demonstrate value for money and are valued using the provisions of the contract
- *payment and incentive mechanisms*: for example, that financial or non-financial incentive structures are linked to the contract's intended deliverables (outcomes), and appropriately applied and managed; that service credits and the like are managed well; that if an open-book approach is adopted the process is managed impartially and expertly; and that commercial terms that could enable the supplier to behave opportunistically during the contract period are avoided

- **Area 7: Risk**: understanding and managing contractual and supplier risk

  Activities addressed under Area 4 are considered under three headings:
  - *processes and plans*: for example, ensuring risk management is instigated, with risks being borne by the party best able to manage them; that there is regular formal risk identification, monitoring and mitigation; that risk governance is accounted for with contingency plans to cover failure of the supplier, and exit strategies in place
  - *contractual terms*: for example, ensuring that they are understood and monitored by the contract manager relating, for example, to termination; to warranties, indemnities and insurance; and to security and confidentiality. Also ensuring that dispute resolution processes, such as mediation, adjudication and arbitration, are adopted.
  - *ongoing supplier risk management*: for example, the supplier's financial health and business performance, and compliance with contractual 'non-performance' issues are monitored by the contract manger

Development

- **Area 8: Contract development**: effective handling of changes to the contract

  Activities addressed under Area 8 are considered under two headings:
  - *change process*: for example, including the regular review of the contract to ensure it continues to meet the evolving needs of the organisation; the establishment of transparent processes that provide governance to contractual change with processes that are proportionate to both minor changes and contract variations and to major contractual changes; where appropriate, the use of benchmarking to test for value for money and market testing; and also the inclusion of dispute resolution processes linked to change management
  - *processes for different types of change*: for example, ensuring that both parties are fully aware of the procedures for agreeing any contract extensions (either in scope or time) or any other related issues; that consistent and equitable processes are in position to deal with changes to the commercial (interpreted as financial) aspects of the contract; similarly that appropriate mechanisms are in place to manage price changes and to test that value for money is attained (these may include competitive tendering, benchmarking or open-book pricing)

- **Area 9: Supplier development**: improving supplier performance and capability

  Activities addressed under Area 9 are considered under two headings:

  - *processes*: for example, that processes are initiated addressing the planning, management and governance of supplier development; and for capturing and measuring benefits derived (so that supplier development drives continuous improvements and the delivery of value for money on behalf of the purchaser). This requires an awareness of what drives and motivates suppliers, together with an understanding of the linkage between the objectives of the supplier and supplier development
  - *improvement activities*: for example, include adopting supplier operational performance improvement activities, such as, '6-Sigma' and 'lean' principles; collaborative working; shared risk reduction; and activities related to wider customer organisational initiatives. Activities may also include the development of the wider supply chain

Strategy

- **Area 10: Supplier relationship management**: having a programme for managing and developing relationships with suppliers

  Activities addressed under Area 10 include, for example: establishing a planned and structured supplier relationship management programme, with appropriate oversight; establishing a benefits realisation plan in respect of supplier relationship management; capturing supplier innovation where appropriate; addressing knowledge management issues; taking an organisation-wide perspective on supplier relationship management; and planning and managing organisational interfaces at board level

- **Area 11: Market management**: managing the wider market issues that impact on the contract, but lie beyond the supplier

  Activities addressed under Area 11 include, for example: establishing processes to review and evaluate 'make or buy' options – that is deliver services in-house or outsource the requirement; utilising market intelligence when benchmarking, contingency planning and developing re-competition strategies; analysing the competence and resources of prospective suppliers; evaluating emergent practice and technology; undertaking 'market making' when necessary; understanding the impact and importance of bidding and switching; establishing a re-competition strategy and plan; and providing feedback to support future procurement processes and procedures and the development of strategy

Contract and commercial management

Adopting a process-based, best-practice perspective, the IACCM book *Contract and Commercial Management: The Operational Guide* (IACCM, 2011) is a practical resource and operational guide to contracting practices and methods for all practitioners engaged in the management or negotiation of contracts. It is based around materials developed for IACCM's online training modules, with significant input from contracting and commercial management practitioners.

The guide is divided into five phases:

- **Initiate**: this seeks to develop an awareness of the relationships between business requirements and objectives, and markets. Further, it explores the value to be gained from aligning contracting arrangements, plus contract terms and conditions, policies and practices with these issues, the objective being to facilitate successful trading exchanges and relationships and to increase the effectiveness of the contracting process
- **Bid**: this deals with the tendering (bidding) and proposal development activities carried out by the parties in order to match customer requirements with supplier capabilities. Legal and regulatory issues associated with these activities are addressed, together with the financial aspects of the intended relationship
- **Development**: this covers the formation of appropriate contracts and outlines the commonly occurring issues and concerns that need to be addressed. It also provides a framework to evaluate the risk involved in particular relationships and acts as a basis for negotiation preparation
- **Negotiate**: this provides a comprehensive guide to contract negotiation: both face-to-face and via 'virtual' technology. The text considers behavioural issues associated with negotiating and deals with common issues and problems encountered by contract negotiators, both internally and with third parties
- **Manage**: this explores post-award management techniques aimed at ensuring the delivery of a project or implementation of a service in accordance with the contract terms. It deals with contract changes and the renegotiation of long-term contracts. Further, it discusses the methods by which project/service deliverables can be evaluated

Commercial function skills framework

The UK MOD has produced a demand-side perspective commercial skills framework to support the development of its defence commercial function (DCF) – see Table 7.8 (MoD, 2008). Adopting a through-life approach it covers the following activities: *shaping the market place, influencing the sourcing strategy,*

Table 7.8 Defence commercial function competencies.

DCF competencies

1. **Market knowledge**: understanding a relevant market sector, the capacity and capabilities of suppliers in that sector and the MODs position in that sector. Competencies include:
   - Spend analysis
   - Market research
   - Supplier assessment
2. **Commercial operations**: commercial and contracting processes that are open, robust and consistent. Competencies include:
   - Roles and responsibilities
   - Commercial strategy
   - Commercial models
   - Commercial enabling
3. **Whole supply network management**: to bring together the whole supply network to review and manage all contributors to the required final contract outcomes. Competencies include:
   - Supply chain analysis
   - Supplier development
4. **Regulatory and legal requirements**: procurement policy. Competencies include:
   - Legality – legislation and regulatory frameworks

DCF skills

These can be used in any phase of the commercial cycle:

- Strategic thinking
- Business planning
- Project management
- Programme management
- Risk management
- Benchmarking
- Negotiation
- Knowledge management
- Influencing skills
- Change management
- Resource planning
- Relationship management
- Conflict management

Source: Adapted from MoD (2008)

capitalising on the supply network, choosing the right commercial approach, exploiting available e-tools, making the deal and managing the contractual relationship for delivery.[23] There is also an associated commercial cycle comprising the following three phases:

1. **Pre-contract**: activities include: requirements management, market research and development, and strategic sourcing
2. **Contract award**: activities include: the tendering process, contract terms and conditions, incentive pricing, cost analysis, performance management, and cost modelling
3. **Contract management**: activities include: supplier performance analysis, supply network management, partnered category management (PCM), value chain management, lean techniques, process improvement, demand management, compliance management, and forensic accounting

The recognised professional qualification for the DCF is the Chartered Institute of Purchasing and Supply (CIPS) Advanced and Graduate Diplomas in Purchasing and Supply.

Similarly, the NAO (2009a) has developed a **commercial skills for complex projects framework**. NAO consider commercial to be a combination of procurement and contract management, and define commercial skills as the ability to interact on equal and professional terms with the private sector (NAO, 2009b). The framework comprises a set of key commercial skills and behaviours (abilities and knowledge) applied across the following five stages of a project/contract: project initiation, feasibility testing, selecting a partner, post-contract award, contract completion. The framework was used to evaluate the commercial skills across UK central government. The review found commercial skills to be 'generally weak' across all government departments, with major deficiencies in contract management, the commissioning and management of advisers, risk identification and management, and business acumen.[24]

The OGC Gateway™ Process

Introduction

Developed by the Office of Government Commerce (OGC) as part of the UK government's modernisation agenda, the Gateway Project Review Process has been operational throughout central civil government since January 2001. Although the process was primarily introduced to promote the delivery of improved public services, ensuring that central government procurement projects and acquisition programmes deliver the benefits anticipated more effectively and with more reliable costs and outcomes, the underlying principles and methodology are equally applicable to private sector organisations. Within UK central government the process is overseen by the newly instigated Major Projects Authority, part of the Cabinet Office*, while Local Partnerships are accredited to offer Gateway Reviews to local government.[25]

The process is designed to be appropriate for a broad range of programmes and projects, including the procurement of construction/property, services, IT-enabled business change and project and programmes employing framework contracts. The principles and process are also applicable to the management of expenditure projects and to the development of policy and change initiatives.

Compliant with the *Gershon Report* on government procurement (Gershon, 1999) and the Cabinet Office report *Successful IT: Modernising Government in Action* (Cabinet Office, 2000), the OGC Gateway™ Process also conforms to the principles of *Achieving Excellence in Construction* (AEC).[26]

The process involves an independent, high-level, evaluation of a programme or project at critical stages during its life cycle (a **peer review**) – in particular, its progress to date and chances of delivering its intended benefits. The aim of the process is to provide assurance to programme/project **senior responsible owners (SRO)** that the undertaking can proceed to the next phase. The **SRO** (in construction projects frequently called the **project owner**) is a generic title, referring to the senior individual who is personally responsible for the successful conclusion of the programme/project. Further, it assists the SRO to deliver their business goals by ensuring that:

- Individuals with appropriate experience and expertise are allocated to the programme/project
- All programme/project stakeholders fully comprehend both the issues entailed in the programme/project and its status
- The programme/project can proceed with confidence to the next stage of its execution or development (additionally, any associated procurement is appropriately undertaken and delivers value for money)
- More realistic cost and time targets for programmes/projects are achieved
- Expertise and competence of government staff are developed via membership of review teams
- Recommendations and guidance are provided to programme/project teams by their peers (OGC, 2007a)

During the life cycle of a project/programme there are five reviews, three before contract award:

- OGC Gateway™ Process **Review 1: Business justification**
- OGC Gateway™ Process **Review 2: Delivery strategy**
- OGC Gateway™ Process **Review 3: Investment decision**

* The following overview of the OGC Gateway™ Process is derived from the following sources: OGC (2007a, 2007b, 2007c, 2007d, 2007e, 2007f, 2007g)

and two which focus on service implementation and the verification of operational benefits:

- OGC Gateway™ Process **Review 4: Readiness for service**
- OGC Gateway™ Process **Review 5: Operations review and benefits realisation**

Recurring over the life of the programme, the OGC Gateway™ Process **Review 0: Strategic Assessment** is a programme-only review which takes a broader corporate or policy perspective. A process model illustrates how the review process relates to the project life cycle.

The review process

OGC Gateway™ Process reviews are carried out on a confidential basis for the SRO. The aim of the process is to determine potential and feasible opportunities to improve a project/programme's governance, management, expertise, or delivery processes, by refocusing key stakeholders interest on the principal risks and essential issues inherent in the programme/project.

The process is based around the assessment of readily available project documentation and non-attributable discussions between the review team and individual key stakeholders, members of the programme/project's management team – and on exceptionally complex or mission critical programmes – government ministers. Review teams may vary in size but usually comprise three or four members, whereas reviews, depending on the scale and risk inherent in the project/programme, generally take between three and five days. The process is, therefore, intended to minimise the burden on project teams and its impact on project/programme duration.

The SRO receives a draft report outlining a small number of proposals and allocating the programme/project either red, amber or green status. While 'red' status does not lead to the termination of the programme/project, it does require immediate remedial action to be taken. Subsequently, the SRO is responsible for putting into practice any suggested corrective action for the progression of the programme/project, and for deciding who should be given access to the report.

Guidance is available to support the OGC Gateway™ Process, including:

- A **manager's checklist** – comprising a series of key questions to help SROs determine the progress and probable success of their programme/project
- **Workbooks** covering each OGC Gateway™ Process Review which contain a comprehensive list of questions to support each review
- A **risk potential assessment**, enabling risk to be assessed prior to each OGC Gateway™ Process Review and for resources and expertise requirements to be determined
- The **Successful Delivery Toolkit** gives access to pdf copies of the workbooks and links to sources of best practice, tools and techniques (source: OGC, 2007a, 2009b)[27]

Gateway™ workbooks

As mentioned above, a series of workbooks have been produced by OGC to facilitate these stage reviews. Due to the uniqueness of projects and programmes, these documents are intended to provide guidance, containing a compilation of suitable questions to raise, together with samples of appropriate evidence to seek/furnish. Moreover, these documents are not considered to be a checklist of compulsory, predetermined items; it is anticipated that they will be tailored to suit individual programme/project needs. They also describe the key project documents to be provided.

OGC Gateway™ Process Review 0: Strategic Assessment

Utilised at the commencement of a programme, the Gateway™ Review 0 is repeated at key stages during its life cycle, the review seeks to satisfy the **programme board** that the programme's scope and rationale have been satisfactorily investigated, that the programme conforms to the priorities, policy and management strategy of the organisation, that the necessary resources to ensure compliant delivery are obtainable, and that the key actors have a shared understanding of its main deliverables. Also, the review investigates whether or not the programme's management structure, monitoring and resourcing arrangements, and projects and subprogrammes are appropriately organised to deliver its stated objectives.

Review 0, therefore, seeks to establish whether or not the stakeholders' view of the programme is realistic, by considering its costs, anticipated outcomes, resource requirements, schedule and overall viability. Later reviews will return to these issues to verify that the major stakeholders have a shared appreciation of the required outcomes and that the programme is set to realise them (OGC, 2007a).

The workbook is divided into the following six sections:

1. Policy and business context
2. Business case and stakeholders
3. Management of intended outcomes
4. Risk management
5. Review of current outcomes
6. Readiness for next phase

Each section contains sample questions (**areas to probe**) and sample sources of evidence (**evidence expected**).

OGC Gateway™ Process Review 1: Business justification

The **project initiation process** generates a justification for the project founded on business needs and an evaluation of the likelihood of it satisfying those needs and its probable costs. The first Gateway™ Review is held following the production of the **strategic business case** (a high-level outline business case), but before any development proposal is considered by a **project board** or similar group and authorisation to continue is given. Focusing on the project's business justification, the review seeks to satisfy the **project board** that the planned scheme meets the business needs, has been satisfactorily investigated and can be realised (OGC, 2007b).

The workbook is divided into the following four sections:

1. Policy and business context
2. Business case and stakeholders
3. Risk management
4. Readiness for next phase: delivery strategy

Again, sample areas to probe and examples of evidence expected are included.

OGC Gateway™ Process Review 2: Delivery strategy

Review 2 seeks to satisfy the **project board** as to the suitability of the project's chosen procurement method (delivery strategy). It assesses the feasibility of the project, the likelihood that it will go according to plan, and whether it is appropriate for the project team to invite tenders or proposals from potential suppliers. The OGC Gateway Review team and project team must be content that all relevant issues have been taken into account, particularly the provision of value for money (OGC, 2007c).

The workbook is divided into the following five sections:

1. Assessment of delivery approach
2. Business case and stakeholders
3. Risk management
4. Review of current phase
5. Readiness for next phase – investment decision

Sample areas to probe and examples of evidence expected are included.

OGC Gateway™ Process Review 3: Investment decision

Before entering into a contract with a supplier or partner (or placing a work order with an existing supplier), Review 3 considers the suitability of the proposed investment decision and verifies the supplier selection procedures. Additionally, it evaluates the effectiveness of the process, the realisation of business needs, the proposed solution (that the supplier and client can manage and execute the project), and that, after placing the contract, appropriate systems are present to ensure project delivery. The review, therefore, examines the **full business case** and the governance of the investment decision. Again, the OGC Gateway Review team and project team must be content that all relevant issues have been taken into account, particularly the provision of value for money. Generally, an OGC Gateway Review 3 will only be carried out once during the project life cycle. However, there are occasions when it may be prudent to repeat the review (OGC, 2007d).

The workbook is divided into the following five sections:

1. Assessment of the proposed solution
2. Business case and stakeholders
3. Risk management
4. Review of current phase
5. Readiness for next phase – readiness for service

Sample areas to probe and examples of evidence expected are included.

OGC Gateway™ Process Review 4: Readiness for service

Review 4 concentrates on establishing the effectiveness of the chosen solution, the preparedness of the organisation to put into practice resulting business changes, that appropriate contract management arrangements are (or are being) prepared, and that apt performance criteria have been developed. In the case of construction/property projects, Review 4 occurs once the project has been certified as ready for use, whereas for IT-enabled projects the review occurs once all testing has been concluded but prior to the products launch (OGC, 2007e).

The workbook is divided into the following four sections:

1. Business case and stakeholders
2. Risk management
3. Review of current phase
4. Readiness for next phase – operations review and benefits realisation

Sample areas to probe and examples of evidence expected are included.

OGC Gateway™ Process Review 5: Operations review and benefits realisation

For some projects one Review 5 should be sufficient, but for most projects, particularly those that incorporate a service contract element, several reviews may occur during its operational service. Moreover, based on whether the project is a long-term service contract or the provision of works, the extent of the review is adaptable. For long-term contracts – for example, PFI and strategic partnering arrangements – four reviews could be held over the contract period, whereas for IT-enabled projects either two or three reviews could be undertaken during a five-year contract.

The initial review is undertaken 6–12 months after the receipt of the asset or the launch of a new service, to enable evidence to be assembled on the operating benefits derived from the service or asset. It focuses on the business case and the effectiveness of service delivery/contract management procedures. An OGC Gateway Review 5 occurs once the organisation has undertaken a post-implementation or comparable review. It utilises the conclusions drawn from the internal review, in conjunction with an evaluation of organisational learning derived from the project. However, the inclusion of a full review of plans for the future is optional.

Subsequent Gateway Review 5s verify whether or not the projected value for money and anticipated benefits outlined in the business case and benefits plans are being delivered. The timing of these additional reviews is determined by the operational business owner (OGC, 2007f).

The workbook is divided into the following six sections:

1. Business case and benefits management
2. Review of operating phase
3. Plans for ongoing improvements in value for money
4. Plans for ongoing improvements in performance
5. Readiness for the future plans for future service provision
6. Review of organisational learning and maturity targets

Sample areas to probe and examples of evidence expected are included.

Summary

This chapter has sought to address the issue of project success and failure, in particular, failures associated with poor client-side project management competence. Best-practice frameworks and guidelines have been presented that seek to develop and improve, both at an organisational and individual level, capabilities in procuring, delivering and managing projects and programmes. Whereas the vast majority of organisations may gain significant benefits from adopting these best practices, they may not be applicable to all projects or programmes. For example, the adoption of an organisational 'standard way' or approach can lead to the imposition of a disproportionate level of bureaucracy, stifle creativity and innovation solutions, and result in practitioners concentrating on individual processes and procedures rather than the specific task to be undertaken. Moreover, due to the sheer volume of applicable best-practice advice, relating to the breadth of activities performed by the commercial function, practitioners may fail to address or see the wider context in which an exchange contract sits.

Exercise 7.6 Best practice management

Evaluate the management practices demonstrated on the Wembley football stadium project (see Case Study A) against the perceived best practice outlined in this chapter.

Endnotes

1 The activities of the Office of Government Commerce (OGC) were subsumed within the Efficiency and Reform Group of the Cabinet Office in 2011
2 In the case of private organisations this would be executive board level personnel
3 See http://www.bis.gov.uk/files/file45270.xls (accessed 17 September 2012)
4 See http://www.offta.org.uk/charts.htm (accessed 17 September 2012); http://www.btplc.com/Thegroup/RegulatoryandPublicaffairs/Ourundertakings/KeyPerformanceIndicators/index.htm (accessed 17 September 2012)
5 The current edition of the code was published in May 2010 and became effective on 29 June 2010. It applies to all companies with a premium listing of equity shares in the UK. See: http://www.frc.org.uk/corporate/ukcgcode.cfm (accessed 17 September 2012)
6 See: https://www.frc.org.uk/FRC/media/Documents/Revised-Turnbull-Guidance-October-2005.pdf (accessed 18 September 2012)
7 See: http://www.icaew.com/en/library/subject-gateways/corporate-governance/codes-and-reports/turnbull-report (accessed 17 September 2012)
8 See http://frc.org.uk/getattachment/9e3ea3c6-6fa0-45e5-aae5-db4f1be84300/Guide-on-use-of-Turnbull-for-section-404.aspx (acccessed 17 September 2012)
9 Responsibility for the Best Management Practice Portfolio passed to the Cabinet Office in 2011
10 PRINCE2®, MSP®, M_o_R®, MoV®, P3O®, and ITIL® are registered trademarks of the Cabinet Office; MoP™ is a trademark of the Cabinet Office

11 OGC Gateway™ is a trademark of the Cabinet Office
12 See APMG Group (2012g) and Best Management Practice (2012g)
13 See OGC (2008b); APMG Group (2012f); and Best Management Practice (2012f)
14 See The Cabinet Office's Common Glossary of Terms and Definitions: http://www.prince-officialsite.com/InternationalActivities/TranslatedGlossaries.aspx (accessed 17 September 2012)
15 See The Cabinet Office's Common Glossary of Terms and Definitions: http://www.prince-officialsite.com/InternationalActivities/TranslatedGlossaries.aspx (accessed 17 September 2012)
16 © Crown copyright 2011. All rights reserved. Material is reproduced with the permission of the Cabinet Office under delegated authority from the Controller of HMSO.
17 See The Cabinet Office's Common Glossary of Terms and Definitions: http://www.mop-officialsite.com/home/InternationalActivities/Translated_Glossaries_2.aspx (accessed 17 September 2012)
18 See The Cabinet Office's Common Glossary of Terms and Definitions: http://www.mov-officialsite.com/home/InternationalActivities/Translated_Glossaries_2.aspx (accessed 17 September 2012)
19 See for example: OGC (2008a); Department of Health (2010); IMF (2010); The World Bank (2011); The Chartered Institute of Purchasing and Supply (CIPS) http://cipsintelligence.cips.org/opencontent/what-is-purchasing-and-supplymanagement
20 Note the current code (OGC, 2010a) is currently under review
21 See, for example, OGC (2002, 2010b); Elsey (2007); 4ps (2007); Capital Ambition (2011); and The Chartered Institute of Purchasing and Supply (CIPS) (2012)
22 See, for example, OGC (2002); Elsey (2007); 4ps (2007); and Capital Ambition (2011)
23 See also the MOD Commercial Toolkit. http://www.aof.mod.uk/aofcontent/tactical/toolkit/content/topics/postcost.htm (accessed 28 June 2010) and MOD (2003)
24 See also NAO (2009b) *Opinion pieces on improving commercial skills for complex government projects*
25 See http://www.localpartnerships.org.uk/PageContent.aspx?id=40&tp= (accessed 18 September 2012)
26 See http://www.nao.org.uk/publications/nao_reports/03-04/0304361-i.pdf (accessed 18 September 2012) http://ctpr.org/wp-content/uploads/2011/03/Successful-IT-Modernising-Government-in-Action-2000.pdf http://ctpr.org/wp-content/uploads/2011/03/Successful-IT-Modernising-Government-in-Action-2000.pdf (accessed 18 September 2012) http://webarchive.nationalarchives.gov.uk/20110601212617/http:/www.ogc.gov.uk/ppm_documents_construction.asp (accessed 18 September 2012)
27 See http://webarchive.nationalarchives.gov.uk/20100503135839/http://www.ogc.gov.uk/resource_toolkit.asp (accessed 18 September 2012)

References

12manage (2008) *Project Management Methodology Focusing on Organization, Management and Control. Explanation of Prince 2.*12manage – The Executive Fast Track. http://www.12manage.com/methods_ccta_prince2.html (accessed 21 February 2012)

Aberdeen Group (2004) *Best Practices in Contract Management: Strategies for Optimizing Business Relationships.* Aberdeen Group Inc., Boston, MA http://v1.aberdeen.com/summary/report/other/BPinCM_092904a.asp (accessed 1 March 2012)

APMG Group (2009) *PRINCE2 Glossary of Terms.* APM Group, High Wycombe. http://www.prince-officialsite.com/InternationalActivities/TranslatedGlossaries.aspx (accessed 21 February 2012)

APMG (2012a) *PRINCE2 – PRojects IN Controlled Environments: The Official PRINCE2® Website.* A Best Management Practice website managed and published by APMG in conjunction with the Cabinet Office (part of HM Government) and TSO. http://www.prince-officialsite.com/ (accessed 21 February 2012)

APMG (2012b) *MSP® – Managing Successful Programmes: The Official MSP®Website.* A Best Management Practice website managed and published by APMG in conjunction with the Cabinet Office (part of HM Government) and TSO. http://www.msp-officialsite.com/ (accessed 21 February 2012)

APMG (2012c) *M_o_R® – Management of Risk: The Official M_o_R®Website.* A Best Management Practice website managed and published by APMG in conjunction with the Cabinet Office (part of HM Government) and TSO. http://www.mor-officialsite.com/ (accessed 21 February 2012).

APMG (2012d) *MoP™ – Management of Portfolios: The Official MoP Website.* A Best Management Practice website managed and published by APMG in conjunction with the Cabinet Office (part of HM Government) and TSO. http://www.mop-officialsite.com/ (accessed 21 February 2012)

APMG (2012e) *MoV® – Management of Value: The Official MoV Website.* A Best Management Practice website managed and published by APMG in conjunction with the Cabinet Office (part of HM Government) and TSO. http://www.mov-officialsite.com/home/home.aspx (accessed 21 February 2012)

APMG (2012f) *P3O® – Portfolio, Programme and Project Offices: The Official P3O Website.* A Best Management Practice website managed and published by APMG in conjunction with the Cabinet Office (part of HM Government) and TSO. http://www.p3o-officialsite.com/ (accessed 21 February 2012)

APMG (2012g) *ITIL®: The Official ITIL® Website.* A Best Management Practice website managed and published by APMG in conjunction with the Cabinet Office (part of HM Government) and TSO. http://www.itil-officialsite.com/ (accessed 21 February 2012)

APMG (2012h) *P3M3®: The Official P3M3® Website.* A Best Management Practice website managed and published by APMG in conjunction with the Cabinet Office (part of HM Government) and TSO. http://www.p3m3-officialsite.com/ (accessed 21 February 2012)

APMG UK (2012a) *PRINCE2® – PRojects IN Controlled Environments.* APMG International. http://www.apmg-international.com/APMG-UK/PRINCE2/PRINCE2Home.aspx (accessed 21 February 2012)

APMG UK (2012b) *MoV® – Management of Value.* APMG International. http://www.apmg-international.com/APMG-UK/MoV/MoVHome.aspx (accessed 21 February 2012)

Baker, BN, Murphy, DC and Fisher, D (1988) Factors affecting project success. In: DI Cleland and WR King (eds) *Project Management Handbook.* 2nd edition. Wiley, New York. pp. 909–919

Best Management Practice (2012a) *Project Management – PRINCE2.* A Best Management Practice website managed and published by TSO in conjunction with the Cabinet Office (part of HM Government) and APMG. http://www.best-management-practice.com/Project-Management-PRINCE2/ (accessed 21 February 2012)

Best Management Practice (2012b) *Programme Management – Managing Successful Programmes.* A Best Management Practice website managed and published by TSO in conjunction with the Cabinet Office (part of HM Government) and APMG. http://www.best-management-practice.com/Programme-Management-MSP/ (accessed 21 February 2012)

Best Management Practice (2012c) *Risk Management – Management of Risk (M_o_R).* A Best Management Practice website managed and published by TSO in conjunction with the Cabinet Office (part of HM Government) and APMG. http://www.best-management-practice.com/Risk-Management-MoR/ (accessed 21 February 2012)

Best Management Practice (2012d) *Portfolio Management – Management of Portfolios (MoP).* A Best Management Practice website managed and published by TSO in conjunction with the Cabinet Office (part of HM Government) and APMG. http://www.best-management-practice.com/Portfolio-Management-MoP/ (accessed 21 February 2012)

Best Management Practice (2012e) *Value Management (MoV).* A Best Management Practice website managed and published by TSO in conjunction with the Cabinet Office (part of HM Government) and APMG. http://www.best-management-practice.com/Value-Management-MoV/ (accessed 21 February 2012)

Best Management Practice (2012f) *Portfolio, Programme and Project Offices – P3O.* A Best Management Practice website managed and published by TSO in conjunction with the Cabinet Office (part of HM Government) and APMG. http://www.best-management-practice.com/Portfolio-Programme-and-Project-Offices-P3O/ (accessed 21 February 2012)

Best Management Practice (2012g) *IT Service Management – ITIL®.* A Best Management Practice website managed and published by TSO in conjunction with the Cabinet Office (part of HM Government) and APMG. http://www.best-management-practice.com/IT-Service-Management-ITIL/ (accessed 21 February 2012)

BSI (1997) *BS EN 1325-1: Value Management, Value Analysis, Functional Analysis Vocabulary – Part 1. Value Analysis and Functional Analysis.* British Standards Institution, London

BSI (2000) *BS EN 12973: Value Management.* British Standards Institution, London

BSI (2004) *BS EN 1325-2: Value Management, Value Analysis, Functional Analysis Vocabulary - Value Management.* British Standards Institution, London

Cabinet Office (2000) *Successful IT: Modernising Government in Action.* Central IT Unit, London

Cabinet Office (2012) Office of Government Commerce (OGC). http://www.cabinetoffice.gov.uk/content/office-government-commerce-ogc (accessed 21 August 2012)

Capital Ambition (2011) *You and Your Contractor: A Manual of Best Practice in Contract and Relationship Management.* 2nd edition. Capital Ambition (London Councils), London. www.capitalambition.gov.uk/srdbestpractice (accessed 1 March 2012)

Chapman, C and Ward, S (2002) *Managing Project Risk and Uncertainty.* John Wiley & Sons, Chichester

CIPS (2012) *Supply Management Guide to Contract Management.* The Chartered Institute of Purchasing and Supply, Stamford, Lincolnshire. http://www.supplymanagement.com/resources/how-to/2010/guide-to-contract-management/. (accessed 21 February 2012)

Cox, A and Townsend, M (1998) *Strategic Procurement in Construction: Towards Better Practice in the Management Of Construction Supply Chains.* Thomas Telford, London

Dallas, MF (2006) *Value and Risk Management: A Guide to Best Practice.* Blackwell Publishing, Oxford

Dallas, M, OGC and Clackworthy, S (2010) *Management of Value.* The Stationery Office, Norwich

Department of Health (DoH) (2010) *Procurement Guide for Commissioners of NHS-funded Services.* Department of Health, London. http://www.dh.gov.uk/prod_consum_dh/groups/dh_digitalassets/@dh/@en/@ps/documents/digitalasset/dh_118219.pdf (accessed 21 February 2012)

Elsey, RD (2007) *Contract Management Guide*. The Chartered Institute of Purchasing and Supply, Stamford, Lincolnshire. http://www.cips.org/documents/CIPS_KI_Contract%20Management%20Guidev2.pdf (accessed 1 March 2012)

4ps (in collaboration with Mott MacDonald) (2007) *A Guide to Contract Management for PFI and PPP Projects*. Public Private Partnerships Programme, London. http://www.partnershipsuk.org.uk/uploads/documents/OTF4ps_Contract Managers_guide.pdf (accessed 1 March 2012)

Financial Reporting Council (FRC) (2004) *The Turnbull Guidance as an Evaluation Framework for the Purposes of Section 404(a) of the Sarbanes-Oxley Act*. Financial Reporting Council, London. http://frc.org.uk/getattachment/9e3ea3c6-6fa0-45e5-aae5-db4f1be84300/Guide-on-use-of-Turnbull-for-section-404.aspx (accessed 18 September 2012)

Financial Reporting Council (FRC) (2005) *Internal Control: revised Guidance for Directors on the Combined Code*. Financial Reporting Council, London. https://www.frc.org.uk/FRC/media/Documents/Revised-Turnbull-Guidance-October-2005.pdf (accessed 18 September 2012)

Gershon, P (1999) *Review of Civil Procurement in Central Government*. HM Treasury. http://archive.treasury.gov.uk/docs/1999/pgfinalr.html (accessed 21 February 2012)

Glenigan (2011) *2011 UK Construction Industry KPIs - Industry Performance Report*. Constructing Excellence in partnership with Department for Business, Innovation and Skills, London. http://www.glenigan.com/PDF/2011_UK_Construction_Industry_KPI_Report.pdf (accessed 21 February 2012)

Graves, A and Rowe, D (1999) *Constructing the Best Government Client: Benchmarking the Government Client, Stage II Study*. HM Treasury/HMSO, London

Graves, A, Sheath, D, Rowe, D and Sykes, M (1998) *Constructing the Best Government Client: The Government Client Improvement Study*. HM Treasury/HMSO, London

Green, P (2010) *Efficiency Review – Key Findings and Recommendations*, Available from http://www.cabinetoffice.gov.uk/sites/default/files/resources/sirphilipgreenreview.pdf (accessed 21 February 2012)

HM Treasury (2004) *Orange Book: Management of risk – Principles and Concepts*. HMSO, Norwich. http://www.hm-treasury.gov.uk/orange_book.htm (accessed 21 August 2012)

HM Treasury (2005) *Government Accounting 2000: A Guide on Accounting and Financial Procedures for the Use of Government Departments: Amendment 4/05*. The Stationery Office, London. http://www.government-accounting.gov.uk/current/frames.htm (accessed 1 May 2007)

IACCM (2011) *Contract and Commercial Management: The Operational Guide*. Van Haren Publishing, Zaltbommel, Netherlands

IMF (2010) *IMF Procurement Guide for Suppliers*. International Monetary Fund, Washington, DC http://www.imf.org/external/np/procure/eng/index.htm (accessed 21 February 2012)

International Organization for Standardization (ISO) (2009) *ISO 31000:2009 Risk management – Principles and Guidelines*. Geneva, Switzerland

Jenner, S, OGC, and Kilford, C (2011) *Management of Portfolios*. The Stationery Office, Norwich

Jones, ME and Sutherland, G (1999) *Implementing Turnbull: A Boardroom Briefing*. The Centre for Business Performance, The Institute of Chartered Accountants in England and Wales, London

Kähkönen, K (2006) Management of uncertainty. In: DJ Lowe (ed.) with R Leiringer, *Commercial Management of Projects: Defining the Discipline*. Blackwell Publishing, Oxford. pp. 211–233

Kelly, J and Male, S (1993) *Value Management in Design and Construction*. E & FN Spon, London

Kelly, J, Male, S.and Graham, D. (2004) *Value Management of Construction Projects*. Blackwell Publishing, Oxford

KPMG International (2005) *Information Risk Management: Global IT Project Management Survey - How Committed Are You?* KPMG International, Switzerland. http://www.kpmg.com/CN/en/IssuesAndInsights/ArticlesPublications/Documents/Global-IT-Project-Management-Survey-0508.pdf (accessed 21 February 2012)

MOD (2003) *Commercial Awareness: A Beginner's Guide for Those New to MOD Acquisition Work*. Commercial Services Group, DSDA (PC) Keynsham, Bristol. http://www.metasums.co.uk/uploads/asset_file/Commercial%20awareness%20-%20a%20beginners%20guide%202003.pdf (accessed 1 March 2012)

MOD (2008) *Defence Commercial Skills Framework*. Defence Commercial Directorate, MOD, London

Murray-Webster, R and OGC (2010) *Management of Risk: Guidance for Practitioners*. 3rd edition. The Stationery Office, London

NAO (2000) *Supporting Innovation: Managing Risk in Government Departments*. The Stationery Office, London. http://www.nao.org.uk/publications/9900/managing_risk_in_gov_depts.aspx (accessed 21 August 2012)

NAO (2001) *Modernising Construction* The Stationery Office, London http://www.nao.org.uk/publications/nao_reports/00-01/000187.pdf (accessed 21 February 2012)

NAO (2004a) *Improving IT Procurement: The Impact of the Office of Government Commerce's Initiatives on Departments and Suppliers in the Delivery of Major IT-enabled Projects*. The Stationery Office, London. http://www.nao.org.uk//idoc.ashx?docId=2f350558-ef70-432c-9cd3-656afb06e767&version=-1 (accessed 18 September 2012)

NAO (2004b) *Getting Value for Money from Construction Projects through Design*, The Stationery Office, London http://www.nao.org.uk/sector/idoc.ashx?docid=9ad62f32-0acf-4a60-aea0-a26a17b8fdca&version=-1 (accessed 21 February 2012)

NAO (2005) *Improving Public Services Through Better Construction*. The Stationery Office, London. http://www.nao.org.uk/publications/0405/improving_public_services.aspx (accessed 1 March 2012)

NAO (2009a) *Commercial Skills for Complex Government Projects*. The Stationery Office, London. http://www.nao.org.uk/publications/0809/commercial_skills.aspx (accessed 1 March 2012)

NAO (2009b) *Opinion Pieces on Improving Commercial Skills for Complex Government Projects*. The Stationery Office, London. http://www.nao.org.uk/idoc.ashx?docId=0b2eabab-cd39-4ce1-b77e-551ce03b801c&version=-1 (accessed 1 March 2012)

NAO/OGC (2008) *Good Practice Contract Management Framework*. NAO Marketing and Communications Team, London.

OGC (2002) *Principles for Service Contracts: Contract Management Guidelines*. OGC, Norwich

OGC (2003) *Building on Success - The Future Strategy for Achieving Excellence in Construction*. OGC, London

OGC (2005a) *OGC Best Practice – Common Causes of Project Failure*. OGC, London. http://webarchive.nationalarchives.gov.uk/20100503135839/http://www.ogc.gov.uk/documents/cp0015.pdf (accessed 21 February 2012)

OGC (2005b) *Managing Successful Programmes: Delivering Business Change in Multi-project Environment.*, The Stationery Office, London

OGC (2006a) *Managing Successful Projects with PRINCE2 Manual 2005*. The Stationery Office, London

OGC (2006b) *Common Minimum Standards*. OGC, London. http://webarchive.nationalarchives.gov.uk/20100503135839/http://www.ogc.gov.uk/documents/Common_Minimum_Standards_PDF.pdf (accessed 21 February 2012)

OGC (2006c) *Making Competition Work for You*. OFT/OGC, London. http://webarchive.nationalarchives.gov.uk/20100503135839/http://www.ogc.gov.uk/documents/CP0144MakingCompetitionWorkForYou.pdf (accessed 21 February 2012

OGC (2007a) *OGC Best Practice – Gateway™ Review 0: Strategic Assessment*. OGC, London. http://webarchive.nationalarchives.gov.uk/20100503135839/http://www.ogc.gov.uk/documents/FINAL_BOOK_0.pdf (accessed 21 February 2012)

OGC (2007b) *OGC Best Practice – Gateway™ Review 1: Business Justification*. OGC, London. http://webarchive.nationalarchives.gov.uk/20100503135839/http://www.ogc.gov.uk/documents/NEW_BOOK_1_APRIL.pdf (accessed 21 February 2012)

OGC (2007c) *OGC Best Practice – Gateway™ Review 2: Delivery Strategy*. OGC, London. http://webarchive.nationalarchives.gov.uk/20100503135839/http://www.ogc.gov.uk/documents/BOOK_2_APRIL.pdf (accessed 21 February 2012)

OGC (2007d) *OGC Best Practice – Gateway™ Review 3: Investment Decision*, OGC, London. http://webarchive.nationalarchives.gov.uk/20100503135839/http://www.ogc.gov.uk/documents/BOOK_3_APRIL.pdf (accessed 21 February 2012)

OGC (2007e) *OGC Best Practice – Gateway™ Review 4: Readiness for Service*, OGC, London. http://webarchive.nationalarchives.gov.uk/20100503135839/http://www.ogc.gov.uk/documents/NEW_BOOK_4_APRIL.pdf (accessed 21 February 2012)

OGC (2007f) *OGC Best Practice - Gateway™ Review 5: Operations Review and Benefits Realisation*. OGC, London. http://webarchive.nationalarchives.gov.uk/20100503135839/http://www.ogc.gov.uk/documents/FINAL_BOOK_5.pdf (accessed 21 February 2012)

OGC (2007g) *A Manager's Checklist*. OGC, London. *http://webarchive.nationalarchives.gov.uk/20100503135839/http://www.ogc.gov.uk/documents/CP0071AEManagersChecklist.pdf* (accessed 21 February 2012)

OGC (2007h) *Achieving Excellence Construction Projects Pocketbook*. OGC, London. *http://webarchive.nationalarchives.gov.uk/20100503135839/http://www.ogc.gov.uk/documents/CP0060AEConstructionPocketbook.pdf* (accessed 21 February 2012)

OGC (2007i) *Achieving Excellence in Construction Procurement Guide 1 – Initiative into Action*. OGC, London. http://webarchive.nationalarchives.gov.uk/20100503135839/http://www.ogc.gov.uk/documents/CP0061AEGuide1.pdf (accessed 21 February 2012)

OGC (2007j) *Achieving Excellence in Construction Procurement Guide 2 – Project Organisation*. OGC, London. http://webarchive.nationalarchives.gov.uk/20100503135839/http://www.ogc.gov.uk/documents/CP0062AEGuide2.pdf (accessed 21 February 2012)

OGC (2007k) *Achieving Excellence in Construction Procurement Guide 3 – Project Procurement Lifecycle*. OGC, London. http://webarchive.nationalarchives.gov.uk/20100503135839/http://www.ogc.gov.uk/documents/CP0063AEGuide3.pdf (accessed 21 February 2012)

OGC (2007l) *Achieving Excellence in Construction Procurement Guide 4 – Risk and Value Management*. OGC, London. http://webarchive.nationalarchives.gov.uk/20100503135839/http://www.ogc.gov.uk/documents/CP0064AEGuide4.pdf (accessed 21 February 2012)

OGC (2007m) *Achieving Excellence in Construction Procurement Guide 5 – The Integrated Project Team*. OGC, London. http://webarchive.nationalarchives.gov.uk/20100503135839/http://www.ogc.gov.uk/documents/CP0065AEGuide5.pdf (accessed 21 February 2012)

OGC (2007n) *Achieving Excellence in Construction Procurement Guide 6 – Procurement and Contract Strategies*. OGC, London. http://webarchive.nationalarchives.gov.uk/20100503135839/http://www.ogc.gov.uk/documents/CP0066AEGuide6.pdf (accessed 21 February 2012)

OGC (2007o) *Achieving Excellence in Construction Procurement Guide 7 – Whole-Life Costing*. OGC, London. http://webarchive.nationalarchives.gov.uk/20100503135839/http://www.ogc.gov.uk/documents/CP0067AEGuide7.pdf (accessed 21 February 2012)

OGC (2007p) *Achieving Excellence in Construction Procurement Guide 8 - Improving Performance*. OGC, London. http://webarchive.nationalarchives.gov.uk/20100503135839/http://www.ogc.gov.uk/documents/CP0068AEGuide8.pdf (accessed 21 February 2012)

OGC (2007q) *Achieving Excellence in Construction Procurement 9 – Design Quality*. OGC, London. http://webarchive.nationalarchives.gov.uk/20100503135839/http://www.ogc.gov.uk/documents/CP0069AEGuide9.pdf (accessed 21 February 2012)

OGC (2007r) *Achieving Excellence in Construction Procurement Guide 10 – Through Health and Safety*. OGC, London. http://webarchive.nationalarchives.gov.uk/20100503135839/http://www.ogc.gov.uk/documents/CP0070AEGuide10.pdf (accessed 21 February 2012)

OGC (2007s) *Achieving Excellence in Construction Procurement Guide 11 – Sustainability*. OGC, London. http://webarchive.nationalarchives.gov.uk/20100503135839/http://www.ogc.gov.uk/documents/CP0016AEGuide11.pdf) (accessed 21 February 2012)

OGC (2008a) *An Introduction to Public Procurement*. The Stationery Office, London https://update.cabinetoffice.gov.uk/sites/default/files/resources/introduction-public-procurement.pdf (accessed 21 February 2012)

OGC (2008b) *Portfolio, Programme and Project Offices (P30)*. The Stationery Office, London

OGC (2009a) *Achieving Excellence in Construction – Successful Delivery Toolkit, Programme and Project Management Resources*. http://webarchive.nationalarchives.gov.uk/20100503135839/http://www.ogc.gov.uk/guidance_achieving_excellence_in_construction.asp (accessed 21 February 2012)

OGC (2009b) *Gateway Review for Programmes and Projects – Successful Delivery Toolkit, Programme and Project Management Resources*. http://webarchive.nationalarchives.gov.uk/20100503135839/http://www.ogc.gov.uk/what_is_ogc_gateway_review.asp (accessed 21 February 2012)

OGC (2009c) *Management of Risk (M_o_R) – Successful Delivery Toolkit, Programme and Project Management Resources*. http://webarchive.nationalarchives.gov.uk/20100503135839/http://www.ogc.gov.uk/guidance_management_of_risk.asp (accessed 21 February 2012)

OGC (2009d) *Managing Successful Programmes (MSP) – Successful Delivery Toolkit, Programme and Project Management Resources*. http://webarchive.nationalarchives.gov.uk/20100503135839/http://www.ogc.gov.uk/guidance_managing_successful_projects_4442.asp (accessed 21 February 2012)

OGC (2009e) *Managing Successful Projects with PRINCE2®*. The Stationery Office, Norwich

OGC (2009f) *Directing Successful Projects with PRINCE2®*. The Stationery Office, Norwich

OGC (2009g) *PRINCE2 – Successful Delivery Toolkit, Programme and Project Management Resources*. http://webarchive.nationalarchives.gov.uk/20100503135839/http://www.ogc.gov.uk/methods_prince_2.asp (accessed 21 February 2012)

OGC (2010a) *The Government Procurement Code of Good Practice: For Customers and Suppliers*. The Office of Government Commerce, London

OGC (2010b) *Contract Management in Long Term Or Complex Projects: Key Commercial Principles to Help Ensure Value For Money*. OGC, London. http://clients.squareeye.net/uploads/east2010/Contract_Management_in_Complex_Procurement(1).pdf (accessed 21 February 2012)

Pinto, JK and Slevin, DP (1988) Critical success factors across the project life cycle. *Project Management Journal*, 19, 67–75

PRINCE2.com (2012) *What is PRINCE2? – The PRINCE2® Definitive Resource*. ILX Group plc http://www.prince2.com/what-is-prince2.asp#intro (accessed 21 February 2012)

Saxena, A (2008) *Enterprise Contract Management: A Practical Guide to Successfully Implementing an ECM Solution*. J Ross Publishing, Fort Lauderdale, Florida

Siegelaub, J (2004) *How PRINCE2 can Complement PMBoK and your PMP*. http://www.prince2.org.uk/nmsruntime/saveasdialog.asp?lID=900&sID=277

Smith, N (2010) *Wave 2 Procurement Capability Reviews (PCR) Launched by OGC*. Press Release, Office of Government Commerce, London

Sowden, R (2011a) *Managing Successful Programmes – 2011 Edition (Overview Brochure)*. The Cabinet Office/The Stationery Office, Norwich. http://www.best-management-practice.com/gempdf/MSP_Overview_Brochure.pdf (accessed 21 February 2012)

Sowden, R (2011b) *Managing Successful Programmes (MSP®): A Basic Overview*. The Stationery Office, Norwich. http://www.best-management-practice.com/gempdf/MSP_A_Basic_Overview_White_Paper_Dec11.pdf (accessed 21 February 2012)

Sowden, R and the Cabinet Office (2011c) *Managing Successful Programmes*. The Stationery Office, Norwich

Sowden, R and OGC (2010) *Procurement | Programmes and Projects: Portfolio, Programme and Project Management Maturity Model (P3M3®)* Version 2.1. The Office of Government Commerce, London. http://www.p3m3-officialsite.com/P3M3Model/P3M3Model.aspx (accessed 21 February 2012)

Thiry, M (2004) Program management: a strategic decision management process. In: PWG Morris and JK Pinto (eds) *The Wiley Guide to Managing Projects*. John Wiley and Sons, Hoboken, New Jersey. pp. 257–287

Wideman, M (2002) *Comparing PRINCE2 with PMBoK*. http://www.maxwideman.com/papers/comparing/comparing.pdf (accessed 21 February 2012)

Winch, GM (2003) Models of manufacturing and the construction process: the genesis of reengineering construction. *Building Research and Information*, 31 107–118

The World Bank (2011) *Procurement Policies and Procedures*. The World Bank, Washington, DC. http://web.worldbank.org/WBSITE/EXTERNAL/PROJECTS/PROCUREMENT/0,,contentMDK:50002392~pagePK:84269~piPK:60001558~theSitePK:84266,00.html (accessed 21 February 2012)

Chapter 8
Commercial Strategies and Tactics

Learning outcomes

When you have completed this chapter you should be able to:

- Understand the relationship between the purchaser's procurement procedures and the supplier's bid cycle
- Explain client purchasing objectives and purchasing decision making
- Evaluate client procurement policy and procurement route assessment criteria
- Identify appropriate tendering processes and bidding procedures
- Appraise UK public sector bidding and bidding within the European Economic area.
- Analyse the decision to bid
- Describe the bid preparation stage
- Determine bid strategy and evaluate bidding tactics
- Evaluate the factors that influence bidding success
- Analyse risk and uncertainty relating to bidding
- Describe the major types of bidding model
- Identify the components of a proposal
- Determine good tendering and bidding practice
- Identify key factors in negotiation
- Demonstrate an understanding of contract management procedures

Introduction

This chapter examines commercial practice across the project life cycle from both the supply and demand perspective. It identifies the interrelationship between the demand-side procurement process (cycle) and the supply-side bidding and implementation cycles. Additionally, it explores how purchasers can effectively articulate their requirements for new assets and services, and then manage the process of asset/service definition and delivery successfully. From the demand side, it addresses how suppliers can identify prospective opportunities, develop and communicate their proposals for the supply of asset and services, manage the deal creation stage and ensure that commitments made are delivery profitably. It is subdivided into three sections:

- **Part A – Intent**: from the demand side, Part A addresses requirement identification, requirement specification and solution selection, while from the supply side it covers opportunity identification, opportunity development and proposition identification

Commercial Management: theory and practice, First Edition. David Lowe.

- **Part B – Deal creation**: Part B focuses on the process and actions undertaken by commercial practitioners that bring an agreement into existence. From the demand side it addresses asset/service procurement, implementing the selected sourcing (procurement) strategy and negotiating the agreement, while from the supply side it covers proposition development, creating, preparing, submitting and negotiating a proposal in respect of the supply of a product and/or service
- **Part C – Execution**: Part C addresses the commercial procedures and tasks involved in implementing, delivering and fulfilling commitments made during the deal creation stage. It outlines the principle aspects of contract management: the monitoring and control of the contract to ensure that the agreement between the purchaser and supplier operates efficiently and effectively, and that all contractual obligations are appropriately discharged. Part C concludes with an explanation of the commercial activities involved in asset disposal and service termination

The chapter is concerned with the procurement/supply of **complex products and services** (projects) (CoPS: see Davies and Hobday 2005 and 2006) and **procuring complex performance** (PCP: see Caldwell and Howard, 2011); as such the construction sector is a primary example. While the case studies and most examples are drawn from the construction sector, the topics addressed are generic and further references are made to the IT/communications and defence/aerospace sectors.[1]

Overview of the transaction process

As outlined in Chapter 1 the transaction process involves the interconnected activities of procurement and bidding.

Procurement

Procurement, the process used to appoint appropriate suppliers, has been defined as:

> '... the systematic process of deciding what, when and how much to purchase; the act of purchasing it; and the process of ensuring that what is required is received on time in the quantity and quality specified.' (Burt, 1984) or
>
> '... the business management function that ensures identification, sourcing, access and management of the external resources that an organisation needs or may need to fulfil its strategic objectives.' (Kidd, 2005)

As mentioned in Chapter 1, the terms procurement, purchasing and supply chain management are frequently used interchangeably. These activities and their associated business functions are increasingly seen as strategically important to successful organisations. Baily *et al.* (2008) summarise the main reasons for this. They include:

- The introduction of innovative ideas and processes – for example, best practice benchmarking, total quality management, supply chain concepts and relationship management, etc.
- Increased specialisation due to technological complexity requiring organisations to purchase goods and services from specialists
- Innovation and the pace of change
- Greater awareness of the finite nature of resources
- A higher amount of external revenue spend
- Concentration of supply in 'fewer but larger' providers
- A raised awareness of environmental issues
- The acknowledgement of purchasing and supply as a professional activity

As the transaction process is invariably instigated by the purchaser, they are responsible for selecting the procurement methodology and for setting the tone of the exchange. It could be said, therefore, that buyers get the suppliers they deserve.

Impetus for change in UK procurement policy has also come from central government: for example, under the banner *Modern Government, Modern Procurement* two separate but complementary reviews have been commissioned on the subject of government procurement: *Government Procurement and the Private Finance Initiative* and a *Review of Civil Procurement*,[2] leading to the establishment of the Office of Government Commerce (OGC) and OGC buying solutions in 2000. Further government reports include, for example:

- *Getting Value for Money from Procurement.* NAO/OGC 2001 (NAO/OGC, 2001)
- *Transforming Government Procurement.* HM Treasury January 2007 (HM Treasury, 2007)
- *From Private Finance Units to Commercial Champions: Managing Complex Capital Investment Programmes Utilising Private Finance – A Current Best Practice Model for Departments.* HM Treasury/ NAO March 2010 (HM Treasury/NAO, 2010)
- *A Review of Collaborative Procurement across the Public Sector.* NAO/The Audit Commission MAY 2010 (NAO, 2010)

The prevailing ethos, which coincided with the Labour administration from 1997 to 2010, was that of collaboration. However, following the formation of a coalition government, made up of the Conservative and Liberal Democrat parties, central government procurement policy is currently under review. An early report commissioned by the new government – the Efficiency Review conducted by Sir Philip Green (Green, 2010) – once again highlighted the issue of inconsistent commercial aptitude of managers across government departments. One of the report's recommendations was for the public sector to '*leverage its credit rating and its scale*' (buying power). This move could lead to fewer collaborative agreements and encourage increased opportunistic and adversarial behaviour by suppliers. Further, changes in the government's approach have led to the activities of the OGC being subsumed within the Efficiency and Reform Group of the Cabinet Office and the establishment of the Major Project Authority (MPA) both in 2011.[3] A partnership between the Cabinet Office and HM Treasury, the MPA's role is to oversee and direct the management of all central government financed and delivered large-scale projects.

Industry sector reports

Since the mid-twentieth century the UK construction industry has been subject to numerous government reports that have called for the sector to improve its performance. Langford and Murray (2003) chronicle these reports, and show how the UK government has attempted to influence the performance and viewpoint of construction industry stakeholders. The most recent include:

- *Constructing the Team: Final Report of the Government/Industry Review of Procurement and Contractual Arrangements in the UK Construction Industry – 'the Latham Report'* (Latham, 1994)
- *Rethinking Construction. The Report of the Construction Task Force* (Egan, 1998)
- *Modernising Construction* (Comptroller and Auditor General, 2001)
- *Rethinking Construction: Achievements, Next Steps, Getting Involved* (Rethinking Construction Group, 2002)
- *Accelerating Change* – (Strategic Forum for Construction, 2002)
- *Improving Public Services through Better Construction* (NAO, 2005)
- *Performance of PFI Construction* (NAO, 2009)[4]

Predominantly, these reports recommend and promote changes to the procurement approaches adopted in the sector and to the nature of exchange relationships.

In terms of telecommunications, OGC published *Procurement Next Generation Networks – Procurement Standards, Guidance and Model Clauses* (OGC, 2007d). Derived from the Next Generation Networks (NGN) Procurement Standards Project, the document was introduced as part of the Cabinet Office's NGN Risk Mitigation Programme. The publication promotes 'best-practice' procurement standards and guidance to assist buyers of NGN-based telecommunication services and sets standards that suppliers are required to meet in order to supply the UK government.

Launched in 1998 as a direct result of the Strategic Defence Review (SDR) published in the same year, the Smart Procurement Initiative (renamed **Smart Acquisition** in October 2000) has had a significant impact on UK MOD procurement. The initiative refocused defence acquisition on acquiring access to capability: the combination of equipment and through-life in-service support. Further, it aimed to deliver defence equipment *faster, cheaper and better.*[5]

Exercise 8.1 Impetus to change

Review the most recent procurement reports and establish:

- The perceived issues affecting your own industry sector
- The measures proposed to improve procurement performance

The bid cycle/process

The bid cycle, the process by which suppliers focus their response on purchaser's tender opportunities, commences when a potential opportunity becomes apparent and concludes with a lessons learnt review following notification by the purchaser as to whether or not their bid/proposal has been successful. Activities involved in the process include influencing purchaser behaviour, collating information, estimating, interacting with the supply chain, developing and producing a proposal document, preparing and delivering presentations and negotiating the 'deal'.

A key prerequisite of the process is the availability of specific knowledge on the potential opportunity. For example, details of:

- **The purchaser**: their precise requirements, the degree of design or innovation required, the needs of the eventual user (or in the case of the supply of intermediate products the requirements of the purchaser's customer), the structure and trends within their market segment, how the contract award will be decided and the criteria used to inform that decision, the composition of the purchaser's '**buying centre**' and their individual biases
- **The supplier**: the bidder's existing experience in providing the required products and services, details of its current offerings (products and services), its current market position, its current workload and the availability of resources to both respond to and deliver the potential opportunity, and the state of its relationship with the purchaser and their advisers
- **The competition**: their existing experience in providing comparable products and services, details of their current offerings (products and services), their current market position and the state of their relationship with the purchaser and their advisers

This information informs the development of the bid strategy.

Interaction between the procurement/bidding processes

Figure 1.7 presented in Chapter 1 provides a framework for the commercial management function, illustrating the through-life interrelationship between the demand-side procurement process and the supply-side bidding cycle and project implementation stages, plus supply-side in-service implementation/ asset maintenance and demand-side service/asset management. It also shows the involvement of the commercial function throughout these interconnected processes. In a business-to-business (b2b) context an organisation will have both a supply and demand side. However, for the purposes of clarity the diagram shows the relationship between two separate organisations. Likewise within a supply chain there will several b2b interactions both vertically and horizontally – similarly these have been excluded.

The framework also identifies the themes covered by the following three parts of this chapter.

Intent

The **intent** stage links a proposed acquisition (project conception) or potential opportunity to supply (pre-request for proposals) with the overall **strategy** (**formulation** and **implementation**) of each organisation. The phase includes the following elements:

- Requirement identification: identifying, analysing and confirming a need or want, a potential opportunity to be exploited or an issue to be resolved. Once a potential need, opportunity or issue has been recognised, a high-level business validation review is required to confirm or otherwise the value and relevance of satisfying the need, exploiting the opportunity or resolving the issue in respect of the organisation's strategy and its implementation. Depending on the capability of the buyer, advice may be sought from solutions providers: consultants and contractors. Following a positive decision the potential purchaser will progress to the next phase of the **procurement process**
- Opportunity identification: identifying potential customer requirements that the organisation is capable of satisfying. This is the initial stage of the **bid cycle** (process), which commences once a potential lead has been identified (the process concludes after winning or losing the opportunity). Ideally, the viability of the lead should be assessed and its appropriateness to the organisation's strategy confirmed via a high-level business validation review. Following a positive decision the potential supplier will seek to promote their capabilities and influence the buyer's decision makers
- Requirement specification: collating and comprehending the internal customer's needs, stating (specifying), invariably in a written form, and a detailed description of precisely what is required. The resulting document – the **specification** – can include: design details, a description of or reference to the materials to be used, service levels or the standard of workmanship to be achieved, and references to acknowledged standards. The phase will also include an evaluation of the viability of the proposed endeavour and the production of a business case. Mapping the entire procurement process (from inception to disposal), as soon as is practical, is recommended as this is considered to aid the realisation of value for money through improved planning and management and by identifying:
  - non-value-adding aspects of the project procurement process
  - potential causes of disruption to the project procurement process
  - alternative means of delivering the required project (contract) outcomes (GCCP, 1999)

It is recommended that this phase concludes with a stage gate review to evaluate the business justification for proceeding with the proposed investment.

- Opportunity development: cultivating a relationship and dialogue with potential customers, and creating, preparing and marketing potential solutions (product/and or services) to them. The activities undertaken by suppliers during this phase seek to influence the buyer choice of solution. Ideally, they would prefer to be selected as the sole supplier (a negotiated solution), or where that is not realistic to be selected as a potential bidder and to influence the content of the buyer's invitation to tender (ITT) or request for proposal (RFP) so that it favours them over the competition. Additionally, during this stage, client managers will attempt to glean as much information as possible about the potential customer's requirements and expectations in order to develop and define a '*win*' strategy in anticipation of an ITT or RFP. Bernink (1995), for example, asserts that '*unidentified RFPs are born losers*'; meaning that if an ITT or RFP is received without prior notice, a competitor has probably influenced the customer to some degree. Again, it is recommended that this phase concludes with a stage gate review to evaluate the business justification for proceeding with the potential opportunity – a continuation of the screening process
- Solution selection: identifying and deciding upon the most appropriate means of satisfying the acknowledged organisational need, that the market is capable of supplying (the **preferred solution**), and

the route to obtaining the selected product(s) or services (the **sourcing strategy**) – the classic '*make or buy*' decision. During this stage the selected solution will be progressed and appraised, potential solution providers assessed, and an appropriate tendering methodology selected; concurrently the business case will be reviewed and revised. It is recommended that this phase concludes with a stage gate review to appraise the selected delivery strategy
- Proposition identification: identifying and refining a potential solution to a customer's identified and specified requirement. During this stage potential bid team members are identified, an outline bid management or 'win' plan is formulated, and processes and procedures established for managing the bid. The phase will conclude with a stage gate review – a screening exercise – often referred to as the bid/no-bid decision

Common 'commercial' activities undertaken during this stage include, for example:

- Building business cases for business needs or requirements
- Selecting appropriate procurement strategy(ies)
- assessing the costs and risks of a particular venture/opportunity
- Constructing a business case for a potential deal
- Obtaining legal and regulatory sign-off for undertaking business opportunities

Deal creation

The **deal creation** stage focuses on bringing an agreement (principally a contract) by two or more parties into existence for their mutual advantage. Colloquially referred to as brokering or 'striking' the deal, the phase contains the following elements:

- **Asset/service procurement**: obtaining or procuring assets, products and/or services through the implementation of the selected **sourcing** (procurement) **strategy**, the **tendering process**, and negotiating the agreement – the precise terms of the exchange. This generally involves the purchaser in preparing tender documentation (comprising both commercial terms and conditions, and a technical specification of the required asset, product or service); issuing ITTs and subsequently distributing tender documents to interested suppliers. Following an appropriate tender period and the receipt of the suppliers' proposals (offers, bids or tenders), the purchaser will evaluate the submissions and either select a 'winning' proposal, preferred bidder or decide to abort the project. Invariably, the purchaser will negotiate with the preferred supplier(s) before awarding the contract. It is recommended that this phase concludes with a stage gate review to inform and support the investment decision
- Proposition development: creating, preparing and submitting a proposal in response to a client's request for proposal, quotation or tender (RFP/RFQ/RFT) in respect of the supply of property, products and/or services. During this stage suppliers will re-evaluate the merits or otherwise of submitting a proposal (bid/no-bid screening exercise), scrutinise the tender documents, assess the resource requirements and inherent risk and opportunity, undertake any required design work, estimate the costs involved in undertaking the work or supplying the asset or service, prepare/publish the proposal documentation, and prepare a presentation to support the bid/proposal if required. The phase will conclude with a stage gate review, sometimes referred to as an **adjudication** meeting, where the bid/no-bid decision is revisited and, if the decision is made to submit a *bona fide* tender, to determining the 'commercial' offer (price) or proposal (a combination of price and design/specification). The objective of the exercise is to be selected as preferred bidder. During the period between the submission of a proposal and the announcement of a preferred supplier or winning proposal, suppliers will seek to keep communication channels open, if allowed under the procurement rules adopted, with the prospective customer. The aim is to gauge the response their submission receives and to propose alternative solutions if necessary
- **Award**: assigning or granting a contract (or commission) to an individual or organisation: typically the winning bidder. Prior to awarding the contract, purchasers will frequently enter into negotiations with a preferred bidder; the aim for both parties is either to agree the most advantageous or mutually beneficial

contract terms possible. The stage concludes either with a decision not to proceed with the 'purchase' or in the award of a contract. It also marks a transition point: the commencement of the implementation stage, requiring the delivery of the commitments made by the contracting parties during the deal creation phase. Following the award of a contract, the purchaser and suppliers (the successful and unsuccessful bidders) should undertake a post-award/bid review to capture any lessons learnt

Common 'commercial' activities undertaken during this stage include, for example:

- Constructing business cases
- Implementing selected procurement strategy(ies)
- Assessing costs and risks
- Dealing with contracts: drafting, negotiating and agreeing complex contracts and agreements
- Preparing bids; forming part of, or leading bid teams (panels)
- Owning the bid authorisation procedures and providing commercial authorisation
- Selecting or developing and publishing standard terms and conditions of contract for products and/or services
- Arranging 'back to back' subcontract terms to mirror main contract provisions
- Obtaining legal and regulatory sign-off for undertaking business opportunities

Execution

The **execution** stage involves implementing, delivering and fulfilling the commitments made during the deal creation stage and incorporated into the contract, concerning the required asset, products and/or service. The term 'execution' has been adopted due to its legal connotation: an executed contract is one where the provisions have been fully performed by both parties. The stage is sometimes referred to as **project implementation** or **contract management**. Depending upon the nature of the purchase (asset/product or service) and associated sourcing strategy the phase comprises the following alternative components.

The purchase of an asset or product involves the following elements:

- Asset supply: designing, manufacturing, constructing, providing and delivering the procured asset or product(s) in accordance with the contract. This entails managing associated supply chains. With regard to future opportunities, it is essential for the implementation team, wherever possible, to maintain good working relationships with the customer, as their performance and behaviour during the contract has an impact, positive and otherwise. For example, a positive impact may lead to follow-on contracts negotiated off the back of the initial agreement. On completion of the contract, it is recommended that the supplier undertakes a post-contract review to capture any lessons learnt
- **Asset receipt**: receiving and accepting the procured asset or product(s) in accordance with the agreed terms and conditions of contract. Following the completion of the contract, it is recommended that the purchaser undertakes a stage gate review to assess an asset's or product's fitness for purpose/readiness for service
- **Asset usage**: applying or utilising the procured asset or product(s). During this phase the customer will be required to maintain and service the asset or product; this is often managed through term contracts (described in Chapter 6). At points throughout the life of the asset/product and immediately prior to its disposal, stage gate reviews are encouraged to evaluate its operational capability, to assess whether or not it is delivering (has delivered) the anticipated benefits, and to capture any lessons learnt
- Asset maintenance: preserving and maintaining the procured asset or product(s). Similarly, at points throughout the life of the asset/product and on completion of a maintenance/service contract, it is recommended that the supplier undertakes a post-contract review to capture any lessons learnt
- **Asset disposal**: divesting of property and assets. The cycle concludes with the disposal of the asset

Alternatively, the acquisition of a service includes the following elements:

- **Service implementation**: executing, effecting and delivering the procured services in accordance with the agreed terms and conditions of contract. At points during the life of the service contract and on its completion, it is recommended that the supplier undertakes a post-contract review to capture any lessons learnt
- **Service management**: receiving, utilising, monitoring and managing the procured services in accordance with the agreed terms and conditions of contract. At points throughout the length of service and immediately prior to its termination, stage gate reviews are encouraged to evaluate its operational capability, to assess whether or not the service is delivering (has delivered) the anticipated benefits, and to capture any lessons learnt
- **Service termination**: bringing a service contract to an end. In this instance the stage concludes with the termination of the service

Termination also refers to the premature end of a supply contract (*termination of a contract*). It occurs when the contracting parties are released from their contractual obligations due to performance of their duties (an executed contract), agreement of the parties, breach of contract or frustration.

Common 'commercial' activities undertaken during this stage focus on the management and administration of complex contracts and agreements (post-award contract management), and include, for example:

- Advising project management teams on contractual issues, providing critical business support to ensure the project is delivered in accordance with the agreed contract terms
- Managing/administering commercial/contractual issues and initiating correspondence:
  - receiving and processing contractual correspondence: variations and claims, etc.
  - valuing, preparing, submitting, assessing, negotiating and agreeing additional payment (fees) and/or extensions of time in respect of contractual changes (variation) and claims
  - producing and maintaining financial reports and forecasts
  - arranging insurance provisions
  - valuing, preparing, submitting, evaluating and agreeing interim payments and ensuring timely receipt of payment
  - resolving disputes and managing conflict
  - valuing, preparing, submitting, assessing, negotiating and agreeing final accounts
- Procuring subcontractors and materials: establishing trading accounts with suppliers and subcontractors:
  - managing/administering subcontracts and suppliers
  - receiving and processing subcontract interim payment claims and final accounts
  - liaising with the accounts section to ensure timely payment of monies due to suppliers and sub-contractors
  - valuing, preparing, submitting, assessing, negotiating and agreeing final accounts

Not all the areas indicated in this framework are addressed in this chapter. For example, strategy was covered in Chapter 3, risk management was addressed in Chapter 4 and financial appraisal techniques were illustrated in Chapter 5.

Exercise 8.2 Overview of the procurement/bidding processes

Evaluate the foregoing framework against the procurement/bidding processes and procedures within your own organisation

Part A: Intent

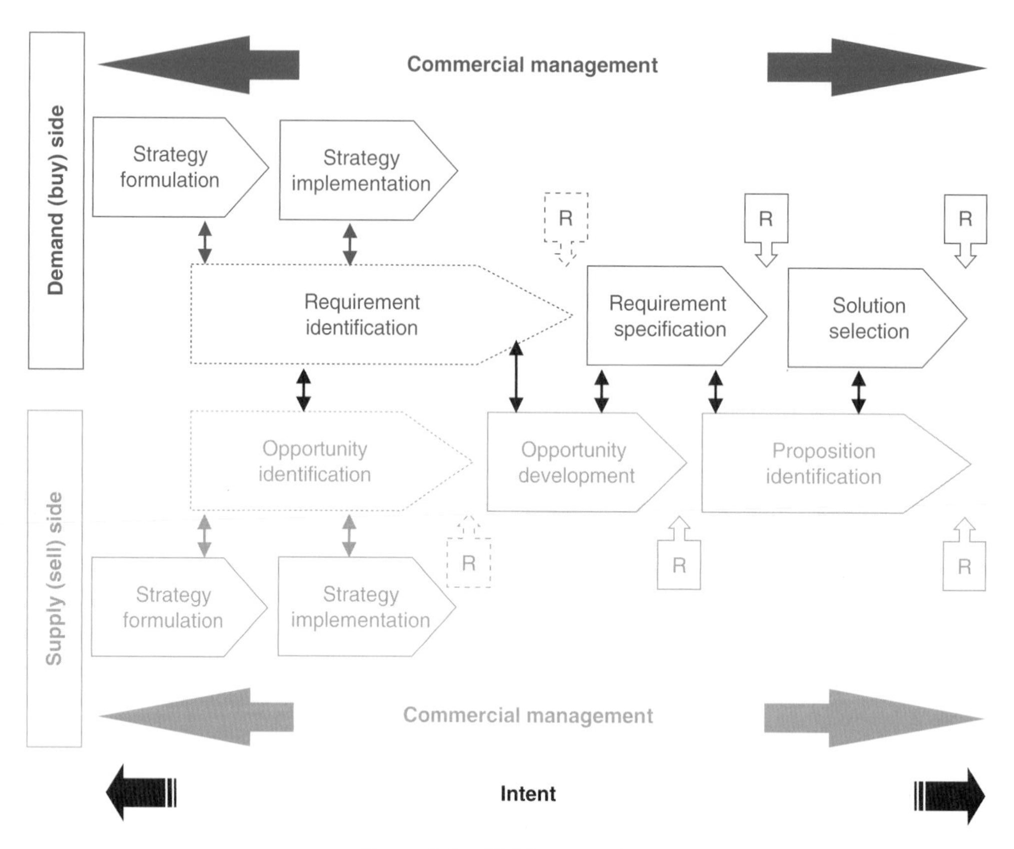

Figure 8.1 CM framework.

Introduction

From a demand perspective, Part A ***Intent*** addresses requirement identification, requirement specification and solution selection, exploring how purchasers can effectively articulate their requirements for new asset and services, and then select an appropriate procurement system to support its/their delivery. From the supply perspective it covers opportunity identification, opportunity development and proposition identification, the aim being to provide greater understanding of how suppliers can create greater purchaser value and screen potential opportunities.

Requirement identification

Ideally, the commercial function is engaged in the exchange process at the point when a requirement (a potential opportunity to be exploited or an issue to be resolved) is identified. Commercial practitioners are then able to contribute to the analysis and high level **business validation review** to confirm or otherwise the value and relevance of satisfying the need, exploiting the opportunity or resolving the issue in relation to the organisation's strategy and implementation plan. Depending on the capability of the buyer, advice may be sought from solutions providers – consultants and contractors. Following a positive decision, the potential purchaser will progress to the next phase of the **procurement process**.

Activity 8.1 Requirement identification

Compare and contrast the client's specific needs outlined in case study A for the Millennium, Emirates and Wembley stadia projects and case study B Heathrow terminal 5 (T5). Consider the motivation for undertaking these projects and the specific constraints faced by each procuring party

Requirement specification

Following the identification and validation of a requirement, the next step is to establish, as clearly, comprehensively and succinctly as possible, the purchaser's **project brief**. From this statement of requirement numerous activities and further decisions emanate. Moreover, it has a direct influence on the ultimate success of the project and is used to evaluate the ultimate success or failure of the undertaking. Sufficient time and resources, therefore, should be allocated to the activity to enable as precise a statement to be developed as possible. Involvement of the commercial function in this activity will enable them to provide appropriate advice concerning the selection of an appropriate procurement methodology and to reflect accurately the purchaser's requirements in the resulting tender documentation and contract. Specification-writing skills are, therefore, crucial.

Importance of the clients' brief

Brief making is defined by Turner (1997) as:

> '... the radical, searching process that a client and his advisers, particularly the in-house executive and/ or the principal adviser, must go through, to explore and to conclude on the nature of the Clients' business in order to decide on an appropriate solution.'

The importance and process of establishing the project brief were covered in Chapter 7. However, the extent to which external consultants and advisers will be involved in this process is usually determined by the relative experience of the client body (**project board**) and the nature of the project, whether or not it relates to a core capability of the organisation. An inexperienced client, for example, would require significantly more assistance from their consultants, in terms of extracting the relevant information about their requirements and how any proposed solution may impact on their business, than an experienced client. Also, time invested in formulating a comprehensive brief is generally time well spent, as introducing changes during the later stages of the design process or during the implementation stage will generally have significant cost and time implications.

The Achieving Excellence in Construction (AEC) procurement guidelines suggest that, irrespective of the procurement route (system) used, all specifications describing the user's needs, except where there are exceptional reasons to the contrary, should be output (outcome) based. Although avoiding prescription, as excessive detail invariably leads to higher costs and impedes innovation, these output specifications must precisely express the required outcome. Also, in order to identify how the project requirements are to be evaluated, in terms of performance and quality, performance measures should be incorporated in the specification (OGC, 2007b).

Procurement objectives

An often-cited definition of **procurement objectives** is:

> 'To buy materials[6] of the right quality, in the right quantity from the right source delivered to the right place at the right time and at right price.' (Lysons and Farrington, 2006)

Baily *et al.* (2008) comment that this rather hackneyed statement is held by certain commentators to be somewhat simplistic and superficial, although they deem the definition to provide a practical framework. The overall aim of the procurement process, therefore, is to meet the purchaser's (and or user's) need, accounting for specific quality or service-level requirements, time constraints, cost-effectiveness and affordability:

- **Quality** or product is generally held to be **fitness for purpose**, as recognised by the client in meeting his/her specific requirements. Purchasers, therefore, assisted by their advisers, should determine the optimal overall quality necessary to meet their needs. Further quality issues relating to the product or service may include the following objectives:
  - defect free on delivery/completion
  - reasonably efficient running cost
  - satisfactory durability
  - a visual statement
  - a value-analysed/engineered solution
  - innovation – the incorporation of original design features
  - the provision of satisfactory guarantees and after-sales service (Latham, 1994; Masterman, 2002)
- **Schedule**: purchaser time-related issues include the following:
  - timely delivery/completion
  - certainty of completion date and other time-related estimates
  - early commencement of work/fabrication/manufacture
  - design proposals to be submitted expeditiously
  - speedy rectification of defects (Masterman, 2002)
- **Cost**: client cost issues include:
  - certainty of cost estimates
  - value for money
  - ease of accountability
  - competition – lowest possible tender
  - cost certainty of product/service or a reduced risk of cost overrun
  - realistic maintenance and running costs (Masterman, 2002)

The resulting project objectives are used to inform the make-or-buy decision and the choice of any subsequent procurement systems. Also, as they may well form the basis of project-specific key performance indicators (KPIs) that will ultimately be used to assess the success or failure of the exchange, they need to be measurable. Additional purchasing objectives may include:

- Continuity and timely provision of components, products and services to meet the organisation's requirements
- Ethical acquisition
- Development and maintenance of cooperative trading relationships
- Access to innovation and new product support
- Supply chain management – the development of key suppliers capability (see for example, Baily *et al.*, 2008)

Business justification stage gate review

Following the selection of appropriate project objectives, a review should be undertaken to confirm their compatibility with the overall objectives of the organisation. The OGC Gateway™ Process Review 1: Business justification best-practice guidance (OGC, 2007e) provides a template that can be used to develop a pro-forma to support the business justification stage gate review. It contains sample questions to be considered relating to the organisation's/project's policy and business context, business case and stakeholders, risk management and delivery strategy (its readiness for the next phase). These questions, for example, include:

- *Does the preferred option meet wider... organisational policies, strategic objectives, standards and business change programmes?*
- *Is there a clear and agreed understanding of business goals and how the project will deliver these?*
- *What are the critical success factors.. and can these... be quantified and measured?*
- *Have all the likely stakeholders been identified and their needs clearly understood?*
- *Has the feasibility study examined a wide enough range of options that will meet the business requirement?*
- *Have contract management issues been considered?*
- *Have the risks for each of the options been evaluated?*
- *Is there an overall project structure for the delivery strategy phase?* (OGC, 2007e)

Illustrations are provided of the anticipated evidence required to confirm compliance/answer each question.

During this phase, commercial practitioners are involved in defining and articulating the precise organisational need – a prerequisite for the development of an appropriate procurement and contracting strategy. They will also engage in evaluating the feasibility of the potential project, applying the risk management, financial decision making and value management techniques described in Chapters 4, 5 and 7 respectively.

Activity 8.2 Requirement specification

Compare and contrast the briefs developed for the Millennium, Emirates and Wembley stadia projects (case study A) with that of T5 (case study B). In particular, consider the implications of the various specific requirements of T5's stakeholders involved in the development of the brief and the key objectives of the client BAA

Solution selection

Once a business requirement has been confirmed, and progression to the next phase of its realisation authorised, a decision has to be made regarding how to resolve the need. Initially, this requires the organisation to address the classic **make-or-buy decision**.

Make-or-buy decision

According to Baily *et al.* (2008) **make-or-buy decisions** entail making choices about the source of assets (components, goods and materials) and services: specifically whether or not an organisation produces the asset and/or provides the associated services internally, or whether it procures a solution from the market: an external supplier. Cousins *et al.* (2008) provide a theoretical perspective on this decision process concerning the boundaries of the firm. They utilise **transaction cost economics** (introduced in Chapter 1 of this text) and the **resource-based view** of the firm (covered in Chapter 3) to explain the interconnectivity of corporate strategy, the make-or-buy decision and supply strategy.

If a procured solution is selected, then a choice has to be made over which procurement system (strategy) should be adopted. To reiterate, this text is primarily concerned with the purchase and supply of complex products (assets/capital goods) and services, either as one-off exchanges or in bespoke, small numbers. It is not primarily concerned with the **purchase of commodities** (natural resources) or **buying goods for resale**.

According to Rathmell (1966), '... *economic products lie along a goods-services continuum, with pure goods at one extreme and pure services at the other, but with most of them falling between these two extremes*'. Bailey *et al.* (2008) differentiates between the two forms of purchasing:

- **Assets and capital goods**: items that are usually purchased to meet specific long-term needs of an organisation and used to produce goods or services. They include, for example, investment in infrastructure, buildings, plant and machinery, computer systems, etc. Generally deferrable, both tax and accountancy procedures deal with capital expenditure differently, while in certain parts of the UK tax allowances or grants may be available to support the purchase of specific capital goods. As a result, tax issues can have significant impact on the decision to purchase
- **Services and capabilities**: services involve the delivery of an organisational benefit, derived from a supplier undertaking a task or activity on behalf of the purchaser. Invariably, the purchase does not include the transfer of ownership of assets or products from the supplier to the purchaser. The acquisition of services has become increasingly strategically important to organisations as the **contracting out** or **outsourcing** of non-core business elements has been more prevalent. Similarly, the contracting out of public sector services has been stimulated by a drive for efficiency and transparency, plus the adoption of competitive purchasing principles. Initially encouraged under **compulsory competitive tendering** legislation, more recently it has been viewed as a means to delivering **best value**

Globally, the boundaries between production and service-oriented organisations have become increasingly blurred. For instance, IBM and the construction company Tilbury Douglas have reinvented themselves as service companies. IBM now concentrates on supplying business solutions, and Tilbury Douglas renamed itself InterServe in 2001, providing advice, design, construction, equipment and facilities management services for infrastructure projects. Davies and Hobday (2005), for example, present the concept of complex products and systems (CoPS) (high technology, high-value capital goods), which they define as:

> '... high-cost, engineering and software-intensive goods, systems, networks, infrastructure and engineering constructs and services, many of which are vital to industrial; growth and the modern economy'.

Box 8.1 Rolls-Royce TotalCare© 'power by the hour'

TotalCare© is the registered name of Rolls-Royce's business model through which it offers clients power capability – 'power by the hour' – where they derive income by making engines available for use. Following the purchase of a fleet of engines, a purchaser enters into a contract with the manufacturer for the supply, as necessary, of additional engines, accessories or components and for the maintenance of the engines. The engine owner pays Rolls-Royce a rate per hour flown by the engine, for which the manufacturer retains responsibility for technical risk and maintenance of the power unit, providing a continuous warranty and guaranteeing engine availability. Additionally, the purchaser obtains budget certainty for operating the engines. In order for Rolls-Royce to capitalise on the opportunity, they need to establish an appropriate hourly charge based on accurate life cycle cost estimates. This requires accurate engine and component life prediction.

Source: Rolls-Royce (1999, 2009); Bagnall *et al.* (2001); Neely (2008). For further information see Anon (2007)

Moreover, they note that project-based CoPS enterprises are a key source of competitive advantage, accounting for a large percentage of the gross value-added of developed countries (see also Davies and Hobday, 2006).

The following procurement approaches are of particular interest:

- **Subcontracting**: in project-oriented sectors, such as construction, main contractors generally delegate a significant portion of the contract works to specialist subcontractors, due to the fragmentation of the sector and increased specialisation. As indicated in Chapter 6, the degree to which main contractors can subcontract elements of projects is governed by the terms of the contract, as the purchaser has no direct contractual relationship with, and therefore control of, these suppliers. In other sectors, the term is used to indicate work a purchaser could have potentially undertaken in-house but, due to the lack of capability, for example, chooses to acquire from an external supplier. Likewise, subcontracting is frequently considered as a mechanism to supplement limited resources – skills, competence, time and capacity – while allowing the supplier to concentrate on their main area of expertise. Specialist subcontractors are better placed to acquire/develop, adopt, retain and exploit developing technologies and innovation
- **Outsourcing**: this is where an organisation contracts out its non-core activities, involving, therefore, a strategic decision process to determine precisely what are its core capabilities (see Chapter 3). Definitions include '*the purchase of a good or service that was previously provided internally*' (Lacity and Hircshheim, 1993, see also Domberger, 1998). Bröchner (2006) views outsourcing as a particular example of contracting, one that includes aspects of a project and process. Further he suggests that it is a special case of the classic make-or-buy decision. Examples include **business process outsourcing** (BPO), **facilities management** (FM), **logistics**, **component production** by third parties in the manufacturing sector, **industrial maintenance**, **public–private partnerships** (PPP) and **off-shoring**.[7] Outsourcing, therefore, may involve the transfer of employees, plant and equipment, and facilities – traditional examples include HR management, communication and computing facilities. More recently some organisations have outsourced procurement activities. Outsourcing requires the outsourcing entity (purchaser) to entering into a long-term contractual relationship with a supplier for the provision of ongoing services. Establishing an appropriate structure/contractual framework for this relationship, together with realistic and measurable service-level agreements (SLAs) is crucial, as changes over the period of the contract are usually inevitable. Moreover, prohibitive switching costs generally result in a power shift between purchaser and supplier. The process of appointing an outsourcing partner is essentially the same as other more general sourcing activities, although due to the transfer of assets and personnel a higher degree of due diligence will often be involved.

Further reading

Outsourcing

Bröchner, J (2006) Outsourcing (Chapter 9). In: DJ Lowe (ed.) with R Leiringer. *Commercial Management of Projects: Defining the Discipline*. Blackwell Publishing, Oxford. pp. 192–206

Increasingly, organisations procure products and services from a global market; **globalisation** therefore impacts many organisations. However, the process of engaging with and appointing an overseas supplier is not essentially that dissimilar from that required to select a local supplier. Remote geographical locations and specific international dimensions of trading relationships can, however, influence the exchange. Chapter 6 addresses specific aspects relating to international law, international commercial terms (Incoterms®) and letters of credit. Further, the commercial function needs to consider the potential impact, for example, of legal jurisdiction and the applicable law governing the exchange contract, differing standards, currency exchange rates, language, communication and cultural issues,[8] and logistics on the purchase.

Supply chain management

Following the publication of several reports into the effectiveness of the UK construction industry (for example: the Latham, 1994; and Egan, 1998 reports – see earlier references), which highlighted the need for organisations within the sector to become more efficient, there has been much interest in the adoption of supply chain management systems as a mechanisms to deliver these efficiency improvements. On the whole, however, there is a lack of agreement on a precise definition of supply chain management; the following are given as examples:

> 'Supply Chain Management is the process of integrating and managing the flows of goods and information in the Supply Chain. [where] ... the Supply Chain is the set of physical links and transformation processes through which goods flow en route from supplier to manufacturer to customer, together with the information links through which data flows from customer to manufacturer to supplier (and vice versa)' (The Institute of Operations Management[9])

> 'Supply Chain Management is the implementation of cross-functional relationships with key customers and suppliers... the integration of key business processes from end user through original suppliers that provides products, services, and information that add value for customers and other stakeholders.' (Supply Chain Management Institute[10])

Additionally, supply chain management includes both inter- and intra-organisational perspectives (Fernie *et al.*, 2000).

Although beneficial within the manufacturing sector, the application of supply chain management principles within construction is problematic due to the project-based context of the industry and the custom of establishing discrete temporary multi-organisations for each project. However, the construction industry, by adopting a more collaborative approach, can deliver performance improvement, which according to Cox and Townsend (1998) is derived through the structured hierarchy of dependency, control and leverage which is intrinsic in any ongoing collaborative supply chain.[11] For a discussion on the introduction of supply chain management in construction, see Cox and Ireland (2006).

Understanding the supply chain process is a fundamental capability of the commercial function within project-oriented organisations (Langford and Murray, 2006).

Further reading

Supply chain management

Cousins, PD, Lamming, R, Lawson, B and Squire, B (2008) *Strategic Supply Management: Principles, Theory and Practice*. Pearson Education Ltd, Harlow

Handfield, RB, Monczka, RM, Giunipero LC and Patterson JL (2012) *Sourcing and Supply Chain Management*, 5th International edition. South Western/Cengage Learning Business Press, Mason, OH

Activity 8.3 Supply chain management

Compare and contrast the approaches to supply change management adopted on the Emirates stadium, Wembley Stadium (case study A) and T5 (case study B)

Procurement oversight

Project management

Project management, as defined earlier in the text, is concerned with the overall planning and coordination of a project, from its inception to completion: the objective is to meet the purchaser's requirements, as discussed above. Alternatively, the management of projects, according to Morris and Hough (1987) and Morris (1994), centres the process on managing the definition and delivery of the project itself in order to deliver a successful outcome. Good project management is crucial to the realisation of the project's goals (OGC, 2007a).

Project management services can be provided by in-house capability or through the appointment of an external organisation or individual. The development of project management maturity and best-practice project management guidance was outlined in Chapter 7.

In-house project executive

Due to the increased prevalence of project-oriented organisational structures, many organisations possess well developed project management experience, capability and authority. On large/complex projects and exchange contracts, Turner (1997) recommends that a nominated in-house project executive (alternatively referred to as **project sponsor**) should:

- Be permanently allocated to the project
- Be the organisation's sole point of contact
- Be empowered to respond quickly and fully to any external queries
- Appreciate and manage the project's internal decision-making processes
- Have delegated authority to act and speak on behalf of the organisation
- Support any externally appointed project manager

Consultant advisers

Where the organisation's internal executive lacks the capacity and/or ability to manage the project, the appointment of an external lead-consultant is crucial. Further, this appointment will probably significantly influence the overall success of the project. Various organisations could provide this service, for example:

- **Consultancies**: Project managers, architects, designers and engineers
- **Main contractors**: Suppliers with appropriate design and management expertise

The principal function of the lead-consultant is to be the purchaser's **agent**.

Procurement systems

Within project environments there are various procurement options for the acquisition of assets and services (**procurement systems**) and several ways of categorising them.* Each has its strengths and weaknesses depending on the context of the purchase. The following is a commonly adopted method of classification.

Separated procurement systems

Separated procurement systems involve the procurement of the design aspects of a project independent of those associated with its production. In the following text the term *production* refers to the processes of assembling, building, constructing, fabricating, implementing and manufacturing, etc. Winch (2010) terms this approach **separated project coalitions**.

A traditional form of contracting (design-bid-build)

A traditional form of contracting is where the client employs a consultant to design the project, select a contractor, supervise the implementation (construction) stage, and certify the completed works. The contractor is selected once the design has been established, usually after some form of competition. A contract is then entered into between the client and contractor, the contractor agreeing to perform the work:

- For a fixed price, irrespective of the final cost of the project
- On a measurement basis, using agreed prices for units of specified work
- For a predetermined fee plus the actual costs of construction

* The following classification is based on a combination of Building EDC (1985); Bennett and Grice (1990); Masterman (2002); and Winch (2010). See also: Turner (1997); Franks (1998); Rowlinson and McDermott (1999); Bowers (2003); and Morledge *et al.* (2006) for a discussion of procurement within a construction context, and Walker and Rowlinson (2008) for a wider projects perspective.

Generally, the contractor's profit margin is not revealed (Franks, 1998).
Bennett and Grice (1990) identify two variants:

- **Sequential design**: where main contractors are invited to tender for projects once the design has been completed and tender documentation prepared by separately appointed advisers
- **Accelerated**: where a main contractor is engaged, either in competition or by negotiation, early in the project life cycle. This may involve the use of a two-stage tender process based on outline documentation

Integrated procurement systems

Alternatively, integrated procurement systems involve the adoption of a procurement approach that combines both the design and delivery (production/implementation) of the project. Winch (2010) refers to this approach as **integrated project coalitions**. Options include one of the following.

Box 8.2 Football stadia case studies

Case study A includes details of the Millennium Stadium, Cardiff, the Emirates Stadium, London and Wembley Stadium, London. In all three cases the client directly appointed designers, architects and engineers to produce concept designs. These designs were subsequently novated to the main contractor following their appointment.

Millennium Stadium: As the stadium's design was not fully developed, the contract enabled Laing, the main contractor, to assume responsibility for its design based on the Welsh Rugby Union's specification (the client's requirements) and the concept design prepared by Hok and Lobb and W.S. Atkins. The New Engineering Contract (NEC) Option C Target Contract with activity schedule standard form was adopted on the project. The contract incorporated risk and reward (pain/gain) sharing, a fixed end date and a guaranteed maximum price (GMP), while performance and progress payments were linked to an activity schedule.

The Emirates Stadium: Arsenal appointed a design team to produce concept designs; these appointments were made on the understanding that the designers (and design) would be novated to the main contractor under a design and build contract once a contractor had been selected. Based on a Joint Contracts Tribunal (JCT) standard form with contractor design, the agreement was converted converted to a GMP contract (a lump-sum, fixed-price, and fixed-date contract) after agreement had been reached with the majority of the works contractors (subcontractors). Following the conversion of the contract and the novation of the professional team, the main contractor, Sir Robert McAlpine, accepted single-point responsibility for both the design and construction of the stadium.

Wembley Stadium: The design for the new stadium was produced by World Stadium Team, a joint venture between Foster and Partners, and HOK Sport, with structural and services design provided by Mott Stadium Consortium. Sir Norman Foster was responsible for the design of the building's arch and roof structure. The design was novated on the appointment of the main contractor, Multiplex, under a design-build contract. Initially, WNSL had intended entering into a 'design, construct and finance contract' with Multiplex. However, Multiplex and WNSL entered into an amended JCT design-build contract, under which Multiplex was responsible for the overall design and construction of the project. In addition to a GMP provision, the contract included a gain share clause, fixed completion date and made provision for Multiplex to claim £12.6 million from WNSL if the contract was terminated.

Source: see Case Study A for more details, particularly concerning the client's motivation for selecting this procurement approach

Design-build

Design and build involves the purchase of an entire product from a single source and where the supplier is responsible for its design and production.[12] Bennett and Grice (1990) present three variants:

- **Direct**: where, following a degree of evaluation but not undergoing a competitive tendering process, a designer–contractor is selected
- **Competitive**: where, based on documentation (client's requirements) prepared by consultants, several contractors submit designs and prices in competition
- **Develop and construct/manufacture**: where independent design consultants are appointed to develop the scheme to a predetermined level, contractors then either complete and warrant the design (using either the client's consultants or their own designers) or novate a completed design, bidding for the work in competition

Package deal

Similar to design-build, a package deal is where the supplier designs and produces the asset/facility under a single contract; however, implicit within the term are standardised or partially standardised solutions.

Turnkey

A turnkey contract is where the client enters into an arrangement with a single organisation, which designs and produces the asset/facility under one contract; normally to a position where the client can take over the facility and use it immediately (see Box 8.3).

Build-Own-Operate-Transfer

Build-Own-Operate-Transfer (BOOT) is a means by which a supplier provides and operates a facility for a set period of time, typically, on behalf of a public client. The facility is transferred to the purchaser at the end of this period. Similar approaches include:

- **Build-Own-Operate (BOO)**: this is where a supplier provides and operates a facility on behalf of a client
- **Design-Fund-Build-Operate (DFBO)**: a further example of where a private sector supplier (frequently a consortium) participates in providing what have previously been considered public services

Design and manage

Design and manage entails the appointment of a sole organisation to both design and deliver the product/project. Subsequently, production and services contracts are awarded to specialist contractors to carry out the work. Bennett and Grice (1990) define two variants:

- **Contractor**: where the contractual risk is borne by a design and management organisation appointed to deliver the product/project to a specified time schedule and agreed price (which may or may not be fixed/guaranteed). The contractor is responsible for appointing designers and specialist subcontractors
- **Consultant**: where the schedule and cost risks are retained by the purchaser, who appoints a project designer/manager as an agent. Under this arrangement it is the purchaser who enters into direct contracts with the specialist contractors

Private Finance Initiative

Public-private partnerships (PPP) may take many forms of which the private finance initiative (PFI) is the most developed model. They are a mechanism that enables the private sector (usually in the form of a consortia) to participate in the provision of public services. They are not, therefore, limited to providing capital assets – for example, buildings – as the supplier generally designs, finances, and constructs the asset/facility. After this the consortium operates and maintains the facility on behalf of a public client for a period of between 15 and 30 years – for example, providing catering facilities and servicing the building. The consortium is recompensed by means of predetermined stage payments over the duration of the contract, provided set performance targets are met. Normally, at the end of this period, the assets generated under the project are transferred to public sector ownership. It is anticipated that efficiencies

Box 8.3 Olkiluoto 3 nuclear power plant project

Background

Rated at 1600 megawatts, Olkiluoto 3 is the fifth nuclear power plant (NPP) to be constructed in Finland. The project, valued at €3 billion, was let in 2003 using a turnkey contract. However, almost from the start, the project has been subject to problems – for example, the reconciliation of disparate operational and communicational practices between participants from different countries. The project's current estimated completion date is after 2014, over five years behind schedule.

Key project participants (actors)

While Teollisuuden Voima (TVO), as the buyer/owner, and AREVA, the main contractor, are the principal actors, the construction of Olkiluoto 3 entails a complex, multi-organisational project network:

Teollisuuden Voima (TVO)	**The buyer and owner**: A Finnish, private limited power company founded in 1969 to produce electricity (at cost) for its shareholders. An experienced participant in the Finnish nuclear sector, TVO already operates two NPPs in Olkiluoto. TVO's principal stakeholders are Fortum Power and Heat and Pohjolan Voima.
AREVA	**The turnkey contractor**: Responsible for designing/engineering and constructing the entire project. An industrial group, created in 2001 following the merger of the nuclear technology divisions of Framatome and Siemens, AREVA is owned by the French State (>90%). Highly experienced and a global player in the nuclear power sector, AREVA is divided into three main divisions: • AREVA NP (nuclear power): responsible for developing and building nuclear reactors • AREVA NC (nuclear cycle): encompasses the entire nuclear fuel cycle: mining to waste disposal • AREVA T&D (Transmission and Distribution): responsible for power transmission and distribution When the contract was let, Framatome had an excellent reputation for building NPPs (having completed 93 plants); moreover, the company saw the Olkiluoto 3 as a demonstration project to highlight its third-generation nuclear reactors. Strategically, the Chinese and Indian markets are important to AREVA
Bouygues	A key subcontractor of AREVA: Bouygues (a French company) is responsible for the construction of civil work. In addition, other companies from AREVA's extensive subcontractor network, both from within and outside Finland, were involved in the project
Finnish state	The Ministry of Employment and the Economy: has the ultimate responsibility for the management and supervision of the Finnish nuclear energy sector
STUK (Radiation and Nuclear Safety Authority)	Accountable to the Ministry of Social Affairs and Health, STUK is a regulatory authority, research centre and source of expertise, and is responsible for the supervision of nuclear safety and the use of radiation. Its mission is to protect people, society, the environment and future generations from the harmful effects of radiation, by minimising exposure to radiation and preventing radiation and nuclear accidents. STUK's role is to oversee radiation safety issues on the project

In addition, numerous other governmental and local organisations participate in the supervision of the nuclear power plant's construction.

Contractual arrangements

As previously mentioned, the Olkiluoto 3 project was let as a lump-sum turnkey project, following a tendering process involving three bidders: AREVA; General Electric (with ABB as a key subcontractor); and Atomstroieksport, a Russian company, that emphasised price-based competition. Atomstroieksport was considered to lack the competence to complete the project, and AREVA was appointed based on its competitive (low) bid. The contract between TVO and AREVA, valued at €3 billion, was signed in late 2003. At the time it was described by a representative of AREVA as:

> '... remote actors in a remote place'.

At time the contract was let, the project's programme, which incorporated innovative nuclear technology, was judged by the press to be very tight. Subsequently, it has been alleged (by the French media) that AREVA undervalued the work in order to win the tender, thereby underestimating the credit insurance given by Coface.

Project objectives

Although officially both parties placed great importance on quality and safety, in order to win the contract (and gain a potential competitive advantage in the international nuclear market) AREVA was willing to accept some risks. Similarly, the Finnish media have suggested that in awarding the contract TVO may have been influenced by the price offered by AREVA at the expense of quality.

Relationships between the project participants

Prior to Olkiluoto 3, TVO and AREVA had not been involved in any significant project together. Similarly, AREVA and STUK did not have a pre-existing relationship, although TVO and STUK had a well-established rapport.

TVO's contracting strategy for the project was to give 'turnkey responsibility' to a single company: AREVA. The resulting ambiguity in the allocation of responsibilities between TVO and AREVA has been criticised in the media (both Finnish and international).

As the project proceeded, severe problems were encountered and the relationship between AREVA and TVO became plainly confrontational, with both parties placing the blame for the ensuing delays on each other. This resulted in claims for economic compensations through international arbitration.

Source: Ruuska *et al*. (2009, 2010)

may be derived through the integrating features of the PFI approach, while the long-term maintenance and servicing aspects incentivise the consortium to consider whole-life costing and to design efficient solutions.

First introduced within the UK in 1992, the use of PFI expanded after the election of the Labour government in 1997. With 642 projects let and a further 61 under procurement as of November 2011, the current UK commitment to PPP/PFI contracts is estimated to be approximately £230 billion (Parker, 2012). This entails an estimated annual payment under PFI contracts for the year 2012–2013 of £9115 million (in nominal terms, undiscounted).[13] These projects include hospitals, housing, offices, prisons, roads and schools. Although generally used to procure new infrastructure, they can be applied to the refurbishment of existing assets and for the provision of services. Globally, there has been much interest in the use of the PFI model.

For the public client, PFIs deliver improved cost certainty as a greater proportion of the inherent project risk is transferred to the private sector. This, coupled with the long-term nature of the approach, results in an extended and involved procurement process and less flexible contracts when compared with traditional forms of procurement. Normally, therefore, PFIs are not applicable for smaller projects.

Box 8.4 The London Underground PPP

The PPP established to upgrade London Underground (LU) was one of the world's largest examples of the approach to delivering public projects. Under the PPP, three infrastructure companies (infracos) were created as delivery vehicles to maintain and upgrade LU's assets:

- BCV: The Bakerloo, Central, Waterloo & City, and Victoria lines
- SSL: The sub-surface lines (District, Metropolitan, Hammersmith & City, and Circle lines)
- JNP: the Jubilee, Northern and Piccadilly lines

The PPP was configured so that LU (the public sector) could focus its activities on the delivery of passenger services (operating trains, signals and stations and overseeing safety), while the infracos (the private sector) concentrated on maintaining and upgrading the infrastructure. The PPP was to run for a 30-year period, at the end of which the assets would be returned to LU.

On 4 April 2003, infracos BCV and SSL were transferred to Metronet, a consortium comprising: Balfour Beatty, Atkins, Bombardier Transportation, RWE Thames Water and EDF Energy (referred to as Metronet Rail BCV Limited and Metronet Rail SSL Limited). Tube Lines, a consortium of Bechtel, Jarvis and Amey, took on infraco JNP on 31 December 2002.

In total, the transaction costs incurred in establishing the PPP amounted to £455 million.

Under the contract, Metronet agreed to invest £17 billion over its 30-year term, while Tubelines committed to spending £4.4 billion over the first seven years (the original estimate was £1.5 billion). On its part, the UK government agreed to provide stable funding, averaging more than £1 billion a year until 2010. The Metronet consortium (shareholders) each invested £70 million of equity in the company, totalling £350 million. A further £1 billion was raised via bank loans, £1billion in bonds and £600 million from the European Investment Bank.

However, despite the perception that PPP projects transfer financial risk from the public sector to the private sector, the Metronet contracts made Transport for London (TfL) liable for the financing of the programme. TfL would be required to repay Metronet's creditors £1.9 billion (95% of the company's £2 billion debt) if Metronet went into liquidation.

Over the term of the contract:

- Approximately 250 stations would be refurbished
- Over 300 new trains would be provided
- 80% of the track would be upgraded

Under its contract, Metronet received a monthly fee of approximately £860 million (as at 2006).

In April 2003, Metronet estimated that its budget would be £8 billion until 2010. However, by November 2006, according to the PPP arbiter, Metronet had only refurbished four stations under the BCV contract (against the 17 it had projected) and 10 under the SSL contract compared with the 18 it had planned to refurbish. Moreover, upgrading the track and tunnels under the BCV contract had cost £5.7 million compared with a budgeted £3 million. The regulator attributed '... *combination of inadequate incentives in the supply-chain contacts and poor implementation*' as the reasons of the poor performance. Likewise, the PPP arbiter in November 2006 concluded that Metronet had not carried out its work in an '*economic and efficient manner*' over the first three years of the contract.

A major criticism of Metronet was its strategy of awarding contracts only to its shareholder companies. In contrast, Tube Lines had avoided cost overruns, which was attributed to its policy of competitive tendering all maintenance work.

In May 2007 Metronet announced a requirement for an additional £2 billion in funding until 2010. When the banks refused to extend Metronet's overdraft the company applied to the PPP arbiter for £551 million (over the next 12 months to cover the overrun) in emergency funding. Metronet argued that TfL and LU were responsible for the increase, having changed the specifications outlined in the contracts. The PPP arbiter determined that there would be no increase until January 2008 and that any additional payments would be conditional on Metronet fulfilling its obligations.

On 18 July 2007 the Mayor of London and LU announced that Metronet Rail BCV Limited and Metronet Rail SSL Limited had entered administration.

As a result, the costs of the failure of Metronet were borne by the government; the two Metronet infracos were transferred to TfL in May 2008.

This required London Underground Limited (London Underground) to purchase 95% of Metronet's outstanding debt obligations in accordance with the contract terms, financed via a grant of £1.7 billion from the Department for Transport (DfT). Subsequently, in October 2009 it was confirmed that responsibility for the former Metronet contracts would be assumed by LU on a permanent basis. This was followed, in June 2010, by Tube Lines becoming a wholly owned subsidiary of TfL.

An NAO report into the failure of Metronet (NAO, 2009) concluded that:

- Overall the direct loss to the taxpayer, as a result of Metronet's entering onto administration, was between £170 million and £410 million (based on 2007 prices)
- The principal reason for Metronet's failure was poor corporate governance and leadership
- Under the PPP arrangement DfT had few formal levers to manage the risk to the taxpayer
- DfT was ultimately exposed to the risk of failure of the supplier under the PPP contracts due to the provision that TfL would be required to repay Metronet's creditors 95% of the company's debt if it went into liquidation

Sources: NAO (2004a, 2004b); NAO/Department for Transport (2009) and http://www.tfl.gov.uk/corporate/modesoftransport/londonunderground/management/1580.aspx (accessed 21 September 2012)

Accordingly, the AEC Guidance Procedures (OGC, 2007b) suggest that it is:

> '... preferable to investigate Public Private Partnerships as soon as possible after a user need has been identified rather than leaving it until a conventional construction project has been selected as the solution. It is possible that a Public Private Partnership may result in a solution (provision of services to meet the user need) that does not require a construction project.'

Approval for a project to be procured using the PFI approach requires the public sector purchaser to establish that the project, accounting for the transfer of risk to the private sector, delivers value for money. Further, the approach (it is held) should only used where it can be shown to meet the government's commitment to efficiency, equity and accountability, and its principles of public sector reform.[14]

Partnerships UK (PUK) was established in 2000 by HM Treasury as a PPP; its role was to:

> '... support and accelerate the delivery of infrastructure renewal, high quality public services and the efficient use of public assets through better and stronger partnerships between the public and private sectors.'

PUK acted exclusively with and for the public sector in:

- Supporting individual infrastructure and complex procurement projects
- Developing public service commissioning models and procurement and investment policies
- Investing in projects and companies

In June 2010, Infrastructure UK was established by the incoming coalition government as a separate unit within HM Treasury; its function was to work in conjunction with the private sector on major infrastructure projects.[15] Consequently, the dissolution of PUK was announced in May 2011. Central government PPP/PFI projects now fall within the remit of Infrastructure UK, while local and community-based public service delivery is overseen by Local Partnerships.[16]

Similarly, Local Partnerships is a joint venture between HM Treasury and the Local Government Association. Its remit is to provide commercial knowledge and expertise to all local public bodies, working in conjunction with them to '*improve their sourcing and commissioning skills, programme and project management capabilities, procurement, negotiating and contract management capacity, and their delivery, funding and partnering abilities*'.[17] Also, established by HM Treasury in 2006 and now located in Infrastructure UK, the Operational Taskforce aims to improve the operational performance of PFI and PPP contracts. It provides guidance and makes recommendations on various issues, such as, developing contract management strategies, market testing, benchmarking, managing change and refinancing.[18]

A National Audit Office study (NAO, 2007) established that:

- A third of all projects (closed between 2004 and 2006) received only two detailed competitive bids
- On average the tender period for PFI projects lasted 34 months (for example, PFI schools taking 25 months, hospitals 38 months, while other PFI projects took 47 months)
- The final negotiation stage, once a preferred bidder had been selected, took on average over a year and occasionally as long as five years
- While negotiating with the preferred bidder, significant changes (positive and negative) were frequently made to supplier's prices, once the competitive tension had been lost
- Systematic processes were not always in place to cascade any lessons learned

In response, the report recommended, for most projects a target procurement period of 18–24 months, which should only be exceeded in exceptional cases. In order to achieve these targets and still deliver value for money, authorities were advised to:

- Obtain early key stakeholder commitment to the project
- Develop more appropriate output specifications
- Establish the affordability of the project before seeking tenders and again before selecting a preferred bidder
- Agree the commercial basis of a deal and the main features of the detailed design before selecting a preferred bidder (NAO, 2007)

Additional recommendations were: ongoing monitoring of projects during the procurement stage; the adoption of a programme approach, where appropriate, to enable experience to be transferred from one project to another – for example, as under Building Schools for the Future (see Box 8.5); and the identification of lessons learnt from recently closed PFI projects.

Under EU directives it is anticipated that public authorities will use the competitive dialogue procedure to procure PFI projects.

The PPP/PFI approach has been criticised on several counts: for example, its failure to transfer risk effectively, the 'creeping' privatisation of public services, and the failure to determine whether or not the approach delivers value for money. The NAO (2009) observed:

> 'Our long-held view on PFI is that it is neither always good value for money, nor always poor value for money. It has the potential to deliver benefits but not at any price or in any circumstances. In practice its value is contingent on a wide range of contract, sector and market specific factors.'

Box 8.5 Partnerships for Schools (PfS)

The Building Schools for the Future (BSF) programme was inaugurated in 2004, with its delivery vehicle Partnerships for Schools (PfS) being launched in 2005. The programme sought to rebuild or renew every state secondary school in England (3500 schools – at a cost of £45 billion) over 15–20 years. PfS's aim was to facilitate the establishment of robust PPPs, allowing the public sector to benefit from private sector expertise and knowledge and obtain greater efficiencies and economies of scale.

The principal focus of PfS was the development and management of:

- The national BSF programme
- Local projects

On a functional level, its remit was to:

- Aid and provide support to individual local authority procurement teams
- Evaluate and co-select private sector participants
- Act as a contributor to each Local Education Partnership (LEP)
- Provide continuing support to local authorities following the creation of LEPs
- Provide a support function to the BSF team in DfES
- To ensure that BSF schools were well designed, delivered on time and on budget, and were appropriately maintained over their lives, and provided value for money

The BSF programme was cancelled in July 2010 following the election of the coalition government, and, since April 2012, PfS has been incorporated into the Education Funding Agency (EFA), a new Executive Agency within the DfE. When all the commissioned projects are completed, over 700 schools will have been funded via the programme.

Sources: http://www.education.gov.uk/schools/adminandfinance/schoolscapital/funding/bsf http://webarchive.nationalarchives.gov.uk/20120202141958/http:/www.partnershipsforschools.org.uk (accessed 21 September 2012)

Further, the NAO (2011) concluded that, following the recent credit crisis, the use of PFI to procure projects may not be as appropriate as it once was. As a result, the adoption of the approach should be challenged more often, as the cost of debt finance has increased by 20% to 33%.

Integrated procurement systems: comment

According to the AEC guidelines (OGC, 2007b), under a design and construct contract the purchaser is most likely to derive the maximum performance benefits from suppliers through innovation, standardisation and integrated supply chains, where suitable output specifications are used. Moreover, there is a risk that the design, performance or quality of the finished asset/service may be compromised where output specifications are not sufficiently developed. To realise the required outcome, therefore, thorough consideration of the specific output specifications is necessary.

Fenn and Lowe (2010) established that '*the interface between employer's requirements and contractor's proposals*' and '*contractor's design liability*' are causes of dispute in design-build projects. Particularly, the absence of any protocol and guidelines was held to be a key feature of many of these disputes, whereas novation and the amendment of standard forms of contract were often cited areas of concern. Finally, although the absence of any protocol and guidelines does not seriously hamper the utilisation of design-build, it does impede the effectiveness of this procurement method, leaving both employers and contractors exposed both commercially and contractually.

In certain conditions it may well be beneficial, when using the design-build procurement route, to extend the agreement to include the maintenance and possibly the operation of the facility for a considerable period of time. By doing this, the supplier has an increased incentive to incorporate innovative solutions that take into account whole life cost and provide greater value for money (OGC, 2007b).

Management-orientated procurement systems

Management methods are where the client appoints design and cost consultants and a contractor or consultant to manage the production/implementation phase for a fee. Winch (2010) labels this approach **mediated project coalitions**. Specialist contractors are appointed to undertake the production work by negotiation or in competition. There are essentially two variants:

- **Consultancy**: where the management of the project is undertaken by a professional organisation and specialist works contractors are appointed to carry out individual work packages under a direct contract with the purchaser (client). Schedule and price risk is held by the purchaser. Within the construction sector this is referred to as **construction management**
- **Contracting**: involves a management contractor entering into a contract to deliver the project to a predetermined schedule and price; they also assume a degree of risk. The management contractor then enters into direct contracts with specialist subcontractors who undertake the work. Under this approach the purchaser generally retains a measure of price/schedule risks. Within the construction sector this is approach is referred to as **management contracting**

Prime contracting

Prime contracting is where an individual supplier (**prime contractor**) acts as the sole point of responsibility between the client and the **supply chain**. As long as the supplier (be it an individual or organisation) has the appropriate competence and experience (for example, the ability to assemble a suitable supply chain to meet the needs of the client successfully), there is no restriction on who can assume the function. It is common for the potential prime contractors, during the pre-qualification stage, to indicate the various participants in their proposed **supply chain**. The prime contractor can, therefore, be a designer, facilities manager, financier or a construction company, for example, who is liable for the management and delivery of the project. In addition to meeting the usual time, budgetary and fitness for the purpose criteria, prime contractors are usually required to show (during the early stage of use of the asset/facility) that pre-agreed performance targets, such as operating costs, have been achieved. Typically, prime contracting incorporates the following:

- **Open-book accounting**: where the actual costs incurred by the supply chain are revealed to the client
- **Target cost pricing**: where the prices paid are established based on value for money for the client and a realistic profit for the supply team
- **Whole-life costing**: where, before construction/implementation begins, a whole-life cost model is generated
- **Pain/gain share**: where both the client and prime contractor benefit financially from reductions in project costs (Franks, 1998; OGC, 2007b)

Collaborative arrangements

Rigby *et al.* (2009), drawing on a Nordic study of partnering (Gottlieb *et al.*, 2007), identify five forms of collaborative arrangement: **project partnering**, **strategic partnering**, **alliances**, **framework agreements** and **construction consortia**. While the first four instances generally involve a degree of client (procurer) participation, construction consortia only entail collaboration between supply-side organisations and are covered later in the chapter. Rigby *et al.*'s report for the EU, lists the following fundamental characteristics of collaborative relationships:

- **Trust and openness**: this is fostered and rewarded between the parties to encourage behaviours that promote reciprocal advantage. The parties are motivated to enter into the arrangements because they believe that it will lead to further potentially more successful projects that will deliver improved business opportunities

- **Forfeiture of benefit**: the parties voluntarily relinquish some advantage or freedom
- **Promotional activities**: these encourage collaboration and establish supportive project environments in conjunction with the formal contract. These measures generally necessitate changes to conventional contractual conditions; in some countries, for example, specific standard collaborative contract forms have been developed

Collaborative arrangements have been found to deliver the following tangible and intangible benefits:

- **For the client**: increased reliability of delivery, including both schedule and budget; superior quality; and increased overall satisfaction with the final product
- **For the supply firms**: increased consistency and transparency in project objectives; an increased focus on project success; a high degree of certainty that payment agreements will be honoured; and increased employee job satisfaction
- **Joint**: improved communication within the project coalition; reduced (or even no) conflict and formal disputes; innovative solutions and fewer design changes. Additionally, where these arrangements extend over several projects, familiarity can result in the following mutual benefits being realised: continuous improvements, increased application of prior knowledge, improved knowledge transfer from previous projects, reduced tendering costs (Rigby *et al.*, 2009)

Partnering

Partnering[19] is a form of collaborative project that is:

> '... based on dialogue, trust, openness and with early participation from all actors. The project is carried out under a mutual agreement expressed by mutual activities and based on mutual economic interests.' (National Agency for Enterprise and Construction, 2004)

Likewise, the Construction Industry Institute of the United States of America defines partnering as:

> '... a long term commitment between two or more organizations for the purpose of achieving specific business objectives by maximizing the effectiveness of each partner's resources. This requires changing traditional relationships to a shared culture without regard to organizational boundaries. The relationship is based upon trust, dedication to common goals, and an understanding of each other's individual expectations and values. Expected benefits include improved efficiency and cost effectiveness, increased opportunity for innovation, and the continuous improvement of quality products and services.' (CII, 1990)

However, as Walker *et al.* (2000) state, '... *the partners still maintain a sense of independence with their own contractual arrangement and a tendering process that may or may not be based on a competitive/cost structure*'. Partnering, therefore, still entails the purchase of an asset/service by means of a procurement process. For example, using the traditional, design-build, or management approach, let either by competition or through negotiation. The advantages of the partnering approach lie in the attitudes and behaviours that govern the commercial process (Walker *et al.*, 2000).

Two forms of partnering are identified:

- **Project partnering**: this is where the purchaser and principal partner (supplier) or coalition partners enter into a formal arrangement to work collaboratively on a single venture. Frequently expressed in the form of a 'partnering charter' or through the adoption of a collaborative form of contract that incorporate collaborative features (see examples reviewed in Chapter 6), such as gain share/pain share clauses and disputes resolution mechanisms such as mediation and arbitration
- **Strategic partnering**: this is where the purchaser and principal partner (supplier) or coalition partners enter into an agreement to work collaboratively on a series of projects. While the full nature of the work to be undertaken may not initially be explicitly defined, the parties commit to quality and performance improvement over the life of the arrangement. Ideally, these objectives are expressed as mutually agreed targets and commitments (Rigby *et al.*, 2009)

A survey by Wood (2005) found that developing **trust** is a key challenge, the transfer of risk from client to supplier without appropriate reward is still prevalent and that clients occasionally regress to cost-driven behaviour. Improved time and cost performance for clients and a guarantee of work for main contractors are seen as the apparent benefits of partnering. Similarly, a report by the Specialist Engineering Contractors (SEC) Group (SEC, 2003) found that local government clients had been slow in developing partnering-style relationships, and that lack of trust and lowest price culture still prevailed.

Further reading

Trust

Swan, W, McDermott, P and Khalfan, M *et al.* (2006) Trust and commercial managers: influences and impacts (Chapter 8). In: DJ Lowe (ed.) with R Leiringer. *Commercial Management of Projects: Defining the Discipline.* Blackwell Publishing, Oxford. pp. 172–191

Box 8.6 Heathrow Terminal 5 (T5) case study

In selecting an appropriate procurement strategy, risk was a significant issue for BAA. Similarly, the approach adopted had to be flexible and responsive to change, enabling the design to evolve during the project's delivery stage. BAA's objectives were:

- To create a project environment to support the delivery of T5
- To achieve and improve upon existing 'best-practice' performance in all stages of the project
- To initiate a blueprint for future major construction projects

To deliver these objectives BAA instigated a wide-ranging strategy to transform its own capabilities and those of its key suppliers. However, the strategy went beyond any previous arrangement implemented in the UK. It redefined BAA's role as project client allowing them to participate fully in the delivery process. The strategy embraced four key principles:

1. The client always bears and pays for the risk
2. BAA retained full liability for all project risks
3. Suppliers' profit levels were to be predetermined and fixed
4. Partners are worth more than suppliers

BAA implemented an 'integrated project team approach' captured in a 'delivery team handbook', which sought to create an appropriate environment for team working. These principles were incorporated into the 'T5 agreement' – a bespoke project-specific contract developed by BAA to govern its relationships with tier-1 suppliers and designed to create a framework to deliver project success.

Under the 'T5 agreement' BAA entered into a direct contractual relationship with all their 'tier-1' suppliers (main suppliers, contractors and consultants). Rather than specifying the precise work required or seeking a commitment from suppliers to carry out a specific amount of work, the T5 agreement required tier-1 suppliers to commit capability and capacity to undertake work on the project. This allowed BAA to compile project teams from the collective expertise within the tier-1 suppliers. Suppliers were then assigned to integrated teams when their capability was required, carrying out discrete parts of the project based on plans of work.

T5 consisted of 16 projects (each valued at between £50 million and £400 million), which in turn were divided into 147 subprojects (comprising approximately 1000 work packages). Each subproject was run by an integrated design and construction team, containing between six and 25 tier-1 suppliers, which was led by a BAA project manager.

The main objective of the agreement was to create a unique contract under which BAA retained all the risk relating to the project. The contract is generally considered to be balanced, facilitating appropriate relationships and behaviours, and creating commercial tension without erecting commercial barriers. The contract, supported by BAA's novel risk insurance policy, was designed to enable all participants to concentrate on:

- The root cause of problems – not their effects
- Working within integrated teams to deliver success in an uncertain environment
- The proactive management of risk rather than the avoidance of litigation (NAO, 2006)

Source: see Case Study B for more details, particularly concerning the client's motivation for selecting this procurement approach

Alliancing

In comparison with partnering, **project alliancing** is more '*all-encompassing*', with alliance partners (the purchaser and key suppliers) combining to create a virtual organisation or in certain instances combining to establish jointly-owned companies. Large complex infrastructure projects have been delivered using an alliancing approach.

Appointed at the beginning of the project, alliance partners are selected based on their competence and expertise to meet the client's procurement objectives and performance standards. Once an alliance team has been formed, an alliance agreement can be formulated. This agreement generally incorporates an alliance charter and performance requirements. Further, project duration and cost targets are determined during the development of the design, incorporating agreed risk/reward sharing arrangements. Value for money and potential cost savings are realised by leveraging the alliance partners' competence and expertise throughout the entire project life cycle to develop and deliver an appropriate solution that meets the client's needs (Walker *et al.*, 2000; Walker and Hampson, 2003).

Walker *et al.* consider the defining features of alliances to include:

- Supplier selection based on general performance criteria requiring firms to display their ability to innovate, manage relationships and deliver world-class performance
- A significant contribution to the development/design stage of the project
- An alliance board, comprising the purchaser and representatives of the key supply partners, which sets and commits to mutual budget and cost/schedule targets
- A pain share/gain share agreement, open-book accounting/costs reimbursement, and a verifiable cost target. Profit is deemed to be 'at risk' as a lever to ensure that established project cost(s) is/are achieved
- Preplanning and establishing project scope before the gain share/pain share mechanism is agreed, which reduces the likelihood of significant changes and variations. All production phase changes are managed by the alliance partners, the cost of changes being paid only when substantial and demonstrable changes in scope arise
- Excellent communication at a personal, operational and organisational level, incorporating shared IT systems and integrated information processing (Walker *et al.*, 2000)

Box 8.7 The Aircraft Carrier Alliance (ACA)

In September 2011 assembly of the first of two Queen Elizabeth Class aircraft carriers commenced; the second beginning the following May. With an initial estimated cost of £3.9 billion, the 65 000 tonne carriers, HMS Queen Elizabeth and HMS Prince of Wales, at 284 m long, 73 m wide and with a draught of 11 m are the largest ships ever to be built for the Royal Navy. Currently, it is anticipated that the HMS Queen Elizabeth will enter service in 2016, followed in 2018 by HMS Prince of Wales. The MOD entered into contracts for the delivery of the vessels in July 2008.

The vessels are being delivered by the Aircraft Carrier Alliance (ACA): a partnering arrangement comprising the UK MOD, Babcock, BAE Systems and Thales UK. Ownership and responsibility for the project are held jointly, with each partner sharing in the risks and rewards.

BAE Systems is responsible for managing the programme and providing overall leadership. Additionally, it is responsible for the design, manufacture and integration of the carrier's mission systems, and is involved in designing and building the ships. Two of BAE Systems' businesses are engaged on the project: BAE Systems Naval Ships and BAE Systems Submarine Solutions. Work undertaken by Babcock includes design and development activities, CAD-based modelling, section and block manufacture, and ship assembly. Thales UK leads the Power and Propulsion suballiance, having previously led the design stage of the programme. The MOD Capital Ships Project Team is responsible for procuring and providing support to the programme.

The alliance is based on collaborative principles. For example, ACA members have committed to:

> '... making decisions in the best interests of the entire programme rather than basing any decisions on their individual best interests alone'.

Based on a culture founded on:

> '... an uncompromising commitment to trust, collaboration, innovation and mutual support'.

The Alliance's current project cost forecast is £5.461 billion, £219 million higher than its agreed target cost.*

For details of the procurement process adopted and the Aircraft Carrier Alliance Charter see: http://navy-matters.beedall.com/cvf1-10.htm

*House of Commons Public Accounts Committee publishes the 56th report of 2010–2012 Session

Sources: http://www.aircraftcarrieralliance.co.uk/en.aspx (accessed 5 March 2012) http:// www.defenseindustrydaily.com/design-preparations-continue-for-britains-new-cvf-future-carrier-updated-01630/ (accessed 21 September 2012) http://www.mod.uk/DefenceInternet/DefenceNews/EquipmentAndLogistics/AssemblyOfNewRoyalNavyAircraftCarriersGetsUnderwayInFife.htm (accessed 5 March 2012)

Framework agreements (including call-off contracts)

Framework agreements have a degree of commonality with strategic partnering. For example, under both approaches the purchaser engages specific suppliers to deliver a level of capability (goods, works or services) for a defined period. There are two formats:

1. **Framework Agreements**: these include a contractual requirement to purchase a specified quantity or value of goods/services

2. **Framework Arrangements**: these include no contractual requirement to either purchase or supply a specified quantity or value of goods/services; however, they generally stipulate the terms and conditions of the contract that would ultimately be applied if and when goods, works or services are purchased

Frameworks can be used in conjunction with prime contracting and design-build procurement approaches, and are particularly suitable for maintenance work. They are not, however, suitable for inexperienced purchasers, who infrequently participate in projects. The significant difference between strategic partnering and frameworks is the requirement to undertake a subsequent selection process to allocate one of the potential suppliers within the framework to a specific project once it has been defined. Framework arrangements are frequently underpinned by both contracts and a commitment to collaborative working.

They are alternatively called **call-off agreements, call-off arrangements** or **call-off contracts**, as orders issued under frameworks are frequently termed **call-offs**. There are three principal forms of framework agreement/arrangement:

1. **Fixed term**: this is regularly used for the supply of goods and services, where the parties enter into a framework for a specified length of time. The agreement/arrangement usually contains an assessment of the overall quantity or value of items to be supplied
2. **Fixed quantity**: this is mostly used for the purchase of inexpensive consumables for which there is a constant demand, or for particular types of service contracts – for example, window cleaning and office equipment maintenance. The agreement offers the supplier more certainty that the predicted volume of goods/services will be purchased
3. **'Insurance' type**: this is generally used for service contracts, where the annual cost of a service is predetermined irrespective of the number of times the service is required. This form of agreement is especially appropriate for contracts covering the maintenance of equipment

The AEC guidelines (OGC, 2007b) assert that framework agreements with either a sole supplier or a few suppliers can produce considerable cost savings for both parties. The anticipated benefits and savings derive from:

- Expeditious procurement and reduced levels of bureaucracy
- The elimination of the requirement to tender each separate project, leading to reduced administration (cost and effort)
- Continuity of work, enabling suppliers to offer competitive prices due to the anticipated overall value derived from the long-term agreement (although, this has to be tempered by the potential for opportunistic behaviour post award). Contracts, therefore, need to incorporate a mechanism to inject/retain a degree of competitive tension
- The transfer of learning across projects supporting continuous improvement
- A less adversarial approach, leading to a reduction in the number of disputes and level of conflict
- The establishment of longer-term trading relationships with the wider supply chain (DTI E.9 Frameworks and OGC, 2007b)

Term contracting

Term contracting is generally taken to refer to a form of agreement that enables work to be executed over a given time frame. It is commonly used to procure a specified capability – for example, the provision of a service, such as repair and maintenance work. The approach can be utilised when the general nature of the work is recognised but its scope is not. Cost reimbursement under term contracts can take two forms:

- Measure and value based on a schedule of rates (prices)
- The prime cost of carrying out the work/providing the capability

In both instances individual orders are raised under the contract, which then becomes a contract in its own right. It is at this point that the terms of the contract become enforceable/binding on the parties. As a result, a term contract could potentially be terminated, without incurring any sanction, if the purchaser decides not

Box 8.8 Procure21+

Building on the ProCure21 National Framework, which ran from 2003 to September 2010 and delivered 647 schemes worth £4.2 billion, the ProCure21+ National Framework and procurement process is a framework agreement for NHS capital investment construction schemes in England. Six principal supply chain partners (PSCPs): Balfour Beatty Group, Integrated Health Projects, Interserve Project Services, Kier Regional, Miller hps, and Willmott Dixon Holdings (selected via an OJEU tender process) have entered into a framework agreement with the Secretary of State for Health. It is scheduled to run until 2016.

Any NHS client or joint venture may select a ProCure21+ supply chain, provided the project has a health component, without the need to instigate a further OJEU tender process. Potential applications include:

- Major works schemes (or refurbishments)
- Minor works programmes (provided the value of each task is less than £1 million)
- Refurbishments
- Infrastructure upgrades and non-health buildings, for example, car parks
- Feasibility studies
- Service planning or reconfiguration reviews

The ProCure21+ procurement principles and process for the design and construction of proposed schemes are detailed in the ProCure21+ NEC3 Contract Template and associated guidance. ProCure21+ espouses 16 key features: tested, flexible, required, educated, assured, accountable, committed, integrated, reviewed, recycled, transparent, challenging, innovative, scaled, streamlined, and evaluated.

Currently, there are 90 active Procure 21+ schemes with an estimated value of £1,386.22 million.

For more details of Procure 21+ see DoH (2011)

Source: http://www.procure21plus.nhs.uk/introduction/ (accessed 21 September 2012)

to issue any further instructions or by the contractor declining to undertake any subsequent orders. In reality, term contracts invariably include termination clauses. Term contracting can, however, also refer to long-term service contracts where sanctions for termination or specific exit provisions are commonly applied.

e-Procurement

> 'Electronic procurement (eProcurement) is the use of electronic tools and systems to increase efficiency and reduce costs during each stage of the procurement process.' (OGC, 2005)

e-Procurement covers a range of solutions that enable both buyers and suppliers to trade electronically over the internet. It is claimed that e-procurement will enable companies to gain a competitive advantage through e-commerce by creating profitable communities of buyers and suppliers. Further, e-procurement will allow organisations to increase significantly the control, speed, efficiency and cost saving around their purchasing and selling processes. The EU public and utility contracts directives (2004/18/EC and 2004/17/EC) encourage the use of e-procurement.[20]

Constructionline provides e-procurement and pre-qualification as a combined solution. It is a PPP between the Department for Business Innovation & Skills and Capita, a major process outsourcing (BPO) and professional services company. Its function is to assist both purchasers and suppliers exploit the benefits of e-procurement and on-line data exchange.[21] However, the most recent RICS Survey of Contracts in Use (Davis Langdon, 2010) found *'little evidence of the use of electronic tendering'* – although the report intimates that their results may have been influenced by under-reporting. Also, the figures are now at least five years out of date and e-procurement may well be more prevalent.

Trends in procurement systems

This section is based on the UK construction industry, due to the availability of data on the contracts in use within the sector. The sector has been subject, over the last 20 years, to significant political (PPP) and industry institutional demands to adopt collaborative working practices. As a result, there has been a discernible change in the way that UK construction projects are procured; although the traditional approach of design–bid–build still remains a popular choice, design-build is the dominant procurement method. The most recent RICS Survey of Contracts in Use (Davis Langdon, 2010) indicates that, although the proportion of work undertaken via design-build has dipped slightly in the latest figures it is still the most popular approach when expressed as a percentage of the total workload based on value. Moreover, design-build was one of only three procurement routes sanctioned by the OGC, the others were: PPP/PFI and prime contracting. The results also reveal the uptake of partnering agreements since 2004. Changes over time in the method adopted to procure UK construction projects were shown in Table 6.1.

Although, the most recent figures indicate a slight increase in use of management contracting (Davis Langdon, 2010) the approach has been in decline for the past 10 years. Construction management, in contrast, is still used: primarily on large-scale complex projects. However, the report by the Rt Hon Lord Fraser of Carmyllie QC on his inquiry into the Holyrood project (the Scottish Parliament building) recommended that:

> 'Construction Management as a procurement route should be used sparingly for any public building project. All risk lies with the client and ultimately the taxpayer. Current Treasury Guidance could not be clearer. It is a procurement route of last resort. I recommend civil servants or local government officials contemplating construction management for a public project should reflect long and hard on the advantages and disadvantages of such a route and should set before the political leadership a full evaluation of the risks.'[22]

Exercise 8.3 Trends in procurement routes

Compare and contrast the foregoing with developments within your own industry sector.

Procurement system selection criteria

As mentioned earlier, procurement system selection is closely linked to the purchaser's specific objectives. Each of the foregoing procurement systems satisfies certain of these objectives: each has distinct advantages and disadvantages in respect of aiding the delivery of quality, cost and schedule certainty and fitness for purpose, plus supporting the provision of innovative solutions or bespoke design statements, post-award flexibility; single-point responsibility, teamworking and purchaser participation. Masterman (2002) provides an extensive discussion of the advantages and disadvantages of the main procurement systems used within the construction sector.

Table 8.1, adapted from the Construction Round Table (1995) and Masterman (2002), lists the most common variables that influence the choice of approach. The questions and the subsequent review of the responses enable the purchaser to adopt a systematic approach to procurement route selection.

A key determinant in procurement system selection is the purchaser's attitude to risk, in particular their willingness to retain an appropriate level of risk or desire to transfer as much risk as is possible to their

Table 8.1 Variables that influence the choice of the most appropriate procurement route.

Variable	Question	Response
Schedule	How important is early completion to the success of your project?	• Crucial • Important • Not as important as other factors
Change control	Do you foresee the need to alter the project in any way once it has begun?	• Yes • Possibly • Definitely not
Technical complexity of projects/commodity	Does your product/project need to be technically advanced or highly complex?	• Yes • Moderately so • No, just simple
Product/quality level	What level of quality do you seek in the design and workmanship?	• Basic competence • Good but not special • Prestige
Price/cost certainty	Do you need to have a firm purchase price before you can commit to proceed?	• Yes • A target plus or minus will do • No
Competition	Do you need to choose your supplier/ subcontractor by competition based on price?	• Yes, for all work • Above a specified price or specific work • No, other factors more important
Management	Can you manage multiple suppliers, or do you want just one organisation to be responsible after the briefing stage?	• Can manage separate firms • Must have only one firm for everything
Responsibility/ accountability	Do you want direct professional accountability from your advisers?	• Yes • Not important
Risk acceptance/ avoidance	Do you want to pay someone to take the risk of cost and time slippage from you?	• Yes • Prepared to share agreed risks • No, prefer to retain risk and therefore control
Collaboration	Do you want to collaborate with the appointment supply chain post-award?	• Yes • Possibly • No
Repeat business	Do you envisage letting further similar contracts?	• Yes • Possibly • No

Source: adapted from Construction Round Table (1995) and Masterman (2002)

supply chain. In terms of risk transfer there is a continuum ranging from design-build through to construction management:

- **Design-build**: this provides a high degree of price and schedule certainty provided the purchaser has a well-developed statement of requirements prior to commencing the tender process and that few if any changes are introduced post-award (Compare and contrast the Wembley and Emirates Stadium cases both of which were let using a design-build approach.)

- **Management-orientated systems**: these enable purchasers to appoint and involve partners earlier in the project life cycle and offer greater degrees of flexibility to introduce changes post-award. However, these benefits come at a 'price': the client is required to bear most of the cost and schedule risk

Assessment models

There are several models available that seek to aid clients and their consultants in selecting the most appropriate procurement route for their particular circumstances. Table 8.2, presented by AEC (OGC, 2007b), is a typical approach to assessing the relative effectiveness of the different procurement routes. It functions as a decision-support tool, assisting but not replacing procurement specialists in selecting an appropriate procurement option. Selection of assessment criteria and their associated weightings are dependent on the scope of the project and the specific client. PFI is not shown as an option, as it is deemed to be a procurement strategy. It could be beneficial, however, to assess PFI against the main procurement routes. As with most decision-support tools, by adjusting the assessment criteria, scoring and weightings, the outcome can be manipulated to produce a predetermined solution.

ProCost

An alternative procurement strategy decision support tool is **ProCost**, a neural network model developed using data from 286 UK construction projects (Emsley *et al.*, 2002; Lowe *et al.*, 2006), as a response to the lack of reliable information on the relative costs of using different procurement routes. The model forecasts the total cost of a potential project – construction cost (final account) plus client on-costs (such as professional fees and internal costs incurred by the client) – these costs are combined to give the total cost to the client. Forty-one independent variables are incorporated and categorised as: **project strategic variables** – for example, contract form, procurement strategy and tendering strategy; **site-related variables** – for example, location and site access; and **design-related variables**. The influences of time and geographical location are accommodated through the use of the Building Cost Information Service (BCIS) indices by adjusting all the data sets (projects) to a common location and base date. The predictive capability (accuracy) of the model has been measured at 16.6% (using the mean absolute percentage error [MAPE]). This compares favourably with the traditional methods of cost estimation, which have reported MAPE values of between 21% and 28%.

Contract strategy and type

Having determined that a 'purchased' approach is required and selected an appropriate procurement system, the purchaser has then to either choose a suitable standard form of contract (or company precedent) or arrange for a bespoke contract to be drafted. Chapter 6 addressed the issue of contract selection, providing examples of several standard forms of contract classified in terms of contract strategy (which corresponds with the procurement system types described above) and outlined the various payment terms that can be adopted and the apportionment of risk (contract type). The aim is to select a contract best able to deliver the project's intended outcomes and support the required behaviour of the parties associated with the proposed procurement systems.

Commercial practitioners are either responsible for making this selection or providing specialist advice to the organisation's procurement function. Further, where a bespoke solution is required they either draft the contract, often in conjunction with the in-house legal department, or commission external lawyers to formulate one. Where the purchase is for either an asset or service that is outside the normal trading sphere of the organisation, consultant commercial advice is often sought. Instances of this can include infrastructure investments, such as office or manufacturing facilities, and computer-enabled services.

Supplier selection

Further commercial decisions are required concerning the process by which a suitable supplier is appointed. Although supplier selection is addressed in Part B: **Deal Creation**, the following section considers the purchaser's motivation for purchasing by tender or direct negotiation.

Purchaser's can select from a continuum of options, ranging from open competition to direct negotiation with a single supplier. This choice is influenced by a variety of factors, generally specific to the purchasing organisation and the purchase to be made. Factors include:

Table 8.2 Example of an approach to procurement route evaluation.

Project title: A construction project											
Procurement route:		**Traditional**		**Design and build**		**Design, build and maintain**		**Design, build, maintain and operate**		**Prime contracting**	
Evaluation criteria (Appropriate to the client and project)	**Criteria weight %**	**Score**	**Weighted score**	**Score**	**Weighted score**	**Score**	**Weighted score**	**Score**	**Weighted score**	**Score**	**Weighted score**
Opportunity for supplier to innovate to yield the most cost effective combination of capital construction, maintenance and operation											
Least disruption in project flow due to perceptions and procedures to meet public accountability – minimisation of disputes											
Certainty of whole-life costs											
Flexibility for future changes in client requirements and post-completion changes											
Speed of project delivery to occupation/first use											
Control over detailed design and design quality (a detailed output specification is still required)											
Control over whole-life health and safety issues											
Reduction in disputes and in-house costs through single point responsibility											
Control of sustainability issues											
Requirement to optimise whole-life costs											
Total scores											
Contract strategy ranking by evaluation criteria											
Preferred order (score and rank combined)											
Members of evaluation panel	Panel member 1		Signature			Panel member 2			Signature		

Source: OGC (2007b) Achieving Excellence Guide 6 – Procurement and Contract Strategies. Reproduced under the terms of the Open Government Licence (http://www.nationalarchives.gov.uk/doc/open-government-licence/)

- **Commercial imperatives/procedural requirements**: for example, compliance with mandatory tendering procedures
- **Market conditions**: the degree of choice within a specific market both in terms of potential solutions and suppliers. This is linked to the structure of the market, the distribution of capability across organisations and prevailing economic conditions
- **Requirement for a bespoke solution**: which may make it difficult to establish a fair 'market' price without undertaking a tender competition
- **Product/service specification**: the ability to specify precisely in advance the required product or service

Depending on the precise nature of the competition, a competitive approach has certain advantages. These include, for example, the provision of:

- A recognised methodology for deriving a market price for the required product or service
- An open, transparent, objective and auditable selection/assessment methodology
- Innovation – competition can stimulate innovation with the supply market

Alternatively, it can be argued that in times of reduced competition within a particular market, greater value can be derived through negotiating with a single supplier as opposed to instigating a competition between potentially indifferent suppliers.

Disadvantages of inviting tenders
The most common drawbacks associated with competitive tendering include:

- High transaction costs, incurred as a result of producing tender documentation, the investment of time by the client, expenditure on consultants' fees, and expensive tendering procedures, especially in the case of open tendering (which the client ultimately pays for either openly or concealed within the winning tender)
- Time-consuming procedures, while efforts to accelerate the process can often lead to additional problems such as ill-conceived solutions and contractor pricing errors
- A tendency to focus on the lowest bid, rather than value for money or quality and performance issues
- A potential to stifle innovation in the supply chain
- High abortive costs, when tenders come in significantly higher that the purchaser's budget
- Reduced competition – for example, while few contractors will decline an ITT due to fears of alienating a purchaser or their agent, they may decide to submit an artificially high tender (**cover price**) – the effect is to reduce the number of competing organisations (Tweedley, 1995)

Further reading

Outsourcing

Langford, D and Murray, M (2006) Procurement in the context of commercial management (Chapter 4). In: DJ Lowe (ed.) with R Leiringer. *Commercial Management of Projects: Defining the Discipline.* Blackwell Publishing, Oxford. pp. 71–92

Additional activities undertaken by commercial practitioners
During this phase, in addition to appraising, selecting and formulating a preferred solution delivery mechanism (procurement and contracting strategy and tendering methodology), commercial practitioners can become involved in assessing (pre-qualifying) potential solution providers, forecasting the likely procurement costs associated with each potential delivery solution, undertaking cost planning and value management of the preferred solution, as it is refined and fully articulated, and reviewing and revising the

project's associated business case. In addition to the risk management (Chapter 4) and the financial decision-making techniques (Chapter 5), and the project and programme management, risk and value management, and procurement and contract management best practice outlined in Chapter 7, commercial practitioners need to be proficient in or have an awareness of the available cost-forecasting techniques and their associated accuracy levels specific to each industry sector. Early-stage estimating techniques used within the construction sector include:

- **Unit method**: the project cost is established by selecting a standard unit of accommodation/usage and multiplying this by a suitable cost per unit (for example, cost/bed or cost/seat)
- **Superficial method**: the project cost is established by multiplying the total area of the proposed building by a suitable cost per square metre
- **Approximate quantities**: the project cost is established by measuring and pricing composite items usually derived from combining or grouping typical bill of quantities items
- **Elemental estimating**: the project cost is established on an elemental basis by reference to cost analyses of previous projects
- **Cost models**: the project cost is established by reference to a mathematical model or formula

The most popular techniques used are the cost per square metre, approximate quantities and elemental analysis methods. Accuracy levels of the resulting cost predictions range from ±30% at the feasibility/brief stage (requirement specification phase), ±20% at the definition/ scheme design stage (solution selection phase), to ±10% at the tender/bid stage (deal creation phase); the increase in accuracy is attributed to the availability of more information on the selected solution.[23]

Similarly, techniques for estimating the cost of IT/IS projects at an early stage include:

- **Analogy estimating**, involving the transposition of costs from similar previous projects
- **Function point analysis**, based on the quantification of functions points (system functionality), such as inputs, outputs, interfaces with other systems and data storage; again associated costs are derived from previous, similar projects
- **Mathematical models**[24]

Delivery strategy stage gate review

Following the selection of an appropriate solution to address the organisation's needs, a review should be undertaken to evaluate the selected delivery strategy. The OGC Gateway™ Process Review 2: Delivery strategy best practice guidance (OGC, 2007f) provides a template that can be used to develop a pro-forma to support the delivery strategy stage gate review. Sample questions relating to the assessment of the delivery approach, business case and stakeholders, risk management, review of the current phase and the investment decision (its readiness for the next phase) are listed. These include, for example:

- *Have all the relevant options for delivery been investigated?*
- *Are the business needs clearly understood by the client organisation and likely to be understood by those involved in delivery?*
- *Has the proposed procurement procedure been evaluated?*
- *Is there adequate knowledge of existing and potential suppliers?*
- *Is the contract management strategy robust?*
- *Is the evaluation strategy... accepted by stakeholders and compliant with EU procurement rules?*
- *Does the Business Case continue to demonstrate business need and contribution to the organisation's business strategy?*
- *Are the benefits to be delivered by the project understood and agreed with stakeholders?*
- *Are the major risks and issues identified, understood, financially evaluated and considered in determining the delivery strategy?*
- *Is the project under control?*
- *Are the project's timescales reasonable, and compliant with EU rules?* (OGC, 2007f)

Activity 8.4 **Solution selection**

Compare and contrast the solutions adopted for the delivery of the Millennium, Emirates and Wembley stadia projects (case study A) with that of T5 (case study B). Comment on the appropriateness of these approaches to delivery of the declared objectives of the projects. Also, review the specific factors influencing Wembley National Stadium Limited's (WNSL) and BAA's decisions.

Opportunity identification and development

From a supply perspective, the commercial function should ideally be involved in the exchange process as soon as a potential opportunity (lead) has been identified (**opportunity identification**). Commercial practitioners are then able to support the sales function in developing the opportunity and participate in a high-level business validation review of the opportunity to assess the feasibility of securing the work and its alignment to the organisation's overall strategy. The validation review is the initial application of a recurring project-screening exercise, referred to as either **qualification** or the **bid/no-bid** decision. This activity is described in detail under **proposition identification**. If the opportunity is supported, the organisation's sales function will endeavour to promote the supplier's capabilities and influence the buyer's decision makers.

During the subsequent **opportunity development** phase the supplier's sales team will attempt to cultivate a relationship and enter into a dialogue with the potential customer (from corporate level to end user), creating, preparing and marketing potential solutions (product/and or services) to them. The aim of this activity is to influence the purchaser's solution selection, so that the ITT or RFP includes components that correspond with their organisation's competitive advantage and to ensure that their organisation is invited to tender for the opportunity or if possible to be selected as the sole supplier (a negotiated solution). The client manager will also seek to collate as much information as possible about the potential customer's requirements and expectations in order to develop and define a **bid strategy** (win strategy) in anticipation of receiving an ITT or RFP.

Business justification stage gate review

Again, it is recommended that this phase concludes with a stage gate review to evaluate the business justification for proceeding with the potential opportunity – a continuation of the screening process. The OGC Gateway™ Process Review 1: Business justification best practice guidance (OGC, 2007e), although developed to support the demand perspective, can be used as a template and amended to develop a pro-forma to support a supply-side business justification stage gate review.

During this phase, commercial practitioners are involved in developing an understanding of the purchaser's requirements, drivers and expectations from the proposed exchange – a prerequisite for proposition (bid) development – and for drafting appropriate contract terms and conditions.[25] They will also participate in evaluating the feasibility of the potential opportunity (project), applying the risk management and financial decision-making techniques described in Chapters 4, 5 and 7.

Further reading

Opportunity identification and development

Preece, C, Moody, K and Brown, M (2006) The effectiveness of marketing spend (Chapter 7). In: DJ Lowe (ed.) with R Leiringer. *Commercial Management of Projects: Defining the Discipline*. Blackwell Publishing, Oxford. pp. 155–171

Opportunity development[26]

The principal activities undertaken during this phase of the bid process include:

- Continuing the sales campaign to influence the purchaser's procurement solution (involving, for example, the staging of briefing presentations and/or demonstrations to showcase the supplier's capabilities and experience, site visits, and other promotional activities), plus cultivating in-house champions within the purchasing organisation
- Developing a bid (tender) strategy for the potential opportunity
- Undertaking a formal opportunity screening (qualification/bid/no-bid) exercise

The supplier's bid strategy

According to Tweedley (1995), no successful supplier enters a bidding opportunity without initially developing a plan for 'winning' the work. The bid process is dynamic and proactive, requiring the supplier to continually test and evaluate alternative solutions to the problems and opportunities presented in each bid. Further, he contends that a good bid strategy reflects a good business plan, which comprises both quantifiable and intangible elements. Central to any strategy, therefore, must be the objective of winning new and profitable business for the organisation, while each bid should support the adopted business strategy, which is generally either purchaser or product/service focused.

A good bid strategy integrates both business and competitive strategies into a cohesive '*win*' plan. The competitive forces influencing the bid process include:

- **Purchaser factors**: award criteria, preferences and buying-power
- **Supplier factors**: price, supply chain and sources of alternative work
- **Competitor factors**: economic environment – newcomers, old rivals and emulators

It is widely held that a supplier should apply competitor profiling, so as to develop a thorough understanding of the competition and to inform the establishment of a strategy to counter any threat they may pose. The purpose is to build a profile of each competing bidder that includes information on the nature of their bid, its chance of being successful, their assessment of the threat posed by your bid, and the possible tactics they may adopt. In several surveys – for example, Mochtar and Arditi (2001) – the most common form of assessment undertaken as part of the mark-up[27] decision is that of the competition.

Likewise, suppliers should assess the purchaser's needs, highlighting key values that they will expect or require – for example, does the proposed solution provide the purchaser with a competitive advantage in their own field of operations, or will it make their operations easier? The term 'buying centre' is used to describe the group of individuals directly involved in the purchase of products or services. The identification of these key individuals and the influence they surreptitiously exert is important as it will affect not only the supplier's negotiating style but also the pricing and marketing strategies adopted prior to these negotiations.

According to Mattson (1988), the purchaser's 'buying centre' comprises three dimensions:

- **Height**: the hierarchy within the purchaser's organisation that exert an influence on the 'buying centre'
- **Width**: the number of functional areas or departments that are involved in the decision to purchase
- **Depth**: the total number of individuals engaged in the purchasing decision

In order to develop an understanding of their purchaser's buying centre, in addition to establishing the identity of the key decision makers, Kennedy and O'Connor (1997) suggest that the supplier has to consider the purchaser's business environment; mission, objectives and markets; communication network; decision-making processes; award criteria both declared and hidden; and establish the roles played by the individuals involved in purchasing the product/service. However, they discovered sharp divergences between the least and most successful bidding organisations across a range of activities involved in understanding the purchaser's buying centre. For example, only one in ten of the least successful firms made any attempt to understand the purchaser's mission, objectives and markets, and only a sixth of the least successful organisations rated themselves as very effective in getting a feel for the roles played by the individuals and interest groups that influence the final choice of supplier, compared with over two-thirds of the most successful companies (Kennedy and O'Connor, 1997).

Based on their interpretation of these data, each supplier will respond to a project opportunity in one of the following ways:

- **Passively**: by adopting previously applied strategies and tactics
- **Aggressively**: by attacking the market with the intension of obtaining the work by adopting an aggressive position and/or aggressive pricing
- **Defensively**: by adopting a protective approach to the bid, predominantly when perceived to be threatened by the competition and unable to adopt an aggressive approach

The final stage in the process is to translate this information into an attack plan for the supplier's own bid and incorporate it into the bid strategy. The aim is to challenge the competition on issues that are critical to the purchaser and where the supplier stands the best chance of being awarded the project.

The decision to bid

The d2b process (alternatively referred to as project selection, pre-bid analysis, project screening or the bid/no-bid decision) is both complex and dynamic, involving many factors (Shash, 1993), and interactive and iterative (Odusote and Fellows, 1992). Moreover, the selection of the most appropriate projects for which to bid is fundamental to a successful commercial strategy. It also forms the basis for generating the supplier's marketing approach to the project in terms of bidding and/or solution development (Cova *et al.*, 2000). Moreover, the decision, as with that of determining the project mark-up, is very important as the success or failure of a supplier's business lies in the outcome derived from these decisions. What evidence there is, however, suggests that this decision is usually determined by subjective rather than objective information (Fellows and Langford, 1980; Ahmad and Minkarah, 1988; Shash, 1998).

Odusote and Fellows' (1992) respondents claimed that when invited to bid they submitted bids for an average of 73.1% of projects and that they had an average success rate of 22.41%. Likewise, Tucker *et al.* (1996) report that most major construction contractors are successful in one out of six tenders submitted. Any improvement in the supplier's selection of projects would, therefore, give significant benefit to both the supplier and consequently to their clients. Moreover, a suitable decision-support model would be a strategic tool in determining the most appropriate projects to seek and for which to submit a bid, resulting in fewer but more successful bids.

The d2b process

Commercial organisations are required to be selective, choosing which work they will seek out and bid for from a continually changing array of potential projects, due primarily to the availability of sufficient resources (Smith, 1995). Although few suppliers will actually decline an ITT, there is still the need to make a strategy decision on whether to submit a *bona fide* bid or not.[28] This decision is extremely important to the supplier, as, beside the issue of resource allocation, the preparation of a *bona fide* bid requires the supplier to commit to some outlay, which is only recovered if the bid is successful. For example, for a typical bid within the UK construction industry, Fellows and Langford (1980)[29] estimate this outlay to be 0.25% of annual turnover or alternatively as 1% of the projected contract sum for each bid submitted. This could be significantly higher, especially where considerable design or development work is required prior to the submission of a bid. In addition to achieving strategic objectives, these statistics demonstrate the potential financial benefit that can be realised by organisations through adopting an effective and systematic approach to decision making when deciding whether or not to bid for a project.

The d2b decision is predominantly a group assessment, especially on large complex projects (Odusote and Fellows, 1992), involving several managers each of whom will have a different organisational perspective. When considering the decision, these decision makers make judgements to balance market opportunities and risks (Thorpe and McCaffer, 1991), the primary objective being the continued existence and further development of the company (Skitmore, 1989). Several sources suggest that this evaluation involves assessing a number of readily discernible features, for example: company objectives and policies (the potential contribution of the contract to the company's turnover in a particular market segment, the overhead recovery and anticipated profit); contract conditions/details; workload; type of work; resources needed (the financial resources required to support the project and the availability of resources); bid documentation; the cost of preparing a bid; contract size; location of contract; and the contract buyer/purchaser (Ward and Chapman, 1988; Thorpe and McCaffer, 1991; Chartered Institute of Building, 1997).

Additionally, Skitmore (1989) highlights special objectives, which include: market, supply, production, financial, personnel and organisational aims; he collates these under the headings of monetary, non-monetary and market-related objectives. Likewise, Ward and Chapman (1988) believe that recent information concerning competitors' workload, business strategy, strengths and weaknesses is required to account for non-price factors. Kwakye (1994) considers the concept of competitive advantage, while Park and Chapin (1992) suggest that the number of competitors has a direct effect on the probability of winning the bid. However, one of the most important issues in the d2b decision is the supplier's need for work (Smith, 1995).

The d2b process is, therefore, multi-bid in scope and seeks to position potential project opportunities within the organisation's portfolio of projects. Its primary objectives according to Cova *et al.* (2000) are to:

- **Prioritise** project opportunities and inform internal resource allocation
- **Determine** the optimum mode of entry into the project system and any associated external resource requirements

Cova *et al.* (2000) present a case study of a d2b opportunity for a shipbuilding project. The process involved an assessment by the prospective bidder of its: 'relational position' or 'network position'; 'functional position'; 'competitive strengths'; and the 'attractiveness' of the proposed project. They conclude that, depending upon the competitive position of the company, the process seems to underpin the selling and the offering actions in two different ways:

- **Bidding approach**: where the supplier submits a specification-compliant bid and conforms to the purchaser's decision criteria
- **Solution approach**: where the supplier develops a non-compliant bid based on the risk approach, which assumes that the purchaser will modify their decision criteria for the project

Factors influencing the d2b decision

To inform the development of d2b models, surveys by Ahmed and Minkarah (1988), Abdelrazig (1995), Wanous *et al.* (2000 and 2003), Odusote and Fellows (1992) and Shash (1993) within the construction industry identified and ranked 31, 37, 38, 42 and 55 factors respectively, perceived to influence the d2b decision. For example, Ahmed and Minkarah (1988) identified type of job, need for work, owner, historic profit and degree of hazard; Odusote and Fellows (1992) identified client related factors, such as the ability of the client to pay, and the type of work; Shash (1993) identified the need for work, number of competitors bidding, experience in such projects, current work load, and purchaser (owner/promoter/client) identity; and Wanous *et al.* (2000 and 2003) identified fulfilling the to-bid conditions imposed by the client, financial capability of the client, relations with and reputation of the client, project size, and availability of time to bid as the top five factors influencing a supplier's decision to bid for a project.

Taken collectively, these studies have much in common, and reveal the top 15 factors that influence the decision to be: type of job/project type; current work load; size of contract/project size; need for work; past profit in similar projects; purchaser identity; purchaser's ability to pay; experience in similar projects; competitors – the number bidding; location of project; tendering duration; degree of safety/hazard; labour environment; overall economy (availability of other projects/work); and pre-qualification requirement. Further, they suggest that suppliers are influenced more by project characteristics and company-related issues, moderately by the bidding situation, and less so by the economic situation and project documentation. Alternatively, the following reasons why firms might decide not to bid are offered by Hillebrandt and Cannon (1990): the lack of skills within the company to undertake the work; unsatisfactory payment arrangements; too many competitors; inadequate capacity in the estimating department; and unsatisfactory experience in a particular geographical area. However, unsatisfactory past experiences of a particular purchaser or consultant regarding personality or payment, high cost and inadequate information often resulted in inflation of the bid price rather than a refusal to bid.

Although there is some agreement between the studies on the identity of the major factors considered to be significant, it is unlikely that decision makers would consider all these factors for every bid opportunity. Smith (1995) suggests that for each potential project the supplier makes a set of intuitive and subjective judgements in the light of the prevailing circumstances, derived – according to Ahmad (1990) – from a combination of gut-feelings, experience and guesses.

Cova *et al.* (2000) categorise the main criteria used to assess project opportunities under the following headings: attractiveness of the project to the bidder, and competitive strengths of the bidder in relationship to the project. Similarly, Lin and Chen (2004) classify the main d2b criteria under two categories:

- **Company factors**: these include the company's reputation and mission, and its internal resources
- **Bid opportunity factors**: these include the competitive environment, probability of the project going ahead, and project risk

Lowe and Parvar (2004) established that the decision makers from a UK construction company (these included: sales, marketing and estimating personnel) were only able to discriminate, between the outcomes 'bid' and 'no-bid', for only eight of their 21-item inventory: strategic and marketing contribution of the project; competitive analysis of the bid environment; competency – project size; competitive advantage – lowest cost; resources to bid for the project; feasibility of alternative design to reduce cost; external resources (implementation); and bidding procedures. Further, they establish that only these eight items had a significant linear relationship with the decision to bid.

Additionally, they identified six underlying components: responsiveness to opportunities, project relationships, strategic competitive advantage, project procedures, financial relationships, and project risks. Correlation revealed that only three of these dimensions had a significant linear relationship with the d2b outcome. In order of strength of relationship, these were: strategic competitive advantage, responsiveness to opportunity (both of which have a positive relationship) and project risk (which correlates negatively). The positive relationships are consistent with those established for the primary items, while the negative relationship is perhaps indicative of a risk-aversion strategy adopted by the organisation. Again, tests for differences revealed that the decision-makers were only able to discriminate, between bid/no-bid, for the same three factors.

These results suggest that, when deciding upon whether or not to accept a bid opportunity, decision makers are effectively using far fewer factors to inform this decision than indicated by some sections of the literature (see, for example, Ahmad and Minkarah, 1988; Odusote and Fellows, 1992; and Shash, 1993).

The d2b output

Ansoff (1965) suggested the following classification of bid opportunity decision outcomes:

- Reject the opportunity to bid
- Provisionally accept the project and prioritise as follows:
  - Add it to a reserve list
  - Remove another project from the reserve list and replace it with the current project
- Unconditionally accept the ITT

Alternatively, Fellows and Langford (1980) offer the following five possible outcomes: returning the documents; submitting a 'cover price'[30]; providing detailed estimates and bid conversion; preparing a bid based on approximate estimates; or reworking the bid. They suggest that each of these possible 'outcomes' is evaluated, using a mixture of intuition, past data, research information, etc., against each 'utility criterion' to give an assessment of the desirability of a particular outcome. However, according to Skitmore (1989) the usual options are simply acceptance or rejection of the opportunity, although rejection does not mean that the supplier does not submit a bid.

d2b models

Generally, the objectives behind the development of a d2b model are to:

- Provide a decision support system for the d2b process by systematising the existing 'hard' and 'soft' knowledge within the organisation held by senior management, marketing and technical personnel
- Reduce the overall number of bids submitted but maintain or increase the number of successful bids and overall profitability

Further, any improvement in the selection of more profitable projects for which an organisation has the 'best chance' of submitting successful bids would give that organisation a significant competitive advantage. Although previous investigators have suggested that suppliers could improve the efficiency and effectiveness of bid preparation by adopting a more systematic approach (Ward and Chapman, 1988), as previously stated, this has primarily been directed to the bid mark-up problem. Having said this, the several d2b models have been devised based on the following modelling techniques:

- Deterministic worth-evaluation (Ahmad, 1990), which contains four hierarchical groups: job-related, market-related, firm-related and resource-related objectives of a construction firm
- Parametric approach (Wanous *et al.*, 2000)
- Neural network (Wanous *et al.*, 2003; Lowe *et al.*, 2004)
- Logistic regression analysis (Lowe and Parvar, 2004) predictor variables: competitive analysis of the bid environment, competency – project type and the non-monetary contribution of the project demonstrated a high prediction capability (see also Oo *et al.*, 2007, 2008a, 2008b)
- Fuzzy linguistic approach (Lin and Chen, 2004), where users subjectively assess screening criteria in linguistic terms and fuzzy values are used to weight their importance
- Feasibility analytical mapping (FAM) (Du and El-Gafy, 2011), which utilises a graph-based analysis to map potential bids

Further reading

d2b models

Lowe, DJ and Skitmore, M (2006) Bidding (Chapter 16). In: DJ Lowe (ed.) with R Leiringer. *Commercial Management of Projects: Defining the Discipline.* Blackwell Publishing, Oxford. pp. 363–364

Summary

The decision whether or not to bid for a project is a strategic one requiring the consideration of strategic intent, competency acquisition and the long-term aims and objectives of the organisation. Moreover, the decision is extremely important to suppliers; besides the issues of resource allocation, the preparation of a *bona fide* bid commits the organisation to consid.erable expenditure, which is only recovered if the bid is successful. Analysis of the literature has identified some 85 factors considered to be important in the d2b process. Deliberation and assessment of these factors (or subgroups of these factors) facilitates: a systematic approach to the decision-making process, which can improve the quality of the decision making by ensuring that all relevant items are considered and assessed; consistency in the decision process; increased productivity; and they assist in achieving the strategic objectives of an organisation. Additionally, the application of d2b models in the form of decision support systems (DSS) can enhance these benefits further by: providing the analyst with more convincing and reliable results; speeding up the decision process, thereby generating a cost saving; and improving the communication of the rationale adopted by the decision makers within the bidding organisation.

While artificial neural networks, regression analysis and analytical hierarchy process (AHP) techniques have been used to model the d2b decision, as illustrated in the previous section, there are still relatively few models. Those that have been developed are predominantly theoretical and/or based on the perception of the decision makers as to the relationships between and importance of the various decision criteria. Moreover, there is little evidence to suggest that these models have been adopted in practice, despite protestations by some researchers that their industrial collaborators had received the models favourably.

Bid strategy/project screening stage gate review

Again, it is recommended that this phase concludes with a stage gate review to re-examine the business justification for proceeding with the opportunity, re-evaluating the preliminary bid/no-bid decision and appraising the proposed bid (tender) strategy. The OGC Gateway™ Process Review 2: Delivery strategy best practice guidance (OGC, 2007f), although developed to support the demand perspective, can be used as a template and amended to develop a pro-forma to support a supply-side bid strategy/project screening stage gate review. This will involved incorporating the factors that influence the bid/no-bid decision and the associated decision-support tools discussed above.

During this phase commercial practitioners are involved in further developing their understanding of the purchaser's requirements, drivers and expectations of the proposed exchange, managing the pre-qualification process, and drafting appropriate contract terms and conditions. They will also participate in the development of the bid strategy and the project-screening exercise, once again, applying aspects of strategic thinking (Chapter 3), plus the risk management and financial decision-making techniques described in Chapters 4, 5 and 7.

Activity 8.5 **Opportunity development/proposition identification**

Evaluate the motivation of Laing Construction and Multiplex, respectively, to pursue the opportunity to deliver the Millennium and Wembley stadia projects (case study A).

Additionally, consider the motivation of prospective tier-1 suppliers to become involved in T5 (case study B).

Part B: Deal Creation

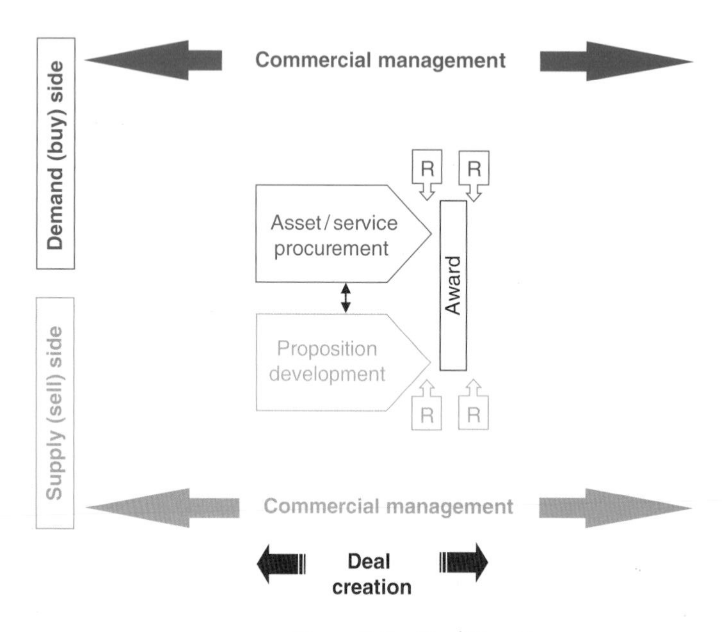

Figure 8.2 Deal Creation.

Introduction

Part B **Deal Creation** focuses on the process and actions undertaken by commercial practitioners that bring an agreement into existence. It is divided into three sections:

1. **Asset/service procurement**: covers implementing the selected sourcing (procurement) strategy. It includes: procurement procedures, key principles of the tendering process (in particular supplier selection); the specific requirements placed on public sector bodies by UK and EU procurement legislation/directives when purchasing assets and services; and negotiating the agreement – the terms and conditions of the exchange
2. **Proposition development**: addresses bid management. It includes creating, preparing, and submitting a proposal in respect of the supply of a product and/or service, and in particular bid price determination
3. **Award**: deals with the negotiation of the agreement and the subsequent contract award

Asset/service procurement

Once a 'purchased' approach to satisfying the purchaser's needs has been adopted and an appropriate procurement system selected, the purchaser has next to decide on suitable procedures for selecting and appointing a supplier.

Procurement procedures

There are essentially four options:

1. **Open**: open tendering affords the purchaser maximum competition by not imposing any restrictions (other than perhaps a deposit to cover the cost of printing the tender documentation, which is typically refundable on the submission of a bona fide tender) on who may submit a tender/proposal for a potential project. Any potential supplier, therefore, can respond to a public advertisement inviting tenders, which are usually placed in trade journals and the national press. Additionally, public sector tenders are advertised in the *Official Journal of the European Community*, and aid-funded projects are publicised in *Development Business*, a United Nations publication
2. **Restricted**: restricted tendering, also termed selective or closed tendering, is where the purchaser performs a selection exercise prior to inviting the chosen suppliers to tender for a project. The purchaser may have an established list of preferred suppliers, or might generate a specific shortlist, in conjunction with their consultant advisers, of known suppliers that have the capacity, competence and experience to undertake the project. In certain circumstances, the purchaser could place an advertisement as with the open procedure, and then select a shortlist from the responding suppliers. When compiling these shortlists, the HM Treasury Procurement Policy Guidelines (HM Treasury, 2005) recommend that the choice should be made using objective criteria, accounting for the evidence allowed under the EC rules where they apply. Under the restricted procedures, tenders may only be submitted by those invited to do so
3. **Negotiated**: the negotiated procedure is where a purchaser approaches either one or more suppliers and negotiates the terms and conditions under which the project will be undertaken. This approach is generally adopted where there are relatively few suppliers competent to undertake the project, where the market is saturated (insufficient suppliers to meet the current demand), or where the client has an existing relationship (contract) with a supplier and wishes to place a further contract with them
4. **Competitive dialogue**: competitive dialogue is a relatively new procurement procedure introduced under the EU Public Sector Procurement Directive (2004/18/EC). It is an adaptable procedure designed to be applied on complex projects where the purchaser needs to enter into detailed discussions with the potential suppliers covering all aspects of the proposed contract. The procedure has been developed to as an alternative to the negotiated procedure (as under the EU legislation the negotiated approach should only be used in 'exceptional cases') and is particularly intended to be used with PPP and PFI projects.

   Following a pre-qualification procedure, the purchaser discusses the form of contract to be adopted and the project's technical requirements with potential suppliers (normally a minimum of three) before issuing the main tender documents. The purchaser then conducts a bilateral dialogue with these suppliers covering all relevant issues; this stage concludes when a suitable solution (or solutions) is identified. Based on the solution(s) derived from the dialogue tenders are submitted. Under competitive dialogue, tender evaluation must be made using the most economically advantageous tender (MEAT) principles. Further, the precise award criteria including their relative weightings should be made available to the tenderers. Once the final tenders have been submitted, the procedures preclude any further negotiation with the bidders

There is some concern, however, that the competitive dialogue approach will lead to higher tendering costs for suppliers and purchasers, although this has not as yet been verified.[31]

Key principles

Although the tendering process is frequently considered to be time consuming, the actual organisation of the process is reasonably uncomplicated. Additionally, to aid the purchaser through the process, exemplars and guides are available, for example, the Construction Industry Board: Code of Practice for the Selection of Main Contractors (CIB, 1997). The CIB Code provides good practice guidelines for the appointment of suppliers and states that, although the precise approach may vary, irrespective of the procurement procedures adopted for the selection of a supplier, the process should be undertaken in the spirit of good practice.

The code considers the key principles of good practice comprise:

- *Clear procedures should be followed that ensure fair and transparent competition in a single round of tendering consisting of one or more stages*
- *The tender process should ensure receipt of compliant, competitive tenders*
- *Tender lists should be compiled systematically from a number of qualified contractors*
- *Tender lists should be as short as possible*
- *Conditions should be the same for all tenderers*
- *Confidentiality should be respected by all parties*
- *Sufficient time should be given for the preparation and evaluation of tenders*
- *Sufficient information should be provided to enable the preparation of tenders*
- *Tenders should be assessed and accepted on quality as well as price*
- *Practices that avoid or discourage collusion should be followed*
- *Tender prices should not change on an unaltered scope of works*
- *Suites of contracts and standard unamended forms of contract from recognised bodies should be used where they are available*
- *There should be a commitment to teamwork from all parties* (CIB, 1997)

If applied consistently the process is objective, unbiased, and precise. However, it is seldom the procurement cure-all the client anticipates.

Pre-qualification

A variety of techniques are available to aid the **pre-qualification** process. According to Smith (1995), pre-qualification, at its fundamental level, comprises the selection of an appropriately sized group of suppliers (from all the potential suppliers) who are deemed capable of satisfying the purchaser's requirement(s), by evaluating them based on a set of predetermined criteria. This requires the purchaser and/or their lead-adviser to establish precisely what the necessary criteria for pre-qualification are. The process, however, has its critics – for example, Drew and Skitmore (1992) are especially derisive of the process asserting that frequently it is merely no more than a '... *crude subjective assessment of bidders' capabilities based on the prequalifier's first or second-hand knowledge of the bidders*'. Further, they suggest the practice is:

> '... naturally rather unreliable and may result in the selection of bidders that are either not interested or not able to provide competitive bids for a contract. There is also the possibility that other ready, willing and able potential bidders may be neglected'.

Therefore if pre-qualification is undertaken, a wide selection of potential suppliers should be carefully evaluated using appropriate techniques.

When selecting a short list, the following issues should be considered:

- The firm's financial standing and record
- Whether the firm has had recent experience of building at the required rate of completion over a comparable contract period
- The firm's general experience, skill and reputation in the area in question
- Whether the technical and management structure of the firm, including the management of subcontractors, is adequate for the type of contract envisaged
- The firm's competence and resources in respect of statutory health and safety requirements
- The firm's approach to quality assurance systems
- Whether the firm will have adequate capacity at the relevant time (NJCC, 1995)

A global construction survey (KPMG, 2007) found that the top ten most important factors used by purchasers to select suppliers are, in rank order: strength of project personnel, quality of reputation, industry expertise, price, delivery on time, delivery on price, balance sheet strength, flexibility, historical customer relationship, and innovation.

Single-stage tenders

Single-stage tendering is where proposals/bids are submitted under a single-stage competition (see for example, NJCC 1996).

Two-stage tenders

The adoption of a **two-stage tendering** process is recommended for projects where the early participation of the contractor is deemed appropriate, for example, to contribute to the development of the solution/design (see NJCC, 1995). The process entails the client's lead consultant conducting individual, comprehensive negotiations with a select number of potential suppliers (generally three or four) on all facets of the project. Following these discussions, price competition can be introduced by means of a schedule of prices or an approximate/notional bill of quantities, with supplementary award criteria applied to select a suitable contractor. It is considered that the majority of clients, where design-build has been selected as the procurement route of choice, would benefit from a two-stage procedure.

Formal presentations

Bidding organisations are frequently required to make formal presentations of their bid proposals to the purchaser and their advisers. Increasingly, decision makers expect the supplier's presentation team to include key members of its proposed implementation team.

Supplier selection[32]

It is suggested that competition promotes economy, efficiency and effectiveness in expenditure (HM Treasury, 2007), while contributing to the competitiveness of suppliers in both domestic and overseas markets. Assisting the competitiveness of suppliers will also assist purchasers to obtain future value for money and security of supply in a competitive market. The UK government's Procurement Policy Guidelines (HM Treasury, 2007) require goods and services to be acquired by competition unless there are compelling reasons to the contrary. They also recommend that whether or not there is any legal requirement for it, advertising proposed contracts could be a useful means of ensuring that the potential of the market is fully tested.

Traditionally, in many market sectors, if suppliers have been pre-selected on their capability to meet the purchasers' procurement objectives, tendered on the same information (detailed design and/or specifications) and submitted a complaint bid, then the bidder with the lowest bid price would usually be awarded the contract: the 'lowest price wins'. The following statement succinctly illustrates this practice:

> '... for every contract there is the optimum bidder who is not only capable of fulfilling the Clients' requirements in terms of time, quality and risk but also in respect of cost is also willing and able to submit a bid lower than any competitor. A fundamental goal of any competitive bidding system is to reveal the identity of this optimum bidder and determine the bid price.' (Drew, 1993)

Although tendering procedure often included clauses such as the following:

> 'Nothing in this Code should be taken to suggest that the employer is obliged to accept the lowest or any tender, although if the procedure advocated in the Code is followed, the successful tenderer will normally be the one offering the lowest price, on a lump sum basis. If it is the employer's intention to allow bids based on alternative criteria, this must be made clear to all potential tenderers at the outset ...' (NJCC, 1995)

Lowest price

Generally, although it is recognised that an element of competition is necessary, the application of the low bid criterion as the sole criterion for discriminating between bidders is believed to be unsound in several ways. For example, a recent review of the UK construction industry, which has a long history of appointing suppliers by competitive bidding, concludes:

'... too many clients are undiscriminating and still equate price with cost, selecting designers and constructors [*suppliers*] almost exclusively on the basis of tendered price. This tendency is widely seen as one of the greatest barriers to improvement. The public sector, because of its need to interpret accountability in a rather narrow sense, is often viewed as a major culprit in this respect.' (Egan, 1998)

'Lowest price' does not guarantee the overall lowest project cost upon project completion, while there is an increased possibility of financial collapse of the supplier, bad performance, delay in completion, time and cost over-runs. Moreover, Hatush and Skitmore (1998) assert that it is a somewhat risky and short-sighted approach, as a supplier may inadvertently submit an unrealistically low, or suicidal, bid – a well-documented phenomenon known as the '*winner's curse*'. Economist William Vickrey has long held that the low bid criterion is not economically efficient and in 1961 he was able to show what has become known as the Vickrey Auction (Vickrey, 1961), in which the low bid criterion is retained for supplier selection, but at the contract price of the second lowest bidder; this is theoretically better, as the price obtained in this way is closer to the market consensus.

Further, this approach is not suitable where suppliers are responsible for elements of design, requiring the evaluation of more than one criterion. At the next level of simplicity, evaluation will be based on two variables – price and quality – while the evaluation of the 'hard issues' (price, delivery, quality and performance, etc.) rapidly becomes difficult, requiring purchasers to develop rating techniques. Principally, therefore, there has been an acceptance that relying on lowest price is not always in the purchasers' best interest and as a result there has been an increasing transition from 'lowest price wins' to 'multicriteria selection' practices in the supplier selection process, the motivation being to attain best value (for money) for the purchaser (Wong *et al.*, 2000).

Best value/value for money

In response to these and other criticisms of competition based solely on price, the Procurement Policy Guidelines (HM Treasury, 2007) stipulate that all public procurement of goods and services, including works, within the UK is to be based on **value for money**, and that the criteria for determining the award of contracts should rarely rely on price alone. Further guidelines assert that:

'... sound practice for the pursuit of value for money will contribute to the competitiveness of suppliers, contractors and service-providers (collectively "suppliers") in both domestic and overseas markets. Assisting the competitiveness of suppliers will also assist purchasers to obtain future value for money and security of supply in a competitive market.' (HM Treasury, 2005)

HM Treasury (2006) define 'value for money' as the optimum combination of whole-life cost and quality (or fitness for purpose) to meet the user's requirement.

In most cases value for money (MEAT principles in EC terms) will involve other factors such as whole-life cost, quality and delivery against price. Procurement Policy Guidelines recommend that appropriate investment appraisal techniques should be used in assessing which compliant bid offers best value for money (see Chapter 5). Also, irrespective of the procurement route adopted, they consider value for money to incorporate an appropriate distribution of risk.

To promote value for money, it is suggested that purchasers should:

- Minimise bidding costs to encourage effective competition
- Remove barriers to participation to encourage the involvement of SMEs and the self-employed, while not victimising larger organisations

Further, the use of purchasing power to pursue alternative goals is incompatible with value for money policy.

Bid evaluation criteria

The purpose of **bid evaluation** is to determine each bid in a manner that permits a meaningful comparison. It is often thought that if the proposal variables can be reduced then like for like comparison will be easier. Therefore, despite the previous comments, bid price is a key factor in determining the award of a contract.

Other project specific factors, however, affect the success of a project, and, therefore, should also be considered (Mahdi *et al.*, 2002).

Usually, an evaluation panel is responsible for evaluating the proposals (bids) and for recommending which organisation, if any, should be offered the contract. The criteria by which they evaluate bids can be categorised as follows:

- **Commercial**: compliance with the ITT, delivery requirements, tender validity period; acceptance of the purchaser's terms and conditions of contract, and payment arrangements; and length of time that product range/design will be on the market
- **Technical**: consideration of: performance and productivity standards; quality (fitness for purpose); inspection requirements; operational and maintenance costs; professional competence; technical/professional support; standardisation; after-sales service; cost and availability of spares and/or consumables; provision of manuals and training; and warranties, etc.
- **Financial**: life-cycle costing comparisons; quantifiable financial benefits; fixed or variable pricing; cost of components, spare parts, consumables and servicing; financial qualifications to fulfil the contract; risk analysis and financial appraisal, etc.

The criteria for evaluating bids should be established before an ITT or quotation request is issued in order to ensure effective and transparent bid evaluation. Purchasers are advised, if they are not to adopt the lowest bid principle, to set out in detail the criteria for evaluation in the tender documentation for the information of bidders[33] (see Box 8.9).

Increasingly, partnering is used to create long-term and durable relationships. For example, the **Smart Acquisition**[34] initiative (previously called the Smart Procurement Initiative, SPI), has transformed how the MOD and industry conduct business. The initiative advocates the pursuit of partnering arrangements in suitable situations. Partnering has had an impact on supplier selection, moving selection criteria from 'low bid' to 'best value', as the evaluation of so-called 'hard issues' will give only a limited indication of the potential for a supplier to be a durable partner in any long-term commercial relationship. Likewise, longer-term partnering arrangements are unlikely to be fully pre-defined and as a result cannot be fully priced at the outset. In such circumstances, price will be less of a determining factor in contract award, requiring other markers to future performance, such as, the commercial and strategic processes within organisations – 'soft issues'. Public accountability requires that the evaluation of these soft issues must be as rigorous and auditable as the more objective assessment of 'hard issues'. A Soft Issues Bid Evaluation Tool (SIBET) has therefore been developed by MOD for this purpose (MOD, 2002).

The implications on the bid preparation process of the transition from 'lowest price' to 'multicriteria' evaluation requires the bidder to signal and substantiate their ability to meet these criteria within their bid documentation.

Multicriteria selection

Studies by Hatush and Skitmore (1998), Wong *et al.* (2000, 2001) – see Box 8.10, Fong and Choi (2000) and Mahdi *et al.* (2002) have identified 24, 37, 68 and 127 criteria involved in supplier selection respectively. The most common criteria considered by purchasers during the bid evaluation stage have been found to include: bid price, financial capability/stability, technical ability (past performance, past experience, and resources), management capability (current workload, current capabilities; past purchaser/supplier relationship and work strategy), health and safety performance of suppliers and reputation (Hatush and Skitmore, 1997, 1998; Fong and Choi, 2000; Mahdi *et al.*, 2002). However, bid price was found to be the only criterion considered by all purchasers in the bid evaluation phase (Hatush and Skitmore, 1997).

Wong *et al.*'s (2000) analysis of the responses of both public and private construction clients revealed much similarity between the two subgroups. Overall, the following were deemed to be important: ability to complete on time, ability to deal with unanticipated problems, maximum resource/financial capacity, actual quality achieved for similar works, quality and quantity of managerial staff, site

Box 8.9 Harmon CFEM Facades (UK) Limited v The Corporate Officer of the House of Commons 2000

This was the first case of any real significance in which the courts adjudicated on a claim by a party whose tender for a public works project was unsuccessful as a result of a public body's failure to apply objective criteria in the tender process. It was also the first case in which an unsuccessful tenderer was awarded damages for breach of the Public Works Contracts Regulations 1991.

The lessons for the public sector and its advisers are summarised as follows:

- If intending to award a public works contract on MEAT principles, an authority must set out detailed criteria for the award of the contract in the OJEC notice (merely stating 'overall value for money' is grossly insufficient)
- According to the Regulations, the authority must put in place internal procedures to ensure that all tenderers are treated fairly and placed on an equal footing. An unsuccessful tenderer may otherwise argue a lack of transparency (the House of Commons project task force was hampered by being unable to agree an approach, selection criteria or an evaluation procedure)
- Under English law the court found there to be an implied contract that, by issuing an ITT, the House of Commons would treat a prospective tenderer fairly and equitably. It was held that, by negotiating with a rival bidder post-tender, the House breached this implied contract. Advisers should be aware that public bodies may not negotiate with one bidder over another after the tender process, but before a preferred partner is chosen

If the authority does not conduct the tender selection process correctly, the unsuccessful tenderer may have a claim for:

- The amount of the profit margin that the tenderer would have earned had it been awarded the contract
- Any tender costs
- Possibly, exemplary damages

In turn, this is the measure of damages that the public authority might seek to recover from its legal advisers if the process were mishandled under their supervision.

Source: Don Stokes, Masons Solicitors. Reproduced by permission of Don Stokes

organisation, rules and policies (health and safety, etc.), training or skill level of craftsmen, comparison of client's estimate with tender price, amount of key personnel for the project and quality and quantity of human resources. In addition, they found an increasing use of project-specific criteria (PSC), while 'lowest price' was not necessarily the purchaser's principal selection criterion. They established a growing realisation that cost has to be tempered with evaluation of PSC in any attempt to identify value for money.

They also established that, while PSC criteria are used in bid evaluation, public sector supplier selection was still dominated by the principle of 'lowest bid price'. However, private clients favoured an evaluation strategy where bid price was equally as important as PSC criteria. Wong *et al.* (2000) put forward a possible explanation for this difference: that the public sector behaviour could be attributed to the need for financial accountability and deficiencies in public procurement systems.

Box 8.10 Project-specific evaluation criteria

Manpower resources
- Quality and quantity of human resources
- Quality and quantity of managerial staff
- Amount of decision-making authority on site
- Amount of key personnel for the project

Plant and equipment resources
- Type of plant and equipment available
- Size of equipment available
- Condition and availability of equipment
- Suitability of the equipment

Project management capabilities
- Number of professional personnel available
- Type of project control and monitoring procedures
- Availability of project management software
- Cost control and reporting systems
- Ability to deal with unanticipated problems

Geographical familiarities
- Contractor's familiarity with weather conditions
- Contractor's familiarity with local labour
- Contractor's familiarity with local suppliers
- Contractor's familiarity with geographic area
- Relationship with local authority

Location of home office
- Home office location relative to job site location
- Communication and transportation: office to job site

Capacity
- Current workload
- Maximum resource/financial capacity
- Finance arrangements

Project execution of the proposed project
- Training or skill level of craftsmen
- Productivity improvement procedures and awareness
- Site organisation, rules and policies (health and safety, etc.)
- Engineering coordination

Other project-specific criteria
- Actual quality achieved for similar works
- Experience with specific type of facility
- Proposed construction method
- Ability to complete on time
- Actual schedule achieved on similar works

Technical-economic analysis
- Comparison of client's estimate with tender price
- Comparison between proposal and average tender prices
- Comparison for client's and proposed direct cost
- Contractor's errors – proposed construction method/procedure
- Proposals review – unit price/labour cost/resources schedule

Source: Compiled from Wong *et al.* (2001)

Further reading

Multicriteria decision support systems

Lowe, DJ and Skitmore, M (2006) Bidding (Chapter 16, p. 369). In: DJ Lowe (ed.) with R Leiringer. *Commercial Management of Projects: Defining the Discipline*. Blackwell Publishing, Oxford

Legal obligations
All clients, but especially those within the public sector, are accountable for making certain that they comply with their legal obligations. The legal framework for public procurement within the UK includes:

- EC and other international requirements
- Specific domestic legislation, such as unfair contract terms, TUPE and competition law
- General contract and commercial law
- Domestic case law

For more information on competition see Chapter 6 and the OGC/Office of Fair Trading 2007 publication *Making Competition Work for You: A Guide for Public Sector Procurers of Construction.*[35]

Comment

Despite moves towards partnering and open-book arrangements, competitive bidding, especially for the appointment of subcontractors, remains prevalent (Fu *et al.*, 2002), and current EU legislation promotes the use of competition. Also, there is, perhaps, an indication within the UK that partnering does not deliver 'best value' – as indicated by the decision by Network Rail to suspend their partnering arrangements and return to traditional competitive bidding.

The objectives of the tendering process and in particular supplier evaluation can be illustrated by the following:

> 'The principal aim of the tendering process is to select the goods and/or services which offer best value for money in performing the outputs required. Therefore, it is not appropriate to accept the lowest price without full evaluation of the total offer. Purchase price is only one consideration when selecting a supplier. As the value and/or complexity of products or services increase, it becomes more important to consider whole of life costs. Moreover, meeting user requirements, quality and service are critical and can be as or more important than price'. (Victoria Treasury and Finance Department, Australia)

The rationale for using an objective bid evaluation method is that purchasers may accomplish most of the objectives, that is, reduce *ex-post* cost and minimise contract failure. The advantage of these alternative criteria, from a purchasers' perspective, is that they attempt to safeguard against the acceptance of unrealistically low bid prices and the resulting claims, disputes and adversarial relationships during the project. It can of course be argued that the use of non-low bid criteria will result in bidders adjusting their prices upwards to try and find the criterion level and that the incentive to develop more efficient methods of production will be lost. The counter-argument is that contrary to first impressions, innovation, technology development and cost reduction will not be discouraged as suppliers will bid at what they believe to be the market price, with any such cost savings made by one supplier still producing significantly higher profit margins in contracts won. Of course, when such savings are available throughout the industry bid prices would be expected to gradually fall and the savings eventually passed on to the purchaser.

The findings show that although competition is still the most common approach to supplier selection, purchasers (particularly those in the construction sector) have been influenced to some extent, either by good practice documentation and/or industry commentators, so that '*bid price*' and '*multi-evaluation criteria*' have equal status when evaluating bid submissions. This transition towards multicriteria selection has implications for the development of supplier's bidding models (discussed later in this chapter).

Exercise 8.4 **Supplier selection**

Compare and contrast the findings of Wong *et al.* (see above) with data available within one specific area of your organisation on the multicriteria selection approaches adopted by its clients.

UK public sector procurement policy

Central government departments are required to comply with a policy framework, which currently state that the government's procurement policy is:

> '... to buy the goods, works and services that it needs under a fair and open procurement process, guarding against corruption and seeking to secure value for public funds with due regard to propriety and regularity. EU law and World Trade Organisation (WTO) agreements underpin these principles.' (HM Treasury, 2007)

The need for probity

Historically, because of the need for public sector clients to be publicly accountable and open to public scrutiny tenders have normally been let under competitive tender with the lowest tenderer being awarded the project. However, as mentioned earlier, Egan (1998) believes that contracts let exclusively on the basis of tendered price to be one of the greatest barriers to improvement. He suggests that the public sector, with its requirement for accountability, is one of the 'prime culprits'.

Compulsory competitive tendering (CCT)

CCT, originally introduced to ensure effective use of public resources, is now no longer seen as the most effective way of procuring services and activities, instead government favouring the adoption of **value for money** (VfM) principles. On a construction project, for example, VfM should be realised over the entire life of the asset, while at the same time incentivising suppliers.

Value for money (VfM)

The Local Government Act passed in 1999 laid down the requirements of **best value**. Since April 2000 all UK local authorities (plus police and fire services, the London Development Agency, park authorities and passenger transport authorities) have been required to conform to the principles of and to obtain best value. Specifically:

> '... a Best Value authority must make arrangements to secure continuous improvement in the way in which its functions are exercised, having regard to a combination of economy, efficiency and effectiveness'.

In order to deliver this, the government anticipated the development of partnerships with the private sector, providing access to high-quality, cost-effective services from external providers. An essential element of VfM is the regular monitoring of performance via a cyclical review process. Performance indicators have been established to support this.

The primary consideration of public clients, therefore, when selecting an appropriate procurement strategy, is the need to obtain VfM over the whole life of the service/facility. This is facilitated by the early involvement of all the project stakeholders: those involved in the use, design, production (construction), operation and maintenance of an asset or service project. Guidelines suggest that VfM is best delivered via the following procurement options:

- PPP
- Design and construct (and where appropriate maintain and operate)
- Prime contracting
- Framework agreements

It is considered that the traditional procurement routes restrict the potential to eradicate inefficient activities and achieve VfM. Their use, therefore, should be restricted to when it can be clearly established, in terms of whole-life costs and overall performance, that they will deliver better VfM than the procurement options listed above (GCCP, 1999).

European and UK procurement regulations

As described in Chapter 6, procurement procedures for the award of public works, supply and service contracts are regulated by the Public Contracts Directive 2004/18/EC within the European Union (EU).[36] This legislation is implemented within the UK under the Public Contracts Regulations 2006,[37] which regulates the tendering procedures for building and civil engineering works contracts offered by public authorities – for example, central government, local and regional authorities, health authorities and similar bodies. Works in the utilities sector (water, energy, transport and postal services sector) are covered by the Utilities Contracts Regulations 2006,[38] which implements the EU legislation: Utility Contracts Directive 2004/17/EC.[39] Only contracts for works exceeding a threshold value of €5,000,000 (approximately £4,348,350) are covered by these regulations, while certain contracts are excluded – for example, works that are declared to be secret by a Member State, works covered by special security measures, and contracts entered into in accordance with an international agreement.

The regulations state the procedure for the award of the contract (open, restricted, negotiated or competitive dialogue) and specify the awarding criteria.

Qualifying projects

Not all procurement is required to undergo this process. There is a minimum value above which the procedures apply and, therefore, projects below this figure are exempt. See Box 6.8 for the current position.

Procurement procedures

Under the UK Public Contracts Regulations the open and restricted procurement procedures (described earlier) are the preferred approaches, although in specific circumstances purchasers may use the competitive dialogue procedure and in exceptional circumstances the negotiated procedure where the purchaser (contracting authorities) can justify its use. Generally, purchasers must publicise their intention to obtain tenders for a public contract; the purchaser is then restricted to inviting tenders from those suppliers (**economic operators**) who respond to their advertisement. Further, under the restricted procedures the minimum number of bidders is five, while for the publicised negotiated procedure and the competitive dialogue procedure the minimum number is three.

Contract notices

As mentioned above, all compliant public contracts have to be publicised in the *Official Journal* and in *Tenders Electronic Daily*, its electronic counterpart.[40] The directives stipulate the precise format and content of the advertisement and specify time limits for the invitations to tender stage. Further, purchasers must provide appropriate notice, issue periodic notices on product and work groups and describe any qualification methodology to be employed. Additionally, *Tenders Electronic Daily* and the *Official Journal* publishes preliminary information, invitation to tender announcements, contract award results and advance notices of larger supply purchases.

Time limits

The directives provide minimum time limits for the various stages of the procurement process – for example, publicising the bid and bid preparation, which differ depending on the type of tendering method adopted. However, when establishing limits, purchasers are required to allow for the complexity of the contract and the time required to compile a tender. The minimum time limit for the receipt of tenders is 52 days under the open procedures, 40 days under the restricted procedures and 37 days under the negotiated procedures and competitive dialogue. However, in the case of the open and restricted procedures where the purchaser has published a prior information notice, these limits may be reduced to 37 days, the absolute minimum being 22 days. Also, where notices are drawn up and transmitted electronically these time limits may be reduced by seven days, plus if the purchaser provides unrestricted and full electronic access to all contract documentation the limits may be reduced by five days. Further, where the tenders are deemed to be urgent the minimum time limit for the receipt of tenders is 15 days or 10 days if the notice was sent electronically under the negotiated procedures and 10 days in the case of the restricted procedures, again commencing from the date of the ITT/receipt of requests to participate.

Contract award criteria

The EU directive requires purchasers to treat suppliers *equally and non-discriminatorily and act in a transparent way* and determine the award of a tender based on either:

- Lowest price
- MEAT principles

Where the MEAT principles are adopted, the purchaser must publish details of the specific award criteria to be used, together with their relative weightings (where practical). Such criteria may include: quality, price, technical merit, aesthetic and functional characteristics, environmental characteristics, running costs, cost-effectiveness, after-sales service, technical assistance, delivery date and delivery period and period of completion.

The directives also provide mechanisms for responding to abnormally low tenders.

Other items

Purchasers are required to notify prospective suppliers of their decisions in writing within 15 days, when requested to by a supplier.

The directives also cover the establishment of framework agreements, and the use of dynamic purchasing systems, electronic auctions, and design contests.

As a result of the decision of the European Court of Justice in the Alcatel case, the new UK regulations have introduced a mandatory 10 day standstill period between the announcement of the award decision and the contract being concluded. The purpose of the new provision is to allow time for any unsuccessful bidder to consider their options and, if appropriate, institute a legal challenge before a contract is signed.

Pre-award negotiation

After instigating an appropriate process and receiving the suppliers' proposals (offers, bids or tenders), the purchaser will evaluate the submissions and either select a 'winning' proposal, preferred bidder or decide to abort the project.

Once a **preferred bidder** has been selected, the parties generally enter into **pre-award negotiations**, with the aim of agreeing the precise terms of the exchange contract, before formally awarding the contract. This may involve reaching agreement over specific amendments to the standard form of contract stipulated by the purchaser in the ITT, or alternatively, it can entail a protracted 'contest' to impose a company's precedent document on the business deal. These precedent documents are generally biased towards the originating party and the outcome of the negotiations is closely linked to the relative power of the trading parties. The term '**battle of the forms**' refers to the legal resolution of a dispute – for example, where the parties acknowledge a contract has been entered into but fail to agree on the precise contract terms that govern the exchange.

In the case of complex projects, pre-award negotiation can take several months if not years to conclude. For example, a NAO study (NAO, 2007) found the pre-award negotiation phase on UK PFI projects took, on average, over a year and occasionally as long as five years. Further, it established that significant changes (both positive and negative) were frequently made to supplier's prices during this period. It is generally recommended that once a preferred bidder has been selected, reductions to the contract price should only be made when accompanied by corresponding reductions to the scope of the project.

The most commonly negotiated contract terms between large global organisations, based on data collected by IACCM (2011), are presented in Table 8.3 These can be contrasted with the top 20 terms considered to be more productive in supporting successful relationships during the implementation phase of contracts.

Negotiation best-practice guidance and training is endemic. This text does not attempt to address the topic in any depth, but merely to highlight some basic principles pertinent to both transaction and dispute negotiation.

Table 8.3 IACCM top 20 negotiated terms 2007–2011.

	The terms that are negotiated with greatest frequency		Terms that would be more productive in supporting successful relationships
1	Limitation of liability	1	Change management
2	Indemnification	2	Scope and goals
3	Price/charge/price changes	3	Responsibilities of the parties
4	Intellectual property	4	Communications and reporting
5	Payment	5	Performance/guarantees/undertakings
6	Liquidated damages	6	Limitation of liability
7	Performance/guarantees/undertakings	7	Delivery/acceptance
8	Delivery/acceptance	8	Dispute resolution
8=	Applicable law/jurisdiction	9	Service levels and warranties
10	Confidential information/non-disclosure	10	Price/charge/price changes
10=	Service levels and warranties	11	Audits/benchmarking
12	Warranty	12	Indemnification
13	Insurance	13	Intellectual property
14	Service withdrawal or termination	14	Payment
15	Data protection/security	15	Information access and management
16	Scope and goals	16	Business continuity/disaster recovery
17	Responsibilities of the parties	17	Applicable law/jurisdiction
18	Change management	18	Confidential information/non-disclosure
19	Invoices/late payment	19	Warranty
20	Audits/benchmarking	20	Assignment/transfer

Source: Adapted and derived from the IACCM 2011 survey of the Top Terms in Negotiation (IACCM, 2011)

It is generally held that there are two theoretical and strategic approaches to negotiation:

1. **Positional negotiation**: this is associated with a competitive negotiation stance that results in compromise. The approach centres on the positional power of the parties, where the dominant party seeks to intimidate the other side into accepting their demands, and is based on the premise that only one party can 'win'. It is frequently referred to as the 'win/lose' approach. Essentially, the approach entails a negotiator adopting a 'position' then attempting to reach an agreement as close to this position as possible. Positional negotiation is characterised by aggressive, predictable behaviour, and concentrates on the adoption of extreme opening gambits and the subsequent justification of perceived 'rights'
2. **Principled negotiation**: this is derived from the Harvard Negotiation Project (Fisher *et al.*, 2003), and is associated with an interest-based, cooperative, collaborative and problem-solving approach to negotiation. Fisher *et al.* (2003) consider the key components of principled negotiation to comprise: separating the parties (individuals) from the problem; focusing on the parties interests and not their positions; inventing options to deliver mutual gain; and selecting a solution, using objective criteria, from amongst these options. They suggest tactics to address the problems that arise when a party '*refuses to play*', uses '*dirty tricks*' or where one party is significantly more powerful than the other. Features of the approach include: focusing on the '*needs, concerns and goals*' of each party; and the development of a '*best alternative to a negotiated agreement*' (BATNA). Other buzzwords include: WATNA (Worst Alternative To a Negotiated Agreement) and PATNA (Probable Alternative To a Negotiated Agreement)[41]

Successful negotiation, best-practice guidance[42] suggests, is based on exhaustive preparation; positioning an appropriate opening offer (or waiting for the other party to make the first move); a clear understanding of the stages through which a negotiation will go through; and an awareness of negotiating behaviours, tactics, strategies and ethics. Critical is the ability to facilitate the bridging of the 'last gap', alternatively termed the 'last dance', to enable the parties to reach an agreement.

Effective negotiation requires participants to have an eclectic mix of negotiating skills and tactics derived from both a positional and principled perspective. A **contingency approach** is generally required based on a perception of what is deemed appropriate to counter and complement the approach adopted by a particular '*negotiating partner*'. Ultimately, however, the relative **power** of the parties is a significant factor in negotiation.

Investment decision stage gate review

Prior to formally entering into an exchange contract, a stage gate review to inform and support the investment decision is recommended. The OGC Gateway™ Process Review 3: Investment decision best practice guidance (OGC, 2007 g) provides sample questions relating to the assessment of the proposed solution, business case and stakeholders, risk management, review of the current phase and readiness for service (its readiness for the next phase). These include, for example:

- *Have the suppliers or partners proposed any alternatives or other options in addition to a fully compliant bid?*
- *Has the proposed solution affected the expectations of business benefits?*
- *Are there resources available for the business to fulfil its obligations within the contract/agreement?*
- *Does the commercial arrangement represent value for money, with an appropriate level of quality over the whole life of the project?*
- *Have all major risks that arose during this phase been resolved?*
- *Does the contract reflect standard terms and conditions and (where applicable) the appropriate allocation of risks between the contracting parties?*
- *Is the project under control?*
- *Have all the required organisational procurement and technical checks been carried out?*
- *Is the working relationship likely to succeed?*
- *Are the supplier's project, risk and management plans adequate and realistic?*
- *Are the service management plan, administration and service level arrangements complete?* (OGC, 2007 g)

During this phase commercial practitioners are involved in preparing tender documentation (including drafting complex bespoke contracts and agreements or selecting standard terms and conditions, writing technical specifications for the required asset, product or service, and in some cases compiling a bill of quantities[43]; they may also implement the selected procurement strategy, issuing ITT and subsequently distributing tender documents to interested suppliers. They are frequently responsible for evaluating the tender submissions, for reporting to the purchaser/buying centre on the relative merits and concerns associated with each bid (for example, assessing costs and risks associated with each bid), and advising the purchaser/buying centre on and/or participating in the supplier selection decision. Additionally, the commercial function is generally responsible for negotiating with the preferred supplier(s), particularly in respect of any proposed changes to the preferred contract terms and conditions, and obtaining legal and regulatory sign-off prior to formally awarding the contract.

These tasks will require the application of the risk management tools and financial decision-making techniques described in Chapters 4 and 5, and the procurement, risk and contract management best practice described in Chapter 7. It will also require a detailed understanding of the legal and contractual issues described in Chapter 6.

Activity 8.6 Asset/service procurement

Evaluate the procurement procedures adopted on the Millennium, Emirates and Wembley stadia projects (case study A) and T5 (case study B). Comment on the conformity of these procedures with current perceived 'best practice' (described elsewhere in this text) and the motivation of the procuring party where actual practice deviates from this.

Proposition development[44]

This phase of the bid cycle entails the supplier generating and submitting a proposal for the supply of an asset or service in response to a client's request for proposal, quotation or tender (RFP/RFQ/RFT). During this stage the supplier will scrutinise the tender documents, assess any potential resource requirements and inherent risk and opportunity, and re-visit the bid/no-bid screening exercise. Assuming the opportunity is still attractive to the supplier, the purpose of the proposition development stage is to produce a compliant proposal, based on the RFP/RFQ/RFT and any additional information obtained from the customer and other project stakeholders. The overall objective is to generate a persuasive proposition that convinces the purchaser/buying centre to award the contract to the supplier. This generally involves undertaking additional design work, estimating the costs involved in carrying out the work or supplying the asset or service, preparing/publishing the proposal documentation, and, if required, preparing a presentation to support the bid/proposal.

As identified in Chapter 1, bid management is an integral part of the commercial role. Commercial practitioners lead or participate in multifunctional bid teams along with customer relationship managers, proposal/bid managers, representatives from finance, design/technical authority and project/implementation managers, plus, where necessary, support from business managers, taxation/treasury, procurement and legal functions.

The activities undertaken during the proposition development stage include:

- Preparing a proposal development plan
- Holding a kick-off meeting involving all the proposition development participants
- Gathering further information on the purchaser, their agents, any potential subcontractors and the likely competition
- Revising the initial 'win plan' themes and strategies
- Developing the proposition in terms of its implementation, financial, contractual, and performance proposals. Both formal and informal reviews of the proposed solution should be undertaken, for example, instigating a value management exercise (often referred to as 'red teaming'). The draft proposal should be subsequently revised to incorporate any feedback obtained
- Discussing the proposed solution, if possible, with the purchaser and other stakeholders
- Drafting the proposal document
- Staging a post-publication review meeting to capture lessons learnt[45]

An early decision concerns the proportion of the project that will be subcontracted. This is a further example of the 'make or buy' decision discussed earlier in relation to the purchaser. The construction sector, for example, particularly in the UK, is highly fragmented with main contractors predominantly providing project and contract management services and subcontracting the 'production' aspects of projects to specialist contractors. Similarly, other project-oriented sectors rely on the contribution of third-party suppliers and partners. A further allied decision concerns the assessment of the organisation's capability and capacity to manage the project/undertake the contract and whether or not the creation of a joint venture (JV) or teaming agreement (TA) with one or more partners, or a consortium with a group of companies to carry out the project/supply the asset or service, would be beneficial. Rigby *et al.* (2009) define **construction consortia** as follows:

> '... where a group of supply interests come to an agreement on the joint development and marketing of their services. The aim of the consortium is therefore to enhance the overall market competitiveness of its member firms.'

A decision to form a JV or consortium, or enter into a teaming agreement to deliver the project will involve commercial practitioners in vetting appropriate partners and drafting, negotiating and agreeing agreements to establish the JV, TA or consortium.

Activities undertaken during the proposition development stage of major outsourcing projects are not too dissimilar from those undertaken for a major infrastructure project, particularly ones that adopt a

PFI delivery approach. One aspect of divergence, however, is the need to undertake due diligence in respect of the transfer of assets from the purchaser (employer) to the supplier, in particular any warranties provided by the employer in respect of these assets, and the implications arising from the transfer of stafffrom the employer to the supplier in respect of the TUPE regulations (discussed Chapter 6).

The following section focuses on bid price determination and the bid submission/adjudication stage gate review. Contractual issues have been addressed in Chapter 6.

Bid price determination

The success or failure of a bid is significantly influenced by the formulation of the bid price. Competitive bidding, however, unlike traditional sales, does not allow an organisation to market test the price it sets for its products and services.

Pricing policies and systems

The strategy and tactics employed by an organisation to implement its pricing policy should reflect the level of profitability it is trying to achieve. As previously stated, mark-up, in addition to profit, also includes additions for running costs or overheads. Further, the costs for all unsuccessful bids have to be recouped from those that are successful.

Generally, pricing policies contain the following components:

- **Business policy**: for most large organisations, policy is predetermined and documented; examples include lowest price and best value, high-quality price leader, or a mid-range policy
- **Business objectives**: the strategic objectives and mission statement of the organisation
- **Pricing strategy**: a set of bid-specific objectives developed by the bid team

Determine bid price strategy

The bid price should be prepared and structured to benefit from the bidder's strengths and weaknesses, accounting for the competition and the purchaser's assessment criteria. Bid-price strategy should consider: **price objectives**, derived from the outcome of previous analysis; an appropriate balance between profitability and the offer price; and the **winning price**. Determining the winning bid price is a challenge – often referred to as the bidder's dilemma – too high a profit margin and the bid will be unsuccessful, too low and despite winning the bid the project could easily return a reduced profit or no profit at all. The process is linked to the purchaser's criteria for awarding the contract discussed earlier. However, the prime factor in deciding upon a price strategy is the consideration of the winning price. Strategies include submitting the lowest price, profit maximisation, submitting a loss-leader bid, gaining market entry, or responding to competitor, purchaser and market factors.

In the construction sector, Drew and Skitmore (1990) established two very successful methods of acquiring contracts:

1. Consistently bidding very competitively for specific types of work, and thereby having a comparatively low bidding variability relative to other bidders
2. Being inconsistently competitive and having a comparatively high bidding variability relative to other bidders

Further, they established a significant correlation between competitiveness and bidding variability: less competitive bidders were more variable in bidding (Drew and Skitmore, 1992), while differences in supplier competitiveness are greater for different contract sizes than for different contract types. The most competitive bidders appear to be those with a preferred contract size range (Drew and Skitmore, 1997).

Pricing systems

Within the context of construction, building and construction work is predominantly subcontracted, with the main supplier primarily providing management services. In terms of pricing, the contractor solicits, selects and compiles, with judgement, quotations from subcontractors into a single bid. Projects are undertaken by temporary coalitions of firms with successful completion of the project balanced against profit and their long-term interest in survival and growth (Winch, 2001, 2006). They often form a loosely organised set of subcontractors who work from time to time for a main contractor – a relationship that tends to be essentially long-term and rarely based on price competition. More often than not the subcontractors do not have to bid to win the work, although main contractors often 'test the market' every few years by holding a tender competition between subcontractors (Eccles, 1981).

This suggests that main suppliers (contractors) within the construction sector belong to one of the service industries, and service industries are not known for their sophistication in pricing (Hoffman *et al.*, 2002).

The following pricing systems exist:

- **Cost-related systems**: including standard-cost pricing (covers standard variable cost and fixed cost per unit, plus profit, adjusted on the basis of competitor pricing); cost-plus-profit (standard mark-up applied to total cost of each product); break-even analysis or target profit pricing (determines the price that will yield the required profit); and marginal pricing (refers to the marginal cost of manufacturing each unit)
- **Market-related systems**: including perceived value pricing (pricing based on assumptions of purchaser beliefs of 'value'); psychological pricing (price used as a tool to condition purchaser beliefs, for example, quality or value); promotional pricing (discounts offered to generate high turnover); and skimming (a high price is bid to 'skim the cream' off the market)
- **Competitor-related systems**: including competitive pricing (tackling the price leader in a particular segment); discount pricing (set artificially high prices and offer discounts to attract purchasers); and penetration pricing (significantly undercutting competitors' prices to generate turnover)

The pricing strategies of service industries have been classified as either cost based or market-oriented (Gabor, 1977), with full-cost pricing, in practice, being the most popular pricing policy. In general, 'real-world' pricing practices in service industries essentially differ in the emphasis placed on production costs and prevailing prices – which mirrors most manufacturing organisational structures with separate, and often conflicting, production and marketing departments – with '*... the ideal pricing policy being simultaneously profit based, cost conscious, market-oriented and in conformity with any other the aims the businessman may have*' (Gabor, 1977, p. 43). Clearly, though, the amount of available knowledge concerning costs and prices depends on the products involved, with services costs generally being harder to calculate than those incurred in the production of commodities.

Mochtar and Arditi (2001) confirm the widely held belief that pricing strategy in construction is predominantly cost based. However, Skitmore *et al.* (2006) have examined the tenability of two mutually exclusive accounts of construction bid pricing:

1. A full-cost pricing policy is used
2. Classical microeconomic theory holds

They demonstrate that the nature of construction firms' marketing activity and price movements in general make (1) highly unlikely. Likewise, Runeson (1996, 2000) has also shown that the predominant form of construction contract pricing is unlikely to be an absorption, or full-cost, pricing policy – once again offering support for the tenability of the neoclassical position. An alternative approach has been suggested in a recent 'comment' by Weverbergh (2002), advocating 'the economic theory of auctions' to underpin empirical tests of statistical bidding models with construction contract bidding data.

However, what is apparent is that:

> 'Businesses that use price as a strategic tool will profit more than those who simply let costs or the market determine their pricing.' (Kotler, 2000, p.459)

Bidding and pricing tactics

Suppliers may adopt various bidding and pricing tactics. These include:

- **Tactical withdrawal of bid**: however, bidders need to consider the implications and impact that this action may have upon existing and future relations with the purchaser
- **Seeking an extension to the submission date**: although it is better to request an extension rather than risk missing a submission date or submitting a poor or incomplete bid, Tweedley (1995) cautions against this tactic unless there are compelling reasons to do so
- **Submitting a non-compliant offer**: where cost saving for both the supplier and the purchaser can be shown, for example, through value engineering, a non-compliant (or innovative) offer may give the bidder a competitive advantage

Likewise, they may implement pricing tactics. These include:

- **Overpricing**: where suppliers do not want the work on offer, but are concerned that a refusal to submit a tender will lead to exclusion from future tender opportunities, they may submit an inflated price. A variant of this, referred to as cover pricing (a form of 'collusion'), is where the supplier obtains an artificially high price from a competitor. Found to be illegal and in breach of the 1988 Competition Act by the Office of Fair Trading (OFT) and the Competition Appeal Tribunal, the practice was widespread in the UK construction sector prior to 2004. It can lead to a reduction in the number of available bidders, an increase in the average bid price and a reduction in bid variance, potentially resulting in artificially increased prices (see Box 8.11). Zarkada-Fraser and Skitmore (2000) consider that the decision to participate in some form of collusion is primarily focused on the individual. Their investigation established that there is a minority of decision makers that admit they would consider participating in some form of collusive tendering agreement under certain circumstances
- **Unbalanced allocation of profit**: where suppliers are required to submit a detailed breakdown of their bid the bid team has a variety of methods for distributing the profit element within the bid

Box 8.11 OFT investigation into illegal 'bid-rigging'

In September 2009, following one of the UK's largest Competition Act investigations, the OFT imposed fines totalling £129.2 million on 103 English construction companies; the OFT concluding that the firms had participated in 'illegal anti-competitive bid-rigging activities' on 199 tenders between 2000 and 2006. This activity predominantly took the form of 'cover pricing' with 86 companies admitting to obtaining cover prices. The breaches included both public and private sector projects valued at more than £200 million; for example, in 11 instances the lowest bidder faced no real competition. Moreover, the OFT claimed to have evidence that over 4000 tenders were affected.

Under the Competition Act 1998 and Article 81 of the EC Treaty firms discovered to have been involved in bid rigging can be fined up to 10% of their worldwide turnover.

However, in 2011 the Competition Appeal Tribunal (CAT) determined in 13 instances the fines imposed by the OFT to have been disproportionate and excessive. Consequently, it reduced the fines covered in the judgments made to date by over 80%, from around £64.3 million to approximately £10.6 million. The tribunal found that the OFT was mistaken in deeming cover pricing to be as serious an infringement as bid rigging and had failed to adequately take into account mitigating factors relating to the practice.

Source: Office of Fair Trading (2009); Adkins and Beighton (2011)

- **Tactical pricing**: where payment is staged there is an opportunity for manipulation of the bid by the supplier (King and Mercer, 1988). Examples include front-end loading, item spotting, backend loading, and maximising the net present worth of the cash flow. Although there is evidence of the use of tactical pricing or unbalanced bidding, it would seem that front loading and item spotting are by far the most common in market sectors such as construction (Green, 1989). This perhaps is not the case in service industries

There is limited research in this area both in investigating practice and model development to systematise the process. However, it is apparent that the techniques used in practice are both subjective and opportunistic.

Price formulation

After deciding to submit a bid, there usually follows a two-stage price formulation process consisting of a baseline estimate and mark-up. Formulated at the operational level, the baseline estimate is forwarded to the business strategy level, where senior management decides the appropriate bid level; the baseline estimate is usually combined with a mark-up to form the bid. Different bidders utilise distinct mark-up policies. Bidding strategy involves determining the mark-up level to a value that is likely to provide the bestpay-off. Standard bidding models, according to Male (1991), assume that bidders endeavour to maximise their expected profit. The bidder, however, may be attempting to fulfil other objectives, including minimising expected losses, minimising profits of competitors, or obtaining a contract in order to maintain production (Drew and Skitmore, 1997).

Cost estimating

All bidders, irrespective of the pricing system they adopt, need to know the likely cost components of a bid, if only as an indication of their bottom line. Cost estimating, therefore, plays an important part in pricing a bid and is critical to the overall success of the resulting project.

Akintoye and Fitzgerald (2000) found the most commonly used cost-estimating techniques within the UKconstruction section to be, in order of preference: standard estimating procedures (the standard text book approach of establishing the costs of construction – labour, material, plant, subcontractors – and project management to which allowances for overheads, contingencies, inflation – if appropriate – and profit margin are added, sometimes referred to as bottom-up estimating); comparison with similar past projects based on personal experience; comparison with similar past projects based on documented facts; establish standards; and intuition.[46] The estimating activity also involves obtaining and analysing quotations from subcontractors and suppliers, and accounting for the ramifications of global sourcing and the delivery of international projects.

Fine (1987) likens traditional cost-estimating techniques to compiling a grocery list. However, he considers large complex projects to be so dissimilar from grocery lists that '*... we may even forget the list content with equanimity*'.[47]

Tender stage cost-estimating accuracy is variable across industry sectors, although average accuracy levels are thought to be in the range of ±2 to 5% (Turner and Remenyi, 1995). Similarly, estimators within the construction sector (supply-side) consider their performance to be approximately ±5%; however, research indicates that this is overly optimistic. Exact details are difficult to obtain because of the confidential nature of the information and as the implementation team have the opportunity to intervene to overcome potential losses. Estimators within the UK construction sector attributed this inaccuracy, again in order of perceived influence, to: insufficient time for estimating, poor tender documents, insufficient tender document analysis, lack of understanding of project requirements, poor communication between project team, and low participation in estimating by site (implementation) team (Akintoye and Fitzgerald, 2000).

Skitmore and Lowe (1995) suggest that the quality of estimates can be improved by:

- **Selecting the most appropriate estimating technique**: several studies have found estimators to be reluctant to adopt new techniques, preferring to select estimating technique based on their familiarity with the method

- **Collating reliable cost information**: to ensure that decisions are made using up-to-date data from a sufficiently large enough data sample
- **Obtaining sufficient design information**: the bid team should ensure that appropriate levels of information are provided by in-house designers, etc. and by the purchaser/purchaser's advisers. Where more information is required this should be requested from the client
- **Ensuring effective feedback is collected and used**
- **Developing the estimator**: this invariably requires the estimator to learn from experience

However, in contrast with the usual assumptions of auction theory, in construction contract auctions bidders have both different costs and imperfect estimates of them.

Setting the bid price

Bidding performance is a reflection of the strategic process (Drew and Skitmore, 1997), and is concerned with the competitive relationships between the bids submitted to the purchaser. Because a bid is an estimate of the unknown market price, most bidders submitting a genuine bid are attempting to submit a bid that is low enough to win the contract but high enough to make a profit (Park and Chapin, 1992).

> 'Essentially, the bid price problem is one of optimising the price of the bidder's proposal against the requirements of the client and the performance of competitors' bids in order to maximise the bidder's profits while ensuring that it maintains a reasonable chance of winning the bid.' (Bussey *et al.*, 1997)

At the time of submitting the bid, the maximum level of competitiveness can be taken to be the lowest bid; the bid price will become the optimum bid. The optimum bid has been defined as:

> '... the lowest priced evaluated bid which has undergone a process of assessment to identify and, where necessary, to price the consequences inherent in the submission.' (Merna and Smith, 1990)

As previously mentioned, suppliers will often base their bid price on an estimate of the cost of the inputs to the process. The procedure of transforming the estimate into a bid is a management activity sometimes referred to as tender adjudication, and is essentially about submitting a winning bid at the best possible price.

> 'Tender adjudication is concerned with getting the job at the best price; securing it in competition by the smallest possible margin; and on the best commercial terms procurable ....' (Tassie, 1980)

To arrive at the bid price, a mark-up (additions for overheads, project financing costs, required profit and risk margin) is added to the net cost estimate. The technical authority will provide the bid team with an analysis of the net estimate, indicating its key components, together with clarification on how risks identified during the estimating process have been resolved. Generally, an organisation's annual business plan will specify its overheads budget and baseline profit requirement. The only item, therefore, left for consideration is the risk margin: the supplier's assessment of the special risks and/or the commercial attractiveness presented by the project under review. This decision will be made taking into consideration any knowledge of the purchaser's available budget for the project, approaches likely to be adopted by the competition, and an appraisal of the potential (realistic) future revenues and margins obtainable through contract variations.

Mark-up estimation, however, is a nebulous decision problem rendering analysis and formulation of a satisfactory solution mechanism problematic. Identification of all the related factors that inform rational decision making is both protracted and complex – likewise the analysis of their individual influence and the measurement of their collective bearing on the decision. As a result, conventional practice is to form bid decisions on the basis of intuition, derived from a mixture of gut feeling, experience and guesses. According to Moselhi *et al.* (1993), this implies the application of some degree of pattern recognition rather than computation or deep reasoning about the components of the problem. Likewise, Fayek (1998) asserts that the margin-size decision process predominantly involves making qualitative and subjective assessments and that, despite the problems involved, a need exists to structure and formalise the process. Many different theoretical approaches to competitive bidding have been proposed and tested with varying results.

Factors influencing the bid price decision

Flanagan and Norman (1982) identified five key factors as influencing contractors' bidding behaviour: market conditions, current and projected workload, the size and complexity of the work, client type and project size. Likewise, research within the construction industry has sought to determine the order of importance of the factors considered by suppliers when setting the bid price. Surveys by Ahmed and Minkarah (1988), Shash and Abdul-Hadi (1992), Dulaimi and Shan (2002), Eastham (1987) and Shash (1993) identified and ranked 31, 37, 40, 52 and 55 factors respectively. Again, taken collectively, these studies have much in common. They reveal that the top 15 factors that influence the bid price decision are: risk involved in the investment; degree of difficulty; size of contract/project size; need for work; uncertainty in (reliability of) cost estimate; current work load; past profit in similar projects; owner (private/public); type of contract; type of job/project type; degree of safety/hazard; anticipated rate of return on project; overall economy (availability of other projects/work); risk involved owing to the nature of the work; and project cash flow. Further, they suggest that bidders are predominantly influenced by project characteristics and company-related issues, moderately so by the economic situation and project documentation, and to a lesser extent by the bidding situation.

The supplier's need for work and risk have been shown to significantly affect bid mark-up (De Neufville and King, 1991), while bidding decision making has been found to be influenced by the results of both previous bidding attempts and construction experience derived from undertaking projects (Fu *et al.*, 2002, 2003). Fu *et al.* found that suppliers that bid more frequently are more competitive than those that bid occasionally. Further, they established the existence of 'experienced market players' who display the traits of a more competitive and consistent bidding performance.

As with d2b, the bid mark-up decision is again a dynamic one where the factors considered vary over time, from project to project and from organisation to organisation.

Risk and uncertainty

In principle, Chapman *et al.* (2000) assert that risk analysis should be incorporated into the process of bid preparation to assess uncertainty inherent in the obligations required by the contract, and also to assist in the formulation of bids that provide a suitable balance between the risk of not being awarded the contract and the risk associated with potential profit and losses if the contract is acquired.

Risk and uncertainty relating to the bid process can be categorised under the following headings:

- **Factors relating to the influence of the decision makers**: as risk assessment decisions are predominantly made intuitively and subjectively, the decision-making process itself gives rise to additional risks and uncertainties
- **Factors relating to the bidding process**: according to Smith (1995), these may be divided into three distinct groups:
  - risks relating to the project itself – for example, procurement system and intended contractual arrangements, adequacy of the tender documentation, planned project time scale and the implications of delayed or non-completion, fixed or fluctuating (variable) price, level of prime cost sums, and the degree of technological difficulty
  - risks relating to the purchaser and the professional team – for example, uncertainty over whether or not it will be paid on time, underpayment of payments on account, underpayment for variations to the contract, delay in agreeing the final account, and purchaser's advisers interpretation of the specification
  - risks relating to the bidding process – for example, the selection of appropriate work rates and price stability

As cited in several influential contributions in the construction literature (for example, Raftery, 1991; Hillebrandt, 2000; Runeson, 2000), the extent of the uncertainties involved in forecasting future costs, as well as the behaviour of competitors and the market in general, requires contract bidders to devote a far greater amount of energy and resources to marketing than is currently admitted in the economic theory of auctions.

Bidding models

Over the past 50 years extensive research has been undertaken into the development of analytical bidding models, with the aim of improving a bidder's likelihood of submitting the 'optimum' bid. These mathematical models can be dated from Friedman's innovative work (Friedman, 1956) and include statistical bidding models, approaches based on historical data, multiple regression analysis and artificial neural network techniques. Other researchers have taken a less mathematical approach, attempting to analyse the human decision-making process as it applies to forecasting. However, most of the bidding literature is concerned with setting a mark-up, m, so that the probability, $Pr(m)$, of entering the winning bid reaches some desired level. Several models have been proposed for calculating $Pr(m)$ (for example, Friedman, 1956; Gates, 1967; Carr, 1982; Skitmore and Pemberton, 1994).

The problems involved in monitoring the performance of competitors

All organisations monitor, to a degree, the bidding performance of their competitors. The use of these bidding patterns to model and predict bidding performance requires the acceptance of a series of interrelated assumptions, which influence the reliability of the results obtained and can produce the following problems:

- **Availability of information**: there is an assumption that there is an adequate supply of information on competitors' bids. However, depending upon the market sector this may be erroneous – for example, in some sectors, relatively few submitted bids are published and the bidder may only able to allocate the 'winning bid' to the successful bidder. In others, even this information remains confidential. Obtaining information required by quantitative models can, therefore, be problematic, particularly as Hillebrandt (1974) recommends that data used should not span a period of more than 3 months. Alternatively, in other sectors, there may be a large number of potential bidders in which case the collection and assessment of data are both time consuming and expensive to undertake
- **Continuity of bidding behaviour**: the use of historic data assumes that competitors will continue to bid as in the past: this is held to be unrealistic (Chapman *et al.*, 2000)
- **Conformity to statistical independence**: predominantly, modelling approaches require that for each competition every rival bid is statistically independent
- **Homogeneity of underlying costs**: for organisations to be able to compare their submitted bid (or underlying estimate) with the bids of others, or to treat all rival bidders as one 'average' competitor, necessitates the assumption that there is no significant difference between the competitors' cost estimates. Unfortunately, this is not accurate due to the variability of each supplier's performance

Taken collectively, these assumptions imply that competitors do not discriminate between contracts, rather they randomly select values from a constant bidding pattern, oblivious to any deviation in attractiveness of contracts. Also, increasingly, as indicated earlier in this chapter, tenders are not always expressed or assessed in terms of the 'lowest' price; in these circumstances where alternative award criteria are applied it can be difficult to obtain information concerning the purchaser's subjective decision making. Further, these approaches do not attempt to predict the outcome of any particular bid: they simply attempt to predict a trend over a period of time, and it may be that most organisations do not submit enough bids in any given period of time for the trend to become established or useful. Despite these comments, used cautiously, bidding models can give suppliers a competitive advantage.

Classification of bidding models

Currently, there are in excess of 1000 papers associated with modelling the bid-price decision. Generally, these models fall into the following main categories:

- **Models based on probability theory**: most bidding models based on probability theory have been derived from the work of Friedman (1956) and Gates (1967), whereas others have sought to improve the precision of probability models, for example Hillebrandt (1974), Carr (1982), King and Mercer (1990) and Skitmore (1991). Skitmore (2004), using real and typical sets of construction bid data, compared Friedman's (1956) model, Gates' (1967) model, Carr's (1982) model and two versions of Skitmore's (1991) model against pure chance. He established that at best they produced only a marginal improvement on chance

- **Multiple regression models**: examples of the application of this approach include Carr and Sandhal (1978) and Seydel and Olson (2001)
- **Neural network models**: artificial neural networks (ANN) models have been developed to aid the prediction of winning bids (McKim, 1993), predict mark up (Dozzi *et al.* 1996) and for optimum mark-up estimation (Moselhi *et al.*, 1991; Moselhi and Hegazy, 1993; Li and Love, 1999; Li *et al.*, 1999)
- **Novel pricing approaches**: most bid price models have attempted to either determine the probability of a bidder winning a bid based only upon the price of its proposal or to estimate the optimum mark up. As shown earlier, in practice, however, purchasers also evaluated bids on the basis of multiple criteria, that they perceive to have value. More recently, bid strategy models have been developed to address these multi-award criteria. For example, Cassaigne and Singh (2001) and Bussey *et al.* (1997) used multi-attribute utility theory to capture the bid selection behaviour of the purchaser. Similarly, Dozzi *et al.* (1996) and Seydel and Olson (2001) have used utility theory. Additionally, the AHP has been applied by Seydel and Olson (1990), Cagno *et al.* (2001), and Marzouk and Moselhi (2003), where competing bids are evaluated on a multiple-criteria basis, whereas Fayek (1998) has developed a competitive bidding strategy model based on the techniques of fuzzy set theory, and Shen *et al.* (1999) and Drew *et al.* (2002) have used an optimal approach. Finally, Chua *et al.* (2001) have developed a case-based reasoning approach: the output derived is the optimal mark-up

Further reading

Bidding models

Lowe, DJ and Skitmore, M (2006) Bidding (Chapter 16, pp.377–381). In: DJ Lowe (ed.) with R Leiringer. *Commercial Management of Projects: Defining the Discipline*. Blackwell Publishing, Oxford

Exercise 8.5 Bid price/supplier's dilemma

Discuss the following statement:

> 'The contractor's dilemma has been stated before, and will be stressed again many times: if he bids high enough to make a profit, he cannot get the job; and if he bids low enough to get the job, he will not make a profit' (Park and Chapin, 1992)

Adoption of models by practitioners

Despite the wealth of research activity over the last 50 years in developing bidding models, in practice there has been relatively little use made of these models. Moreover, in many organisations the decision on risk margin is a very subjective one. Surveys of the application of bidding models within the construction sector reveal the limited use of any kind of mathematical or statistical bidding model. Additionally, within the context of Singapore, Chua and Li (2000) found that 80% of their respondents had never used any statistical model to assist their bidding decisions, whereas almost 95% of Dulaimi and Shan's (2002) respondents were comfortable with the way they made the mark-up decision.

There appears to be an issue concerning the apparent apathy with which these theoretical bidding models have been received in practice. This raises the question: considering the academic interest in this area, why are bidding models not more widely used? Chapman *et al.* (2000), suggest that this may reflect a lack of understanding on the part of practitioners and/or a lack of organisation effort in collecting, collating and interpreting relevant information. They assert that this is a result of failure on the part of the model developers to persuade practitioners that investing their time and resources is worthwhile, and of theorists to convince practitioners that theoretically sound approaches are practical propositions. Alternatively, Fayek (1998) believes that current models do not correspond with the actual practice of bidders.

eAuctions and game theory

e-Procurement refers to the electronic enablement of the purchasing process. Of particular interest to the topic of bidding are electronic **reverse auctions** (eAuctions): online exercises in which suppliers vie against each other with open bids for the entitlement to provide the purchaser with assets or services. The OGC (2004) maintain that during an eAuction the impression of competition is intensified as suppliers compete in real time by submitting lower bids as the auction develops. The process can accommodate either lowest price or MEAT award criteria. Also, in addition to submitting a lower bid, a supplier can introduce new or enhanced value aspects to their bids, which are open to all the competing suppliers. Increasingly used in public and private sectors, the OGC (2004) reports efficiency improvements in public sector contracts in the range of 20–25%.

Linked to reverse auctions is the application of game theoretic techniques. An explanation of game theory falls outside the scope of this chapter – for more information see, for example, Binmore *et al.* (1993) and Binmore (1994, 1998). Probably the most famous application of Binmore's work is the sale of the British 3G telecom licences (Binmore and Klemperer, 2002), claimed to be the biggest auction ever.

Summary

This section has shown that, notwithstanding the increased use of partnering and open book arrangements, competitive bidding is still the most commonly used approach to supplier selection. Moreover, EU legislation advocates the use of competition. However, purchasers (particularly those in the construction sector) have been persuaded, either by good practice documentation or industry pundits, so that currently within the UK, 'bid price' and 'multi-evaluation criteria' have equal status when evaluating bid submissions, the motivation being to attain best value. This transition towards multicriteria selection has implications both for the bid preparation process, requiring bidders to signal and demonstrate their ability to meet these criteria within their bid documentation, and on the development of bidding models.

As with the assessment of whether or not to bid for a project – determining the bid submission, the interplay between bid price and value aspects of the submission – is a strategic decision. Likewise, bidding performance is an expression of a strategic process. Factors believed to influence the bidding decision include risk involved in the investment, degree of difficulty, size of contract/project size, need for work, and uncertainty in (reliability of) the cost estimate. These are held to be important, suggesting that bidders are primarily influenced by project characteristics and company-related issues. The prominence of risk implies that risk analysis should be central to the decision process (see Chapter 4).

For over 50 years, extensive research has been undertaken into the development of analytical bidding models, with the aim of improving a bidder's likelihood of submitting the 'optimum' bid price. Originating from the work of Friedman (1956) they include statistical bidding models, approaches based on historical data, multiple regression analysis and ANN techniques. More recently, bid strategy models have been produced to take account of multi-award criteria, employing utility theory, AHP, fuzzy set theory, and case-based reasoning approaches. In practice these models have not been widely adopted. Furthermore, in many organisations conventional practice is to form bid decisions on the basis of intuition, derived from a mixture of gut feeling, experience and guesses.

Bid submission/adjudication stage gate review

Prior to submitting a bid/proposal, a stage gate review, in some sectors termed an **adjudication** meeting, is recommended. The remit of the review is to re-examine, once again, the business justification for proceeding with the opportunity, revisiting the bid/no-bid decision, and, if the decision is made to submit a *bona fide* tender, determining the 'commercial' offer (price) or proposal (a combination of price and design/specification). The aim is the submission of a bid/proposal that will result in the firm being selected as the preferred bidder but at the 'right' price in relation to the purchaser's scope of work, quality expectations, time scale and proposed terms and conditions. The OGC Gateway™ Process Review 3: Investment decision best practice guidance (OGC, 2007 g), again while developed to support the demand perspective, can be used as a template and amended to develop a pro-forma to support a supply-side bid submission/adjudication stage gate review. This will involve incorporating the factors that influence the bid/no-bid decision, bid price determination and the associated decision support tools discussed above.

During this phase, depending on the allocation of tasks across the organisation's business functions, commercial practitioners are involved in either leading or participating in bid management teams. Activities undertaken include: reviewing and analysing the purchaser's tender documentation (ITT/RFP/RFQ); capturing the customers' requirements; assessing resource requirements, the inherent risk and potential opportunities presented by the venture; where the response requires the formation of a joint venture or project alliance, preparing teaming and joint venture agreements; drafting complex bespoke contracts and agreements or selecting standard precedents; defining and selling the customer proposition, writing technical specifications that define the supplier's capability and proposed solution (products and/or services); estimating the costs involved in undertaking the work; undertaking value management exercises; preparing the commercial response and developing the commercial shape of the deal; establishing pricing and payment policy; obtaining requisite bonds/guarantees/insurance/ export control licences; overseeing the publication of the proposal documentation; owning and implementing the bid authorisation process and providing commercial authorisation; arranging 'back to back' subcontract terms to mirror main contract provisions; obtaining legal and regulatory sign-off for undertaking business opportunities; providing an assurance role, in relation to the submission of bids/proposals and the application of effective CRM techniques; and preparing and participating in bid/proposal presentations. They will invariably play a prominent role in the adjudication meeting and have responsibility for evaluating and reporting on the success of bids and the effectiveness of bid processes and procedures.

The tasks listed above will require the application of the risk-management tools and financial decision-making techniques described in Chapters 4 and 5, plus the risk, value and contract management best practice described in Chapter 7. It will also require a detailed understanding of the legal and contractual issues described in Chapter 6.

Exercise 8.6 Adjudication process

A. Compare and contrast the factors to be considered by a supplier's management during the adjudication process with the output produced by the mathematical models currently available to assist this decision-making process.
B. Critically appraise the use of bidding models within an integrated tendering strategy

Activity 8.7 Proposition development

Compare and contrast the strategies and tactics adopted by Laing Construction, Sir Robert McAlpine and Multiplex when bidding for the Millennium, Emirates and Wembley stadia projects (case study A). In particular, consider how each contractor responded to/utilised delays in the procurement process and what effect their responses had on the ultimate success or failure of the projects.

Additionally, comment on implications of the T5 procurement process on the pre-award activities of prospective tier-1 suppliers.

Following the submission of a bid/proposal and before a preferred supplier (or winning proposal) is announced, suppliers will attempt to, if permitted by the procurement procedures, maintain a dialogue with their contacts in the purchasing organisation. The objective is to establish the buying centre's reaction to their bid/proposal and if deemed appropriate to put forward alternative solutions.

As described under asset/service procurement, once a preferred bidder has been selected, the parties frequently engage in pre-award negotiations to determine the precise terms of the exchange contract. Frequently discussed terms and conditions include the limitation of liability, indemnification, cost/price structures, intellectual property rights, payment terms, liquidated and ascertained damages, and performance guarantees, etc. Alternatively, the negotiations can centre on one or both parties attempting to impose their in-house precedent document on the other.

Award

Assuming that the pre-award negotiation phase leads to the establishment of a common understanding (a '*meeting of minds*' or '*mutual agreement*') on the formation (terms and conditions) of the contract, the contract can then **awarded**, that is, assigned or granted to the successful individual or organisation. The stage marks a **transition point** – the commencement of the **implementation stage** – requiring the exchange parties to deliver on the **commitments** made during the deal creation phase. Following the award of the contract the purchaser and suppliers (the successful and unsuccessful bidders) are advised to carry out a **post-award/bid review** to capture any lessons learnt. Prince2 provides a template: the **lessons report** and the process **managing stage boundaries** can be utilised at this point.

Activity 8.8 Award

Assess the appropriateness of the contract negotiation strategies and tactics, and the award procedures adopted on the Millennium, Emirates and Wembley stadia projects (case study A) and T5 (case study B). Comment on the conformity of these procedures with current perceived 'best practice' and the motivation of the parties where actual practice deviates from this.

Evaluate the suitability of the resulting contracts to deliver the stated objectives of each project.

Summary

Procurement best practice

This section has addressed strategic (asset/service) procurement from a demand-side perspective, the aim being to identify the purchaser's motivation for adopting a specific procurement and contracting strategy and recognise good procurement/contracting practice. From a supply perspective it has focused on proposition development, the preparation and submission of a proposal for the supply of a product and/or service, and in particular bid-price determination. It culminates in the contract award.

Effective procurement and supply is dependent to a large degree on the role adopted, both in terms of attitude and participation, by the purchaser (client). Bennett and Grice (1990) suggest that clients adopt the following key principles, which are still relevant today:

- **Commitment**: clients' initiate projects, set the tenor and manner of the exchange and are key team members
- **Role definition**: clients need to define their own role and be capable of performing it. For example, client involvement in the design and management of projects requires appropriate resourcing and expertise

- **Realism**: priorities and expectations, for example, in relation to schedule and price, should be realistic and fair
- **Briefing**: a clear brief is essential to establish exactly what is required, including the level of service, specification or product
- **Negotiate**: a willingness to negotiate with advisers, partners and suppliers is required to maintain effective project teams
- **Variations (changes)**: restrict changes to the design, specification or product to the essential. If changes are necessary a systematic approach to dealing with them is crucial
- **Communication**: communicate expectations both internally and externally. Ensure there is a clear chain of communication and decision-making, and avoid confusion by communicating via a single voice

To the above we can add:

- **Uncertainty**: an acknowledgement that projects are subject to unexpected occurrences which delay completion and can lead to disagreement

Additionally, purchasers should seek to be *fair, efficient and courteous*, adopting the following measures:

- Choosing their advisers, partners and suppliers judiciously
- Implement an appropriate procurement methodology that best fits the project's priorities
- Adhering to fair tendering procedures

The latter should include:

- The publication of procurement opportunities
- The preparation of appropriate tender documentation
- The identification and selection of an appropriate number of suitable tenderers
- An appropriate period of time for the preparation of tenders
- A method of dealing with errors within the tender documentation
- A consistent procedure for the submission and inspection of tenders
- A method for dealing with errors within tenders
- The provision of feedback to all tenderers on the outcome of bids
- The application of the highest professional standards in the management of contracts
- The adoption of procedures to respond to suggestions, enquiries and complaints
- The adoption of prompt payment procedures

Further, public clients are instructed to do their best, in all dealings with suppliers and potential suppliers, to preserve the highest standards of honesty, integrity, impartiality and objectivity.

Part C: Execution

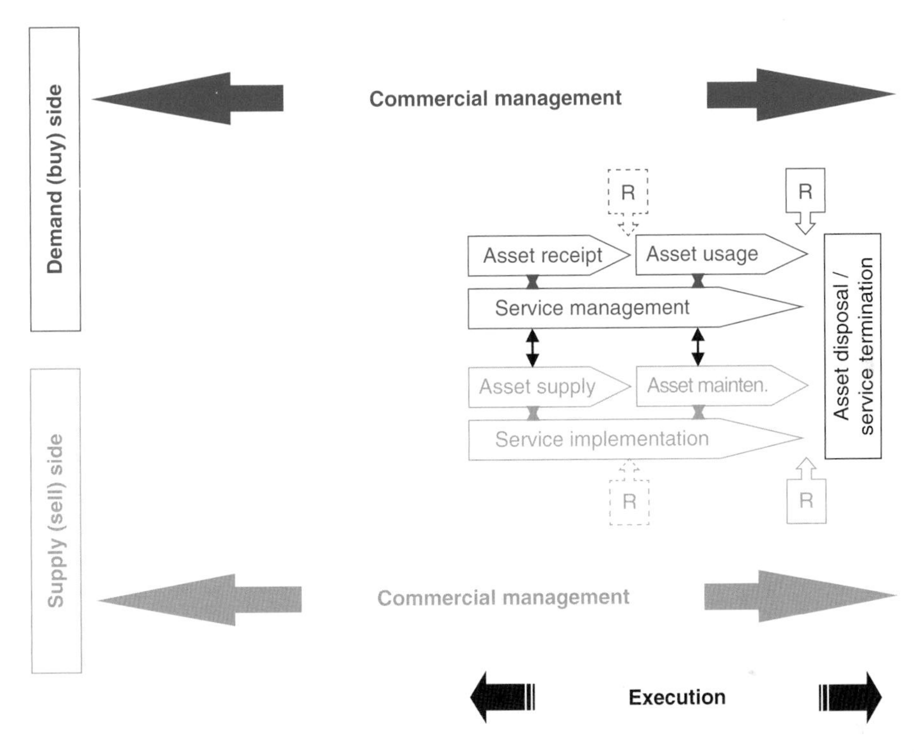

Figure 8.3 Execution.

Introduction

Part C **Execution** addresses the commercial procedures and tasks involved in implementing, delivering and fulfilling **commitments** made during the deal creation stage. It outlines the principle aspects of contract management: for example, the monitoring and control of the contract to ensure that the agreement entered into by the purchaser and supplier operates efficiently and effectively, and that all contractual obligations are appropriately discharged. Part C concludes with an explanation of the commercial activities involved in asset disposal and service termination.

Post-award delivery and maintenance of assets and services

The post-award activities carried out in supporting the supply/ **receipt**, **usage**/ maintenance and **disposal** of assets, and the implementation/ **management** and **termination** of services by commercial practitioners from both the demand-side and supply-side have much in common. These activities are present where appropriate as generic capabilities.

Post-award commercial activities

'Commercial' activities, undertaken during this execution phase by both sides of the exchange transaction, focus on the management and administration of complex contracts and agreements (post-award contract management).

Post-award contract management
These activities include, for example:

- **Provision of specialist advice**: advising project management teams on contractual issues, providing critical business support to ensure the project is delivered in accordance with the agreed contract terms
- **Managing/administering commercial/contractual issues**: specific tasks in respect of the main contract include:
  - receiving, processing and initiating contractual correspondence for example in relation to change management (variations) and claims, etc.
  - valuing, preparing, submitting, assessing, negotiating and agreeing additional payment (fees) and/or extensions of time in respect of contractual changes (variation) and claims
  - producing and maintaining financial reports and forecasts
  - arranging insurance provisions
  - valuing, preparing, submitting, evaluating and agreeing interim payments and ensuring timely receipt of payment
  - resolving disputes and managing conflict
  - valuing, preparing, submitting, assessing, negotiating and agreeing final accounts

Post-award subcontract management
- **Procuring subcontractors and materials**: establishing trading accounts with suppliers and subcontractors. These activities mirror the procurement process between the purchaser and the main contractor/supplier described earlier in this chapter
- **Managing/administering commercial/contractual issues**: specific tasks in respect of multiple subcontracts and supply contracts include:
  - valuing, preparing, submitting, evaluating and agreeing interim payments
  - liaising with the accounts section to ensure timely payment of monies due to suppliers and subcontractors
  - valuing, preparing, submitting, assessing, negotiating and agreeing additional payment (fees) and/or extensions of time in respect of contractual changes (variation) and claims
  - producing and maintaining financial reports and forecasts
  - resolving disputes and managing conflict
  - valuing, preparing, submitting, assessing, negotiating and agreeing final accounts

Again, the management and administration of the subcontract and supply contracts mirror to a high degree the management of the exchange transaction between the purchaser and the main contractor/supplier

The areas to be addressed in this section, therefore, relate to best-practice contract management outlined in Chapter 7

Contract management

Chapter 7 identified and defined best-practice contract management and acknowledged the responsibility of the purchaser's and supplier's contract management capability. Additionally, it identified five common aspects of contract management:

1. Establishing the contract management team
2. Contract administration
3. Performance measurement
4. Relationship management
5. Risk management

Contract administration

Contract administration aspects of execution stage (post-award) contract management, primarily fall into two components:

1. **Managing the contract**: in the physical sense – for example, establishing a repository for 'hard copy' contracts and maintaining a timetable for key decisions-making action points and the administration of the contract closure/termination. This component can also include the utilisation of IT/IS enabled enterprise contract management (see, for example, Saxena, 2008)
2. **Payment and incentives**: ensuring that interim payments are made in accordance with the contract and that agreed incentive mechanisms are in place and well managed (OGC/NAO, 2008; see Chapter 7 for more detail)

It is common practice for a fresh execution/implementation phase contract management team to assume responsibility for all commercial aspects of the exchange transaction post-award. The transfer of oversight has, however, to be well managed with pre-award decisions and agreements well documented in order to minimise the potential for misunderstandings to arise later in the project.

Further reading

Contract administration

Walker, I and Wilkie, R (2002) *Commercial Management in Construction.* Blackwell Science, Oxford

Performance measurement

Managing the performance of suppliers and subcontractors is a key activity for both demand-side and supply-side contract management teams. The aim is to ensure the service/project is delivered in accordance with the contract terms. The NAO/OGC Good Practice Contract Management Framework identifies two key areas:

1. **Service delivery**: for example, ensuring the provision of timely information and the existence of a performance management framework and that SLAs are in position
2. **Feedback and communications**: for example, providing feedback to suppliers on their performance and instigating formal performance reviews (OGC/NAO, 2008; see Chapter 7 for more detail)

Benchmarking involves the measurement of the effectiveness and efficiency of an organisation's strategies, procedures, portfolios, programmes, projects, products and services, followed by an evaluation of the results in comparison with those of their peers/industry standards. The purpose of the exercise, according to Baily *et al.* (2008), is to reveal **best practice** *wherever it might be found, and to attempt to identify and isolate the variables that accompany or are part of this best practice.* The belief is that these variables can then be utilised within an organisation as key indicators (benchmarks), with the objective of drawing attention to how performance could be equalled (or surpassed). However, within the context of outsourcing and managed services, the term has become synonymous with cost cutting. This is a misrepresentation of the function of benchmarking as described above.

The concept of benchmarking is closely linked with that of **continuous improvement**. **performance measurement** and is a key feature of the AEC guidelines (OGC, 2007c), which state:

> '... it is essential to measure performance in construction projects to determine whether planned improvements in the efficiency and quality of facilities are being achieved, and to learn lessons for future projects.'

For example, performance improvements on unique projects can be encouraged by incorporating a target cost incentive payment provision.[48]

Refer to the sections on performance measurement and cash flow management in Chapter 5.

Further reading

Performance measurement

Horner, M (2006) Performance measurement (Chapter 12). In: DJ Lowe (ed) with R Leiringer. *Commercial Management of Projects: Defining the Discipline*. Blackwell Publishing, Oxford. pp. 270–297

Ross, A and Hugill, D (2006) Signals from site – embodied logic and management accounting (Chapter 18). In: DJ Lowe (ed.) with R Leiringer. *Commercial Management of Projects: Defining the Discipline.* Blackwell Publishing, Oxford. pp. 417–439

Relationship management

Relationship management involves the development of solid relationships with all the key participants and stakeholders to facilitate successful project/service delivery. The OGC/NAO (2008) framework identifies two aspects to relationship management:

1. **Roles and responsibilities**: for example, ensuring that all project participants are aware of their roles and responsibilities
2. **Continuity and communications**: for example, clearly communicating the customer's expectations, ensuring effective communication between stakeholders; defining and applying appropriate dispute resolution processes (OGC/NAO, 2008; see Chapter 7 for more detail)

Contractual mechanisms for dispute resolution and conflict management were addressed in Chapter 6, while negotiation was briefly introduced earlier in this chapter.

Further reading

Relationship management

Fenn, P (2006) conflict management and dispute resolution (Chapter 11). In: DJ Lowe (ed.) with R Leiringer. *Commercial Management of Projects: Defining the Discipline*. Blackwell Publishing, Oxford. pp. 234–269

Fenn, P (2012) *Commercial Conflict Management and Dispute Resolution*. Spon Press, Abingdon

Activity 8.9 Relationship management

Compare and contrast the relationship management strategies and tactics adopted by the parties involved in the Emirates and T5 projects with those of the Millennium and Wembley Stadia projects.

Activity 8.10 Dispute resolution

In respect of the various disputes between Multiplex and WNSL, and Multiplex and its supply chain:

(a) How were the underlying issues resolved?
(b) Evaluate the negotiation tactics adopted by the parties
(c) To what extent did the project's main contract aided or hinder the resolution of the issues?

Risk management

Post-award risk management requires an understanding of and the ability to manage both contractual and supplier/purchaser risk. This involves, for example:

- **Processes and strategy**: for example, an acceptance that risks should be borne by the party best able to manage them, and the instigation of a formal risk identification, monitoring and mitigation process
- **Contractual terms**: for example, are understood and monitored by the contract manager
- **Ongoing supplier/purchaser risk management**: for example, measuring the supplier's/purchaser's financial status and business performance (OGC/NAO, 2008; see Chapter 7 for more detail)

Also, refer to Chapter 4.

Further reading

Risk management

Kähkönen, K (2006) Management of uncertainty (Chapter 10). In: DJ Lowe (ed.) with R Leiringer. *Commercial Management of Projects: Defining the Discipline*. Blackwell Publishing, Oxford. pp.211–233

Activity 8.11 Risk management

Compare and contrast the risk management strategies and tactics adopted by the parties involved in the Emirates and T5 projects with those of the Millennium and Wembley Stadia projects.

Gateway reviews

During the post-award period a series of interrelated stage gate review are recommended.

Readiness for service stage gate review

Carried out by the purchasing organisation on completion of the supply contract, a readiness for service stage gate review is advised to assess the asset's/products' fitness for purpose/readiness for service. The OGC Gateway™ Process Review 4: Readiness for service best practice guidance (OGC, 2007h) provides sample questions relating to the business case and stakeholders, risk management, review of the current phase and readiness for service (its readiness for the next phase). These include, for example:

- *Is the Business Case still valid?*
- *Are there any changes between award of contract and completing of transition/testing that affect plans for business change?*
- *Is the organisation ready for business change?*
- *Are risks and issues associated with the implementation phase being properly identified and managed?*
- *Does the total service or facility meet the acceptance criteria?*
- *Is the project under control? Is it running according to plan and budget?*
- *Have all parties accepted the commissioning/test results and any action plans required?*
- *Are there workable and tested business contingency, continuity and/or reversion plans for rollout, implementation and operation?*
- *Is the organisation ready to manage the contract in the operational environment?*
- *Is the long-term contract management process in place?*
- *Is there a process to manage and measure performance … and … benefits?* (OGC, 2007h)

Post-contract stage gate review

On completion of both supply and service contracts a stage gate review is recommended to capture any lessons learnt. Again Prince2 provides a template, the **Lessons Report**, and the process, **Closing a Project**, can be utilised at this point.

> **Activity 8.12 Asset supply/asset receipt**
>
> Compare and contrast the challenges faced by the parties in implementing the Millennium, Emirates and Wembley stadia (case study A) and T5 (case study B) projects. Comment on the different approaches adopted on each project to deliver its stated objectives.

Operations review and benefits realisation stage gate review

Carried out by the procuring organisation, the operations review and benefits realisation review is potentially a repeatable exercise, particularly on those projects that incorporate a service contract element, where several reviews may occur during its operational life. Moreover, based on whether the project is a long-term service contract or the provision of works, the extent of the review is adaptable. For long-term contracts, for example, PFI and strategic partnering arrangements, four reviews could be held over the contract period, whereas for IT-enabled projects either two or three reviews could be undertaken during a five-year contract. An initial review is undertaken 6–12 months after the receipt of the asset or the launch of a new service, to enable evidence to be assembled on the operating benefits derived from the service or asset. It focuses on the business case and the effectiveness of service delivery/contract management procedures.

The OGC Gateway™ Process Review 5: Operations review and benefits realisation best practice guidance (OGC, 2007i) provides sample questions relating to the business case and stakeholders, risk management, review of the current phase and readiness for service (its readiness for the next phase). These include, for example:

- *Is the service/facility operating to defined parameters?*
- *Are the governance and contractual relationships satisfactory?*
- *Is change management effective?*
- *Is relationship management effective?*
- *Are the business benefits being realised as set out in the Business Case?*
- *Are the users satisfied with the operational service?*
- *What is the scope for improved value for money?*
- *Are commercial mechanisms providing appropriate incentives?*
- *Is the organisation setting realistic targets for continuous improvement year-on-year from this service? Are the targets Specific, Measurable, Agreed, Realistic and Timely (SMART)?*
- *Are the client and partner working together actively to identify opportunities for improvement through innovation?*
- *Are performance measures that relate to delivery or capability improvement tracked against an existing baseline?*
- *Has there been a review of how well the project was managed?*
- *Are there any major issues with the current contract that could affect the approach to re-competition where relevant?* (OGC, 2007i)

An OGC Gateway Review 5 occurs once the purchasing organisation has undertaken a post-implementation or comparable review. It utilises the conclusions drawn from the internal review, in conjunction with an evaluation of organisational learning derived from the project. However, the inclusion of a full review of plans for the future is optional. Subsequent Gateway Review 5s verify whether or not the projected value for money and anticipated benefits outlined in the business case and benefits plans are being delivered. The timing of these additional reviews is determined by the operational business owner (OGC, 2007i)

Activity 8.13 **Project success/failure**

Generate a lessons learnt log for the Millennium, Emirates and Wembley stadia (case study A) and T5 (case study B) projects.

Asset disposal/service termination

As defined earlier, **asset disposal** involves the divestment or disposal by other means of property and assets at the end of their economic life. The commercial framework concludes with the disposal of the asset. Similarly, **service termination** refers to the process that brings a service contract to an end. In the case of the delivery of a service the commercial framework concludes with the termination of the service. As noted earlier, termination also refers to the premature end of a supply contract (termination of a contract), occurring when the contracting parties are released from their contractual obligations due to performance of their duties (an executed contract); agreement of the parties; breach of contract; or frustration.

At the end of a service contract the purchaser should reassess the business case that underpins the rationale for the service to determine whether or not there is a continuing commercial requirement for the service (this should form part of the operations review and benefits realisation review). If there is, the purchaser then has to determine whether to renew, renegotiate, or terminate the agreement and seek to appoint a replacement provider or, in the case of an outsourced capability, bring the service back within the boundaries of the firm.

The commercial function is involved in supporting this decision process – for example, through the provision of supplier performance data over the life of the contract – and advising the decision maker(s) on the contractual position surrounding the termination of the service and on the implications of bringing the service in-house. If a decision is made to appoint an alternative provider, then the commercial function will participate in the 'rerun' of the procurement/implementation process. Additionally, they will oversee the transfer of the service from one provider to another. Alternatively, if the service is to revert back to an organisational capability, they will be involved in overseeing the contractual process that facilitates this.

Where the disposal of an asset is involved the commercial function may provide advice on the sale or disposal of the asset – for example, the demolition of a facility. Where specialist legal advice is required they may liaise with the organisation's in-house legal team or commission external legal advisers.

Epilogue

Consider the following: an allegory for the future of NHS provision.

Case Study NuevoJoints

Background

Established in 2015 by Sir Neil 'Chubby' Checker, following the merger of Hips-R-Us and Knees4U, NuevoJoints is the largest joint replacement franchise operating within the UK. In 2016 the organisation performed in excess of 110 000 procedures in England and Wales,[1] accounting for approximately 60% of all hip and knee replacement surgery. Encompassing both the public (NHS) and private sector provision, the company has a presence in over 200 hospitals.[2]

The company's success has been built upon its innovative approach to hip, knee and ankle surgery, which incorporates minimal invasive surgical approaches, components made from pioneering composite materials, the latest surgical equipment and patient-specific blood harvesting both before and during operations. Patients' physiotherapy regimes start approximately 4 weeks before an operation and resumes the day after the procedure has been performed. Moreover, NuevoJoints' consultants, surgeons, nurses and theatre technicians all undergo extensive in-post training, and work exclusively on hip, knee and ankle conditions.

Overall, these developments have reduced average operation times to below one hour, while most patients remain in hospital for only 48 hours post operation. KPIs within NuevoJoints' SLAs with NHS commissioners reflect these improvements, although it is acknowledged that 15% of all operations may fall outside these measures due to inherent complications. However, irrespective of this, NuevoJoints receives a fixed fee per patient that includes pre-operation consultancy and physiotherapy, the procedure and all post-operative care and physiotherapy, but excludes accommodation, meals and specialist rehabilitation.

NuevoJoints' business model

NuevoJoints lease theatre space within existing hospitals (NHS and private sector) with NHS-funded procedures generally being performed during off-peak periods. Specialist equipment, such as the revolutionary twist knee and hip arthoplasty table and robotic equipment, is leased on a 'power-by-the-hour' basis from manufacturers, while ongoing physiotherapy is delivered through a partnering agreement with Entire Health & Fitness Ltd utilising its on-site purpose-built facilities, which include beauty clinics, gyms, physiotherapy, saunas, steam rooms, studios and swimming pools.

NHS changes that makes this possible

Prompted by the recommendations of Dame Emerald Callow's review and changes in NHS commissioning procedures, the current trend is for hospitals to transfer the majority their functions to concessionaires. Based upon established commercial management principles developed, for example, in department stores and airport terminals, around 70% of hospital floor space is currently allocated to franchises and concessionaires. This a continuation of the practice of introducing high street brands, such as Burger King, Caffè Ritazza, Costa and Upper Crust franchises into hospitals begun by Medirest (part of the Compass Group). Medirest also provides catering, cleaning and portering services across the health sector. Concessionaires pay a rental per square foot of floor space occupied plus a percentage of their annual turnover to establish a presence within a hospital.

Overnight accommodation within hospitals, for both patients and visitors, is provided through the Panache Hotel chain, which has a 10-year deal with the NHS. Depending upon the size of the hospital and its location, Panache offers a range of accommodation: Debris (budget brand), Peach (economy brand), Modal (midscale brand) and Boutique (luxury brand). NHS patients are entitled to beds within the Debris range, although they are able to upgrade their accommodation upon payment of a top-up fee. Panache is also responsible for checking in all patients, including day cases. Similarly, Wellington Coalition GmbH has been awarded a five-year franchise to provide both out-patient pharmacy and hospital-wide pharmacy provision

Importantly, the fusion of hospital and airport design – for example, the adoption of interconnected satellite buildings linked by concourses – has enabled a marked increase in retail and restaurant provision. It is common to find florists, designer nightwear outlets, mobility specialists, home care nursing and support services, 'Office Naturale' (providers of office space/facilities on a pay-by-the-hour basis and bedside mobile phone/internet access services), taxi services for patients and visitors, and personal injury lawyers as concessionaires within hospitals. Increasingly, high street names such as Mamas & Papas, Mothercare, Carphone Warehouse, Toys 'R' Us, Waterstones and W.H. Smith are to be found in major hospitals. Also, as with airport terminals, space within hospitals is increasingly zoned – for example, pre-check-in (retail stores, restaurants and hotel accommodation, etc.), post check-in (waiting lounges, retail stores, restaurants, etc.) and service areas (operating theatres, mortuaries, haematology, etc.).

Notes:

[1]Since 2010 there has been a 15% increase in the number of procedures undertaken each year.

[2]Presently, 298 facilities within England and Wales are licensed to perform Hip, Knee and Ankle surgery (compared with approximately 400 in 2010).

Source: Lowe (2010)

Exercise 8.7 NuevoJoints

Highlight the development issues faced by the commercial function in order to support this vision of the future. Consider the risks and opportunities that that might arise from adopting such an approach and the barriers that need to be overcome in order for this to become a reality.

Endnotes

1 Note: numerous web links are provided to additional sources of information; these are provided as a further resource for those who are particularly interested in a topic or require additional help to understand a particular concept or procedure, etc.
2 See: http://archive.treasury.gov.uk/docs/1999/proclaunch.html (accessed 23 September 2012) http://webarchive.nationalarchives.gov.uk/20060214043311/ogc.gov.uk/index.asp?id=377& (accessed 23 September 2012)
3 See: http://www.cabinetoffice.gov.uk/content/major-projects-authority (accessed 19 September 2012)
4 See: *Rethinking Construction*: http://www.constructingexcellence.org.uk/pdf/rethinking%20construction/rethinking_construction_report.pdf (accessed 21 September 2012)
Modernising Construction: http://www.nao.org.uk/publications/nao_reports/00-01/000187.pdf (accessed 21 September 2012) *Rethinking Construction: Achievements, Next Steps, Getting Involved*: http://www.rcne.org.uk/Documents/Rethinking_Construction_2002.pdf (accessed 21 September 2012)
Accelerating Change: http://www.strategicforum.org.uk/pdf/report_sept02.pdf (accessed 21 September 2012)
Improving Public Services through Better Construction: http://www.nao.org.uk/publications/nao_reports/04-05/0405364.pdf (accessed 21 September 2012)
5 See: MoD (2001); and http://www.contracts.mod.uk/competition-policy/ (accessed 23 September 2012)
6 Materials could be replaced by the terms products, components, assets or services
7 See Bröchner (2006) for a detailed discussion of the various applications of outsourcing
8 For a discussion of potential cultural issues see Fellows (2006) and Tijhuis and Fellows (2012)
9 The Institute of Operations Management: http://www.iomnet.org.uk/knowledge-bank/s/supply-chain-management-scm.aspx (accessed 5 March 2012)
10 Supply Chain Management Institute. See: http://scm-institute.org/ (accessed 20 September 2012)
11 For a discussion on the introduction of supply chain management in construction, see Cox and Ireland (2006)
12 For details on best practice tendering for design-build projects see Griffith *et al.* (2003)
13 See: http://www.hm-treasury.gov.uk/ppp_pfi_stats.htm (accessed 21 September 2012)
14 For more information on the PFI approach, see: Akintoye *et al.* (2001) and http://webarchive.nationalarchives.gov.uk/+/http://www.dti.gov.uk/files/file26067.pdf (accessed 21 September 2012)
15 See: http://www.hm-treasury.gov.uk/ppp_infrastructureuk.htm (accessed 21 September 2012)
16 See: http://www.partnershipsuk.org.uk/What-PUK-Do.aspx (accessed 21 September 2012)
17 See: http://www.localpartnerships.org.uk/ (accessed 21 September 2012)
18 See: http://www.hm-treasury.gov.uk/ppp_operational_taskforce.htm (accessed 21 September 2012)
19 OGC guidance on effective partnering can be found at: http://webarchive.nationalarchives.gov.uk/20110822131357/http://www.ogc.gov.uk/documents/cp0078_Effective_partnering.pdf (accessed 21 September 2012)
20 See, for example: OGC (2005) and National e-Procurement Project (2005) http://library.sps-consultancy.co.uk/documents/guidance-policy-and-practice/implementingeprocesourcingv11.pdf (accessed 27 September 2012) http://www.epractice.eu/files/media/media_513.pdf (accessed 21 September 2012)
21 See: http://www.constructionline.co.uk/ (accessed 21 September 2012)
22 For more information on Lord Fraser's report on the Holyrood project see: http://www.scottish.parliament.uk/SPICeResources/HolyroodInquiry.pdf (accessed 21 September 2012)
23 For more information on design stage estimating techniques in the construction sector see: Ashworth and Hogg (2007); Kirkham (2007); Ashworth (2010); Winch (2010)
24 For more information on early stage estimating techniques in the IT/IS sector see: Turner and Remenyi (1995); McConnell (2006); Jones (2010)
25 See CIPS (2012)
26 This section is based upon an extended and updated version of material first published in Lowe and Skitmore (2006)
27 Mark-up, in addition to profit, also includes additions for running costs or overheads
28 'Non-bona fide tenders' are a commercial reality derived from a perception that suppliers will be penalised in the future for not responding positively to an invitation to tender
29 Due to the commercial nature of this information, it is difficult to obtain precise figures on the costs incurred by suppliers when responding to a bid opportunity. However, despite the age of this reference there is no evidence to suggest that these costs have significantly reduced
30 A non bona fide tender (Beeston, 1983, p. 112)
31 For more information see OGC (2006a, 2006b) and OGC/HM Treasury (2008) and Freshfields Bruckhaus Deringer (2005): http://www.freshfields.com/publications/pdfs/2005/13494.pdf (accessed 21 September 2012)

32 This section is based upon an extended and updated version of material first published in Lowe and Skitmore (2006)
33 Harmon CFEM Facades (UK) Limited v The Corporate Officer of the House of Commons (2000) 67 Con LR 1
34 See: http://www.majorprojects.org/pdf/seminarsummaries/093SmartProcurement.pdf (accessed 27 September 2012) Review of Acquisition for the Secretary of State for Defence: An independent report by Bernard Gray (2009) http://www.mod.uk/NR/rdonlyres/78821960-14A0-429E-A90A-FA2A8C292C84/0/ReviewAcquisitionGrayreport.pdf (accessed 27 September 2012) For information on the MoD's current acquisition policy see: http://www.mod.uk/DefenceInternet/AboutDefence/WhatWeDo/FinanceandProcurement/AOF/ (accessed 27 September 2012)
35 See: www.oft.gov.uk/shared_oft/reports/comp_policy/oft892.pdf (accessed 21 September 2012)
36 See: http://eur-lex.europa.eu/LexUriServ/LexUriServ.do?uri=CELEX:32004L0018:En:HTML (accessed 21 September 2012)
37 See: http://opsi.gov.uk/si/si2006/20060005.htm (accessed 21 September 2012)
38 See: http://opsi.gov.uk/si/si2006/20060006.htm (accessed 21 September 2012)
39 See: http://eur-lex.europa.eu/LexUriServ/LexUriServ.do?uri=CELEX:32004L0017:en:HTML (accessed 21 September 2012)
40 See: http://ted.europa.eu/TED/main/HomePage.do (accessed 21 September 2012)
41 For a glossary of negotiating terms see the Centre for Effective Dispute Resolution (CEDR) Glossary of terms: http://www.cedr.com/about_us/library/glossary.php (accessed 21 September 2012)
42 See for example: Fisher *et al.* (2003); Fenn (2006, 2012); Donaldson (2007); Peeling (2007); Lewicki *et al.* (2010)
43 For example, the detailed measurement of proposed construction and infrastructure works in accordance with the RICS New Rules of Measurement (NRM2): Detailed measurement for building works (RICS, 2012) in conjunction with Building Information Modelling (BIM)
44 This section is based upon an extended and updated version of material first published in Lowe and Skitmore (2006)
45 See for example: Tweedley (1995); and Lewis (2012)
46 See also Ashworth and Skitmore (1983)
47 See 'Cost determination and cost behaviour' in Chapter 5
48 For details on performance improvements see Achieving Excellence in Construction Procurement Guide 8 – Improving Performance: project evaluation and benchmarking: http://webarchive.nationalarchives.gov.uk/20110601212617/http://www.ogc.gov.uk/documents/CP0068AEGuide8.pdf (accessed 22 September 2012)

References

Abdelrazig, AA (1995) Unpublished MSc dissertation. King Fahd University of Petroleum and Minerals, Saudi Arabia

Adkins, B and Beighton, S (2011) Construction bid-rigging: implications for the OFT's fining policy. Practical Law Publishing Limited http://construction.practicallaw.com/1-505-7463 (accessed 1 March 2012)

Ahmad, I (1990) Decision-support systems for modelling bid/no-bid decision problem. *ASCE Journal of Construction Engineering and Management*, **116**, 595–608

Ahmad, I. and Minkarah, I. (1988) Questionnaire survey on bidding in construction. *ASCE Journal of Management in Engineering*, **4**, 229–243

Akintoye, A, Beck, M and Hardcastle, C (2001) *Public-Private Partnerships: Managing Risks and Opportunities*. Blackwell Science, Oxford

Akintoye, A and Fitzgerald, E (2000) A survey of current estimating practices in the UK. *Construction Management and Economics*, **18**, 161–172

Anon (2007) *'Power By The Hour': Can Paying Only for Performance Redefine How Products are Sold and Serviced?* Knowledge@Wharton, Wharton School of the University of Pennsylvania http://knowledge.wharton.upenn.edu/article.cfm?articleid=1665 (accessed 5 March 2012)

Ansoff, HI (1965) *Corporate Strategy*. Penguin, London

Ashworth, A (2011) *Cost Studies of Buildings*, 5th Edition. Pearson Education Limited, Harlow

Ashworth, A and Hogg, K (2007) *Willis's Practice and Procedure for the Quantity Surveyor*. Blackwell Publishing, Oxford

Ashworth, A and Skitmore, RM (1983) Accuracy in estimating, *CIOB Occasional Paper*, **27**, Englemere

Bagnall, SM, Shaw, DL and Mason-Flucke, JC (2001) *Implications of 'Power by the Hour' on Turbine Blade Lifing*. Defense Technical Information Center OAI-PMH Repository (United States) http://ftp.rta.nato.int/public//PubFulltext/RTO/MP/RTO-MP-037///MP-037-12.pdf (accessed 5 March 2012)

Baily, P, Farmer, D, Jessop, D and Jones, D (2008) *Purchasing Principles and Management*, 10th Edition. Financial Times/ Pitman Publishing

Beeston, DT (1983) *Statistical Methods for Building Price Data*. E & F N Spon, London

Bennett, J and Grice, T (1990) Procurement systems for building. In: PS Brandon (ed.) *Quantity Surveying Techniques – New Directions*. Blackwell Scientific Publications, Oxford

Bernink, B (1995) Winning contracts. In: JR Turner (ed.) *The Commercial Project Manager*. McGraw-Hill Book Company, London

Binmore, K (1994) *Game Theory and the Social Contract – Vol 1: Playing Fair*, The MIT Press, Cambridge, MA

Binmore, K (1998) *Game Theory and the Social Contract – Vol 2: Just Playing*. The MIT Press, Cambridge, MA

Binmore, K, Kirman, A and Tani, P (eds) (1993) *Frontiers of Game Theory*. The MIT Press, Cambridge, MA

Binmore, K and Klemperer, P (2002) The biggest auction ever: the sale of the British 3G telecom licences. *Economics Journal*, **112**, C1–C23

Bower, D (ed.) (2003) *Management of Procurement*. Thomas Telford, London

Bröchner J (2006) Outsourcing. In: DJ Lowe (ed.) with R Leiringer. *Commercial Management of Projects: Defining the Discipline*. Blackwell Publishing, Oxford. pp. 192 – 206

Building EDC (1985) *Thinking about Building – A Successful Business Customer's Guide to Using the Construction Industry*. National Economic Development Office, London

Burt, DN (1984) *Proactive Procurement: The Key to Increased Profits, Productivity, and Quality*. Prentice-Hall, Englewood Cliffs, NJ

Bussey, P, Cassaigne, N and Singh, M (1997) Bid pricing – calculating the possibility of winning. *IEEE International Conference*, **4**, 3615–3620

Cagno, E, Caron, F and Perego, A (2001) A multi-criteria assessment of the probability of winning in the competitive bidding process. *International Journal of Project Management*, **19**, 313–324

Caldwell, N and Howard, M (eds) (2011) *Procuring Complex Performance: Studies of Innovation in Product-Service Management*. Routledge, Abingdon, Oxon

Carr, RI (1982) General bidding model. *ASCE Journal of Construction Division*, **108**, 639–650

Carr, R and Sandahl, J (1978) Competitive bidding strategy using multiple regression. *ASCE Journal of Construction Engineering and Management*, **104**, 15–26

Cassaigne, N and Singh, MG (2001) Intelligent decision support for the pricing of products and services in competitive markets. *IEEE Transactions on Systems, Man, and Cybernetics, Part C*, **31**, 96–106

Chapman, CB, Ward, SC and Bennet, JA (2000) Incorporating uncertainty in competitive bidding. *International Journal of Project Management*, **18**, 337–347

Chua, DKH and Li, D (2000) Key factors in bid reasoning model. *ASCE Journal of Construction Engineering and Management*, **126**, 349–357

Chua, DKH, Li, D and Chan, WT(2001) Case-based reasoning approach in bid decision making. *ASCE Journal of Construction Engineering and Management*, **127**, 35–45

CIOB (1997) *Code of Estimating Practice*. 6th edition. Chartered Institute of Building, Ascot.

CIPS (2012) *Purchasing and supply management: Writing Contracts*. CIPS Position on Practice. http://cipsintelligence.cips.org/opencontent/cips-purchasing-supply-management.-writing-contracts (accessed 1 April 2012)

Comptroller and Auditor General (2001) *Modernising Construction*. The Stationery Office, London. http://www.nao.org.uk/publications/0001/modernising_construction.aspx (accessed 1 March 2012)

Construction Industry Board (CIB) (1997) *Code of Practice for the Selection of Main Contractors*. Thomas Telford Ltd, London

Construction Round Table (1995) *Thinking about Building*. The Business Round Table, London

Cousins P D , Lamming R , Lawson B and Squire B (2008) *Strategic Supply Management: Principles*, Theory and Practice. Pearson Education, Harlow.

Cova, B, Salle, R and Vincent, R (2000) To bid or not to bid: screen the Whorcop Project. *European Management Journal*, **18**, 551–560

Cox, A and Ireland, P (2006) Strategic purchasing and supply chain management in the project environment: theory and practice In: DJ (ed.) with R Leiringer. *Commercial Management of Projects: Defining the Discipline*. Blackwell Publishing, Oxford. pp. 390–416

Cox, A and Townsend, M (1998) *Strategic Procurement in Construction: Towards Better Practice in the Management of Construction Supply Chains*. Thomas Telford, London

Davies, A and Hobday, M (2005) *The Business of Projects: Managing Innovation in Complex Products and Systems*. Cambridge University Press, Cambridge

Davies, A and Hobday, M (2006) Strategies for solutions. In: DJ Lowe (ed.) with R Leiringer. *Commercial Management of Projects: Defining the Discipline*. Blackwell Publishing, Oxford. pp.132–154

Davis Langdon (2010) *Contracts in Use: A Survey of Building Contracts in Use during 2007*. RICS (Royal Institution of Chartered Surveyors), London. http://www.rics.org/site/scripts/download_info.aspx?downloadID=4748&fileID=5853 (accessed 25 July 2011)

Department of Health (2011) *The ProCure21+ Guide: Achieving Excellence in NHS Construction*. The Department of Health, Leeds http://www.procure21plus.nhs.uk/resources/downloads/ProCure21Plus%20Guide%20v2.2%202011.pdf (accessed 1 March 2012)

Domberger, S (1998) *The Contracting Organization: A Strategic Guide to Outsourcing*. Oxford University Press, Oxford

Donaldson, MC (2007) *Negotiating For Dummies*. Wiley Publishing Inc, Hoboken, NJ

Dozzi, SP, AbouRizk, SM and Schroeder, SL (1996) Utility-theory model for bid markup decisions. *ASCE Journal of Construction Engineering and Management*, **122**, 119–124

Drew, D (1993) A critical assessment of bid evaluation procedures in Hong Kong. *Professional Builder*, Chartered Institute of Building (Hong Kong Branch), June

Drew, D, Shen, LY and Zou, PXW (2002) Developing an optimal bidding strategy in two-envelope fee bidding. *Construction Management and Economics*, **20**, 611–620

Drew, DS and Skitmore, M (1990), Analysing bidding performance; measuring the influence of contract size and type. In: *Building Economics and Construction Management: Management of the Building Firm*, International Council of Construction Research Studies and Documentation, CIB W-65 Sydney, Australia. pp. 129–139

Drew, DS and Skitmore, RM (1992) Competitiveness in bidding: a consultant's perspective. *Construction Management and Economics*, **10**, 227–247

Drew, D and Skitmore, M (1997) The effect of contract type and size on competitiveness in bidding. *Construction Management and Economics*, **15**, 469–489

DTI E.9 Frameworks – Procurement, Buyers' guides, DTI procurement manual section E: part 3 – managing the procurement process. http://www.dti.gov.uk/about/procurement/buyers-guides/page22748.html accessed (accessed 1 May 2007)

Du, J and El-Gafy, M (2011) Feasibility Analytical Mapping (FAM) for the bidding decision: a graphic bidding decision making model based on multidimensional scaling and discriminant analysis. *International Journal of Construction Education and Research*, **7**, 198–209

Dulaimi, MF and Shan, HG (2002) The factors influencing bid mark-up decisions of large and medium–size contractors in Singapore. *Construction Management and Economics*, **20**, 601–610

Eastham, RA (1987) The use of content analysis to determine a weighted model of the contractor's tendering process. In: P.S. Brandon (ed.) *Building Cost Modelling and Computers*. E & FN Spon, London

Eccles, RG (1981) The quasifirm in the construction industry. *Journal of Economic Behaviour and Organisation*, **2**, 356–357

Egan, J (1998) *Rethinking Construction: The report of the Construction Task Force*. The Stationery Office, London

Fayek, A (1998) Competitive bidding strategy model and software system for bid preparation. *ASCE Journal of Construction Engineering and Management*, **124**, 1–10

Fellows, R (2006) Culture. In: DJ Lowe (ed.) with R Leiringer. *Commercial Management of Projects: Defining the Discipline*. Blackwell Publishing, Oxford. pp. 40–70

Fellows, RF and Langford, DA (1980) *Decision theory and tendering. Building Technology and Management*, October, 36–39

Fenn, P (2006) Conflict management and dispute resolution. In: DJ Lowe (ed.) with R Leiringer. *Commercial Management of Projects: Defining the Discipline*. Blackwell Publishing, Oxford. pp. 234–269

Fenn, P (2012) *Commercial Conflict Management and Dispute Resolution*. Spon Press, Abingdon, Oxon

Fenn, P and Lowe, D (2010) Employer's Requirements/Contractor's Proposals: Design and Build Best Practice. *Findings in Built and Rural Environments (FiBRE) SERIES*. RICS, London

Fernie, S, Root, D and Thorpe, T (2000) Supply chain management – theoretical constructs for construction. In: Serpell, A. (ed.) *Construction Procurement*. Pontificia Universidad Catolica de Chile

Fine, B (1987) Kitchen sink economics and the construction industry. In: PS Brandon (ed.) *Building Cost Modelling and Computers*. E & FN Spon, London. pp. 25–30

Fisher, R, Ury, W and Patton, B (2003) *Getting to Yes: Negotiating Agreement without Giving In: The Secret to Successful Negotiation*, 2nd edition. Random House Business Books, London

Flanagan, R and Norman, G (1982) Making use of low bids. *Chartered Quantity Surveyor*, **14**, 226–227

Fong, PS and Choi, SK (2000) Final contractor selection using the analytical hierarchy process. *Construction Management and Economics*, **18**, 547–557

Franks, J (1998) *Building Procurement Systems: A Clients' Guide*, 3rd edition. Addison Wesley Longman Limited

Freshfields Bruckhaus Deringer (2005) *Competitive Dialogue: The EU's New Procurement Procedure*. Freshfields Bruckhaus Deringer. http://www.freshfields.com/publications/pdfs/2005/13494.pdf (accessed 1 March 2012)

Friedman, L (1956) A competitive-bidding strategy. *Operations Research*, **4**, 104–112

Fu, WK, Drew, D and Lo HP (2002) Competitiveness of inexperienced and experienced contractors in bidding. *Construction Management and Economics*, **20**, 655–666

Fu, WK, Drew, D and Lo, HP (2003) The effect of experience on contractors' competitiveness in recurrent bidding. *ASCE Journal of Construction Engineering and Management*, **129**, 388–395

Gabor, A (1977) *Pricing: Principles and Practices*. Gower, Aldershot

Gates, M (1967) bidding strategies and probabilities. *ASCE Journal of Construction Division*, **93** (CO1), 75–107

Gottlieb, SC, Haugbølle, K and Larsen JN (2007) *An Overview of Partnerships in Danish Construction: State of the Art Report Prepared for Nordic Innovation Centre* (Draft). Danish Building Research Institute, Aalborg University, Aalborg

Government Construction Clients' Panel (GCCP) (1999) *Guidance Procedures No. 5 Procurement Strategies*. http:/www.hm-tresury.gov.uk/pub/html/gccp/ (accessed 1 May 2007)

Green, P (2010) Efficiency Review - Key Findings and Recommendations. Available from http://www.cabinetoffice.gov.uk/sites/default/files/resources/sirphilipgreenreview.pdf (accessed 21 February 2012)

Green, SD (1989) Tendering: optimisation and rationality. *Construction Management and Economics*, **7**, 53–63

Griffith, A, Knight, A and King, A (2003) *Best Practice Tendering for Design and Build Projects*. Thomas Telford, London

Hatush, Z and Skitmore, M (1997) Criteria for contractor selection. *Construction Management and Economics*, **15**, 19–38

Hatush, Z and Skitmore, M (1998) Contractor selection using multicriteria utility theory: an additive model. *Building and Environment*, **33**, 105–115

Hillebrandt, PM (1974) *Economic Theory and the Construction Industry*. Macmillan Press, Basingstoke

Hillebrandt, PM (2000) *Economic Theory and the Construction Industry*, 3rd edition. Macmillan Press, Basingstoke

Hillebrandt, PM and Cannon, J (1990) *The Modern Construction Firm*. Macmillan, Basingstoke

HM Treasury (2005) *Government Accounting 2000: A Guide on Accounting and Financial Procedures for the Use of Government Departments: Amendment 4/05*. The Stationery Office, London. http://www.government-accounting.gov.uk/current/frames.htm (accessed 1 May 2007)

HM Treasury (2006) *Value for Money Assessment Guide*. HMSO, Norwich. http://www.hm-treasury.gov.uk/d/vfm_assessmentguidance061006opt.pdf (accessed 5 March 2012)

HM Treasury (2007) *Managing Public Money*. The Stationery Office, Norwich. http://www.hm-treasury.gov.uk/d/mpm_whole.pdf (accessed 5 March 2012)

HM Treasury/NAO (2010) *From Private Finance Units to Commercial Champions: Managing Complex Capital Investment Programmes Utilising Private Finance – A Current Best Practice Model For Departments*. National Audit Office, London. http://www.hm-treasury.gov.uk/d/ppp_managing_complex_capital_investment_programmes.pdf (accessed 1 March 2012)

Hoffman, KD, Turley, LW and Scott, WK (2002) Pricing retail services. *Journal of Business Research*, **55**, 1015–1023. http://webarchive.nationalarchives.gov.uk/20100503135839/ http://www.ogc.gov.uk/documents/CP0062AEGuide2.pdf (accessed 21 February 2012)

International Association for Contract and Commercial Management (2011) *2011 Top Terms in Negotiation*. IACCM, Ridgefield, Connecticut http://www.iaccm.com/admin/docs/docs/top%20terms%202011%201.pdf (accessed 23 September 2012)

International Association for Contract and Commercial Management (IACCM) (2011) *Contract and Commercial Management: The Operational Guide*. Van Haren Publishing, Zaltbommel, Netherlands

Jones, C (2010) *Software Engineering Best Practices: Lessons from Successful Projects in the Top Companies*. McGraw-Hill, Columbus, OH

Kennedy, C and O'Connor, M (1997) *Winning Major Bids: The Critical Success Factor*. Policy Publications, Bedford

Kidd, A (2005) *The Definition of Procurement*. CIPS Australia Pty Ltd, Melbourne. http://www.cips.org/documents/Definition_Procurement.pdf accessed (accessed 1 May 2007)

King, M and Mercer, A (1988) Recurrent competitive bidding. *European Journal of Operational Research*, **33**, 2–16

King, M and Mercer, A (1990) The optimum markup when bidding with uncertain costs. *European Journal of Operational Research*, **47**, 348–363

Kirkham, R (2007) *Ferry and Brandon's Cost Planning of Buildings*, 8th edition. Blackwell Publishing, Oxford

Kotler, P (2000) *Marketing Management*. Prentice Hall International, Upper Saddle River, NJ

KPMG (2007) *Construction Procurement for the 21st Century: Global Construction Survey 2007*. KPMG International Cooperative, Switzerland. http://www.kpmg.com/global/en/issuesandinsights/articlespublications/pages/globalconstruction-survey-2007.aspx (accessed 25 July 2011)

Kwakye, AA (1994) *Understanding Tendering and Estimating*. Gower, Aldershot

Lacity, MC and Hirschheim, RA (1993) *Information Systems Outsourcing; Myths, Metaphors, and Realities*. John Wiley & Sons, Inc., New York

Langford, D and Murray, M (2003) *Construction Reports 1944–98*. Blackwell Science, Oxford

Langford, D and Murray, M (2006) Procurement in the context of commercial management. In: DJ Lowe (ed.) with R Leiringer. *Commercial Management of Projects: Defining the Discipline*. Blackwell Publishing, Oxford. pp.71–92

Latham, M (1994) *Constructing the Team: Final Report of the Government/Industry Review of Procurement and Contractual Arrangements in the UK Construction Industry*. HMSO, London

Lewicki, RJ, Barry, B and Saunders, DM (2010) *Essentials of Negotiation*, 5th edition. McGraw-Hill Higher Education, New York

Lewis, H (2012) *Bids, Tenders and Proposals: Winning Business Through Best Practice*, 4th edition. Kogan Page, London

Li, H and Love, PED (1999) Combining rule-based expert systems and artificial neural networks for mark-up estimation. *Construction Management and Economics*, **17**, 169–176

Li, H, Shen, LY and Love, PED (1999) ANN-based mark-up estimation system with self-explanation capacities. *ASCE Journal of Construction Engineering and Management*, **125**, 185–189

Lin, CT and Chen, YT (2004) Bid/no-bid decision-making – a fuzzy linguistic approach. *International Journal of Project Management*, **22**, 585–593

Lowe DJ (2010) Bringing a commercial perspective to public sector managers. *Transforming Management*, 9th November. http://tm.mbs.ac.uk/features/bringing-a-commercial-perspective-to-public-sector-managers/ (accessed 1 December 2010)

Lowe, DJ, Emsley, MW and Harding, A (2006) Predicting construction cost using multiple regression techniques. *ASCE Journal of Construction Engineering and Management*, **132**, 750–758

Lowe, DJ and Parvar, J (2004) A logistic regression approach to modeling the contractor's decision to bid. *Construction Management and Economics*, **22**, 643–653

Lowe, DJ, Parvar, J and Emsley, MW (2004) Development of a decision support system (DSS) for the contractor's decision to bid: regression and neural networks solutions. *Journal of Financial Management of Property and Construction*, **9**, 27–42

Lowe, D and Skitmore, M (2006) Bidding. In: DJ Lowe (ed.) with R Leiringer. *Commercial Management of Projects: Defining the Discipline*. Blackwell Publishing, Oxford. pp. 356–389

Lysons, K and Farrington, B (2006) *Purchasing and Supply Chain Management*. Pearson Education Limited, Harlow

Mahdi, IM, Riley, M, Fereig, SM and Alex, AP (2002) A multi-criteria approach to contractor selection. *Engineering, Construction and Architectural Management*, **9**, 29–37

Male, S (1991) Strategic management for competitive strategy and advantage. In: SP Male and RK Stock (eds) *Competitive Advantage in Construction*. Butterworth-Heinemann, Oxford. pp. 1–4

Marzouk, M and Moselhi, O (2003) A decision support tool for construction bidding. *Construction Innovation*, **3**, 111–124

Masterman, JWE (2002) *An Introduction to Building Procurement Systems*. 2nd edition. E&FN Spon, London

Mattson, MR (1988) How to determine the composition and influence of the buying centre. *Industrial Marketing Management*, **17**, 205–214

McConnell, S (2006) *Software Estimation: Demystifying the Black Art*. Microsoft Press, Redmond, Washington

McKim, RA (1993) Neural network applications for project management: three case studies. *Project Management Journal*, **24**, 28–33

Merna, A and Smith, NJ (1990) Bid evaluation for UK public sector construction contracts. *Proceedings of the Institution of Civil Engineers, Part 1*, **88** (February), 91–105

Mochtar, K and Arditi, D (2001) Pricing strategy in the US construction industry. *Construction Management and Economics*, **19**, 405–415

MOD (2001) *Ministry of Defence Policy Paper No 4, Defence Acquisition*, MoD, Directorate of Corporate Communication – DCCS London. http://www.mod.uk/NR/rdonlyres/1B07C74B-F841-4E78-9A13-F4A0E0796061/0/polpaper4_def_acquisition.pdf (accessed 1 March 2012)

MOD (2002) Soft Issues Bid Evaluation Tool (SIBET) User Manual. Ministry of Defence, London. http://www.ams.mod.uk/ams/content/docs/toolkit/gateway/guidance/linkdocs/sibet.rtf (accessed 1 May 2008)

Morledge, R, Smith, A and Kashiwagi, DT (2006) *Building Procurement*. Blackwell Publishing, Oxford

Morris, PWG (1994) *The Management of Projects*. Thomas Telford, London

Morris, PWG and Hough, GH (1987) *The Anatomy of Major Projects*. John Wiley & Sons Ltd, Chichester

Moselhi, O and Hegazy, T (1993) Markup estimation using neural network methodology. *Computing Systems in Engineering*, **4**, 135–145

Moselhi, O, Hegazy, T and Fazio, P (1991) Neural networks as tools in construction. *ASCE Journal of Construction Engineering and Management*, **117**, 606–625

Moselhi, O, Hegazy, T and Fazio, P (1993) DBID: analogy-based DSS for bidding in construction. *ASCE Journal of Construction Engineering and Management*, **119**, 446–479

NAO (2005) *Improving Public Services through Better Construction*. The Stationery Office, London. http://www.nao.org.uk/publications/0405/improving_public_services.aspx (accessed 1 March 2012)

NAO (2006) *Case Study: BAA Terminal 5 Project.* NAO Defence Value for Money. http://www.nao.org.uk/idoc.ashx?docId=920DDE14-0756-4219-B5EB-455FBE9CE572&version=-1 (accessed 21 December 2012)

NAO (2007) *Improving the PFI Tender Process.* The Stationery Office, London. http://www.nao.org.uk/publications/0607/improving_pfi_tendering.aspx (accessed 1 March 2012)

NAO (2009) *Performance of PFI Construction.* National Audit Office, London. http://www.nao.org.uk/publications/0809/pfi_construction.aspx (accessed 1 March 2012)

NAO (2011) *Lessons from PFI and Other Projects.* National Audit Office, London. http://www.nao.org.uk/publications/1012/lessons_from_pfi.aspx (accessed 1 March 2012)

NAO/OGC (2001) *Getting Value for Money from Procurement.* National Audit Office, London. http://www.nao.org.uk/idoc.ashx?docId=B0E6F6DE-A132-4DDC-B723-22986DD05CC3&version=-1 (accessed 1 March 2012)

NAO/The Audit Commission (2010) *A Review of Collaborative Procurement Across the Public Sector.* National Audit Office, London. http://www.nao.org.uk/publications/0910/collaborative_procurement.aspx (accessed 1 March 2012)

National Agency for Enterprise and Construction (2004) *Guidelines for Partnering.* National Agency for Enterprise and Construction, Copenhagen

National e-Procurement Project (2005) *Desktop Guide to e-Procurement – Part 4: What is e-Tendering.* Local e-gov, National e-Procurement Project. http://www.localtgov.org.uk/webfiles/NePP/Guidance/8.0%20e-Sourcing/8.2.1.7.pdf (accessed 1 March 2012)

Neely, A (2008) Exploring the financial consequences of the servitization of manufacturing. *Operations Management Research*, **1**,103–118

Neufville, R de and King, D (1991) Risk and need-for-work premiums in contractor bidding. *ASCE Journal of Construction Engineering and Management*, **117**, 659–673

NJCC (1995) *Code of Procedure for Selective Tendering for Design and Build.* RIBA Publications, London

NJCC (1996) *Code of Procedure for Single Stage Selective Tendering.* RIBA Publications, London

Odusote, O and Fellows, RF (1992) An examination of the importance of resource considerations when contractors make project selection decisions. *Construction Management and Economics*, **10**, 137–151

OFT (2009) *Construction Firms Fined for Illegal Bid-Rigging.* Office of Fair Trading, London. http://www.oft.gov.uk/news-and-updates/press/2009/114-09 (accessed 1 March 2012)

OGC (2004) *eAuctions.* Office of Government Commerce, Norwich. http://webarchive.nationalarchives.gov.uk/20110601212617/http:/ogc.gov.uk/documents/OGC_Guidance_on_eAuctions.pdf (accessed 1 March 2012)

OGC (2005) *eProcurement in Action: A Guide to eProcurement for the Public Sector.* Office of Government Commerce, London. http://www.epractice.eu/files/media/media_513.pdf (accessed 1 March 2012)

OGC (2006a) *Procurement Policy – Practical Guidance on the use of Competitive Dialogue.* Office of Government Commerce, London. http://www.ogc.gov.uk/documents/ProcurementPolicyCompetitiveDialogue.pdf accessed 01/05/07) (accessed 1 March 2012)

OGC (2006b) *Competitive Dialogue Procedure.* Office of Government Commerce, London. http://webarchive.nationalarchives.gov.uk/20100503135839/ http://www.ogc.gov.uk/documents/guide_competitive_dialogue.pdf (accessed 21 February 2012)

OGC (2007a) *Achieving Excellence in Construction Procurement Guide 2 – Project Organisation.* OGC, London. http://webarchive.nationalarchives.gov.uk/20100503135839/ http://www.ogc.gov.uk/docurr

OGC (2007b) *Achieving Excellence in Construction Procurement Guide 6 – Procurement and Contract Strategies.* OGC, London. http://webarchive.nationalarchives.gov.uk/20100503135839/http://www.ogc.gov.uk/documents/CP0066AEGuide6.pdf (accessed 21 February 2012)

OGC (2007c) *Achieving Excellence in Construction Procurement Guide 8- - Improving Performance.* OGC, London. http://webarchive.nationalarchives.gov.uk/20100503135839/ http://www.ogc.gov.uk/documents/CP0068AEGuide8.pdf (accessed 21 February 2012)

OGC (2007d) *Procurement Next Generation Networks – Procurement Standards, Guidance and Model Clauses.* OGC, London. http://webarchive.nationalarchives.gov.uk/20110822131357/http://www.ogc.gov.uk/documents/Next_Generation_Networks(1).pdf (accessed 21 February 2012)

OGC (2007e) *OGC Best Practice – Gateway™ Review 1: Business Justification.* OGC, London. http://webarchive.nationalarchives.gov.uk/20100503135839/http://www.ogc.gov.uk/documents/NEW_BOOK_1_APRIL.pdf (accessed 21 February 2012)

OGC (2007f) *OGC Best Practice – Gateway™ Review 2: Delivery Strategy.* OGC, London. http://webarchive.nationalarchives.gov.uk/20100503135839/http://www.ogc.gov.uk/documents/BOOK_2_APRIL.pdf (accessed 21 February 2012)

OGC (2007g) *OGC Best Practice – Gateway™ Review 3: Investment Decision.* OGC, London. http://webarchive.nationalarchives.gov.uk/20100503135839/http://www.ogc.gov.uk/documents/BOOK_3_APRIL.pdf (accessed 21 February 2012)

OGC (2007h) *OGC Best Practice – Gateway™ Review 4: Readiness for Service*. OGC, London. http://webarchive.nationalarchives.gov.uk/20100503135839/http://www.ogc.gov.uk/documents/NEW_BOOK_4_APRIL.pdf (accessed 21 February 2012)

OGC (2007i) *OGC Best Practice – Gateway™ Review 5: Operations Review and Benefits Realisation*. OGC, London. http://webarchive.nationalarchives.gov.uk/20100503135839/http://www.ogc.gov.uk/documents/FINAL_BOOK_5.pdf (accessed 21 February 2012)

OGC/HM Treasury (2008) *Competitive Dialogue in 2008 – OGC/HMT Joint Guidance on Using the Procedure*. Office of Government Commerce, Norwich. http://www.hm-treasury.gov.uk/d/competitive_dialogue_procedure.pdf (accessed 1 March 2012)

OGC/ NAO (2008) *Good Practice Contract Management Framework*. NAO Marketing and Communications Team, London

OGC/OFT (2007) *Making Competition Work for You: A Guide for Public Sector Procurers of Construction*. The Office of Government Commerce, London www.oft.gov.uk/shared_oft/reports/comp_policy/oft892.pdf (accessed 1 March 2012)

Oo, B-L, Drew, DS and Lo, H-P (2007) Applying a random coefficients logistic model to contractors' decision to bid. *Journal of Construction Management and Economics*, **25**, 387–398

Oo, B-L, Drew, DS and Lo, H-P (2008a) A comparison of contractors' decision to bid behaviour according to different market environments. *International Journal of Project Management*, **26**, 439–447

Oo, B-L, Drew, DS and Lo, H-P (2008b) Heterogeneous approach to modeling contractors' decision-to-bid strategies. *Journal of Construction Engineering and Management*, **134**, 766–776

Park, WR and Chapin, WB (1992) *Construction Bidding: Strategic Pricing for Profit*. John Wiley & Sons, New York

Parker, D (2012) *The Private Finance Initiative and Intergenerational Equity*. The Intergenerational Foundation, London. http://www.if.org.uk/wp-content/uploads/2012/02/PFIs-and-Intergenerational-Equity.pdf (accessed 5 March 2012)

Peeling, N (2007) *Brilliant Negotiations: What the Best Negotiators Know, Do and Say: What Brilliant Negotiators Know, Say and Do*. Pearson Education Limited, Harlow

Raftery, J (1991) *Principles of Building Economics*. BSP Professional Books, Oxford

Rathmell, JM (1966) What is meant by services? *Journal of Marketing*, **30**, 33–34

Rethinking Construction Ltd (2002) *Rethinking Construction: Achievements, Next Steps, Getting Involved*. Rethinking Construction Ltd, London. http://www.constructingexcellence.org.uk/pdf/rethinking%20construction/Rethinking_Construction_2002.pdf (accessed 5 March 2012)

RICS (2012) *New Rules of Measurement (NRM2): Detailed Measurement for Building Works*. RICS, London

Rigby, J, Courtney, R and Lowe, D (2009) *Study on Voluntary Arrangements for Collaborative Working in the Field of Construction Services (Parts 1–3)*. Manchester Business School, University of Manchester. http://ec.europa.eu/enterprise/sectors/construction/studies/collaborative-working_en.htm (accessed 5 March 2012)

Rolls Royce (1999) *The Power of Partnerships. The Times* Newspaper Limited and MBA Publishing Ltd. http://www.rolls-royce.com/Images/partnerships_tcm92-11186.pdf (accessed 5 March 2012)

Rolls Royce (2009) Total Care. Rolls Royce, Derby. http://www.rolls-royce.com/Images/TotalCare_A4FINAL_tcm92-15424.pdf (accessed 5 March 2012)

Rowlinson, S and McDermott, P (1999) *Procurement Systems: A Guide to Best Practice in Construction*. E & FN Spon, London

Runeson, G (1996) *The Tenability of Neoclassical Microeconomic Theory and Tendering Theory*. PhD Thesis, School of Construction Management and Property, Queensland University of Technology, Australia

Runeson, G (2000) *Building Economics*. Deakin University Press, Geelong, Australia

Ruuska, I, Ahola, T, Artto, K, Locatelli, G and Mancini, M (2010) A new governance approach for multi-firm projects: lessons from Olkiluoto 3 and Flamanville 3 nuclear power plant projects. *International Journal of Project Management*, **29**, 647–660

Ruuska, I, Artto, K, Aaltonen, K and Lehtonen, P (2009) Dimensions of distance in a project network: exploring Olkiluoto 3 nuclear power plant project. *International Journal of Project Management*, **27**, 142–153

Saxena, A (2008) *Enterprise Contract Management: A Practical Guide to Successfully Implementing an ECM Solution*. J Ross Publishing, Fort Lauderdale, Florida

SEC (2003) *Construction Procurement: Is Local Government Applying Best Value?* Specialist Engineering Contractors Group, London

Seydel, J and Olson, DL (1990) Bids considering multiple criteria. *ASCE Journal of Construction Engineering and Management*, **116**, 609–623

Seydel, J and Olson, DL (2001) Multicriteria support for construction bidding. *Mathematical and Computer Modelling*, **34**, 677–702

Shash, AA (1993) Factors considered in tendering decisions by top UK contractors. *Construction Management and Economics*, **11**, 111–118

Shash, AA (1998) Bidding practices of subcontractors in Colorado. *ASCE Journal of Construction Engineering and Management*, **124**, 219–225

Shash, AA and Abdul-Hadi, NH (1992) Factors affecting a contractor's mark-up size decision in Saudi Arabia. *Construction Management and Economics*, **10**, 415–429

Shen, L, Drew, D and Zhang, Z (1999) Optimal bid model for price-time biparameter construction contracts. *ASCE Journal of Construction Engineering and Management*, **125**, 204–209

Skitmore, M (1989) *Contract Bidding in Construction: Strategic Management and Modelling*. Longman Scientific and Technical, Harlow

Skitmore, M (1991) An introduction to bidding strategy. In: SP Male and RK Stock (eds), *Competitive Advantage in Construction*. Butterworth-Heinemann, Oxford. pp. 139–162

Skitmore, M (2004) Predicting the probability of winning sealed bid auctions: the effects of outliers on bidding models. *Construction Management and Economics*, **22**, 101–109

Skitmore, RM and Lowe, DJ (1995) Human factors in estimating. In: NJ Smith (ed.) *Estimating Capital Cost of Projects*. Institution of Civil Engineers/Thomas Telford, London. pp. 91–100

Skitmore, M and Pemberton, J (1994) A multivariate approach to construction contract bidding mark-up strategies. *Journal of Operational Research Society*, **45**, 1263–1272

Skitmore, M, Runeson, KG and Chang, X (2006) Construction price formation: full cost pricing or neo-classical micro-economic theory. *Construction Management and Economics*, **24**, 773–783

Smith, AJ (1995) *Estimating, Tendering and Bidding for Construction*. Macmillan, Basingstoke

Strategic Forum for Construction (2002) *Accelerating Change*. Construction Industry Council, London. http://www.strategicforum.org.uk/pdf/report_sept02.pdf (accessed 1 March 2012)

Tassie, CR (1980) Aspects of tendering: converting a net estimate into a tender. In: *The Practice of Estimating*. Chartered Institute of Building, London. pp. 6–8

Thorpe, T and McCaffer, R (1991) Competitive bidding and tendering policies. In: SP Male and RK Stock (eds) *Competitive Advantage in Construction*. Butterworth-Heinemann, Oxford. pp. 163–194

Tijhuis, W and Fellows, R (2012) *Culture in International Construction (Cib)*. Spon Press, Abingdon Treasury Corporation of Victoria, Australia (2002) *Bid Analysis Evaluation Policy*. http://wwwvicgovau/treasury/treasuryhtml (accessed on 10 January 2002)

Tucker, SN, Love, PED, Tilley, PA, Salomonsson, GS, MacSporran, C and Mohamed, S (1996) *Perspective's of Construction Contractors Communication and Performance Practices, Pilot Survey*, DOC 96/29 (M), DBCE, CSIRO Press

Turner, A (1997) *Building Procurement*, 2nd edition. Macmillan, London

Turner, R and Remenyi, D (1995) Estimating costs and revenues. In: JR Turner (ed.) *The Commercial Project Manager*. McGraw-Hill Book Company, London. pp. 31–52

Tweedley, N (1995) *Winning the Bid: A Manager's Guide to Competitive Bidding*. Pitman Publishing, London

Vickrey, W (1961) Counterspeculation, auctions, and competitive sealed tenders. *Journal of Finance*, **16**, 8–37

Walker, D and Hampson, K (eds) (2003) *Procurement Strategies: A Relationship-based Approach*, Blackwell Science, Oxford

Walker, DHT, Hampson, K and Peters, R (2000) Project alliancing and project partnering in information and communication. In: A Serpell (ed.) *Construction Procurement*. Pontificia Universidad Catolica de Chile

Walker, D and Rowlinson, S (eds) (2008) *Procurement Systems: A Cross-Industry Project Management Perspective*. Taylor and Francis, Abingdon

Wanous, M, Boussabaine, AH and Lewis, J (2000) To bid or not to bid: a parametric solution. *Construction Management and Economics*, **18**, 457–466

Wanous, M, Boussabaine, AH and Lewis, J (2003) A neural network bid/no bid model: the case for contractors in Syria. *Construction Management and Economics*, **21**, 737–744

Ward, SC and Chapman, CB (1988) Developing competitive bids: a framework for information processing. *Journal of the Operational Research Society*, **39**, 123–134

Weverbergh, M (2002) A comment of 'predicting the probability of winning sealed bid auctions: a comparison of models'. *Journal of the Operational Research Society*, **45**, 1156–1158

Winch, GM (2001) Governing the project process: a conceptual framework. *Construction Management and Economics*, **19**, 331–335

Winch, GM (2006) The governance of project coalitions – towards a research agenda. In: DJ Lowe (ed.) with R Leiringer. *Commercial Management of Projects: Defining the Discipline*. Blackwell Publishing, Oxford. pp. 324–343

Winch, GM (2010) *Managing Construction Projects: An Information Processing Approach*, 2nd edition. Blackwell Publishing, Oxford

Wong, CH, Holt, GD and Cooper, PA (2000) Lowest price or value? Investigation of UK construction clients' tender selection process. *Construction Management and Economics*, **18**, 767–774

Wong, CH, Holt, GD and Harris, P (2001) Multi-criteria selection or lowest price? Investigation of UK construction clients' tender evaluation preferences. *Engineering, Construction and Architectural Management*, **8**, 257–271

Wood, G (2005) Partnering practice in the relationship between clients and main contractors. RICS Research Paper Series, **5**. RICS, London. http://www.rics.org/site/download_feed.aspx?fileID=4906&fileExtension=PDF (accessed 1 March 2012)

Zarkada-Fraser, A and Skitmore, M (2000) Decisions with moral content: collusion. *Construction Management and Economics*, **18**, 101–111

Part 4

Case Studies

Case Study A	Football Stadia	381
Case Study B	Terminal 5 (T5) Heathrow	455

Case Study A
Football Stadia

Introduction

Recently, several major stadia developments have been constructed in the UK, forming the centrepiece for urban redevelopment. The procurement and implementation of these projects exemplify elements of both 'best' and 'poor' practice, as illustrated by the following case studies: the Millennium Stadium, Cardiff; the Emirates Stadium, Islington, London; and Wembley Stadium, London.

Table A.1 Football stadia comparison of key features, etc.

	The Millennium Stadium, Cardiff	The Emirates Stadium, Islington, London	Wembley Stadium, London
Capacity	74 500 UEFA 5-Star rating	60 000 UEFA 5-Star rating	90 000 UEFA 5-Star rating
Cost	£126 million	£220 million	£792–£827 million
Construction started	April 1997	March 2004	30 September 2002
Construction completion	October 1999	July 2006	12 March 2007
Special features	Retractable roof	Low slung roof	Sliding roof and 133 m high landmark arch
Sponsor/client	Millennium Stadium plc (a subsidiary company of the Welsh Rugby Union [WRU])	Ashburton Properties Ltd (a member of the Arsenal group of companies)	Wembley National Stadium Ltd (WNSL)
Architect	HOK & LOBB Sport Architecture	HOK Sport Architecture	World Stadium Team, a joint venture between Foster and Partners and HOK Sport Architecture
Structural engineer	W S Atkins	Buro Happold	Mott Stadium Consortium
Project manager / quantity surveyor	Bute Partnership/OBK	AYH plc	(Capita) Symonds
Main contractor	Laing Construction	Sir Robert McAlpine	Multiplex

The Millennium Stadium, Cardiff

The prime function of Cardiff's Millennium Stadium, the Welsh national stadium, is to host international rugby union and football fixtures. When opened in June 1999 it was the UK's largest sports venue with seating for 74 500 spectators; this capacity, however, has since been exceeded by Old Trafford, Twickenham and the new Wembley Stadium. The venue was developed and is operated by Millennium Stadium (MS) plc, a subsidiary company of the Welsh Rugby Union (WRU), although the project received significant public funding from the Millennium Commission.

Featuring the first retractable roof in the UK, the stadium is a multipurpose, UEFA five-star rated, world-class venue. To date, five major sports organisations have used its facilities, with fixtures including two Rugby World Cups, an historic Grand Slam victory, six FA Cup Finals and numerous headlining concerts. Further, since opening it has attracted in excess of 1.3 million visitors each year.

In 1995, Russell Goodway was appointed leader of Cardiff County Council (CCC). Goodway wished to raise the profile of Cardiff and developed a strategy to encourage economic growth and development in the city. A key feature of this policy was the support of major sporting and cultural events that would give Cardiff an international presence: this included backing the WRU's proposal in 1995 to hold the 1999 Rugby World Cup in the city. Both Goodway and the then treasurer of the WRU, Glanmor Griffiths, are jointly considered to be the project champions of the Millennium Stadium. CCC estimated that the stadium development would generate annual revenue of £19 million for the city through staging events, plus £38 million through the employment of local construction workers during its building phase, and, based on South Africa's experience in 1995, a percentage of the revenue from staging the Rugby World Cup (overall, the South Africa Rugby World Cup grossed £800 million).

The opening of the stadium coincided with the inauguration of the Welsh Assembly in 1999.

Requirement identification

In 1994 a committee was established to investigate the viability of rejuvenating the Welsh National Stadium (Cardiff Arms Park) as part of the regeneration of west Cardiff. Although the existing stadium was relatively new, having been completed in 1983, it was based on an outdated design developed in 1962. The capacity of Cardiff Arms Park was initially 53 000, which included standing room for 11 000 spectators. However, with the introduction of new safety regulations, this had been reduced to 48 000. Crucially, the stadium lacked the prerequisite amenities to stage major international events: its only facilities were toilets. The committee concluded that the stadium had '*long since*' been surpassed by those available in England and Scotland, both countries having developed stadia with capacities of 75 000 (Twickenham Stadium) and 67 000 (Murrayfield Stadium) respectively. Further, France was on the verge of constructing its 80 000 capacity Stade de France.

Requirement specification

In 1995, the WRU were awarded the franchise to stage the 1999 Rugby World Cup.

Due to the significant disruption of the Durban Rugby World Cup in 1995, the result of poor weather conditions, Cardiff's selection was conditional on the provision of a covered stadium at Cardiff Arms Park. In order to fulfil this obligation, plus create a multifunctional venue and provide a natural grass pitch (a necessity for staging rugby fixtures), a retractable roof was included in the design brief along with a removable pitch. This requirement would entail an innovative design, as at this point in time the only retractable roofed stadium in Europe was Ajax Football Club's 50 000-capacity Amsterdam Arena.

The brief, therefore, was to provide a versatile venue, which incorporated a retractable roof and removable pitch, allowing income to be raised by staging a variety of events at the stadium.

Solution selection

Site location

Several diverse development solutions were reviewed: one option included the addition of a third tier to the existing stadium, and another involved relocating the stadium to a completely new site outside the city centre. The latter option, however, was ultimately rejected following pressure from local traders, and because the current site was considered to be an excellent location, easily accessible by public transport via the adjacent bus and rail terminals. Ultimately, it was decided to redevelop the stadium in its existing location as part of a wider regeneration of the city centre.

Site constraints

Redevelopment of the Cardiff Arms Park site necessitated rotating the orientation of the stadium by 90 degrees so that the pitch faced north–south as opposed to east–west as in the old stadium. (See Figure A.1 which shows the configuration of the new stadium in relationship to the existing venue.) As can be seen from Figure A.1, the site was restricted on all sides, with the River Taff flowing down its west side, Westgate Street to the east (where a number of residential apartments and offices, including the BT tower, were located), and to the south and east a number of buildings had to be removed to make way for a wide concourse to the stadium and the River Walk. These buildings included The Crown public house, Department of Social Security (DSS) offices, the Territorial Army headquarters, the Olympic-sized Empire Swimming Pool and a BT telephone exchange. Compensation payments for relocating the occupants and demolishing the existing buildings were reported to have been approximately £23 million.

As the River Taff is tidal, a further problem to be considered was the potential for flooding of the Cardiff Bay area, which could affect the water table and, therefore, the foundations of the new stadium. To alleviate this, a barrage was constructed on the river, and the Welsh Assembly installed a pumping station to control water table levels and indemnified the WRU from flooding above a set datum.

However, the constraint that proved to be the most significant for the project involved a wall shared by the adjacent Cardiff Rugby Football Club's (CRFC) south stand and Cardiff Arms Park's north stand.

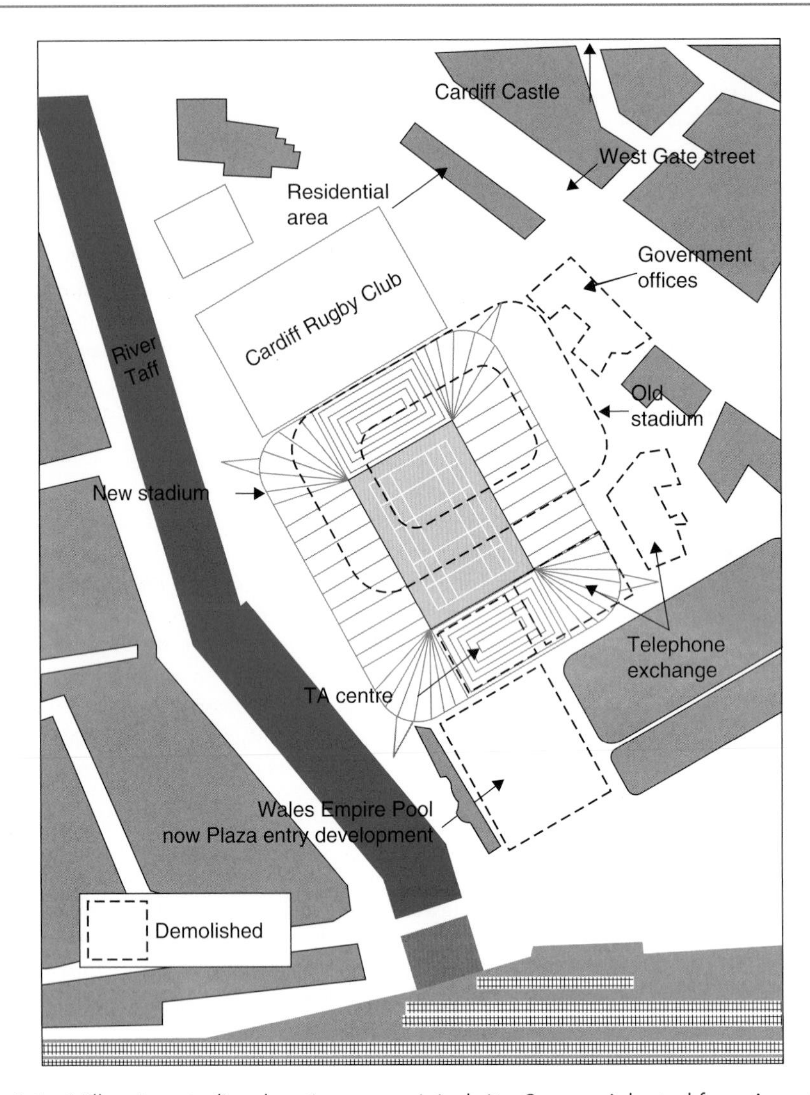

Figure A.1 Millennium stadium location over original site. Source: Adapted from Anon (1999a).

Project scope

Rather than undertaking a £40 million refurbishment of the existing stadium, the need for a retractable roof meant that funding of over £100 million would be required to build a new stadium.

Stadium design solution

The WRU appointed a conceptual design team with responsibility for the concept design, contracting strategy and bidding process. The team comprised: architect Rod Shear of Hok & Lobb (formerly Lobb Partnership), who had recently finished the award-winning design of the 1995 RIBA Award-winning McAlpine Stadium in Huddersfield and who subsequently went on to design the Sydney 2000 Olympic stadium; structural engineers, WS Atkins; and the Bute Partnership who provided quantity surveying and project management services (however, their appointment was subsequently subsumed within the appointment of project managers O'Brien Krietzberg [OBK]). The Bute Partnership was responsible for formulating and negotiating the contract with Laing Construction (Laing) which very effectively protected the interests of WRU.

The resulting design for the 222.4 m long by 181 m wide, 74 500 seat capacity, three tier stadium with its fully retractable roof was novated to Laing.

Box A.1 Key statistics

The Millennium Stadium provides:

- 74 500 seats overall
- 17 400 debenture seats and 1500 club seats
- 128 hospitality boxes
- 6 restaurants, 15 public bars, 25 food and beverage outlets, 13 programme and merchandise outlets, 13 parenting rooms and 17 first aid rooms
- 760 toilet pans and 640 wash band basins
- 15 communications rooms
- A grassed area of 9636 m^2 and pitch size of 120 m × 79 m

Design features included:

- **Roof**: the stadium's key feature is its retractable Kalzip roof; comprising a 26 862 m^2 'fixed' area and an 8960 m^2 moveable section that opens to form a 105 × 80 m aperture. The opening can be opened or closed in 20 minutes to form a fully enclosed arena; when closed, air is removed via extractor fans in the roof. The roof is suspended by cables from four 90 m high pylons, which support two main steel trusses running the length of the stadium: each 13 m deep at the centre and 4 m deep at its ends. Spanning between these main trusses, each sliding roof section contains five lateral trusses. The design avoids the need for any view-obstructing columns
- **Cantilevered walkways**: access to the stadium is by means of a cantilevered timber-decked walkway over the adjacent River Taff, formed using cable-stayed cantilevered plate girders and secondary steel beams
- **Pitch**: the pitch utilises the GreenTech® Turf System based on 150 mm deep 1.3 m^2 modular trays. The stadium incorporates 7412 of these modules, which can be individually removed to enable localised repairs to be made both easily and inexpensively. Further, the system allows the playing surface to be removed so that other events, for example, concerts and exhibitions, can be staged on the underlying asphalt surface
- **Recycling system**: the stadium's roof drainage system was designed to minimise its discharge into the River Taff, incorporating a pitch irrigation system fed by recycled water. When the roof is open, the system manages each roof section independently, but when closed it integrates to drain the entire roof area, with a maximum capacity of 820 litres/s
- **Security and spectator safety**: included within the security systems work package (valued at £2.8 million) was the supply and installation of security and CCTV, fire detection, and public address voice alarm systems, plus radio, television, data and telephone distribution and a system for the hard of hearing

Development programme

The key milestone dates for the Millennium Stadium are shown in Table A.2.

Project budget

The initial budget estimate for the stadium development was £121 million; however, this figure rose to £126 million on the appointment of the main contractor Laing Construction. The construction contract accounted for £99 million with the remaining £27 million made up of consultants' fees and compensation payments to the owners of four buildings that had to be demolished to achieve a clear concourse for the stadium.

Financing the project

Based on this initial budget, it was apparent that financial support would be required from the public sector. The only viable mechanism to dispense this funding, however, was the National Lottery, which had been established in 1994. Funded by the National Lottery, the Millennium Commission had been set up to

Table A.2 Millennium stadium project milestones.

1995	WRU win the bid to host the 1999 Rugby World Cup
1996	
March	The Millennium Commission decides to support the Cardiff Stadium project. The Millennium Stadium (MS) plc is established to finance and operate the stadium
June	Laing and Taylor Woodrow are declared the preferred bidders out of 14 formal bids received, which included Ballast Needham, Bovis and Sir Robert McAlpine
July/August	Laing is appointed main contractor in preference to Taylor Woodrow. A dispute breaks out with CRFC over a shared wall
1997	
March	Laing signs a £99 million GMP contract. Laing's 'A' team, who had been responsible for building the second Severn Bridge, are brought in to construct the Millennium Stadium. The existing west stand is demolished
26 April	The last game (the Welsh Cup Final) is played in the old stadium. The demolition of Cardiff Arms Park begins shortly afterwards
June	Demolition works continue. CRFC and Ebbw Vale win an injunction in the High Court against WRU over the issue of 'golden' shares and their threat of expulsion from the Premier Division
July	The south stand is demolished and the new foundations are commenced. The existing east stand is initially retained and used by Laing as offices OBK appoint Dick Larson as programme manager
September	CRFC claims that work on the Millennium Stadium has resulted in flooding to three of its function rooms
1998	
July	The theft of a confidential unsigned letter from Dick Larson of OBK, which criticised the management structure at WRU and stated his wish to resign, is reported. Also, a letter from Leo Williams of Rugby World Cup to Glanmor Griffith is leaked to the press, expressing his concern that the stadium may not open on time
September	Laing proposes working 24 hours a day from Monday to Saturday and an 8 hour shift on Sundays. Laing installs double glazing in nearby homes and pays compensation to residents for noise pollution
October	Problems occur between Laing and Cimolai its Italian steelwork subcontractor, resulting in delays (a claim is settled in Cimolai's favour for £14 million). Local residents object to Laing's proposal to extended working hours but CCC approves the measures. A dispute arises with disabled associations and CCC who reject Laing's plans for non-compliance with disability regulations. Laing announces losses of £26 million on the project, due mainly to the decision not to demolish the north stand
November	Laing's project manager, Simon Lander, resigns and is replaced by Martin Foster
December	Laing announces that it will not be taking legal action to recoup its losses incurred on the project
1999	
January	Erection of steelwork commences
February	A new project director (the project's third), Trevor Hodges, is appointed as a replacement for Martin Foster
March	Laing commence negotiations with WRU to increase its GMP in an attempt to recoup its £26 millon overspend. Roof sections, including two 220 m span steel trusses are in position
16 June	Contract completion date

(Continued)

Table A.2 (*cont'd*).

23 June	The first game is held at the stadium to an audience of 27 000 (when the stadium was only three-quarters complete). After the game, the turf pallets are removed to enable the work to continue
1 July	Roof cladding panels are fitted and temporary support steelwork replaced by cables threaded through the masts. Governance of Wales devolved to the National Assembly of Wales
29 August	The Millennium Stadium is awarded a crowd safety certificate
September	Laing announce a £31 million loss on the project
1 October	Rugby World Cup opening ceremony is staged
6 November	Rugby World Cup Final
31 December	At a cost of £300,000, additional steel supports are added to reinforce the stadium's middle tier, to withstand the additional load of rock fans at a Manic Street Preachers 'Manic Millennium' Concert
2000	
May	Plans are drawn up to demolish the retained north stand that adjoins CRFC's south stand, as the latter is found to have 'concrete cancer'

Source: Adapted from Morris (2005)

fund projects across the country to celebrate the millennium. In order to qualify for Millennium Commission funding, the stadium development had to meet the following criteria:

- Attract public support
- Make a substantial contribution to the community
- Look back on the past Millennium and forward to the new one
- Denote an important movement in history
- Be of a high architectural design and environmental quality
- Involve collaboration with the local community

Finally, projects eligible for funding from other sources would not be considered.

Only one significant project would be financed in each of the eight areas identified by the Commission (one of which included Cardiff). The stadium development was in direct competition for funding with a proposal to develop the Cardiff Bay Opera House, which had attracted significant political support. However, in March 1996, having initially rejected WRU's first proposal, the Millennium Commission decided in favour of a revised stadium proposal, awarding the scheme a grant of £ 46,360,000 (following this award, the proposed new stadium was renamed the Millennium Stadium). As previously mentioned, despite receiving substantial public funding (which accounted for approximately 38% of the initial budget), the ownership of the stadium was retained by the WRU.

The WRU was, therefore, required to raise the balance from commercial sources. In conjunction with CCC, the WRU established Millennium Stadium (MS) plc (incorporated in March 1996) to raise this additional finance and to design, build and operate the stadium for 50 years on behalf of the WRU. WRU retained 50 000 shares of MS plc and appointed six members of the board, and CCC had one 'golden share' and appointed five members. Glanmor Griffiths, initially the treasurer and then chairman of MS plc and a former director of the Midlands Bank, was actively involved in raising the additional funds: a £60 million loan was obtained from Barclays Bank, with a further £20 million generated through debentures.

An extra £4 million was provided by CCC for the Riverside Walk.

Project management

Pat Thompson, operations manager of CCC, was appointed project coordinator for the Millennium Stadium. His first key decision was to appoint OBK, the largest project management consultancy in the US, who had been involved in the provision of facilities for the Atlanta Olympic Games and in numerous major sporting stadia throughout the US. In March 1996 OBK appointed a project controls manager to assist in establishing a tender process and contract strategy for the project. Following this, in July 1996 they assigned a project manager to oversee the engineering aspects of the project and then in July 1997 a programme manager.

Project procurement (contractor appointment)

As the project had received a substantial amount of public funding, MS plc adopted a two-stage tender process, implemented in accordance with European Union procurement policies for open tendering. Initially 30 companies expressed an interest in the project with 14 contractors submitting formal bids. In June 1996, these were reduced to five companies comprising Ballast Nedham, Bovis, Laing Construction, Sir Robert McAlpine and Taylor Woodrow (their bids ranged from £92 million to around £110 million). Subsequently, two companies entered the second stage: Laing Construction and Taylor Woodrow. Laing's bid was £92 million, and Taylor Woodrow's tender was approximately £95 million.

The bids from Laing Construction and Taylor Woodrow (who reportedly had submitted an impressive technical bid) were subjected to thorough evaluation. However, as Laing had recently successfully completed the second Severn Bridge (a major local project) it utilised the construction team from this project to support its proposal; this impressed the decision makers, Russell Goodway and Glanmor Griffiths. Therefore, having submitted a slightly lower bid, Laing was selected as the preferred bidder.

In March 1997 the construction contract for the Millennium Stadium was awarded to Laing Construction based on a guaranteed maximum price (GMP) of £99 million. Whether or not Laing had undertaken sufficient engineering work to enable a reliable cost estimate to be prepared is unclear. Also, there were indications that Laing's bid had been based on the 'value' or 'price' acceptable to the client, rather than on an estimation of the likely cost of the development plus an acceptable level of profit (Morris, 2005). Porter's 'Bargaining Power of the Buyer'(Porter, 1985) may have resonance with Laing's pricing strategy. MS plc had revealed that an 'acceptable price' was in the region of £100 million, due to its own tight financial constraints: therefore, if the resulting bids were any higher, the project would probably have been aborted.

Contract

The project coordinator identified three requisites for the contract. It should:

1. Identify a project manager
2. Be a design-build contract
3. Include a guaranteed maximum price clause

Based on these criteria, the New Engineering Contract (NEC) Option C Target Contract with activity schedule was considered to be the most appropriate standard form and, therefore, adopted on the project. Additional reasons for its selection included:

- Responsibility for the design could be novated to Laing. As the stadium's design was not fully developed, the contract enabled Laing to assume responsibility for its design based on the WRU's specification (client's requirements) and the concept design prepared by Hok & Lobb and WS Atkins
- Financial risk and rewards (pain/gain) could to be shared between the parties
- A GMP could be agreed, thereby limiting the financial exposure of the client (an acknowledgment of WRU's severe funding constraints)
- Performance and, therefore, progress payments could be linked to the activity schedule. The completion date was immovable: the stadium had to be completed in time for the Rugby World Cup in October 1999, with a game scheduled for June 1999 between Wales and South Africa

A number of astute revisions were made to the contract by OBK and Bute Partnership on behalf of MS plc. The following are examples:

> 'If the Price for Work Done to Date exceeds the Guaranteed Maximum Price, the Contractor is liable and bears the amount by which the Guaranteed Maximum Price is exceeded.'

> 'The existing north stand will be retained in its existing condition and there will be no time or cost implications arising from this.' (Morris, 2005)

The latter insertion proved to be extremely costly for Laing.

Laing had a year to review its price, from March 1996 when the Millennium Stadium was given the 'green light' to March 1997 when contracts were signed; however, it appears that its engineering definition at this point was not as advanced as its rival Taylor Woodrow. A Laing representative was later to admit that its price was based on '*sketchy concepts*', while its estimates fell short of the customary 70–80% cost certainty. Similarly, an observer considered that:

> 'Their [*Laing's*] design was 20% progressed, not the 85% progressed that you would expect as a basis to sign the contract... ...Taylor Woodrow knew more about the scheme in the 12 months having never touched it compared to Laing having not looked at it.' (Morris, 2005)

Taylor Woodrow, on the other hand, was deemed to have spent considerably more on front-end engineering, compared with Laing Construction. However, neither Laing nor Taylor Woodrow had received either a letter of intent or memorandum of understanding during this period. Any work undertaken, therefore, would have been at the contractor's risk.

> 'At the end of the day, Laing failed because during the 12 months, they did not put any commitment into understanding what they were building.' (Morris, 2005)

Generally, contractors are reluctant to enter into GMP contracts due to concerns over potential 'risk dumping' by the client. Further, the tender process generally does not allow sufficient time for the contractor to carry out appropriate risk assessment and/or peer review to determine a suitable price premium to cover this commitment, and time constraints create a greater risk of human error (Minogue, 1998).

On the Millennium Stadium, Laing agreed to accept the risk inherent in a dispute between the WRU and the neighbouring CRFC, despite there being limited time to assess its potential impact on the stadium's design. During negotiations between MS plc and Laing (between late 1996 and early 1997), the relationship between CRFC and the WRU had completely broken down, with the parties resorting to litigation (the case was eventually heard in June 1997). Around February 1997, Laing was informed that because of this dispute, two of the stadium's four masts would have to be redesigned so as not to enter CRFC's airspace, that the north stand could not be demolished as it would affect the CRFC's south stand, and that its cranes would not be permitted to over-swing CRFC's airspace. Laing was given a month to consider the cost implications of these changes and constraints. It was reported that Laing's assessment was that:

> '... the north stand would compensate for any disruption that they would have... [and that overall there would be] a neutral cost and time impact.' (Morris, 2005)

The contract established a target price of £91 million below which Laing and MS plc would divide any savings 50/50. However, should Laing's costs exceed the GMP of £99 million then Laing would be solely liable for the additional expenditure (see Figure A.2). Interestingly, initially Laing believed that it could reduce the target price by 15% through value engineering. Laing was to be paid on an open-book basis.

Although Laing's representatives negotiated and agreed to every clause in the contract, it was apparent that Tony Evans (MD of Laing Civil Engineering) was keen to 'win' the contract.

> 'The Millennium Stadium is a prestigious development, which, when completed, will host the finals of the 1999 Rugby World Cup. John Laing has the expertise to handle such a complex operation. As a Welshman, with rugby in my blood, it is an honour to be involved in this project.' (Tony Evans quoted in Anon, 1996)

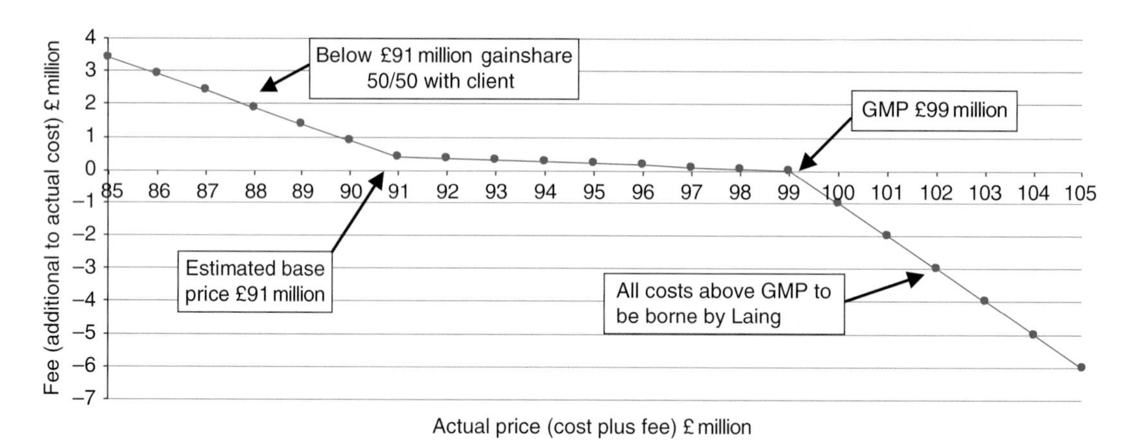

Figure A.2 Fee incentives versus actual price for Laing Construction. Source: Adapted from Morris (2005).

The question to be asked is:

> Why did Laing agree to accept the risk inherent in the dispute with CRFC?

It seems that Tony Evans, a senior member of the CRFC, believed he would be able to use his influence within the club to mitigate and manage this risk. However, as it transpired he had no influence at all; in fact, CRFC continued to be a source of difficulty for Laing throughout the life of the project. David Barry, managing director of OBK, was to comment:

> 'I don't think Laing fully understood the implications of the dispute.' (Quoted in Cook, 1998)

Asset supply

As mentioned above, the outbreak of a dispute between WRU and CRFC over a shared wall resulted in significant revisions to the proposed design of the stadium, for example:

- Two of the four corner masts had to be redesigned so as not to over-sail CRFC's airspace, requiring the masts to be repositioned and strengthened
- This also had a knock-on effect on the stadium's roof structure; altering the loadings relating to the roof's position

These recalculations and design changes had significant cost implications for Laing.
The dispute also impacted on Laing's working methods, for example:

- Tower cranes were not permitted to over-sail CRFC's airspace
- The cladding subcontractors were required to work from within the site limits
- Access to and from the site was severely restricted

Again, due to inefficient working, etc, these increased Laing's costs.
Work on the scheme commenced in March 1997 with the demolition of the existing west stand. Laing's first work package for demolition works, however, came in at £5 million compared with a budgeted £3 million. Overspend on the project averaged £1 million a month against earned value; so right from the outset Laing was encountering schedule and cost overruns.
Ian Jamieson, Laing's operations manager on the project, later commented that there had been three key elements of risk to be addressed, relating to:

1. The demanding time constraints placed on the project
2. Environmental issues particularly associated with noise control, and dust and water pollution
3. The restrictions imposed by the confined site

As an indication of the scale of the project, the quantities of materials/components used on the stadium included: 1350 foundation piles, 4000 tonnes of steel reinforcement, 40000 tonnes of concrete, 12000 tonnes of structural steel, 4 masts (90341 m high), 212520 km of tendons, 24000 m^2 of wall cladding and 650 smoke/heat detectors.

OBK introduced systems to monitor all potential changes and to log correspondence on the project, thereby ensuring that all issues raised received an appropriate formal response: a requirement of the contract. Also, budgetary control (based on a 14-day cycle) was attained by measuring and reconciling earned value against the project's budget. Further, procedures highlighted potential slippages in the programme.

According to Caletka and Merrifield (2000):

> '... if there was a dispute regarding un-agreed issues, the aggrieved party was then required to "put up or shut up", that is accept the project management team's decision or seek immediate recourse through adjudication. It may seem harsh, but the great merit of such a procedure is that it forces issues out into the open at which time the parties can seek resolution. This approach proved good protection for the client later in the project when Laing's design, construction and budget difficulties were realised... Although threatened a few times, there have been no adjudications on this project.'

In contrast, it was considered by observers that Laing did not prepare an appropriate implementation programme against which progress could be measured. Neither, apparently, did they have an adequate construction plan – for example, little thought had been given to where pre-assembly activities for the steelwork would take place. Laing had assumed that all the works could be carried out from inside the bowl of the stadium, but this proved not to be possible (Morris, 2005).

Due to the reorientation of the pitch, piling works for the new east stand could begin immediately as it was positioned on the original pitch (see Figure A.1). Work then continued anti-clockwise around the new pitch in stages, as the site was handed over to Laing. The final stage, however, required the demolition of a BT telephone exchange. This section was due to be handed over to Laing in August 1998; however, it was apparent that the contractor knew that it would be impossible to achieve the completion date given its current progress. To illustrate just how far behind schedule Laing was at this point, the piles to the south stand were only installed during September 1998, a year before the Rugby World Cup in October 1999 and only nine months before the scheduled match against South Africa. Rumours that Laing would not meet its completion date emerged in the spring of 1998.

Despite concerns over the lack of progress on the stadium raised by Leo Williams, chairman of Rugby World Cup Ltd (concerns echoed internally by WRU), Laing's management did not appear to respond to the seriousness of the situation until September/October 1998, effectively 18 months into the project. It is unclear precisely what happened during this period, apart from demolition works and laying the stadium's foundations. Perhaps part of the problem was that Laing had been slow in appreciating the consequences of having to reposition two support masts requiring the provision of asymmetric support to the roof. This resulted in:

- The shear walls being increased to twice their original size
- Revisions to the steel reinforcement in concrete slabs to maintain rigidity
- An additional 6000 tonnes of steel due to the need to redesign the roof structure

Subcontractor dispute and 'a step change'

The probable trigger that led senior managers at John Laing plc to intervene in the delivery of the project was the dispute that arose between Laing Construction and Cimolai, its Italian steel subcontractor. Laing and Cimolai had worked successfully together on the recent Severn Bridge project but commentators questioned why Laing had insisted on using Cimolai again on the Millennium Stadium. It was originally anticipated that a British steel maker would supply and fabricate the steel – for example, Mowlem or Watson. In the event, British steel was shipped to northern Italy where it was fabricated and then transported back to the UK. At the time the Yugoslavian war was in progress and at one point the US Navy was threatening to close the Adriatic Sea.

Reportedly, Laing's budget for the combined steelwork package was £20 million, a relatively low figure. Subsequently, Cimolai submitted an £18 million claim against Laing for 'providing insufficient and late design information'. Apparently, this claim was settled privately by Laing for a reported £14 million in order to meet the crucial completion date. As Laing had entered into a traditional contract with Cimolai, none of the risk/reward incentives were passed on to the subcontractor. Laing, therefore, had to absorb the increased steelwork costs (Morris, 2005).

Johnson *et al.*'s/Mendelow's 'power interest matrix'[1] is used to allocate the relevant parties involved in the Millennium Stadium project to four categories according to their power to influence/level of interest (Figure A.3). These categories are labelled: key stakeholders, keep satisfied, keep informed and minimal

POWER TO INFLUENCE	LEVEL OF INTEREST: LOW	LEVEL OF INTEREST: HIGH
HIGH	**KEEP SATISFIED** **Millennium Commission:** Chairman: Virginia Bottomley Project Monitor: David Baird **John Laing plc:** Chairman: Sir Martin Laing Deputy Chairman: Robert Wood **Cardiff Rugby Football Club:** Chairman: Peter Thomas **Barclays Bank:** £60 million loan **Debenture holders:** £20 million **Steel subcontractor:** Cimolai	**KEY STAKEHOLDERS** **Client:** Cardiff County Council and Welsh Rugby Union under Millennium Stadium plc **Welsh Rugby Union** Chairman: Vernon Pugh QC Treasurer and then Chairman: Glanmor Griffiths (Project Champion – main fund raiser) **Cardiff County Council:** Council Leader (Labour): Russell Goodway (Project Champion) Chief Executive: Byron Davies Project Coordinator and Operations Manager: Pat Thompson **Main Contractor:** Laing Construction Chairman: David Blair (left Oct 1998) replaced by James Armstrong Managing Director Civils: Tony Evans (left Oct 1998) Project Director: Simon Lander (left Oct 1998) replaced by Martin Foster Project Manager: Trevor Hodges Site Construction Manager: Ian Jamieson **Project Manager:** O'Brien Krietzberg Programme Manager: Dick Larson Project Manager: Todd Staley replaced by Simon Turner Project Controls Manager: Anthony Caletka
LOW	**MINIMAL EFFORT** **Local Community** **Associated Rugby Clubs**	**KEEP INFORMED** **Appointed by WRU/CCC:** Architect: Hok & Lobb Civil and Structural Engineer: WS Atkins **Subcontractors (for example):** Turf System Supplies: Inturf Roofing Specialist: Melvyn Rowberry Welding Services Ltd Roof Material: KALZIP 400 by Hoogovens Aluminium Building Systems Ltd Monitoring Systems: Honeywell Control Systems Ltd Climbing Framework: Doka UK Formwork Technologies Ltd **Local residents:** 147 flats Westgate Street **Disabled associations** **Local press:** Western Mail, The Echo

Figure A.3 Power/Interest Matrix Millennium Stadium Project. Source: Morris (2005) – developed from Johnson *et al.* (2005) and Mendelow (1981).

effort. Based on this analysis, Cimolai is shown as having considerable power to influence the delivery of the project despite its apparent subsidiary role as a subcontractor.

In October 1998, Laing's parent company John Laing plc announced losses of £26 million on the Millennium Stadium project. This loss was blamed on the consequential changes to the stadium's design arising from WRU's dispute with CRFC. There followed a significant reorganisation of Laing's management structure with three key personnel leaving the company: Tony Evans, Managing Director of Laing Civil Engineering, Simon Lander, Project Director and David Blair, chairman of Laing Construction. The following statement made by Simon Lander is particularly telling:

> 'While I regret having to leave the Millennium Stadium project before its completion, I have taken a great deal from my experiences on such a challenging project. One of the areas I have learnt valuable lessons from relates to the need to ensure that agreed budgets are capable of delivery and are in line with client's requirements.' (Quoted in Anon, 1998e)

Laing appointed a new project manager whose background was in building construction. This appointment was opportune as the project was at the point of transformation from a 'civil engineering' to a 'building' project, requiring building project management skills.

In November 1998 it was revealed that, despite their outward expression of confidence in Laing, WRU had a contingency plan to stage their match with South Africa on 26 June at Le Stade, Paris in the event of the Millennium Stadium not being completed on time.

Finally, with less than a year of the contract period left, Laing acknowledged that it had problems and that all the parties involved with the project would have to collaborate if they were to meet the completion date. As an example of how Laing attempted to promote a spirit of cooperation, before commencing the roofing contract senior management and key personnel from both Laing and their subcontractor Kelsey took part in a conference to discuss the principles of lean construction and collaboration, in particular how they could be applied to the project. Moreover, Sir Martin Laing, chairman of John Laing plc, visited Cardiff to assure the client that Laing 'would deliver'. However, in order to deliver this undertaking a recovery plan was required, as, based on Laing's current progress, the stadium wouldn't be completed until May 2001.

Significantly, Laing took the decision not to consider litigation, as a means of recouping their losses, until after the project was completed, preferring to concentrate on finishing the works, believing such a course to be a detraction and divisive.

Accelerating the works

From October 1998 to October 1999, Laing and their subcontractors worked 24 hours a day, Monday to Saturday, and 8 hours on Sundays. However, to enable this and to obtain appropriate permissions, Laing incurred costs of £400,000 through installing double glazing and compensating local residents.

Based on predictions generated by OBK, additional cranes were required to help improve progress. At times, 17 cranes were operating within the stadium requiring detailed programmes to prevent jib strikes, while over 2000 people were employed on the site. Also, logistics crews were set up to control deliveries through the night, such as steel, which came in as exceptional loads from the docks with police escort.

> 'Laing mobilised all their best people from all over the country. They performed wonders. Once they had accepted where they were and what a mess they were in. And what they achieved that year! They performed wonder: absolutely phenomenal.' (Morris, 2005)

However, problems persisted, with the first section of temporary steelwork only being erected in January 1999. Further, to facilitate this, Laing required all the area surrounding the stadium, causing delays to the construction of the nearby Westgate Plaza (simultaneously being built by Taylor Woodrow) and requiring Laing to negotiate the use of CRFC's pitch on which to layout their steelwork prior to its erection. Apparently, Laing had to pay a substantial premium for this facility, the cost of which has not been disclosed.

Project completion

On 26 June 1999, Wales were scheduled to play South Africa. It was important that this fixture was held at the stadium to generate confidence that the venue would be ready for the October World Cup. Although the game went ahead, attended by 27 000 spectators, the stadium was only '*sufficiently complete*' and when it was over, the pitch was removed so that construction works could continue. The stadium was finally awarded a safety certificate on the 29 August 1999.

Although Laing did complete the stadium in time for the Rugby World Cup in October 1999, it incurred a loss, at the time believed to be in the order of £34 million. Surprisingly, Laing made no formal claim against MS plc to recoup these costs. In one article, David Barry of OBK was reported as saying:

> 'We have not had a notification from Laing, nor do we expect any.... [due to] clear and clever revisions to the NEC contract that protect us.' (Quoted in Cook, 1998)

Further, a client representative stated:

> 'Laing had not one basis for having to claim, whilst the client had significant claims, not just for damages, delays compensation and counter claims to the order of £20 million, which they never pursued..' (Morris, 2005)

In September 1999, a disappointed Laing announced that it had reached an agreement with WRU over the cost of the project. Laing were to be paid their original GMP, plus the cost of works re-specified by the client. Further, John Laing plc's finance director, Adrian Ewer, admitted that while Laing had delivered a '*super project*' it had been a '*financial disaster*'. He attributed the losses incurred to errors in their tendering process, stating: '*... we got it wrong in the bidding*', and suggested that Laing had been mistaken to bid when there were risks involved that they were unable to manage.

Following the Rugby World Cup, Laing continued working for a further six months to complete the project and for another year after that to rectify all defects.

Reasons for the project's difficulties

So what contributed to the delivery problems encountered on the Millennium Stadium?

The contractor

The cost overrun on the Millennium Stadium was mainly attributable to the following issues:

- Pre-award:
  - poor project definition during the bid stage: including insufficient design/engineering work, which led to inaccurate assessment of the likely cost of the project
  - inadequate planning of the execution of the works
  - failure to appreciate the potential implications of the dispute between WRU and CRFC
- Post-award:
  - increased demolition costs
  - additional works, as a result of the dispute with CRFC, which included, for example, the strengthening of steel trusses to the retractable roof due to the requirement for an asymmetric support structure
  - having to work 24 hour shifts for a year
  - the inability to manage relationship issues with CRFC
  - the dependence on subcontractors, particularly Cimolai who appear to have exploited their position

Laing's management of the project raises several questions.

When dealing with their subcontractors, Laing had a reputation for being tough negotiators (a trait highlighted in *Harmon CFEM Facades (UK) Limited* v *the Corporate Officer of the House of Commons*[2], where Mr Justice Humphrey Lloyd had criticised Laing's procurement practices which had led to Harmon entering into liquidation).

So why, therefore, had Laing not been able to control the costs incurred by their subcontractors on the Millennium Stadium, particularly those of the steelwork subcontractors Cimolai?

Cimolai were able to exert a dominant position over Laing by influencing the rate of progress of the works and making significant claims. Why and how Cimolai came to be in this position is unclear, although one possible reason why Laing so readily agreed to the pay their subcontractor a reported additional £14 million, recounted by Morris (2005), was that Laing had elected to settle an outstanding debt from the second Severn Bridge project through the Millennium Stadium contract (although widely rumoured this explanation is unconfirmed).

While the dispute with Cimolai provoked John Laing plc to intervene, the question that must be asked is:

Why did it take the parent company 18 months to respond when the project had been losing on average £1 million a month?

Finally:

Why did Laing not withdraw from the project in October 1998?

Perhaps Laing were looking to gain potential future work within the Cardiff area, particularly the £1.2 billion development of the Cardiff Bay area, and the loss of reputation if the Rugby World Cup had to be held elsewhere would have had far-reaching consequences for Laing internationally. Also, apparently the liquidated damages on the project were far less onerous than those reported for the Wembley Stadium project, and WRU let it be known that provided the match against South Africa took place on 26 June liquidated and ascertained damages would not be imposed.

The client

At one point MS plc considered replacing Laing. However, their switching costs would have been considerable without necessarily guaranteeing any improvement in the progress of the works. Moreover, during the project the client had their own financial problems to resolve, including increased compensation demands from BT, the discovery of a huge underground tunnel system below the concourse, and not being able to sell a number of the hospitality boxes upon which their business plan depended – all of which increased WRU's debt burden.

Relationships

Welsh rugby turned professional in 1995, which completely changed the relationship between the WRU and its 220 clubs (plus a further 175 affiliated clubs). The WRU had demanded 51% golden shares in each of the leading rugby clubs in return for revenue rights earned through televising games, and participation in both the premier division and European competition. Although most of the major rugby clubs such as Swansea acquiesced to the deal, the two most powerful clubs, CRFC and Ebbw Vale, refused.

This disagreement became personal between two prominent personalities, Vernon Pugh QC, chairman of WRU, and Peter Thomas, chairman of CRFC, with the WRU threatening to expel the two clubs from the premier division. The dispute was resolved by the courts in June 1997 with CRFC and Ebbw Vale successfully obtaining an injunction against WRU. However, the disagreement led to unforeseen consequences for Laing. But it was assumed that Laing had accepted the inherent risk in the dispute as Tony Evans, the then MD of Laing Civil Engineering, was a senior member of the CRFC, who reportedly felt his personal relationships within the club would help mitigate the dispute.

C	Customer	Who would be victims/ beneficiaries of the purposeful activity?	• WRU and its associated/affiliated clubs through staging rugby events and generating additional revenue from other events (football, concerts, etc.) • CCC in helping to promote economic growth in Cardiff and Wales • Millennium Stadium plc – a company set up by WRU and CCC to lease, fund, build and operate the stadium on behalf of WRU for 50 years • Local trades and businesses who benefit from events held at the stadium
A	Actors	Who would do the activities?	• O'Brien Krietzberg acting as Project Managers on behalf of WRU/CCC (MS plc) • Laing Construction as the main contractor • Laing's supply chain
T	Transformation Process	What is the purposeful activity?	• To provide a stadium with a retractable roof and removable pitch making it a facility that can be used for multipurpose entertainment in all weather conditions with seating for 74 500 • To provide facilities that would encourage economic growth in the Cardiff area • To provide stadium facilities more appropriate for the capital city of Wales during devolution to the Welsh Assembly • To modernise facilities as required for the 21st century • To bring an international focus to Wales by holding major sporting and cultural events, including hosting the Rugby World Cup and FA Cup
W	Weltanschauung (worldview)	What view of the world makes the definition meaningful?	• The seating capacity (only 48 000) and facilities at Cardiff Arms Park had become unacceptable, and increased safety requirements and its vulnerablility to the weather conditions had become issues • The stadium, although finally completed in 1983, was based on a 1962 design, and needed to be modernised • The WRU wanted to build a new facility which could compete with other major rugby venues in the world such as NewZealand and South Africa • The stadium formed part of an overall strategy to promote economic growth and regeneration in Cardiff which included the £1.2 billion development programme of the Cardiff Bay area
O	Owner	**Who could stop this activity?**	• WRU, CCC and Millennium Commission
E	Environmental factors	What are the environmental constraints?	• *Financial constraints*: The Millennium Commission provided a £46 million disbursement, but all funding above this had to be raised by the WRU. It proved difficult to raise funding through the sale of 125 hospitality boxes, and the associated rugby clubs were cash poor. The WRU had to raise most of the funds through a loan with Barclays Bank (£60 million) and debentures (£20 million) • *Timescale constraints*: The stadium had to be completed at the latest by October
			1999 to host the Rugby World Cup. The construction schedule was fast tracked to 28 months whereas a construction project of this size would normally have taken 3½ years. • *Stakeholders*: See Figure A.3 • *Technical constraints*: Site access was restricted by the River Taff to the west, and by various buildings to the south and east, which had to be demolished to make way for the concourse and River Walk (these included: a BT Telephone Exchange, DSS building, Territorial Army HQ, Empire Swimming Pool and The Crown public house). However, the major restriction resulted from the actions of CRFC who prevented the north stand being demolished, cranes working in their air space and support masts overhanging their airspace – leading to considerably more steelwork in the redesigned roof

Figure A.4 Millennium stadium CATWOE description.

This dependence on personal relationships and Tony Evans' desire as a patriotic Welshman and rugby enthusiast to build the stadium could potentially have adversely influenced contractor's decision making. In the case of the Millennium Stadium, there appears to have been a determination by Laing Construction to win this landmark project at a cost acceptable to the client.

Using Checkland and Scholes' (1990) CATWOE mnemonic (Checkland and Scholes, 1990), Morris (2005) performed an analysis of the problems that occurred on the Millennium Stadium (see Figure A.4). The analysis puts into context the rationale for rebuilding the stadium, identifying the purposeful activity in constructing the venue; that is, what has been achieved in the transformation process of building the stadium and what the worldview is in making it meaningful. It can be seen in practical terms, such as increased seating requirements, or in terms of business growth to allow the WRU to compete globally to stage major sporting events, or socio-political, that is the bigger picture of building the stadium to help promote economic growth in Cardiff.

Post-project outcome

The Millennium Stadium has focused international attention on Cardiff; in addition to providing a unique, multipurpose venue in the heart of the city, it has become the most visited attraction in Wales and one of the top ten attractions in the UK. Further, it is credited with bringing substantial economic benefits to the area, far more than the publicly funded Opera House. An independent survey carried out for WRU found that the stadium had contributed £725 million to the Welsh economy. It also reported that the stadium annually:

- Generates £105 million (of which £88 million is spent in and around Cardiff)
- Supports the equivalent of 2400 jobs
- The WRU pays more than £900,000 a year in rates (Anon, 2007)

For Laing, involvement in the Millennium Stadium project had significant repercussions. Between 1998 and 2001 Laing Construction incurred losses exceeding £196 million, including a reported £34 million loss on the Millennium Stadium project. This led to the sale of Laing Construction to O'Rourke in April 2002. While Laing Construction had assets valued at £27 million, these were transferred to O'Rourke for a nominal consideration of £1. Moreover, it was later reported that O'Rourke was able to extract a further £35 million plus from the projects it took over from Laing.

As a postscript to the dispute with CRFC: the adjacent club subsequently made enquiries into replacing their south stand (the source of many of Laing's problems) due to the discovery of concrete cancer, although this was never undertaken as estimates for the work were in the region of £15–20 million. However, in May 2008, it was announced that the rugby club would share a new 25 000-capacity Cardiff City Stadium – then currently under construction in the Leckwith area of the city – with Cardiff City Football Club from the beginning of the 2009–2010 season. The Cardiff City Stadium was officially opened on 22 July 2009.

Acknowledgements

This case study was prepared by David Lowe and Shan Morris from the following published sources. It draws heavily on the latter's MSc dissertation undertaken at the University of Manchester (Morris, 2005).

Endnotes

1 Johnson *et al.* (2006) adapted Mendelow (1981)

2 For details of the case see Box 8.9 in Chapter 8

References and sources

Anon (1996) Builder chosen for Millennium project. *South Wales Echo*, 3 July

Anon (1998a) Behind-the-scenes chaos at the Millennium Stadium... *Western Mail*, 2 July

Anon (1998b) Wales' 1999 Rugby World Cup may be held in France, with Laing failing to convert. *Construction News*, 4 November. http://www.contractjournal.com/Articles/1998/11/04/18930/wales-1999-rugby-world-cup-may-be-held-in-france-with201-laing-failing-to.html (accessed 4 June 2008)

Anon (1998c) The growing rift between the Welsh Rugby Union and Millennium Stadium contractor Laing... *Western Mail*, 5 November

Anon (1998d) Cardiff's £121m Millennium Stadium project has been hit by the third top-level departure in two weeks... *South Wales Echo*, 12 November

Anon (1998e) The former project manager for the £121m Millennium Stadium, Simon Lander... *Western Mail*, 11 December

Anon (1998f) Construction giant Laing will not consider legal action... *Western Mail*, 12 December

Anon (1999a) Ready for the world: the construction of Cardiff's Millenium Stadium. *New Civil Engineer Supplement* 30 September, pp. iv–viii

Anon (1999b) Cardiff's £121m Millennium Stadium will not be completed by the date promised... *South Wales Echo*, 20 March

Anon (1999c) With only a day to go to the big kick-off at the Millennium Stadium... *Western Mail*, 25 June

Anon (1999d) A game of two halves at Cardiff. *Contract Journal*, 7 July. http://www.contractjournal.com/Articles/1999/07/07/16641/a-game-of-two-halves-at-cardiff.html

Anon (1999e) Jubilant Welsh Rugby chairman Glanmor Griffiths has hit back... *Western Mail*, 30 August

Anon (1999f) UK: Wales – Cardiff stadium 'financial disaster'. *BBC News*, 9 September http://news.bbc.co.uk/1/hi/wales/442978.stm (accessed 28 December 2012)

Anon (1999g) The Millennium Stadium might be magnificent for rugby fans but for John Laing... *Western Mail*, 10 September

Anon (2000) Pulling off the wow factor. *Master Builder*, June. http://www.fmb.org.uk/publications/masterbuilder/june00/19.asp (accessed 4 June 2008)

Anon (2002) Laing lost out on £40m in sale of division. *Construction News*, 21 March. http://www.cnplus.co.uk/news/laing-lost-out-on-40m-in-sale-of-division/868204.article (accessed 28 December 2012)

Anon (2007) Scrap £900k rates, says stadium, BBC News, 6 November. http://news.bbc.co.uk/1/hi/wales/7081067.stm (accessed 28 December 2012)

Cardiff Research Centre (1998) *The Economic Impact of the Millennium Stadium.* Cardiff County Council.

Caletka, A and Merrifield, D (2000) Cardiff's Millennium Stadium – the final account, NEC User Group No, 12

Cook, A (1998) Penalty time. *Building*, 13 September, pp. 18–19.

Cook, A (1999) Millennium Stadium, Cardiff – spectacular comeback. *Building*, 17 September, pp. 42–45. http://www.building.co.uk/buildings/architecture/spectacular-comeback/4255.article (accessed 28 December 2012)

Johnson, G, Scholes, K and Whittington, R (2005) *Exploring Corporate Strategy*. 7th edition. Financial times Prentice Hall, Harlow. Adapted: Mendelow, A (1981) *Proceedings of the Second International Conference on Information Systems*. Cambridge, Massachussetts

Jones, C (2002) The stadium and economic development: Cardiff and the Millennium Stadium. *European Planning Studies*, 10, 819–829

Macalister, T (2001) Laing sells ailing business for £1. *The Guardian*, 28 September. http://www.guardian.co.uk/business/2001/sep/28/5 (accessed 28 December 2012)

Millennium Stadium – Millennium Stadium Information – Background to the Millennium Project, Millennium Stadium plc. http://www.millenniumstadium.com/3473_3557.php (accessed 4 June 2008)

Millennium Stadium – Millennium Stadium Information – About Millennium Stadium, Millennium Stadium plc. http://www.millenniumstadium.com/information/about.php (accessed 28 December 2012)

Millennium Stadium (2007) Millennium Stadium delivers huge boost to Welsh economy, Millennium Stadium plc. http://www.millenniumstadium.com/news/21176.php (accessed 28 December 2012)

Minogue, A (1998) What became of GMP? *Building*, 15 May, p. 38

Morris, SL (2005) *Insolvency in Construction: The Collapse of Laing Construction plc in 2001*. MSc Commercial Project Management Dissertation, University of Manchester

Porter, M (1985) *Competitive Advantage: Creating and Sustaining Superior Performance*. Free Press, New York

Sports Venue Technology – Millennium Stadium, Cardiff, United Kingdom. http://www.sportsvenue-technology.com/projects/cardiff/specs.html (accessed 26 February 2005)

Thompson, PD (1998) The Millennium Stadium – Cardiff Arms Park. In: PD Thompson, JJA Tolloczko and JN Clarke (eds) *Stadia Arena and Grandstands*. E. & F.N. Spon, London. pp. 221–228

The Emirates Stadium, Islington, London

Completed in July 2006, Arsenal's Emirates Stadium, named after its principal commercial sponsor, is England's third largest football ground after Wembley and Old Trafford with a seating capacity of 60000. The stadium project has been acclaimed an *'exemplar of good construction practice'*, demonstrating an *'outstanding contribution to the built environment'*.

In 2007 Sir Robert McAlpine, as part of the team responsible for delivering the Emirates Stadium, were awarded both the 'Major Project prize' at the Quality in Construction Awards and the inaugural 'Building Project of the Year' at the Annual Building awards. The 'Major Project prize' was presented in acknowledgment of *'... exceptional project management, masterful team-building and an inspiring client'*, while the 'Building Project of the Year' judges described Emirates Stadium as a *'phenomenal project'* and added:

> 'The way that this team tackled design changes should be a lesson to the industry on how to move forward. ' (Quoted in Sir Robert McAlpine, 2007a)

Additionally:

> 'It was the project that most integrated design and construction. It came with severe planning constraints, but it was still designed and built with the minimum of fuss.' (Robin Nicholson, Chairman of the judges, quoted in Spring, 2007)

The judges added:

> 'The whole process was an example of how important teamwork should be to a project, with everyone from the client to the contractor and subcontractors coming together and working successfully to ensure the project was completed on time and under budget. The way that this team tackled design changes should be a lesson to the industry.' (Quoted in Spring, 2007)

Also at the 2007 Annual Building Awards, Sir Robert McAlpine was proclaimed 'Major Contractor of the Year', received the 'Constructing Excellence' award, and director Benny Kelly was named 'Personality of the Year'. When announcing the award, the judges stated that Sir Robert McAlpine's performance on the Emirates Stadium project had '... *done much to restore the reputation of the whole contracting sector*' (quoted in Sir Robert McAlpine, 2007a).

The project, however, was particularly complex, requiring the integration of numerous specialisms and entailing the negotiation of a series of complex commercial deals. The scheme was more than just a conventional construction project, forming the nucleus for the regeneration of the surrounding area, becoming a focal point for the community and generating employment for local residents.

Requirement identification

Arsenal had occupied its existing stadium at Highbury since 1913, although significant alterations to the ground had been made over the years, in particular the creation of an all-seated facility. By the 1990s, these changes had resulted in a reduction in the ground's capacity to approximately 38 000. Decisively, at this time there were ten Premiership teams with larger stadia than Highbury: the largest was Old Trafford, which seated 67 000 fans, enabling Manchester United to generate an additional £1 million per home game when compared with Highbury.

Further, Arsenal had aspirations to become a global presence in football. Therefore, in order to achieve this status; a new high-class stadium with a significantly increased spectator capacity was seen as a necessity, enabling Arsenal to:

- Create a venue worthy of a world-class football club (Arsenal became a member of the G14 European Football Clubs Grouping in 2002)
- Boost their income from ticket sales to a level comparable to Manchester United, allowing the club to compete financially with the largest teams in Europe
- Provide state-of-the-art facilities for their fans
- Continue to attract top players to the club

However, redevelopment of Highbury was not viable, due to both financial and physical constraints. Highbury Stadium was located in a heavily populated residential area, prone to congestion on match days, difficult spectator management and poor access for emergency vehicles. The football club, therefore, took the decision to relocate to a new stadium.

Requirement specification

Arsenal's requirements, therefore, necessitated the construction of a state-of-the art stadium that met the needs of all the club's stakeholders, providing, for example:

- Spectators with a safe environment, excellent views of the pitch – irrespective of their position in the stands, comfortable seating and a remarkable atmosphere
- Players with an outstanding pitch, first-class changing facilities, etc.
- Shareholders with a commercially viable venue – an iconic stadium to strengthen the Arsenal 'brand' and signal the club's international aspirations
- A solution that reduced the ground's impact on the adjacent neighbourhood and which met the community's objectives

Additionally, the club required a discrete source of funding for the project that would allow them to construct the stadium without impacting on its ability to continue investing in world class players.

Solution selection

Site location

Arsenal investigated a number of alternative locations, including tenancy of the new Wembley Stadium. However, by the late 1990s it chose to develop a site at Ashburton Grove, in the London

Borough of Islington, near to Holloway Road tube station. Approximately one mile away from Highbury, the site allowed the club to retain its links with the area it had been connected with for more than 90 years.

Comprising a 17-acre triangular plot of land, access to the site was restricted: bounded between two railway tracks – the Kings Cross mainline and a suburban line – and Queensland Road. The site, owned by Islington Council, was a light industrial estate, comprising numerous small businesses in properties leased from the council; Islington Council offices, main depots and workshops; and the Islington waste transfer station. Although willing to sell the site to Arsenal, the council imposed three main conditions:

1. The relocation of some 80 businesses based on the site, requiring Arsenal to enter into separate negotiations with each concern and to acquire alternative properties for them in Islington. The relocation issue was a subject of controversy as not all the businesses were content to move
2. The relocation, with no disruption in provision, of the council's service providers located on the site. Again, Arsenal met with some local resistance to this requirement, particularly concerning the repositioning of a waste disposal unit
3. That the new stadium would form part of a wider urban regeneration scheme

Additionally, the stadium would have to be at least 100 m from surrounding principal residential areas.

In order to comply with the above conditions, the club was involved in significant land and property acquisitions during the planning application process. These acquisitions, however, did not run smoothly – for example, an essential eight-acre plot of land at the centre of the proposed site was placed on the open market by its owners Sainsbury's plc.

Formulating the planning application for the scheme took two years, and required the issuing of a compulsory purchase order. Arsenal presented its first planning application in November 2000 and, following the submission of three substantial revisions, its final revised scheme was submitted in August 2001. Islington Council implemented an extensive public consultation exercise which resulted in over 100 modifications to the plans. These included the incorporation of a large new community health facility and the reconfiguration of proposed housing at Highbury Stadium so that a large horse chestnut tree could be retained.

Following this, two-and-half years were spent negotiating the legal arrangements to facilitate the relocation of the council's services and other existing businesses based on the site. In total, it took Arsenal four-and-a-half years and several million pounds to achieve vacant possession of the site.

Site constraints

The proposed brownfield site at Ashburton Grove, therefore, imposed several significant constraints/challenges on the project. These included:

- The removal of existing buildings and other structures before construction work could commence
- The regeneration of contaminated industrial land – a full environmental impact assessment of the site necessitated the removal of 25 000 m^3 of contamination
- Working within the constraints of existing utilities, the adjacent railway lines and restricted access to the site
- Providing appropriate pedestrian access to and from the site, and to tube and rail lines
- Providing adequate access to the venue for emergency services
- Relocating the waste transfer station prior to the closure of the existing facility on the site

Problems encountered in achieving vacant possession of the entire site resulted in a phased handover of the land to the contractor. Delays in relocating the waste transfer station, in particular, impacted on project planning, requiring the stadium to be constructed in parallel with its removal. Attaining vacant possession also had financial implications for the developers, as it was a prerequisite condition to borrowing the money to fund the construction phase of the project.

When compared with Manchester United's 100-acre venue, the Ashburton site appears very compact; however, unlike United's supporters who predominantly arrive by car, approximately 70% of Arsenal's fans use public transport. As stated above, this imposes different constraints on the venue: appropriate public transport provision.

Project scope

At the time, the relocation of Arsenal Football Club to Ashburton Grove was one of the UK's largest development schemes, valued at £390 million. However, Arsenal's proposals went beyond the construction of a stadium, entailing an ambitious regeneration project, comprising three elements:

1. A 60 000 seat state-of-the-art stadium at Ashburton Grove designed by HOK Sport Venue & Event Ltd. This element of the scheme also included two pedestrian/service access bridges across the adjoining railway lines, the club's head office, a residential development – Ashburton Triangle (plans included the construction of 730 homes, with 40 flats allocated to teachers working in Islington) – and commercial and office premises in Queensland Road. Plans also included a 'Red Learning Zone' where courses in computer skills would be provided
2. Regeneration of Lough Road with residential properties design by architects CZWG, a replacement sealed waste and recycling plant designed by Sheppard Robson and the relocation of Islington Council's vehicle depot and civic services to a facility in Hornsey Street. Also, in Hornsey Street the club planned to build 230 units
3. A 500-unit residential development on the site of Arsenal's existing Highbury Stadium designed by architects, Allies and Morrison. This element was to commence once the club had relocated to Ashburton Grove

In addition to these three elements, Arsenal also agreed to fund a number of community projects. These included:

- Nursery facilities at both Highbury Stadium and Lough Road
- Four community health facilities at Queensland Road, Lough Road, Drayton Park and Arsenal Stadium
- Office space for the local health authority within the Ashburton Grove scheme

Further, Arsenal agreed to finance a multimillion pound public transport improvement scheme, centred on Holloway Road, Drayton Park and Finsbury Park stations.

In total, Arsenal's stadium redevelopment project and the regeneration of the area surrounding it sought to generate approaching 2500 new residential properties (25% with 'affordable housing' designation), long-term employment for between 1800 and 2600 people and lever approximately £400 million into the area.

Stadium design solution

In 2000, Arsenal appointed a design team comprising HOK Sport as architects and the engineering consultancy Buro Happold to produce concept designs. The appointments were made on the understanding that the design team would be novated to the main contractor under a design-build contract once a contractor had been selected. The scheme design for the project was presented in 2001.

Design features included:

- **Site constraints**: to overcome the limitations of the confined site, the stadium was designed so that it maximised the available space. This entailed reducing the secure area of the arena to within the building's structure, while parts of the surrounding podium circulation space were cantilevered over adjacent railway land. Also, two bridges were required to facilitate rapid spectator egress from the stadium and to link it to the nearby underground system, and a third bridge was incorporated for emergency vehicle access only

- **Podium**: the stadium is surrounded by an elevated external pedestrian-only concourse (the podium), which, in addition to assisting spectator access/egress and servicing, and separating pedestrian from vehicle access, reduces the height within the building fans have to climb to get to the upper terrace. The stadium's entrance plaza is surrounded by the stadium's box office, disabled entrance, Arsenal's offices and its 10 000 m^2 megastore
- **Structure**: a glass and steel construction with areas of smooth concrete and curtain walling, the stadium forms a large ellipse, rising 41.4 m from ground level at its highest point, 35 m high at its eaves and extending 250 m lengthwise: in compliance with planning restrictions concerning its height and location on the site – over 100 m from the nearest residence. The design, also, enables noise levels within the stadium to remain similar to those encountered at Highbury and to reduce the impact of artificial lighting. Principally, accommodation is spread over five levels with 60 000 seats arranged in three uninterrupted tiers that surround the pitch in the form of a saddle. Private boxes are contained within a mezzanine level, while additional amenities include restaurants, bars, conference rooms and merchandise shops
- ***Roof***: a clear-span roof that slopes downwards towards the pitch, creating the illusion of a dish, enhances the pitch's microclimate and maximises the amount of natural light reaching the pitch. The roof design promotes natural air circulation around the seating and increases spectator comfort. Support for the 27 200 m^2 roof is provided by two 204 m long tubular steel trusses, triangular in section, spanning the pitch north to south with two smaller 100 m long trusses spanning east to west. All the triangular-shaped trusses were 'self-stable' while being erected resulting in shorter assembly times and a reduction in temporary support. This was an important feature bearing in mind that construction of the stadium ran simultaneously with demolition of the waste transfer station
- ***Mechanical and electrical (M&E) installations***: M&E installations contributed significantly to the construction cost of the stadium. Special requirements included a ventilation system that could handle the influx of spectators prior to a game, especially during rain; car park ventilation; communications networks including voice and data links to all areas; CCTV and public address networks. Following a deal with Sony, Arsenal's stadium became the first in England to install HDTV.
- ***Support services***: support services for the stadium are contained within two levels below ground level. These comprise: changing rooms, commercial kitchens, education and press centres, plus parking spaces and loading areas for both the stadium and businesses located in the nearby Queensland Road
- ***Sustainability***: the stadium design integrated a number of environmentally sustainable features: a passive and mixed-mode ventilation system, skylights and glass panelling to maximise natural lighting, and photovoltaic solar power. Also, the scheme incorporated rainwater collection and storage to be used in irrigation and toilet flushing. An aspiration for the new stadium, compared with current stadium construction practice, was that construction waste should be reduce by a minimum of 50%, for example, through the reuse and recycling of all demolition waste, and the building should be more energy efficient and cost less to maintain.

Box A.2 Key statistics

The Emirates Stadium provides:

- 60 000 seats (21 581 more than at Highbury)
- 7000 club level seats with attached restaurants
- 150 executive boxes (Highbury had only 48 boxes) – with a capacity to serve 2000 meals on match day
- 250 catering serving points (positioned around the Stadium) and over 900 WCs – plus 370 m of urinals and 113 disabled toilets
- Capacity for up to 250 wheelchair users and 250 ambulant disabled spaces
- 41 TV camera positions and 215 seats for media
- A grassed area of 113 m × 76 m and pitch size of 105 m × 68 m

Table A.3 Emirates Stadium project milestones.

2000	Appoint concept design team
January	Project commences
November	Planning application submitted for three sites (approx. 370 000 m^2)
2001	
August	Third and final revised planning application submitted
September	Shortlisted contractors Taylor Woodrow and Sir Robert McAlpine submit tenders
December	Planning assessment by the London Borough of Islington, amendments and resubmission
2002	
January	Design development, second-stage tender, and site acquisition commences
February	Sir Robert McAlpine selected as main contractor
2003	Funding; waste and recycling centre and other relocations; demolition on stadium site
April	Revised project end date of July/August 2006 announced
2004	
February	Funding secured and Sir Robert McAlpine take possession of site
March	Stadium construction begins
May	Launch of North Bridge
August	Launch of South Bridge
August	Waste and Recycling Centre complete - final vacant possession of Emirates site
October	Announced that the stadium will be known as the Emirates Stadium
2005	
June	Roof steelwork completed (one month ahead of programme)
2006	
January	Works commence on pitch drainage and under-soil heating
February	Installation of stadium's major internal fittings commences, including the seats
March	Pitch seeding starts and roof is completed
May	Pitch seeding and installation of turnstiles completed
July	Stadium achieves practical completion ahead of schedule and on budget

Development programme

Relocation to Ashburton Grove was first proposed in 1999 and, initially, Arsenal had anticipated moving into its new stadium in time for the start of the 2005–2006 season. However, in April 2003, due to a number of delays, the project completion date had to be amended to coincide with the start of the 2006–2007 season. These delays were attributed to the scale and complexity of the project and covered a variety of issues – for example, legal, financial and transport infrastructure problems. They also related to the relocation of existing local businesses which occupied the site and the requirement to provide affordable housing. These issues, in particular, the club's inability to provide their main contractor with vacant possession of the site, as specified in their contract, made the August 2005 completion date unattainable. Additionally, as with the main contractor, the release of funding for the project was dependent on achieving vacant possession.

The key milestone dates for the Emirates Stadium are shown in Table A.3

Project budget

A budget of approximately £390 million was set for the stadium relocation scheme, a significant financial outlay for the club. The figure comprised:

- £200–220 million for the construction of the stadium (a relatively high figure to cover the required high specification and distinctive design, and the constraints imposed by constructing within an urban residential area)

- The remaining £180 million accounted for fees (architects, planning, legal, finance, insurance, etc.), costs related to the neighbouring infrastructure developments (the compulsory purchases of several business premises, relocating a waste recycling centre, and upgrading Holloway Road Underground station), regeneration of the surrounding area and expenditure in compliance with planning requirements

The budget was based on the club's overall business plan, which was linked to the anticipated income and running costs of the development.

Financing the project

To enable the financing of the complex stadium development, Ashburton Properties (a subsidiary of Arsenal Holdings plc) was created; and following months of negotiation it obtained a £260 million senior loan facility from a stadium facilities banking group, consisting of the Royal Bank of Scotland plc, Espirito Santo Investment, the Bank of Ireland, Allied Irish Banks plc, CIT Group Structured Finance (UK) Limited and HSH Nordbank AG. The loan was based on non-recourse financing, under which lenders could only call upon the anticipated revenues generated by the new venue as opposed to the club and its assets. Initially, interest on this debt was set at a commercial fixed rate over a 14-year term.

The shortfall in the project's funding was to be met via revenues derived from Granada, Nike, the sale of surplus land assets relating to the stadium site, and following the relocation of the football club in 2006 the residential development of Highbury. Additionally, Delaware North negotiated a 20-year exclusive contract to run the stadium's catering facilities, which contributed £15 million towards the capital cost of the stadium. A deal with Emirates Airline, announced on 5 October 2004, generated a further £100 million from naming the stadium 'the Emirates Stadium' for, at least, the first 15 years of operation and an eight-year shirt sponsorship deal starting in the 2006–2007 season.

The construction contract, however, was the principal mechanism by which Ashburton Properties raised the debt finance to construct the stadium.

Project procurement (contractor appointment)

The building contract comprised the construction of the stadium, three bridges over the adjacent railway lines, roads and site buildings; a new waste recycling centre; a council depot, offices and a community sports centre. It also incorporated the novation of the concept design team.

Arsenal adopted a two-stage tender process for the selection of its main contractor. Within the tender documentation, specification of the employer's requirements was predominantly performance-based; however, in respect of the pitch and changing rooms, for example, the client retained a degree of control over elements of the design and specification. The client also stipulated a demanding project time constraint.

Following expressions of interest, Taylor Woodrow and Sir Robert McAlpine were shortlisted for the role of main contractor on the project, and submitted their final bids in September 2001. In February 2002 it was announced that Sir Robert McAlpine had been selected. The contractor had previously built Hampden Park Stadium in Glasgow and West Ham's Centenary Stand at Upton Park. The proposal included a lump sum to cover the contractor's preliminaries, and their overheads and profit were recovered by means of a fixed percentage.

Sir Robert McAlpine allocated a discrete team to the project. Prior to binding the contractor to a GMP, the team had to attain a high degree of cost certainty. To achieve this, definitive design solutions and firm prices had to be negotiated with the individual package contractors (subcontractors), of which there were 110 on the stadium portion of the contract, and 87 involved on the new waste recycling centre. As numerous works packages incorporated a significant contribution to the stadium's design, the involvement of the design team in these negotiations was essential to assess and clarify the design input and to delineate the interfaces between the separate works packages. This involvement enabled efficiencies to be achieved in the construction process and reduced project risk.

As mentioned earlier, delays were experienced during this phase of the project. However, effective use of this postponement was made, enabling a re-evaluation of the design, programme limitations, and construction methodology to occur. The agreement was converted to a GMP contract after agreement had been reached with the majority of the works contractors (subcontractors). In 2004 Ashburton Properties Ltd signed a fixed-price design-build contract with Sir Robert McAlpine, which included £220 million to build the stadium.

Originating from a JCT standard form with contractor design, the agreement was converted into a lump-sum, fixed-price, and fixed-date contract with the majority of the construction and programme risk held by Sir Robert McAlpine. On conversion of the contract to a GMP and the novation of the professional team, the main contractor accepted single-point responsibility for both the design and construction of the stadium. The appointment of the concept design team had anticipated this course of action, and afforded Sir Robert McAlpine a remedy against them if a claim was made under the construction contract due to their actions.

Each component of the project had its own defined completion date and apportioned liquidated and ascertained (L&A) damages (a genuine pre-estimation of the probable loss incurred by the client due to the contractor failing to complete the project within the contract period). In respect of the football stadium, the L&A damages provision comprised:

- The interest charged to support the debt
- The loss of revenue experienced by the club in the event of delay (this element of the L&A damages accounted for the variability in the club's income stream, that is, whether or not the delay occurred during the football season)
- The consequential loss arising from the delay in commencing the conversion of the existing stadium at Highbury into flats, which was dependent on the completion of the new stadium

Asset supply

Sir Robert McAlpine finally took possession of the site in February 2004. The onerous programme imposed by the contract (project completion in time for the 2006–2007 football season), required the adoption of a fast-track construction approach, which involved commencing the groundwork for the stadium in one part of the site while simultaneously demolishing the existing waste recycling centre in another. This fast-track approach, together with the complexity of the project, required the implementation of integrative, collaborative working practices. Again, this was a challenge for the project team with more than 120 separate companies involved in the project. Sir Robert McAlpine utilised 4Projects, a project collaboration solution, that allowed over 450 individuals based in several locations to access and collaborate on project documentation. Over the duration of the project, in excess of 55 000 drawings were uploaded using the software, with over 600 000 documents downloaded from it.

Significant risks on the stadium project included:

- The construction of the bridges across the adjacent railway lines: In addition to the possibility of damaging the rail infrastructure, fixed possession dates had to be negotiated with Network Rail, which imposed severe penalties in the event of disruption to the network
- The pitch: In order to retain single-point accountability, Sir Robert McAlpine procured the pitch through a subcontract package

The challenges facing the implementation team included:

- A highly restricted site: the triangular site was bounded by railway lines on two sides
- A demanding time constraint: the stadium had to be completed in time for the start of the 2006–2007 football season
- Working around a number of businesses (including a waste processing site that needed to be relocated without disruption to its operation) that already occupied the site

To give an impression of the scale of the project, Sir Robert McAlpine employed over 1000 operatives on the stadium and its associated structures, while over 9300 construction personnel participated in its construction. In terms of the quantities of materials used, the project incorporated 55 000 m^3 of concrete, 10 000 tonnes of steel reinforcement, 3000 tonnes of steel in the main roof, 100 flights of stairs, 4500 m of metal hand railing, 15 000 m^2 of glazing and over 2000 doors.

Likewise, to illustrate the construction methods adopted to erect the stadium, the main roof trusses were prefabricated on site from tubular steel. Two primary trusses, measuring 204 m × 15.5 m, were fabricated in two sections, lifted into position using two crawler cranes and bolted together at their midpoint. Once connected, they were jacked down into their permanent position and the temporary supports removed.

Project completion

Despite the inclusion of significant additional works, the stadium was completed and handed over to Arsenal Football Club on 14 July 2006, on budget and just over two weeks ahead of schedule; fully operational for the start of the 2006–2007 football season. The first match in the new stadium was a testimonial for Dennis Bergkamp, held on Saturday 22 July.

As indicated earlier, the stadium project has been acclaimed a major success.

Arsenal Director, Ken Friar, commented:

> 'We are delighted that Emirates Stadium has reached practical completion ahead of time… The project has gone extremely well and we would like to thank Sir Robert McAlpine, and the whole project team, who have done an absolutely fantastic job for us and delivered probably the best club stadium in Europe.' (Quoted in Sir Robert McAlpine, 2006a)

Concerning Arsenal's objective of raising their income generation from home matches, Arsène Wenger (Arsenal's manager), has stated that the new venue should increase the club's revenue by 50% to £150 million a year. In terms of providing a state-of-the-art facilities for their fans, Wenger, commented:

> 'It has been seven years since the club first announced the initial plans for the new stadium but I'm sure the fans will agree it has been worth the wait when they experience the fantastic new facilities and great views as they take their seats for the first time.' (Quoted in Arsenal.com, 2006)

Additionally, the Emirates Stadium was commended by Jack Lemley (at the time Chairman of the Olympic Delivery Authority), who stated that it was '*a project the London 2012 team could learn from*' (quoted in Sir Robert McAlpine, 2006b).

Reasons for success

So what contributed to the successful delivery of the Emirates Stadium?

When asked on Radio 4's Sports News how they had managed to build the stadium on time Keith Edelman, a director of Arsenal Football Club, stated:

> 'Well, I'd say there were three reasons… The first is we chose a very good contractor, Sir Robert McAlpine, who have done an absolutely fantastic job for us, planned it extremely well and built it to plan. I think the second two reasons are that we've made very quick decisions – there are always a lot of changes when you build a stadium… this is a very complicated building and we have to make very quick decisions in terms of difficulties and changes that are made to the actual design. And the third I think is the partnership between all the parties which has been very strong.' (Quoted from BBC, 2006)

The contractor

A significant factor in the success of the project, therefore, was the contractor – Sir Robert McAlpine – a company with a 150 year reputation in construction and particularly accomplished in managing its supply chain. For example, by using Sir Robert McAlpine's existing supply chain, subcontractors were appointed earlier in the construction process, commonly using a negotiated approach with prices benchmarked against up-to-date cost data.

The client

The role adopted by the client – Arsenal Football Club – was also influential, for example:

- As they were also the end user, the club assumed a proactive role in the delivery of the project
- They empowered their 'employer's' representatives – they appointed two main board directors to assume responsibility for the delivery of the project
- They provided prompt access to their key decision makers
- They facilitated timely decision making – the employer's representatives were able to make key decisions when required

Relationships

Other features of the project that were believed to have contributed to its success included:

- The adoption of an integrated team approach – promoting collaboration between the client, the professional advisers (architects, engineers, project managers and quantity surveyors, etc), and the construction team
- The co-location of the main actors in the design and construction team
- Transparency and change control
- Allocating of sufficient time to deal adequately with issues – for example, effective use of the project's postponement was made, enabling a re-evaluation of the design, programme limitations, and construction methodology to occur
- The convening of frequent meetings of the project's stakeholders

Theses features were underpinned by the development of relationships based on communication, empowerment and mutual trust. Notably, there were no formal disputes on the project, indicating that problems were resolved through discussion and negotiation.

Other factors

The success of the Emirates Stadium was also influenced by the incorporation of:

- **Buildability**: while the designers succeeded in producing a state-of-the-art stadium, constructability was an important feature of the project. Unlike Wembley the construction of the Emirates Stadium was not impeded by the incorporation of untried iconic design features
- **Realistic cost budgets**: the stadium cost approximately £3670 per seat, as predicted (in contrast Wembley cost approximately £9000 per seat compared with its budget of £2056 per seat)

Although the project was not without its own 'problems and challenges', being a private commercial undertaking (Sir Robert McAlpine is a privately owned company), these issues were resolved away from the scrutiny of the press and public. Further, it is believed that the project team consciously hid behind the very public failures of Wembley.

Finally, a key influence on the project team was the tight contract programme imposed in order to ensure that the stadium was completed in time for the 2006–2007 football season. The drive to deliver the stadium on time and within budget was acknowledged by the whole project team; this formed and inspired the ethos of the entire project, underpinning the project's success.

Post-project outcome

Sir Robert McAlpine's successful implementation of the Emirates Stadium project enabled them to negotiate a £160 million deal with Highbury Holdings (a subsidiary of Arsenal Football Club) to redevelop the club's previous venue – Highbury Stadium. The club had planning permission and listed building consent to provide 711 residential units, a general medical practitioners' surgery, a nursery, a gym/health club and a retail unit, along with associated areas of parking and landscaping on the site. Construction works began on the Highbury Square development in July 2006 and were completed on 24 September 2009.

The successful delivery of the Arsenal project resulted in Sir Robert McAlpine being the only viable contender for the role of main contractor on the London Olympics' main stadium.

Acknowledgements

This case study was prepared by David Lowe from the following published sources.

References and sources

Anon (2002) Arsenal awards stadium deal. http://news.bbc.co.uk/1/hi/business/1763759.stm (accessed 28 December 2012)

Anon (2006) Sir Robert wins second major deal with Arsenal. *Contract Journal*, 20 September, p. 10

Arsenal.com (2004) Contract signed for Gunners Stadium. http://www.arsenal.com/emiratesstadium/article.asp?article=205705&Title=Construction+work+has+commenced&lid=the+stadium+-+Latest+News (accessed 27 November 2012)

Arsenal.com (2006) Emirates Stadium – countdown to kick off. http://www.arsenal.com/155/unhoused-import-pages/latest-stadium-news/emirates-stadium-countdown-to-kick-off (accessed 28 December 2012)

AYH plc - Emirates - The home of Arsenal FC more than a world class stadium. www.ayh.com/resources/EmiratesStadium Casestudy.pdf (accessed 27 November 2012)

BBC (2006) Radio 4 interview with Arsenal director. http://www.bbc.co.uk/radio4/(accessed 27 November 2012)

Broughton, T (2001) Two in running to build new Arsenal stadium. *Building* Issue 35. http://www.building.co.uk/news/two-in-running-to-build-new-arsenal-stadium/1010907.article (accessed 28 December 2012)

Buro Happold (2000–2006) Emirates Stadium, London UK. http://www.burohappold.com/BH/BHTemplate8.aspx?ID=2725268C8EB77CCFABE785B57DB2C36E (accessed 27 November 2012)

Design Build Network – Ashburton Grove Football Stadium, London, United Kingdom. http://www.designbuild-network.com/projects/ashburton/ (accessed 28 December 2012)

Edwards, S (2005) How Arsenal moved home, *Building*, Issue 25. http://www.building.co.uk/professional/legal/how-arsenal-moved-home/3069251.article (accessed 28 December 2012)

Exceptional Performers (2007) Relationships key to success of Emirates Stadium for Sir Robert McAlpine. http://www.exceptionalperformers.com/general/relationships-key-to-success-of-emirates-stadium-for-sir-robert-mca.html (accessed 27 November 2012)

Harris, N (2006) Football: multiplex denies 'Wembley is sinking' claim. *The Independent*, 4 April. http://www.independent.co.uk/sport/football/news-and-comment/multiplex-denies-wembley-is-sinking-claim-472692.html (accessed 28 December 2012)

JCT (2007) Hong Kong students see best of British. http://www.jctltd.co.uk/stylesheet.asp?file=06092007103517 (accessed 27 November 2012)

Liddell, I (2006) Pitch perfect. *Ingenia*, Issue 28, September. http://www.ingenia.org.uk/ingenia/articles.aspx?Index=385 (accessed 28 December 2012)

Phillips, N. New century, new stadium. http://www.arsenaltrust.org/newsletter02.html#stadium

Sir Robert McAlpine (2006a) Emirates Stadium completed early. http://www.sir-robert-mcalpine.com/news/?page=24&id=1606 (accessed 28 December 2012)

Sir Robert McAlpine (2006b) ODA chairman praises Emirates Stadium. http://www.sir-robert-mcalpine.com/ (accessed 27 November 2012)

Sir Robert McAlpine (2007a) Four trophies won at building awards. http://www.sir-robert-mcalpine.com/ (accessed 27 November 2012)

Sir Robert McAlpine (2007b) Quality in Construction Award Winner. http://www.sir-robert-mcalpine.com/ (accessed 27 November 2012)

Sports Venue Technology - Ashburton Grove, Arsenal Stadium, London, United Kingdom http://www.sportsvenue-technology.com/projects/arsenal/index.html (accessed 27 November 2012)

Spring, M (2006) Gunning for glory. *Building*, June. http://www.building.co.uk/buildings/gunning-for-glory/3068562.article (accessed 28 December 2012)

Spring, M (2007) They are the champions. *Building*, Issue 16. http://www.building.co.uk/they-are-the-champions/3085081.article (accessed 28 December 2012)

The Emirates Stadium. http://www.4projects.com/Page.aspx?CAT=328 (accessed 27 November 2012)

Wembley Stadium, London

Completed in March 2007, Wembley Stadium, located in North London, is the National Stadium of England. The 90000 seat venue, operated by Wembley National Stadium Ltd (WNSL), hosts key football, rugby league and music events, and can, if required, be adapted to stage major athletic competitions. A wholly owned subsidiary of the Football Association (FA), WNSL was established in 1997 to design, build, finance and run the stadium.

The redevelopment of Wembley Stadium was an immense and complex project, generating controversy and fascinating both industry observers and the general public. It also provided the impetus for the regeneration of Wembley and its surrounding area.

It is projected that the stadium will make a significant contribution to both the local (Brent and surrounding area) and the UK economy. For example, it is anticipated that the project will generate approximately:

- £229 million annually from visitor expenditure, both at the stadium and in the shops, bars, hotels and leisure facilities in and around the site and elsewhere in London

- £40 million annually for the UK government in taxation of activities linked to the venue
- 7500 new permanent job opportunities

During its construction phase the project employed around 3500 people.

Requirement identification

Taking only 300 days to construct and costing £750,000, the original 120 000 capacity Wembley Stadium opened in 1923 – a key feature of the British Empire Exhibition. The same year the FA signed a 21-year contract with Wembley. The first major sporting event to be staged at the venue was the 1923 FA Cup Final, only four days after its completion.

The stadium hosted many memorable events including track and field athletics during the 1948 Olympics, the 1966 FIFA World Cup Final and Live Aid in 1985. However, by the mid 1990s the stadium had reached the end of its useful life; renovation had been discounted due to limitations in its capacity – 85 500 spectators – and the belief that it could not be brought up to the requisite standards of UEFA for hosting major events. A replacement National Stadium for England was therefore mooted.

Closed in 2000, the old Wembley Stadium was demolished in 2002.

Requirement specification

The aim of the project was to design-build a distinctive, state-of-the-art national stadium: a world-class home for English football. Also, in addition to hosting major football events, such as the FA Cup final and England international matches, the stadium was to be capable of staging major athletics and music events. Its design was to be both functional and architecturally significant: an iconic replacement for the old Wembley stadium, world famous for its twin towers. A key prerequisite of the new venue was spectator comfort – for example, the provision of comfortable seats, generous leg-room, unobstructed views of the pitch, and outstanding catering facilities.

The new stadium would generate an important new income stream for the FA with a proportion of the profits being reinvested in football.

Solution selection

Site location

In 1995, the National Lottery panel of the English Sports Council (now Sport England) was assigned the task of allocating Lottery funds for the construction of a new English National Stadium. The new venue was to be capable of staging football, athletics and rugby league events.

The English Sports Council issued a national, open invitation for proposals to develop a National Stadium for England in March 1995. Bids were submitted by Birmingham, Bradford, Manchester, Sheffield and Wembley, from which Wembley and Manchester were shortlisted for further review in October 1995. Manchester and Wembley submitted their detailed proposals in September 1996, which were subject to detailed evaluation by the Sports Council. Subsequently, Wembley was selected as the preferred location for the English National Stadium in December 1996.

Initially, the scheme was to be developed by Wembley plc (owners of the original Wembley Stadium) using a special purpose vehicle, the English National Stadium Trust (ENST). However, in May 1998 ENST announced that it had been unable to secure a lease from Wembley plc, following which the FA withdrew from the project, only to return in July 1998 as the lead organisation under a revised project structure.

To ensure that Wembley would be the location for the new English National Stadium, Arsenal FC had expressed an interest in the site; it was decided that the Lottery grant would primarily be used to purchase the existing stadium. The new stadium would be designed, built, financed and operated by Wembley National Stadium Ltd (WNSL), a wholly owned subsidiary of the FA.

The Sports Council agreed, in principle and subject to exacting conditions, to allocate Lottery funding to the project up to a sum of £120 million. The grant was made through three Lottery Funding Agreements (LFAs). The final LFA dated 12 January 1999 required WNSL to:

- Design and build a new stadium with a minimum capacity of 80 000 seats for football and rugby matches and 65 000 seats for athletics events
- Run the stadium
  - principally as a venue for 'flagship' sporting events, and
  - secondly for music, conference and ancillary purposes
- Promote the reputation and integrity of English sport and acknowledge the contribution made by Sport England
- Operate the stadium in accordance with specific policy requirements, for example, ensuring non-discriminatory public access, ample transportation and community use (Select Committee on Culture, Media and Sport [SCCMS], 2002b)

Sport England was responsible for monitoring WNSL's compliance with these LFA requirements.

In March 1999 WNSL acquired the existing Wembley stadium for £103 million from Wembley plc.

However, when WNSL encountered difficulties in raising the required finance for the project (discussed later in the case study) the FA approached the Birmingham consortium (comprising the Birmingham and Solihull Councils and the National Exhibition Centre) in May 2001 regarding an alternative proposal to locate the national stadium in the Midlands. Despite the consortium investing upwards of £700,000 (plus approximately £300,000 in 'in-kind' from the private sector), the preferred location remained Wembley.

Stadium location

Wembley Stadium, located in north-west London, falls within the London Borough of Brent, and was built on the site of the original Wembley Stadium.

Site constraints

WNSL's freehold title to the Wembley site was split into two sections:

1. Twenty-four acres (approximately) – the site of the old stadium
2. Seven acres of development land to the north of this

Although owned by WNSL, Wembley plc held an option on the second section of land:

- Failure by WNSL to commence construction of the new stadium by 31 December 2002 would result in the land reverting in full to Wembley plc
- WNSL were required to give Wembley plc three months notice of its intention to use the option land. Failure to give notice by 30 September 2002 would result in WNSL losing the land
- WNSL were obliged to hand back any part of this plot not used for the development of the new National Stadium to Wembley plc

In addition, further restrictions were placed on the use of the site:

- A lease structure (on the whole site) restricted the use of the land to the provision of a national stadium and included a means of disbursing 1% of WNSL's revenue to charitable purposes once the stadium had been in operation for five years

- A covenant held by Hilton Hotels, on the whole Wembley site, prevented the development and operation of a hotel in competition with Hilton. The initial scheme included a hotel; however, this was later discarded
- Under the freehold transfer document, Wembley plc were entitled to approve the planning consent for the development; it also regulated WNSL's rights over adjoining land owned by Wembley plc during the construction of the stadium
- Further covenants imposed additional restrictions on the use of the stadium site to safeguard Wembley plc's ongoing business interests (Select Committee on Culture, Media and Sport [SCCMS], 2002b)

Brent Council granted the project initial planning permission on 1 June 2000, conditional on WNSL providing £17.2 million towards the cost of improving Wembley's transport infrastructure.

Project scope

The initial scheme included a stadium suitable for staging football, rugby league and music events and which could, if required, be adapted to stage major athletic competitions. It also incorporated a hotel, office accommodation and a visitor centre.

The inclusion of a removable platform, rather than a permanent running track, to fulfil Sport England's athletics requirements for the stadium was criticised by Kate Hoey, the sports minister, and the British Olympic Association. They considered the solution to be unsuitable and detrimental to London's bid to stage the Olympic Games; this led the Culture Secretary, Chris Smith, to order the design to be reviewed.

In January 2000, WNSL announced that the stadium would be developed as a 'dual-purpose athletics/football' venue, despite the decision by the Culture Secretary that the stadium would be used solely for football. According to WNSL, the Secretary's decision had 'come too late' to amend the design. Despite an earlier 'gentleman's agreement' to repay Sport England £20 million of the Lottery grant in the event of athletics being removed from the scheme, further controversy ensued when WNSL declined to honour the agreement, arguing that they had fulfilled the Lottery criteria. As it turned out, athletics was reinstated in the scheme after it had been established that a platform solution was both technically feasible and 'cost-effective', although Sport England anticipated that the stadium would only be used as an athletics venue once or twice over a 20-year period. The offer to repay £20 million was withdrawn.

In September 2000 WNSL withdrew its scheme after failing to attract commercial funding for it. It was assumed that the inclusion of a hotel and office space in the original proposal had been the 'deal breaker' for the financial institutions. However, preliminary approval for a scaled-down stadium was given by the FA and the government in December 2001. On 20 March 2002 planning approval was granted to a revised stadium design, which excluded the hotel, office accommodation and visitor centre, that formed part of the earlier scheme.

Infrastructure improvements and the regeneration of Wembley

The new stadium was conceived as a public transport destination, requiring the existing transport infrastructure serving the venue (roads, rail, bus and pedestrian routes) to be rebuilt and revitalised, to meet the increased demand placed on it on major event days. WNSL, therefore, worked directly with all the key stakeholders: Transport for London (TfL), the London Development Agency (LDA), the Department for Culture, Media and Sport (DCMS), WNSL, the London Borough of Brent, the Metropolitan Police and the main train, tube and coach operators. Over £70 million was invested, via Section 106 monies, to deliver the required improvements.

As a key component of the project, Wembley Park, Wembley Central and Wembley Stadium station, the three principal stations serving the stadium, were all upgraded and refurbished to:

- Improve passenger management – for example, the capacity of Wembley Park station was increased by 70% to accommodate 37 500 passengers/hour

- Provide access for mobility impaired passengers: lifts were installed in all stations
- Provide faster and safer public transport journeys: approximately 100 trains/hour service the stadium on event days

Additionally, a new access road (referred to as the Stadium Access Corridor) was created linking the stadium's car park with the M40, M1, and A1(M) by means of the North Circular Road.

The then Mayor of London, Ken Livingstone, said:

> 'As well as being an enormous asset to sport in this country, the new Wembley Stadium will make a substantial economic contribution to London's economy, bringing visitors, jobs and better transport to a very deprived area of North West London. That's why I am pleased to have played my part by putting the case for Wembley to Government, opinion formers and general public, and by securing £21 million for the stadium itself as well as funding for the major transport improvements that will make this a world class project.' (Quoted in Sport England, 2002)

The project also provided the catalyst for the regeneration of the area surrounding the stadium by Brent Council and the developer Quintain Estates.

Following the identification of Wembley as a nationally important 'opportunity area' for leisure and related development, Brent Council launched their 'vision' for a rejuvenated Wembley in November 2002, presenting a far-reaching plan for the area, which had as its focal point the new Wembley Stadium. In addition to the stadium, the scheme included the following key components:

- A community focus for Brent
- A national, regional and local leisure destination
- The London Convention Centre
- A centre for work
- An accessible development
- A cultural and educational centre
- High quality commercial and retail facilities
- A mixture of housing types and tenures (Brent Council, 2002)

In September 2003 the council approved the Wembley Development Framework, which outlined the council's aspirations for the area: the transformation of Wembley into a 'world-class' destination. Proposals were published in June 2004 by both Brent Council and Quintain to improve local amenities and provide thousands of new homes and employment opportunities in the area.

Stadium design solution

In 1996 Foster and Partners were appointed to produce a master plan for redeveloping Wembley Stadium. Ultimately, the design for the new stadium (based on a 50-year design-life) was produced by World Stadium Team, a joint venture between Foster and Partners, and HOK Sport, with structural and services design provided by Mott Stadium Consortium consisting of Mott MacDonald, Connell Wagner and Sinclair Knight Merz and Weidlinger Associates. Sir Norman Foster was responsible for the design of the building's impressive arch and roof structure. The design was novated on the appointment of a main contractor under a design-build contract.

The design solution adopted required the demolition of the existing Wembley Stadium, including its famous 30 m high twin towers.

Design features included:

- **The Bowl**: a key design aspect was the creation of a single bowl to accommodate spectators and to provide uninterrupted views of the pitch. To achieve this, the designers produced an imposing, saucer-shaped stadium with a 90 000-seat capacity (double the size and four times the height of the existing stadium) distributed on three tiers of the eight-floor structure. Acoustic modelling of the existing stadium was used to recreate the famous Wembley atmosphere

- **Retractable roof**: currently, the stadium is the world's largest football venue where all seats can, if required, be protected from the elements. This is achieved through the inclusion of retractable panels in the 50 000 m^2 roof. The process (although initially envisaged to take 15 minutes) takes one hour and should only be undertaken when the stadium is empty. Retracting the panels also enables sunlight to reach all parts of the pitch. A lattice arch with asymmetric catenary cable net and stayed trusses spans the stadium bowl, supporting 5000 tonnes of roof structure
- **The Arch**: the stadium's iconic steel arch, referred to as a 'tiara' by Norman Foster, forms a highly visible landmark on the London skyline by creating a 'tube of light'. Positioned above the north stand, the lattice arch projects 52 m above the stadium, measuring 133 m at its highest point. It is the world's longest single-span roof structure, having a diameter of 7.4 m at its base, weighing 1750 tonnes and spanning 315 m. At an angle of 68° from the horizontal, the arch contains 41 steel diaphragms, linked by 500 spiralling steel tubes to form thirteen 20.5 m long modules which are connected to 70 tonne hinges. The loading of the arch is transferred through concrete bases to 35 m deep piles
- **Pitch**: a key challenge for the design team was the requirement for a high-quality grass pitch, a pre-requisite to the stadium acquiring a UEFA five-star rating. To achieve this, the stadium's retractable roof exploits the optimum conditions for the pitch in terms of sunlight and air movement (computer modelling of the existing pitch was used to inform the design) while between events the roof will be left open. Also, the design incorporates a sub-air system capable of warming the pitch via ducts below the playing surface and extracting moisture when the grass it is too damp. The pitch retains its east-west orientation
- **Seating**: the stadium design incorporates seating superior to that of its predecessor, which is as near the pitch as possible and provided in higher, more steeply raked tiers that offer unobstructed views of the pitch. The central tier of the stadium, 17 000 seats, is assigned to Club Wembley in the form of corporate boxes (incorporating bar, kitchen, lavatories, TV screen and luxury seating) and ten-year seat licences, with access to four of the largest restaurants in London. The stadium retains its Royal Box for the presentation of trophies, etc. positioned, as before, in the middle of the north stand and reached from the pitch by climbing 107 steps, 68 more than in the old stadium
- **Concourses**: an external concourse surrounds the stadium, providing refreshment facilities for approximately 40 000, thereby, reducing the impact of spectators on the local environment. It also encourages spectators to turn up early for events and facilitates easy and safe access to and egress from the stadium. This feature required the stadium to be repositioned 30 m north of the existing venue, nearer to Wembley Park Station. Internally, spacious concourses encircle the stadium with 26 lifts and 30 escalators providing easy transit through the building
- **Athletics**: although its principal function is to host football, rugby and music events, the stadium can be adapted, if required, to stage international athletic events to IAAF standards (a requirement of the Lottery funding agreement). This is achieved by using an innovative elevated athletics platform, which covers the pitch and part of the lower bowl. The removable modular steel and concrete platform, endorsed by the IAAF, is designed to be erected and dismantled in a matter of weeks (approximately 11 weeks to install and 6 weeks to remove). When configured for athletics, the stadium has a capacity of 68 400, although this can be increased to 70 000 by utilising temporary seating

The stadium is connected to both the London Underground (Wembley Park Station) by means of Olympic Way and Wembley Central Station via the White Horse Bridge.

Development programme

Initially, it was anticipated that construction of the stadium would commence in December 1999 and be completed in time to stage the 2003 FA Cup final. However, the redevelopment of Wembley was

Box A.3 Key statistics

Wembley Stadium provides:

- 90 000 seats
- 17 000 Club Wembley seats
- Four of the largest restaurants in London including the capital's largest banqueting hall with capacity for 2000 guests
- Space to cater for 10 500 seated covers per event
- 98 kitchens, the biggest one being one-third the size of the Wembley pitch
- 688 food and drink service points, 2618 toilets, 19 parenting rooms, and 20 first aid rooms
- Capacity for up to 310 wheelchair users and companions and 150 toilets designed for disabled access
- Three interior concourses, each 1 km in circumference
- Two giant screens each the size of 600 domestic television sets
- A pitch size of 105 m × 68 m

protracted and problematic, subjected to scrutiny, including in 2001 a House of Commons Select Committee report, and significant media interest.

The key milestone dates for Wembley Stadium are shown in Table A.4.

Project budget

Initially, the construction budget for the project was £185 million, but this was reduced to £140 million in June 1998 to reflect a revised 'Cardiff-style' development. By January 2000, the estimated cost of redeveloping Wembley stadium stood at £475 million. However, in December of the same year this had risen to £660 million. It was argued that the development costs were appreciably higher than those of similar sports venues because, in addition to the stadium, the project incorporated a broad portfolio of business opportunities, including a hotel and other facilities. However, as mentioned earlier, the hotel and office accommodation were ultimately removed from the scheme.

By September 2002, the basic building cost of the stadium was £364 million but this increased to a design and construction contract sum of £445 million once costs for design, demolition, fit-out, insurance and professional fees, etc. were included. The total project cost was predicted to be £757 million.

Tropus, WNSL's initial project management consultants, were later to comment that WNSL's Senior Management Team had failed to establish a 'realistic budget' for the project and as a result their designers were not required to produce a scheme within '*strict cost parameters*' (SCCMS, 2002c).

Financing the project

In June 1997, the English National Stadium Trust (ENST) was established as a charitable trust to oversee the development of the new stadium; it was also used by Sport England as a mechanism to protect the public interest in the project. ENST then formed a limited company, the English National Stadium Development Company Ltd (ENSDC), to develop, construct and operate Wembley Stadium. ENSDC changed its name to WNSL in March 1999. Initially, WNSL had two shareholders – the FA and ENST – all of the ordinary shares were held by the FA, while ENST retained a single 'A' or 'golden' share. The FA had acquired their shares from ENST in January 1999, when the 'golden' share was created as a condition of Sport England support.

Table A.4 **Wembley stadium pre-contract milestones.**

1995	
April	Competition to select the location for a new English National Stadium instigated by English Sports Council (renamed Sport England)
July	Birmingham, Bradford, Manchester, Sheffield and Wembley submit bids for Lottery funding to build the new national stadium
1996	
December	Wembley is selected as the preferred site for the new stadium by the Sports Council (now Sport England)
1998	
April	The English National Stadium Development Company Ltd (ENSDC – later renamed Wembley National Stadium Limited [WNSL]) purchases Wembley for £103 million – Sport England provides the funding through a £120 million Lottery grant
May	ENSDC appoints Ken Bates (Chelsea Football Club chairman) as chairman
November	It is revealed that Wembley's famous twin towers will be demolished. Initially, English Heritage fights to retain them but then withdraws its opposition
1999	
14 July	WNSL issue an ITT for Wembley stadium; contractors who respond include: Mowlem, HBG, Sir Robert McAlpine and Bovis/Multiplex (BMX)
29 July	Norman Foster's design for the new stadium is revealed
15 November	Planning application for the stadium is submitted to Brent Council
November	Backed by the British Olympic Association, sports minister, Kate Hoey, criticises the proposed stadium design, which features a removable platform rather than a permanent running track, as being unsuitable for an Olympic bid. Following this criticism, the Culture secretary, Chris Smith, calls for the design to be reviewed
22 December	Chris Smith announces that Wembley is to be developed as a stadium for football and rugby only. The FA enters into a gentleman's agreement to repay £20 million of the £120 million Sport England Lottery grant
2000	
22 February	Despite submitting a non-compliant bid, Bovis/Multiplex joint venture announced as the preferred contractor (appointment commenced 6 March). It would appear that neither Mowlem nor Sir Robert McAlpine submitted formal bids
June	Brent Council grant planning permission for the new stadium following WNSL's agreement to fund new road and rail links to the ground
30 August	After failing to agree on a contract sum, WNSL's negotiations with Bovis/Multiplex are terminated
11 September	WNSL appoint Multiplex as preferred contractor based on a GMP of £326.5 million and a 39-month construction programme. Bovis threatens legal action
7 October	Wembley hosts its last competitive fixture: Germany defeating England 1-0.
December	Sir Rodney Walker replaces Ken Bates as chairman of WNSL. Bates is demoted to vice-chairman
2001	
April	FA fails to secure commercial support for the project
1 May	FA announces that the project will fail unless the Government provides additional funding. The request rejected by the prime minister, Tony Blair, and a review of the project by Patrick Carter is authorised
December	Wembley re-surfaces as the likely location for the national stadium, although a final decision is deferred

(Continued)

Table A.4 (*cont'd*).

2002	
11 January	WNSL announces that it plans to overhaul its management team – Michael Jeffries of WS Atkins subsequently appointed chairman
8 February	Ken Bates resigns from the WNSL board
21 May	Reports by Tropus and David James are published; these raise concerns over WNSL's procurement procedures and governance of the project
31 May	The FA announces the conclusion of a loan agreement with WestLB and that enabling work on the stadium will begin within three months
26 September	Final approval for the project is given by the FA, Sport England and the Culture Secretary. In return for further controls, the government provides additional funding of £20 million, while WNSL secures £433 million in bank loans. The cost of the project now stands at £757 million
30 September	Construction of the new Wembley stadium commences. The contract stipulates a 39-month construction programme and a GMP of £445 million

Source: adapted from Bewsey (*2006*), BBC Sport (*2006*), Brent Council (*2007*), Clark (*2002*), Conn (*2006*), Daley (*2006*), Oliver (*2002*), FA, Multiplex and WSNL press releases

Table A.5 Cost breakdown. Based on a projected project cost of £757 million the costs can be broken down as follows.

	£ million	£ million
Multiplex (excluding bank and legal)		
Base construction	364	
Fees, FF&E, inflation for delayed start, insurances, contingency and other	81	445
Other		
Acquisition of stadium (purchase of land, stamp duty and fees)	106	
Construction interest	57	
WNSL costs	39	
Bank arrangement fees/finance costs	22	
Local infrastructure (Section 106)	21	
Project contingency	20	
WNSL management fee	19	
Legal and other professional fees	11	
Other	17	312

Source: Carter (*2002*). Reproduced under the terms of the Open Government Licence (http://www.nationalarchives.gov.uk/doc/open-government-licence/)

WNSL's articles of association included a list of 'restricted matters', actions which WNSL could not take without the agreement of ENST (endorsed by Sport England), or of Sport England itself. These matters include:

- *Anything that would result in the FA ceasing to have control of WNSL*
- *Anything that would alter the constitution of WNSL or the rights of the golden share itself*
- *Initiating certain insolvency proceedings in relation to WNSL*
- *Materially changing the nature of WNSL's business*
- *Acquiring any interest in another company, partnership or joint venture*
- *The sale or other disposal of the stadium*
- *Paying a dividend during the first five years following opening of the stadium* (SCCMS, 2002a)

Further, the 'golden' share allowed ENST to appoint a specified number of directors to the board of WNSL.

As mentioned earlier, the National Lottery contributed £120 million to the project; this was distributed by Sport England to WNSL through three LFAs between November 1997 and May 2000. The money was used as follows:

- £3.26 million on initial design work prior to December 1998 and the legal/due diligence fees required for the purchase of the existing Wembley Stadium from Wembley plc
- £103 million to acquire the existing stadium (on 15 March 1999); the land was valued at £64.5 million. However, the buildings were subsequently written off – due to the need to demolition the old stadium
- Approximately £3 million was paid to the government in stamp duty

The residual funds were used to finance specific expenditure incurred on the project (SCCMS, 2002b).

Although the purchase of the existing stadium was financed by a grant from Sport England, in order for the project to go ahead, the majority of the funding had to be raised by WNSL from private sources. Chase Manhattan Bank was appointed in February 2000 to raise the required capital from the financial markets. However, by October 2000 Chase Manhattan were experiencing severe difficulty in raising the required finance. As a result the demolition works were postponed.

The following April, the FA announced that it had been unsuccessful in securing commercial support for the project, followed on 1 May 2001 by an announcement that the project would fail unless the government injected additional funds into the development. This request was rejected by Tony Blair, the prime minister, and a review of the project by Patrick Carter was instigated.

At this point the project was at risk of being aborted. If that had occurred then WNSL would have been required to reimburse Sport England the £120 million Lottery grant. Although measures had been taken to protect the public sector investment in the project, WNSL's ability to repay the grant in full was not assured.

The funding of the project was subject to controversy. A parliamentary report published by the Culture, Media and Sport Committee on 20 November 2001, seriously criticised Chris Smith's role (as Secretary of State for Culture, Media and Sport) in the Wembley project. It described his transactions with the FA and Sport England as a '*scandalously inept treatment of public money*'. Smith had brokered a deal with the FA, that required them to return an arbitrary £20 million to Sport England following the exclusion of athletics from the Wembley scheme. However, after two years no legal agreement had been signed nor any money repaid to Sport England. Similarly, for permitting this to happen, Sport England was censured for being '*slack and negligent*'. Further, Sport England were criticised for their '*cavalier and egregious use of public funds*' for prematurely dispensing Lottery funds of £120 million for the purchase of the existing Wembley stadium site, prior to the scheme obtaining planning permission.

In December 2001, Wembley resurfaced as the likely location for the national stadium, but a final decision on whether or not to proceed with the project was deferred.

Once appointed, Multiplex (the main contractor on the project), in partnership with WNSL, led the commercial and financial backing of the project. The construction contract was initially conceived as a design, finance and construct agreement. However, the stadium was eventually acquired under a design-build contract. Following the departure of Chase Manhattan, the FA simultaneously entered into discussions with both Barclays and the German bank Westdeutsche Landesbank (WestLB) to coordinate the financing of the project. However, on 26 April 2002, after becoming aware of the FA's parallel dialogue with WestLB, Barclays withdrew from the discussions.

On 8 May 2002, Tessa Jowell, the Culture secretary, extended the FA's deadline to obtain commercial funding for the project. On 23 May she informed the House of Commons that the current financial negotiations represented the 'last chance' for Wembley, and that if they failed Birmingham's bid should be discussed. Also, during May it was revealed that if the negotiations currently being held with WestLB were to fail, then the original stadium would need to be refurbished and reopened to meet WNSL's staging commitments. WNSL had obtained a guaranteed income stream for the stadium by negotiating a 30-year staging agreement with the FA; similarly, 20-year deals were also agreed with the Football League and the Rugby Football League.

Table A.6 Breakdown of funding. Based on a projected project cost of £757 million, the funding sources are as follows.

Source	£ million	%
Bank loans (WestLB)	426	56.3
Senior subordinated debt (Credit Suisse First Boston)	7	0.9
The FA contribution	148	19.5
Sport England contribution	120	15.9
LDA contribution	21	2.8
DCMS contribution (government)	20	2.6
WNSL surplus from operations	15	2.0

Source: Carter (2002). Reproduced under the terms of the Open Government Licence (http://www.nationalarchives.gov.uk/doc/open-government-licence/)

In September 2002 WNSL announced that they had secured the £426 million financial backing for the project through a consortium led by WestLB. WestLB's chairman, Juergen Sengera, commented:

> 'It has been a complex transaction from an arranger's perspective but we are confident that work on the new stadium can go ahead underpinned by solid commercial and financial foundations.' (Quoted in Sport England, 2002)

Further financial support was provided by the FA, Sport England, the LDA and the DCMS, and WNSL contributed £17.2 million towards improving the transport infrastructure to the stadium. A significant part of WNSL's business plan was the sale of corporate boxes and Club Wembley ten-year seat licences; by March 2005, approximately £250 million had been raised through this activity. Credit Suisse First Boston (CSFB) provided £7 million of funding, plus a letter of credit guaranteeing WestLB up to £25 million if there were to be a significant shortfall in the income generated via advance sales of premium seats (before the completion of the stadium). Also, Delaware North negotiated a 25-year concession to run the stadium's catering facilities, which contributed £20 million towards the capital cost of the stadium at practical completion.

Based on the predicted total project cost of £757 million, the contributions towards the funding of the project were as follows (Table A.6).

Project management

Initially, Tropus (formerly McBains Limited) provided project management services on the project between November 1997 and August 2001. However, following a hiatus in the development and a period of review their appointment was not renewed. WNSL then appointed Symonds, who had previously worked on the Millennium Dome, as project managers. Their role involved assisting the client to secure financial support for the project and then, during the construction phase, providing traditional project management services.

Although Tropus prepared a project execution plan (PEP), as the project developed this document was not (according to Tropus) revised; moreover, the procedures contained within it were not imposed. Similarly, Tropus instituted an appropriate structure and procedures to allow the project management team to undertake the routine management of the development, with WNSL's Senior Management Team providing strategic direction. However, in the opinion of Tropus, the senior management team became too involved in project management activities. Further, they commented to a subsequent House of Commons Select Committee that in their opinion the project management team had not been '*allowed to mange roles and tasks efficiently*' (SCCMS, 2002c).

In reaching its decision to allow the project to continue, DCMS considered the potential risks inherent in the delivery and operational phases of the project. It also reviewed the mechanisms adopted by WNSL (in conjunction with the other interested parties) during the contract negotiation stage to mitigate these risks. Table A.7 presents these key risks and the 'risk response' employed.

Table A.7 Project risks.

Risk	Steps taken to mitigate/transfer the risk

Construction risk
 	
The potential for the:
 	
• Construction phase of the project to take longer than anticipated
• Construction costs to exceed the agreed contract sum
 	
resulting in a funding short fall
	The construction contract apportioned risk between WNSL and the main contractor. It also:
	
	• Incentivised the contractor to deliver on time and within budget:
	  ◦ in the event of late delivery (provided the contractor was not entitled to an extension of time) WNSL were entitled to levy liquidated and ascertained (L&A) damages from the contractor to a maximum equal to 9-months' delay
	  ◦ the contract was based on a guaranteed maximum price (GMP) and incorporated a gain-pain share clause. If the contractor was able to construct the stadium below the GMP, they would keep the difference; if the stadium cost more then they would bear the additional cost (except in specific circumstances)
	• Included the provision, by Multiplex, of a £60 million performance bond. In addition, the contract enabled WNSL to hold a retention of up to £40 million until completion has been certified. These provisions provided some protection for WNSL if the main contractor were to fail to deliver the project as expected (due, for example, to contractor failure or insolvency)
	• Incorporated a £30 million contingency sum (WNSL contributing £20 miliion and Multiplex £10 million)
	
	A risk management exercise, which sought to minimise the risk of design changes, established that the most likely circumstances in which Multiplex would be entitled to claim for additional costs and/or time lay within WNSL's control. For example, if WNSL were to issue late instructions or make design changes
	In addition to these provisions, if the project were to encounter difficulties, WNSL could appeal to Sport England for the relaxation of some their funding conditions: for instance those concerning naming rights and anchor tenancy, thus, providing WNSL with a further revenue stream

Operating risk The potential for the stadium's: • Running costs to exceed those anticipated • Revenue falling short of those projected – for example due to failing to sell sufficient premium seats thereby, undermining the viability of the project and WNSL's capacity to meet its liability to WestLB	The overall business case was assessed by the banks and reviewed under the Carter report. It was deemed to be acceptable Primarily, the financial success of the project was based on the sale of premium seats (it was anticipated that this component of the business plan would generate approximately 70% of WNSL's income). To address this WNSL: • Appointed IMG (a sports marketing agency) to sell the premium seats, incentivising them to maximise returns • Obtained a letter of credit from Credit Suisse First Boston, guaranteeing WestLB up to £25 million if there were to be a significant shortfall in the income generated via advance sales of premium seats (before the completion of the stadium) The FA guaranteed WNSL an annual income via an undertaking to stage certain events at the stadium As above, if the project were to encounter operational difficulties, WNSL could appeal to Sport England for the relaxation of some their funding conditions

Source: Adapted from NAO (2003)

DCMS's review of the project (NAO, 2003) concluded that project success would be 'heavily dependent' on WNSL adopting:

- **A strong executive team to manage the project**: in particular its relationship with the main contractor. Additionally, to reduce the risk of construction costs exceeding the agreed fixed price, the establishment of a 'no change' culture was imperative
- **Effective corporate governance to oversee the executive management of the project**: in response to problems encountered earlier in the project and advice from PricewaterhouseCoopers, WNSL strengthened their corporate governance arrangements. Modifications included:
  - reconstituting its Board under a new Chairman, with construction sector experience
  - establishing a committee structure to support the Board; each committee having defined terms of reference and non-executive directors as members
  - requiring board members and senior managers to complete declarations of interest Additionally, WNSL appointed an internal auditor to review their operating systems

Further, to protect the interests of the 'public sector' stakeholders, a range of measures were introduced:

- WNSL appointed a 'compliance officer' (their Chief Executive): responsible for verifying the appropriate allocation of public funds and the attainment of WNSL's public interest obligations
- The FA and WNSL undertook to perform further OGC reviews at key points during the construction of the stadium
- Sport England retained the right to appoint three directors to WNSL's Board

Project procurement (contractor appointment)

Despite receiving £161 million of public sector funding (amounting to over 21% of the anticipated project cost), WNSL took the decision that the project was not subject to EU procurement directives.

In late 1997, ENSDC adopted a procurement strategy recommended by Tropus, under which the project was divided into three phases:

1. Demolition and enabling works (to be let under a lump sum contract)
2. Shell and core (to be let on a lump sum, detail and construct contract)
3. Fit out (to be let either as one or more lump sum contracts, or through negotiation with the main contractor appointed under phase 2)

However, after a preliminary meeting with WNSL, in May 1999, Multiplex submitted a proposal (21 June 1999) whereby Multiplex would be appointed as 'preferred contractor' on an 'open book' arrangement. Under the proposal Multiplex would be paid a fixed fee of 2%, work would be measured against an agreed cost plan (all cost savings would revert to WNSL). Further Multiplex would provide a 100% performance bond. In order to test the market, on 14 July 1999, WNSL asked several UK and European contractors to submit alternative proposals for the construction of the stadium (SCCMS, 2002c).

Between July and October 1999 WNSL carried out a formal pre-qualification exercise involving 16 main contractors, following which, five contractors (Bouyges, Bovis-Multiplex, HBG, Sir Robert McAlpine and Mowlem) were invited to submit detailed 'first-stage' bids. Four bidders were then interviewed by WNSL (after which one contractor withdrew) before it issued amended project documentation to the remaining three contractors, who were required to submit revised proposals by January 2000. Unwilling to adopt a two-stage tender process, WNSL requested bids based on a lump sum, fixed-price contract.

On 18 February 2000, following further negotiation and evaluation, WNSL announced that their preferred bidder was a joint venture between Bovis Lend Lease and Multiplex Construction Limited

(BMX), in preference to HBG and Mowlem, despite their submission of a non-compliant bid. The curriculum vitae of both JV partners included extensive stadium construction experience: Multiplex on Stadium Australia (the Sydney Olympic Stadium), while Bovis had been construction and project managers on both the Atlantic Olympic and Sydney Football Stadia. It was later claimed that Bouyges and Sir Robert McAlpine had withdrawn from the competition due to onerous contract terms, particularly WNSL's insistence on a GMP. This 'preferred contractor' status would be effective for 20 weeks. Dependent on WNSL obtaining planning permission for the development, the formal appointment was due to commence on 6 March 2000.

On 7 September 2000, however, WNSL announced that it had been unable agree a contract sum with BMX and that their appointment as preferred contractor was, therefore, terminated. BMX's initial bid had been £390 million, which following a value engineering exercise had been reduced to £347 million; however, this revised figure was still unacceptable to WNSL who had set a limit of £320 million for the construction works. It was reported in *Building* that WNSL had adopted a robust negotiating stance, as illustrated by the following quote from Bob Stubbs, Chief Executive of WNSL:

> 'We're not meeting anyone halfway. We either get the right numbers [for the price] or we retender the whole thing.' (Quoted in Glackin, 2000)

WNSL were convinced that the overall project costs could be contained within the £600 million budget used throughout the negotiation phase. Chris Palmer, communications director of WNSL, said:

> 'We feel we have a very accurate price for building the new stadium which we feel stands up to robust inspection... We were rather more optimistic in our costing than them [*Bovis*] so we decided to go our separate ways... This will not effect [sic] the scheduled completion date of 2003 although we are not in a position to say whether it will be ready in time for the FA Cup final of that year.' (Quoted in Stone, 2000 and Anon, 2000)

It was later revealed that there had been a disagreement between the two joint venture partners over the cost of construction work.

Despite renewed interest from Bouyges, on 11 September WNSL awarded the contract to Multiplex (as sole contractor) based on a GMP of £326.5 million. However, the completion date was set for early 2004 to enable the FA Cup final to be held in the new stadium (a year later than previously anticipated). It was later alleged that this 'heads of terms' had been signed on behalf of WNSL without appropriate legal advice being sought. Apparently Bovis considered taking legal action against Multiplex to recover its costs incurred in bidding for the project, as the JV partnership was still effective when WNSL and Multiplex entered into separate negotiations, although this did not materialise.

At the time of its appointment, Multiplex was Australia's largest construction company, wholly owned by its Chairman and founder John Roberts. Its successes included delivering Stadium Australia the 110 000-seat centrepiece venue of the Sydney 2000 Olympic Games three months ahead of schedule (the stadium received more than 15 major international construction and engineering awards). In the UK their projects included: 'Knightsbridge Green', a residential development in London; 'Sentinel Point', a luxury office development in Vauxhall; Chelsea Football Club's new West Stand; and a Sports and Leisure Club for Chelsea Village plc.

However, commencing in September 2000, effectively two years were lost due to political infighting over Wembley's function; simultaneously WNSL experienced difficulties in attracting investors and the stadium was redesigned several times.

Contract

Between September 2000 and September 2002 the basic construction costs of the project rose from £326.5 million to £364 million.

Initially, WNSL had intended entering into a 'design, construct and finance contract' with Multiplex. However, on 26 September 2002 Multiplex and WNSL entered into an amended JCT design-build contract, under which Multiplex was responsible for the overall design and construction of the project. In addition to the GMP provision, the contract included a gain share clause, which distributed any savings on the project 80% to the contractor and 20% to WNSL; all costs above the GMP were to be borne by Multiplex. The agreement also made provision for Multiplex to claim £12.6 million from WNSL if the contract was terminated. Contract completion was to be 30 January 2006 (in time for the 2006 FA Cup final), a year later than initially anticipated. L&A damages on the project were set at £120,000/day to a cap equal to nine months delay. In addition, Multiplex provided a Chubb performance bond and was instrumental in obtaining a novel set of project insurances.

WNSL believed that their contract would protect them from the risk of increased costs and construction delays.

Review of procurement process

Subsequently it was claimed that, although there was no evidence of corruption, the procurement process adopted by WNSL failed to follow normal commercial practice, specifically that the Bovis–Multiplex joint venture had received 'preferential treatment' during the tendering period. Further, this treatment continued in respect of Multiplex after the withdrawal of Bovis from the project.

Two documents, the Tropus report and the James study, both criticised WNSL for inappropriately procuring and managing the project.

The Tropus report claimed that:

- BMX and then Multiplex had been given preferential treatment by WNSL during the contractor selection process. During the initial tender phase, BMX had been privy to information and documents not made available to the other bidders. This partiality continued in respect of Multiplex following the withdrawal of Bovis from the project.
- WNSL had entered into a contract with Multiplex with unwarranted haste. It said:

> 'This appointment was, in our opinion and experience, on less onerous and well-defined terms than the original appointment and we believe that the other short-listed contractors may have submitted competitive bids had they been given the same opportunity…
>
> In our professional experience we find it remarkable that this two-and-a-half page letter (with five pages of attachments) was signed on behalf of WNSL without appropriate legal advice being taken…' (SCCMS, 2002c).

Further, it did not consider that:

> '… the Multiplex construction cost offer had been properly interrogated.' (SCCMS, 2002c)

- The project could have been re-tendered. David Hudson, a director of Tropus, later said:

> 'UK and large European contractors would still have been interested in the project. If they had started the process in the summer, demolition could have started at the end of last year. This scheme was capable of being built by a UK contractor. The UK industry was not given a legitimate opportunity to pitch for the scheme.' (Reproduced in Clark, 2002)

Additionally, the James report raised the issue of potential conflicts of interest by member(s) of the WNSL board. The former chairman of WNSL, Ken Bates, had been accused of favouring Multiplex: the contractor was, at the time, constructing a new stand at Chelsea FC (where Mr Bates was chairman). Bates strenuously denied any conflict of interest, asserting that Multiplex had been appointed because they were prepared to undertake the project on a fixed-price basis. The report concluded that:

> 'The process adopted by WNSL is unlikely to satisfy best practice standards as usually deemed necessary in any project involving government or lottery funding.' (Reproduced in Clark, 2002)

Both reports concluded that the procurement process adopted by WNSL had not complied with acknowledged best practice, creating uncertainty as to whether the process had been applied impartially or that it had achieved value for money.

Value for money

In April 2002, international construction and property consultants, Cyril Sweett, produced an independent value for money report for WNSL (Cyril Sweett, 2002). The report confirmed that, while the deal with Multiplex did not constitute the '*cheapest price*' the design-build contract did in fact represent value for money, falling broadly within '*... cost parameters for a project of this type and scale*'.

Cyril Sweett considered that Multiplex's price reflected the particular terms of the proposed contract, which included:

- Demanding contract conditions (for example, that afforded WNSL greater price certainty)
- High design standards (a requirement of the Lottery funding agreement)

Further, the price was potentially higher due to:

- The stadium's seating capacity of 90000 (as large stadia are proportionately more expensive than those with a lower seating capacity)
- The provision of a considerable amount of additional accommodation (included to generate income, as the stadium was required to be financially self-sustainable) (NAO, 2003)

The consultant's report also contained a like-for-like comparison of the projected cost of Wembley with other comparable international stadia (see Table A.8). Although this appears to be the first published cost comparison, it was not presented to WNSL until after the appointment of Multiplex as the main contractor. Similarly, there is no evidence of value for money being considered earlier in the project.

Table A.8 Benchmarking analysis – stadium costs.

Stadium	Location	Capacity (seats)	Accomm-odation area (m^2)	Area/seat (m^2)	Total bench-marked cost	Cost/ seat
Wembley	London, UK	90000	173000	1.92	£352,603,000	£3,918
Stade de France	Saint Denis, France	80000	70000	0.88	£266,597,067	£3,332
Telstra Stadium (Australia)	Sydney, Australia	83500	100000	1.25	£278,897,627	£3,468
Munich (new)	Germany	66000	–	N/A	£248,239,862	£3,761
Arena Aufschalke	Gelsenkirche, Germany	51000	58796	1.15	£180,432,432	£3,538
Sapporo Dome	Japan	42122	53800	1.28	£245,959,091	£5,839
Husky Stadium	Washington, US	72500	–	N/A	£359,642,567	£4,995
Denver	US	76125	–	N/A	£338,503,518	£4,447
Cincinnati	US	66000	–	N/A	£275,875,744	£4,180

Source: Adapted from Department for Culture, Media and Sport (DCMS) (2003)

Additionally, Cyril Sweett reviewed the amended design-build contract in September 2002; they stated:

> 'In broad terms the costs of the design and construct works have not changed since our previous report...we are able to conclude that the new Multiplex design and construct bid of the new Wembley stadium is value for money.'

At the time commentators were sceptical as to whether Multiplex would complete the works on time or generate a profit on the project, bearing in mind the demanding construction programme. It was clear that lowest cost had been a predominant award criterion of WNSL:

> 'We went to the market and looked at various bids and we selected one with a good price.' (Chris Palmer, communications director of WNSL quoted in Cronin, 2000)

However, Paul Gandy, managing director of Multiplex UK, informed *Building* that:

> 'The scheme is very advanced and defined. I think it's one of those schemes that isn't going to be subject to a great deal of change.' (Quoted in Anon, 2002)

Asset supply

Multiplex commenced work on the stadium on 30 September 2002.

Again, to give a sense of scale to the project, at its peak Multiplex employed up to 3500 operatives on the stadium and, in terms of the quantities of materials used, the project incorporated 90000 m^3 of concrete, 23000 tonnes of structural steelwork, 15000 tonnes of steel reinforcement, 4000 individual piles driven to depths of up to 35 m, and 56 km of heavy-duty power cable.

Construction highlights

A key component of the project was the steel signature arch designed by Norman Foster.

Over the weekend of 19–20 July 2003 Multiplex and its subcontractors laid the foundations for the western base of the arch. The continuous concrete pour (4800 m^3 of concrete) lasted for 19½ hours: the longest in UK construction history.

Manufacture of the arch began in 2003: prefabricated off-site by subcontractor Cleveland Bridge UK Ltd (CBUK) it was then brought to site in modular form for final assembly and erection. Initially, the arch was raised into position in four key movements in June 2004. It was finally rotated into its ultimate position of 112° to vertical in the latter part of 2005, supporting the full load of the roof, which exceeds 1300 tonnes, via its permanent cable net and eyebrow catenary cable system.

Delivery issues

During the delivery stage of the project Multiplex encountered numerous issues and problems. These included:

1. An operative died following an accident on site on 15 January 2004 – work on site was suspended temporarily while investigations were undertaken
2. A dispute developed during 2004 between Multiplex and their steelwork subcontractor CBUK. The steelwork work package experienced delays resulting in the postponement of the erection of the steel arch. Then on 2 August 2004 CBUK withdrew from the project, alleging breach of contract and fearing that it would not be paid for work undertaken. (A more detailed account of the dispute is provided later under the heading: Multiplex and their subcontractors.)
3. Labour disputes: a strike commenced on 23 August 2004 when 200 operatives were sacked following a dispute over working hours and rest periods
4. In March 2006 a temporary roof support beam fell by over half a metre during on-site welding, resulting in 3000 personnel being evacuated from the site. The site was closed for a day, plus further delays ensued while inspections were completed and reports compiled
5. Also in March 2006, sewers beneath the stadium collapsed as a result of ground movement, requiring remedial work. Rumours of subsidence were emphatically denied by Multiplex

Table A.9 Wembley stadium post-contract milestones.

2003	
July	Access to the new stadium is improved following the erection of a 60 m footbridge
2004	
15 January	One construction worker is killed and another seriously injured when scaffolding collapses on the site
22 May	Arch lift commenced by steelwork subcontractor CBUK
2 August	CBUK ceases work on Wembley – litigation between CBUK and Multiplex ensues
23 August	Construction workers go on strike when 200 men are sacked following a dispute over working hours and rest periods
5 November	CBUK submitted a claim of approximately £20.9 million plus interest
2005	
28 February	Multiplex announces that the Roberts family has undertaken to indemnify Multiplex up to Au$50 million in respect of any losses incurred on the Wembley Stadium
27/30 May	Multiplex report a significant deterioration in the anticipated margin on the Wembley project – the loss is estimated to be £45 million (excluding the Au$50 million indemnity provided by the Roberts family
14 August	The FA confirms that Cardiff's Millennium Stadium has been reserved as a substitute venue for the 2006 FA Cup Final in the event that Wembley is not completed in time
18 August	Multiplex's construction division post full-year losses of £26.1 million compared with £30.8 million profits in 2004
14 October	WNSL announce the sale of 9000 seats in the new stadium valued at £300 million over the next 10 years
2 November	Multiplex report a potential loss of £180 million on Wembley
8 December	Brian Mawhinney, chairman of the Football League, anticipates that the league play-offs finals, scheduled to take place at Wembley following the 2006 FA Cup final, will be staged at Cardiff
19 December	Multiplex announce that there is a 'material risk' the stadium will not be completed in time for the 2006 FA Cup final
2006	
January	Subcontractor AR Security leaves the site due to payment issues. Also, due to cash flow problems, emergency discussions take place between Multiplex and Hollandia to prevent redundancies
3 January	Following a dispute, work commences on the stadium's main pedestrian access walkway – the work is scheduled to take 13 weeks
13 January	Progress on the stadium is again delayed as Multiplex admits using the wrong grade of concrete in its foundations
30 January	Original contract completion date. Multiplex announce that there is only a 70% chance the stadium will be completed in time for the FA Cup final to be staged at Wembley on 13th May stating that, while the firm was committed to completing the project by 31 March, there was still a risk of non-completion
21 February	FA announces that the 2006 FA Cup final will be held at Cardiff's Millennium Stadium
31 March	Multiplex's revised completion date. FA confirms that all events scheduled for Wembley in 2006 are to be held elsewhere
April	Multiplex announce their intention to submit a claim for £150 million against WNSL due to 560 design changes
12 June	Pitch laying commences

(Continued)

Table A.9 (*cont'd*).

3 July	Multiplex inform WNSL that construction work will not be completed until after September
1 August	Multiplex report that it is unlikely that the stadium will be able to hold a test event until June 2007, and advise WNSL of its entitlement to an extension of time until October 2007. WNSL maintains that the stadium will host the 2007 FA Cup final
17 August	Multiplex announced a Au$364.3 million (before tax – Au$255m after tax) loss on the project
19 October	WNSL announce that all disputes between the FA, WNSL and Multiplex have been resolved
2007	
12 March	Certificate of Practical Completion issued
17 March	First 'ramp-up' event – a community day takes place
24 March	The second 'ramp-up' event is held – an under-21 International against Italy, watched by over 55 000 spectators
27 March	Brent Council award the stadium a General Safety Certificate
19 May	The 2007 FA Cup final played at the new Wembley stadium

Source: Adapted from Clark (2002), Oliver (2002), Bewsey (2006), BBC Sport (2006), Conn (2006), Daley (2006), Brent Council (2007), FA, Multiplex and WSNL press releases

Project completion

As mentioned earlier, the start of the project was delayed by over two years due to complications encountered in raising the required finance to support the project and the ensuing political fall-out. Under WNSL's contract with Multiplex practical completion of the stadium was scheduled for 30 January 2006 to allow WNSL time to complete preparatory and commissioning works prior to staging the FA Cup final at the stadium on 13 May.

During 2004 rumours started appearing in the press that progress on site was falling behind programme, particularly after CBUK withdrew from the project. Although admitting in August 2004 to 'some minor slippage' on the steelwork subcontract package, both Multiplex and WNSL strongly asserted that overall the stadium was on schedule.

In February 2005, the Roberts family announced that it would indemnify Multiplex up to Au$50 million (approximately £21 million) with regard to any potential loss the company might incur on the Wembley project. The same month Multiplex completed a detailed review of the 'programme and costs' associated with the project. This highlighted a significant increase in the cost of finishing the project, prompting Multiplex to announce that they anticipated the project would merely break even.

While admitting that the project had suffered a two-month delay following the departure of CBUK, in March 2005 Multiplex stated that the project was now 'back up to speed' due to the innovative approach adopted by Hollandia (CBUK's replacement). Hollandia had introduced an on-site production line and night working. However, on 30 May 2005 Multiplex announced a predicted loss on the project of £45 million (not including the indemnity provided by the Roberts family) and that they anticipated the project to be completed by the end of March 2006 (not December 2005 as previously predicted). Additionally, they listed the following major risks to achieving this:

- *The ability to successfully recover claims against third parties*
- *The ability to meet the project's programme*
- *The costs associated with completion of the project's steelwork*
- *The costs of project preliminaries and, as required, acceleration*
- *Weather* (Multiplex, 2005)

Shortly after this announcement John Roberts, the founder of Multiplex, resigned as chairman of the company, followed by Multiplex's chief operating officer. Again Multiplex stressed that the stadium would '*absolutely*' be completed in time for the 2006 FA Cup final.

In June 2005 Multiplex introduced an incentive payments scheme, approved by the unions Amicus and the GMB, valued at approximately £1 million. The payments were initiated to motivate the operatives to complete the stadium's roof by mid-October in an attempt to complete the project in time for the May 2006 FA Cup final.

The FA confirmed in August 2005 that, as a contingency, the Millennium Stadium in Cardiff had been reserved as an alternative venue for the 2006 FA Cup final.

By November 2005, Multiplex's predicted loss on the project had risen to £183 million. Then, on 19 December, the contractor announced that there was a 'material risk' of the stadium not being completed in time for the 2006 FA Cup final. This risk of non-completion was assessed by Multiplex (on 30 January, the original completion date) to be 70%. However, the contractor confirmed that it was still committed to completing the project by their revised completion date, 31 March.

According to Multiplex, the key barrier to completing the stadium was the 'de-propping' of a section of the roof, plus the performance of subcontractors, design changes and relations with trade unions. In an attempt to deflect full responsibility for the delay, Martin Tidd (Multiplex's Managing Director in Britain and Europe) was quoted as saying:

> 'It does not all come down to us at the end of the day.' (Anon, 2006a)

From 31 January WNSL were entitled to withhold £120,000/day in L&A damages, subject to a claim by Multiplex for an extension of time.

On 21 February the FA announced that the 2006 FA Cup final would be held in Cardiff. This was followed, on 31 March 2006 by a further statement that all events scheduled to be held at Wembley during 2006 were cancelled, and that a revised opening date would only be given once Multiplex were able to guarantee a completion date with 100% certainty. WNSL's chief executive, Michael Cunnah, commented:

> '.... Their [*Multiplex*] revised construction schedule leaves us with no other choice but to make this decision... We have an over-riding responsibility and duty to ensure that the stadium is completed to the highest standards of safety and quality.' (Quoted in WNSL Press Release 31 March 2006)

Despite having around 3500 operatives on site, Multiplex's outstanding works at this point included: completion of the sliding roof, the internal fit-out and the stadium bowl, which required the installation of seats and scoreboards. Also, the pitch was unfinished with the turf still to be laid and the walkway to the stadium was months behind schedule.

Multiplex subsequently issued several revised completion dates – for example, following the FA's announcement on 31 March, 'substantial completion' was anticipated by the end of June, then in May it announced that it anticipated to finish commissioning, cleaning and other works in September 2006 (interestingly the contractor did not refer to this as practical completion). In June substantial completion was predicted for 13 July.

On 12 June Multiplex commenced laying the pitch; then on 3 July they informed WNSL that practical completion may not be achieved until after September. However, on 1 August Multiplex reported that it was unlikely that the stadium would be in a position to stage a test event until June 2007; it also advised WNSL of its entitlement to an extension of time until October 2007. (A more detailed account of the dispute between WNSL and Multiplex is provided in the next section, and Appendix B contains details of a press release issued by the two parties during this period.)

On 17 August 2006 Multiplex announced an Au$364.3 million (before tax – Au$255 million after tax) loss on the project.

After a period of relative inactivity, WNSL announced on 19 October that all disputes between the FA, WNSL and Multiplex had been resolved. The parties then worked together to complete the project with the stadium attaining its Certificate of Practical Completion on 12 March 2007. This was quickly followed by two successful 'ramp-up' events:

- A Community Day held on 17 March to allow local residents to visit the stadium
- An England under-21 International on 24 March, which attracted over 55 000 spectators

Brent Council awarded the stadium a General Safety Certificate on 27 March, enabling the 2007 FA Cup final to be played at the new Wembley Stadium on 19 May.

Project relationships

WNSL and Multiplex

Following the announcement by the FA that the 2006 FA Cup final would be staged at Cardiff and not Wembley, the relationship between WNSL and their main contractor became increasingly acrimonious. Each blamed the other for the non-completion of the works, in what quickly escalated into a very public disagreement with both parties using media briefing to promote their cause (Appendix B provides examples of press releases issued by Multiplex, WNSL and the FA). These announcements were widely disseminated by all forms of media both within the UK and internationally.

For example, on 1 August 2006, Multiplex declared that the stadium was 'substantially' complete (despite the fact that 10 000 seats were still to be installed following the insolvency of the seating subcontractor). Further outstanding works included the removal of temporary works, commissioning and cleaning. The press release stated that there were a number of critical 'client works' (activities for which WNSL were responsible, for example, the installation of electrical and communication infrastructure – radio and telephony systems) and outside their control that had yet to be started or were still incomplete. Similarly, it declared that there were 'other client works' that must be completed before the stadium could obtain its safety certificates, etc. that were preventing Multiplex from achieving practical completion. Further, the press release announced that Multiplex had prepared its own programme for the completion of the 'client works', which indicated that it was unlikely that the stadium would be completed before June 2007, effectively ruling out Wembley as the venue for the 2007 FA Cup final. Finally, the statement reasserted Multiplex's claim that it was entitled to an extension of time to October 2007 and that it intended to begin legal proceedings against WNSL. Multiplex's position was further expounded in statements to the press, for example:

> 'Our contract with the client clearly sets out the criteria required for us to achieve "Practical Completion"... As WNSL have failed in their obligation to commence certain works critical to this criteria, the position has worsened with another month passing. We still have no clear idea from WNSL about when they intend to begin some of their works.' (A Multiplex spokesperson quoted in Daley, 2006)

Multiplex's press release prompted a swift response from Michael Cunnah, chief executive of WNSL:

> 'The contract with Multiplex has two critical future milestones which Multiplex have deliberately confused. Multiplex is required to hand WNSL a completed stadium which is defined in the contract as 'Practical Completion'. WNSL then has to work with Multiplex to finish certain works and to hold the various test events which will enable the safety certificate to be obtained and achieve 'Operational Completion', the point at which a fully-functioning stadium is delivered...
>
> Multiplex claim that they are substantially complete, however... Multiplex has yet to finish the stadium's roof, the extensive remedial works to the stadium's drainage network, the building management and life safety systems and the installation of approximately 10 000 of the stadium's 90 000 seats, or to hand over any of the c.3400 spaces in the stadium for snagging.
>
> In the absence of a detailed programme of work from the contractor that we can rely on, we estimate that Multiplex will finish their work at some point late this year... If they can achieve that this year, then we will be able to open the stadium in early 2007. We hope that Multiplex will now devote all their

energies to completing the stadium at the earliest opportunity as is their responsibility.' (Michael Cunnah, WNSL Press Release, 1 August 2006)

This was followed almost immediately by a statement from the FA:

> '... Multiplex have to date missed all key deadlines... once they do deliver the stadium they were contracted to hand over several months ago, WNSL will then be in a position to complete a detailed timetable of commissioning work, test events and any other remaining works. They cannot complete this work until the stadium is completed by Multiplex.' (FA Press Release, 1 August 2006)

However, it is unclear why WNSL and Multiplex could not undertake their work, or some of it, simultaneously.

Multiplex were driven, to a degree, by its shareholders (having been listed on the Australian Stock Exchange in December 2003) to initiate steps to recover the losses incurred on the project and to restore its reputation. Similarly, due to its financial constraints, the criticism it received over the selection of Multiplex and the intense media interest in the project, WNSL were resolved not to exceed the GMP.

According to Multiplex, the reasons why the project was late and over budget were primarily a result of WNSL issuing approximately 600 variations (design changes) and their client's indecisiveness. This interpretation of events was firmly disputed by WNSL, who countered that although Multiplex may have encountered some 'bad luck' regarding the performance of its subcontractors, their difficulties were also attributable to poor project management.

It was then reported that Multiplex had submitted a claim, generally considered to be for £150 million (although one source put it at £350 million). Multiplex's actions resulted in Michael Cunnah, chief executive of WNSL, declaring that Multiplex would not be given any additional '*money or time*'. Following this, the project entered into a phase of protracted stand-off between WNSL and Multiplex, the prospect of litigation putting at risk Wembley's ability to stage events for several years to come.

Mediation talks began in September 2006, following the intervention of Lord Carter, the recently retired chairman of Sport England, as intermediary. Lord Carter facilitated a meeting between Alex Horne, chief financial officer of the FA and Andrew Roberts, chief executive of Multiplex. Ultimately, a carefully worded settlement was announced, perceived as a win–win solution.

The terms of the agreement granted Multiplex an extension of time of 130 days, plus £70 million (above the GMP which had already risen to approximately £459 million as a result of previously agreed additions and Multiplex being awarded over £1 million following arbitration proceedings against WNSL) to cover design changes. However, Multiplex acknowledged their responsibility for some of the delays on the project and agreed to reimburse WNSL £34 million of the £70 million settlement. Under the arrangement Multiplex were to achieve practical completion between January and March 2007, allowing WNSL sufficient time to stage two 'ramp-up' events before the 2007 FA Cup final. It also granted WNSL access to the site, enabling them to commence their own installation works. The agreement, subsequently accepted by WNSL's banks, saw the stadium's overall development costs rise to approximately £792 million (some sources suggested that the figure was nearer £827 million).

It was later discovered that in December 2005 WNSL and Multiplex had struck a deal that would have given Multiplex a portion of the proceeds from the 2006 FA Cup final plus a further £25 million provided the contractor completed the stadium in time for the event to be staged at Wembley. However, following the imposition of conditions by Wembley's banks, which Multiplex considered unacceptable, the contractor rejected the deal.

Ultimately it was financial pressure, principally from their banks, that forced WNSL to come to an agreement with Multiplex, driven by the need for the stadium to:

- Start generating a return
- Meet its loan repayment plan (in September 2006 WNSL were forced to renegotiate their repayment schedule when a payment of £40 million fell due)
- Stage the FA Cup final in 2007 – if Wembley did not achieve this, the banks would not sanction a further revision to WNSL's repayment schedule

Multiplex and their subcontractors
Multiplex entered into disputes with several of their subcontractors, some of which led to legal action.[1]

Cleveland Bridge UK Limited
The origins of the disagreement lay in the escalation of the steelwork subcontract costs and the apportionment of blame between the two parties for the delay in the erection of the signature steel arch. As one commentator put it:

> 'The problems were building up and up on Wembley, there was simmering discontent with CBUK and the people in charge at the time weren't willing to conciliate.' (Quoted in Rogers, 2006a)

In September 2002 CBUK entered into a £60 million lump-sum contract with Multiplex for the fabrication and erection of the structural steelwork on the project (Multiplex was subsequently to argue that CBUK had undervalued the work). Delays occurred in the erection of the steel arch; therefore, during 2003 an agreement was reached that allowed the works to be accelerated, although later the parties were unable to agree a mechanism by which the subcontractor would be reimbursed for this.

During this period CBUK were experiencing cash-flow problems and apparently, in January 2004, the company had notified Multiplex that it was close to insolvency. By February (despite reports that CBUK had concerns over their treatment by Multiplex and that Multiplex was worried that CBUK would not adhere to the revised programme) the parties reached a financial agreement. The settlement, covered by a heads of agreement, included a completion date for the erection of the arch (21 April 2004) and payment terms; a further agreement was signed on 14 May 2004. The parties still had difficulties in valuing of the works. However, an oral agreement was reached under which CBUK received £32.66 million.

Despite these agreements, at the end of June it was announced that although CBUK would continue to fabricate and supply steel to the project, they would cease their on-site installation work. At this point Multiplex appointed Hollandia to execute the work. Initially, Hollandia utilised the 200 steelworkers employed on-site by CBUK, but following a dispute over pay and conditions these operatives were laid off; an unofficial picket of the site ensued.

However, by July Multiplex was withholding payments from and looking to recoup monies paid to CBUK. As a result, on 2 August 2004 CBUK pulled out of the project, alleging breach of contract and declaring that it had '*...no faith in Multiplex honouring its commitments and paying the sums due*' (quoted in Morris, 2004).

Multiplex instigated proceedings against CBUK claiming £45 million in damages and for breach of contract, while CBUK, counter-sued for £22.6 million. Commencing on 25 April 2006 at the High Court (Technology and Construction Court), the case was heard by Mr Justice Jackson[2].

During the proceedings, CBUK alleged that in 2003 Multiplex was aware that the project could not be completed within the contract period. Programme delays were due to the construction phase starting before the design had been finalised and because a section of the foundations had to be re-laid following the use of the wrong grade of concrete. As a result, Multiplex 'fast-tracked' the work as it was '*... very keen to have such a high-profile project come in on time*', while CBUK sought a one-year extension of time.

Regarding the payment of £32.66 million, CBUK claimed the figure was an agreed final settlement with Multiplex for work completed prior to 15 February 2004, which the main contractor had reneged on under its 'Armageddon' plan. CBUK contested that the 'Armageddon' plan was a scheme to force the subcontractor into bankruptcy and/or off site in July 2004. The strategy, it was alleged, was driven by Multiplex's need to make savings on the steelwork package, as by May 2004 the cost of the subcontract had increased from £60 million to £90 million. As proof of such a strategy CBUK cited the infamous 'fix and f*** them later?' email of May 2004. Multiplex's failure to adhere to the Supplemental Agreement and its repeated undervaluation of interim payments, CBUK alleged, constituted a repudiatory breach.

Similarly, Multiplex claimed that CBUK had repudiated the contract. This case centred on the law of repudiation; with both parties asserting that the other was in repudiatory breach of contract (their actions indicating that they did not intend to fulfil their duties under the contract, thereby enabling the innocent party to terminate the contract).

The main contractor argued that the delays on the project were due to CBUK's *'disorganisation and incompetence'* – delays which forced Multiplex to make extra payments to other subcontractors on the project due to the consequential disruption of their work. Additionally, Multiplex asserted that CBUK had deliberately held up the works as a means of applying 'commercial pressure' on Multiplex to obtain better terms and conditions (allegedly a tactic later to be used by Multiplex on WNSL). Multiplex claimed that the payment of £32.66 million was only made 'on account' in order to improve the subcontractors 'cash flow' and was subject to counterclaims for damages and deductions for defective work.

Multiplex described the 'Armageddon' plan as follows:

> 'Armageddon is not having a steelwork contractor... There was a plan which I understand was to recover a debt which was owed while at the same time working with CBUK to find a way of going forward.' (Quoted in Anon, 2006f)

It was suggested that, given CBUK's precarious financial position, Multiplex would have been imprudent not to have an alternative steel contractor on hand. However, as a part of the 'Armageddon' plan, Multiplex also instigated a strategy of issuing interim certificates based on '... *the lowest sums which it could properly defend in adjudication*'.

Unsurprisingly, rather than saving Multiplex money, CBUK's exit from the project led to significant increases in the cost of the steelwork work package. During the hearing it emerged that the work package was currently valued at approximately £187.7 million, more than three times over budget.

Although in his ruling Mr Justice Jackson declared there to be no 'outright winner', in fact Multiplex's interpretation of the contract in regards to the two main issues of the case were supported. The judge decided that:

- Under the agreement Multiplex was entitled to adjust its valuations of the subcontract work to 'offset' for what it deemed to be substandard workmanship

The judge concluded that the £32.66 million payment by Multiplex was in effect a valuation for interim certificate (payment on account) purposes and not as a final valuation of the works. He held if it had been the latter it would have been included in the formal agreement – this was not the case. Regarding Multiplex's lack of consultation with CBUK, prior to issuing certificates based on the lowest valuation of the works it believed it could substantiate, Mr Justice Jackson found Multiplex to be in breach, but not repudiatory breach, of the contract. Further, he accepted Multiplex's interpretation of the so-called 'Armageddon' plan – that it was not a strategy to force CBUK into liquidation. Although describing Multiplex's strategy regarding CBUK to have been both '*ruthless*' and '*deplorable*', the judge found that their actions were, however, permissible under the contract and, therefore, lawful.

- CBUK were in breach of the contract by walking off site.

As Multiplex's actions were held to be lawful, CBUK were not entitled to withdraw from the site and were, therefore, in repudiatory breach of the contract. The breach entitled Multiplex to recover damages from CBUK.

The judge arrived at this decision despite CBUK having five adjudication awards made in their favour; these awards required Multiplex to make additional payments to CBUK in excess of £6 million. Moreover, one of these adjudications had established that Multiplex (not CBUK as claimed by Multiplex) had been predominantly responsible for the delay to the erection of the steel arch as a result of defective work by their concrete subcontractors.

The judgment was succinctly summarised by one commentator:

> 'CBUK might have had a legitimate grievance against Multiplex but they were the ones who walked off site. They got it wrong.' (Quoted in Rogers, 2006c)

In his judgment, Mr Justice Jackson encouraged both parties to settle the issue of damages through negotiation.

In respect of Multiplex, the outcome of the dispute enabled the main contractor to recoup some of their losses on the project and to salvage a certain amount of pride, although revelations made during the case did little to erase Multiplex's 'hard-man' image. In response to the judgment, Martin Tidd (Multiplex's Managing Director in Britain and Europe) said:

> 'This enables us to draw a close to this particular chapter. This was an issue with CBUK and it's not our style to pursue all other trade contractors.' (Quoted in Rogers, 2006b)

In January 2007, Mr Justice Jackson pronounced that CBUK were not liable for the cost of temporary roof works on the project, works subsequently undertaken by Hollandia. In the same month the courts granted CBUK leave to appeal against the earlier judgment. However, at the Court of Appeal on 10 May 2007 the initial judgment was upheld.

The first ruling reduced the amount of damages Multiplex were able to claim by approximately £15 million (however, a subsequent ruling on 20 December 2007 reversed the ruling in part stating that CBUK was responsible for the design of the temporary works, but not its fabrication, entitling Multiplex to claim approximately £5 million from CBUK); the second opened the way for the courts to consider the issue of damages.

Although, CBUK's liability for damages was capped at £6 million, Multiplex was entitled to claim additional monies from their former subcontractor in respect of the revaluation of CBUK's works. The dispute was eventually settled in September 2008 when the courts ordered CBUK to pay 20% of Multiplex's costs.

Honeywell Control Systems Ltd[3]

Honeywell was contracted to Multiplex to install communications and control systems on the Wembley project – work that was required to be carried out during the final stages of the project. Due to the delays encountered on the project, the subcontractor was on site a year longer than anticipated. As Honeywell's contract with Multiplex required the work to be completed by a specific date, its failure to comply with this term rendered it liable to a claim for damages by Multiplex. Honeywell claimed that Multiplex had been responsible for the delays on the project, and firmly criticised Multiplex's project management of the works. As a result, Honeywell argued that time was 'at large' on the project; therefore, as the contract completion date was no longer enforceable it was only required to finish its works '*reasonably and at progress with general construction*'. Multiplex would, thereby, lose its entitlement to recover L&A damages.

In March 2007, the courts rejected Honeywell's argument that time was 'at large' and ruled that the subcontractor was bound by its contractual completion date, thereby entitling Multiplex to levy L&A damages. In arriving at this decision Mr Justice Jackson overturned the findings of an earlier adjudication. However, during its submissions Multiplex conceded that the subcontractor had given proper notifications of delay, and therefore it was obliged to consider these notices and award Honeywell an appropriate extension of time. Subsequently, in June 2007, Honeywell issued a court order against Multiplex for damages amounting to £33 million, alleging breach of contract.

PC Harrington Contractors Ltd[4]

Subcontractor PC Harrington was appointed to carry out the concrete works package on the project. Following the issue of an Interim Certificate on 20 April 2007, a dispute arose relating to a pending adjudication.

PC Harrington disputed Multiplex's valuation of their works, and submitted a claim for an extension of the time. It also instigated legal proceeding against Multiplex for additional payments exceeding £17 million. Multiplex counterclaimed that elements of the subcontractor's work were defective or incomplete, resulting in the main contractor having to engage third parties to finish off the concrete work package. Multiplex accounted for its alleged loss and damage in the monthly interim payment process as permitted under the contract.

In calculating the April Interim Certificate Multiplex deducted approximately £13.05 million in respect of incomplete and/or defective work. One element of this deduction concerned £1,658,665 relating to concrete floors poured by PC Harrington. Following correspondence between the parties, PC Harrington declined to accept liability for the defective floors or to reimburse Multiplex the cost of remedial work undertaken. Multiplex valued this at £2,070,110.50 (£1,658,665 plus a claim for consequential losses). In September 2007 Multiplex informed PC Harrington of its intention to refer the issue to adjudication, as permitted under the contract.

PC Harrington appealed to the courts for a ruling, arguing that as Multiplex's claim relating to the concrete floors was less than the amount certified as due in the April Interim Certificate (£2,301,608) the main contractor was not entitled to recoup any further financial award through the adjudication process. However, the subcontractor acknowledged that Multiplex may well have been able to demonstrate its right to damages from PC Harrington.

On 30 November 2007 Mr Justice Clarke declined to grant the declaration sought by PC Harrington.

Mott MacDonald

Mott MacDonald (Mott), as part of a consortium (The Mott Stadium Consortium), were retained as consultants on the project. Mott had been appointed in May 1998 to undertake engineering design work, entering into a consultancy agreement with WNSL for the provision of professional services regarding the design and construction of the stadium on 6 April 1999.

Following the appointment of Multiplex (under a design-build contract), Multiplex, WNSL and Mott entered into two novation agreements on 26 September 2002 regarding the provision of:

- Civil and structural (C&S) engineering services
- Mechanical and electrical (M&E) services

Under the novation agreements the majority of engineering services provided by Mott on the project were novated to Multiplex, who replaced WNSL as Mott's employer. However, a small number of specified services were not novated under the agreement, for which WNSL remained Mott's employer.

Multiplex sought access to Mott's records during the summer of 2006. This included information on the progress of the stadium's design, cost estimates and correspondence between Mott and WNSL. It was understood that Multiplex required the information:

- To pursue a claim against Mott for damages, following delays to the project (During, the CBUK case, Mr Justice Jackson acknowledged that: '*The design was incomplete and Mott were late in issuing information*'.)
- For proceedings against Mott, and/or
- For litigation against third parties

In the opinion of Multiplex, by the beginning of October 2006 four specific disputes had developed and on 4 October it served an adjudication notice on Mott, pertaining to:

1. The meaning given to the words '*all records pertinent to the Services*' within the novation agreement
2. The extent to which the series of specific document requests fell within the words '*all records pertinent to the Services*'
3. The meaning of the words '*at all reasonable times*'
4. The extent to which (if any) such confidentiality obligations as Mott MacDonald may owe to WNSL precluded provision of '*all pertinent records relating to the Services*' by Mott MacDonald to Multiplex

Following the adjudication decision, dated 20 November 2006, Mott afforded Multiplex access to over 100 000 documents. However, the main contractor considered that these were insufficient and that Mott had not fully complied with the adjudicator's ruling. Mott countered that Multiplex had '*not made a proper attempt to inspect them*', suggesting that the main contractor should have undertaken this task prior to seeking an injunction to compel Mott to hand over further documentation. Further, Mott was of the opinion that owing to its confidentiality agreements with WNSL, there were certain documents they were not permitted to make available.

As a consequence, Multiplex commenced proceedings against Mott for the enforcement of the adjudicator's ruling[5].

A Mott spokesperson said:

> 'This matter involves an adjudication about rights under the contract to access project records. We believe we have complied with the adjudicator's decision.' (Quoted in Richardson, 2007a)

Mr Justice Jackson ruled in favour of Mott (with the exception of one minor part); awarding Mott costs, he said:

> 'Multiplex has failed quite spectacularly in all areas of relief other than on the issue of jurisdiction [*of the arbitrator*]... To Mott MacDonald, these have been extremely important proceedings.'

However, Mr Justice Jackson refused to grant Multiplex an order compelling Mott to hand over further documents, stating he was unable to rule on Mott's compliance based on written evidence.

The dispute continued with Multiplex announcing that it was seeking £253 million in damages from Mott relating to breaches of contract and acts of negligence. In response Peter Wickens, chairman of Mott, stated:

> 'The Mott Stadium Consortium (MSC) performed its contractual obligations to Multiplex professionally and diligently. MSC will vigorously defend any claim.' (Quoted in Anon, 2008)

The dispute was finally settled out of court in June 2010.

Further subcontractors involved in the Wembley project experienced financial difficulties. On 28 November 2005 the pipe work subcontractor, **SGD Engineering**, entered into administration, after completing 85% of their £4.5 million work package. SDG alleged that Multiplex had underpaid them by £2 million after the subcontract had incurred a 40% cost overrun as a result of Multiplex introducing specification changes. As with CBUK, Multiplex was adamant that the incident would not impact on its overall programme and initiated a claim for breach of contract against SDG.

In addition to SGD Engineering:

- **AR Security**: responsible for installing security and smoke alarms on the project, entered into voluntary liquidation in April 2006
- **Stadium Seating and Pel Stadium Seating**: contracted by Multiplex to install the stadium's 90 000 seats, both went into administration – Stadium Seating in January 2006 followed by the Pel Group in July 2006
- **Scanmoor**: undertook C&S engineering work valued at £9.9 million and also entered into administration after encountering cash-flow problems on the project (according to the administrator: '*Wembley wasn't the main reason, but it was a factor*')
- **Phoenix Electrical**: the M&E contractor also experienced cash-flow problems on the project leading to 250 electricians withdrawing their labour in May 2006. As a result Multiplex supported the subcontractor with cost and time dispensations, although its owners were later obliged to sell the business
- **Weatherwise**: designers and installers of the stadium's main roof, Weatherwise's contract lasted a year longer than anticipated, but despite Multiplex revising the subcontractor's final account to reflect the delay, its impact on the business as a whole led to the company posting a loss of £157,000 in 2006 (Richardson, 2007c)

Reasons for the project's difficulties

The contractor

An internal review by Multiplex of the Wembley project concluded that problems encountered by the organisation 'spanned' both governance and operational issues. In particular, the contractor reported to shareholders that the majority of its difficulties were associated with 'repeated' issues linked to the substantial and complex steelwork package.

Additionally, industry commentators suggested that Multiplex's problems could also be attributed, in part, to the following:

- The submission of an unrealistic tender price: no UK-based contractor had been prepared to submit a bid (or had withdrawn from the competition). The question was raised: why was Multiplex prepared to take on the project at or near WNSL's budget?
- Establishing the price prematurely, before the design had been sufficiently advanced, while the fixed-price (GMP) aspect of the contract meant that budget overruns were difficult to recoup and had led to subcontractor disputes
- Accepting an unrealistically tight contract programme
- Inexperience of working in the UK (Wembley was Multiplex's first major UK project): the contractor did not have the requisite established and integrated supply chains. It also failed to achieve the on-site efficiencies attained in Australia
- Naivety (alternatively referred to as arrogance and culture shock): the contractor was unprepared for the adversarial mindset of the UK subcontractors. Similarly, they did not appreciate the relative strength of the UK subcontractors compared with the balance of power in Australia. Multiplex was accused of adopting a confrontational stance with its subcontractors and suppliers, at a time when the UK market was embracing a more collaborative approach
- Poor project management: illustrated by the use of the wrong grade of concrete in the foundations of the arch, shortages of steel, and drainage problems, etc. Also, initially Multiplex's information systems were inadequate, failing to keep senior management abreast of developments on the project: mid project Multiplex introduced additional risk management procedures
- Poor expectation management: Multiplex, publicly at least, was overly optimistic of their ability to meet the programme
- Poor public relations: at the beginning, at least, Multiplex did not appreciate the depth of public interest in the project

The client

The problems encountered by WNSL have been credited, by certain construction sector critics, to:

- The adoption of the 'lowest price' award criterion: WNSL's insistence on a low tender price based on an unrealistic cost estimate
- Their contract strategy: WNSL had taken a strategic policy decision to reject the 'partnering/best practice' approach, insisting on a fixed-price (GMP) contract
- Poor project management: the adoption of 'tough' contract terms was seen by many as a poor substitute for appropriate project leadership (alternatively, other commentators suggested that WNSL had abdicated their overall project leadership role or that they had provided weak leadership). It was also considered that WNSL were reluctant to take a more active role in the project for fear that any intervention would result in variations that may have given Multiplex grounds for additional payments, thereby undermining the price certainty of the contract (although, according to Multiplex, WNSL were continually revising their plans). Moreover, once problems emerged on the project, WNSL failed to work with Multiplex to find solutions
- Indecisiveness: one ex-Wembley executive suggested that the project's failure was due to the indecision of WNSL, stating that the senior management team were '*scared of getting it wrong*'. Additionally, it was suggested that client's may be '*intimidated by their own contract terms*'
- Inflexibility: by adopting a GMP contract WNSL effectively '*tied its own hands*', as it did not afford the client the flexibility to intervene where appropriate
- Poor communication (in conjunction with the issues of indecisiveness, inflexibility and poor project management) resulted in a deterioration in the relationship between the client and its main contractor, leading to disputes and further delays
- The stadium's design and issues of buildability: the signature arch, for example, was a major feat of engineering, being the longest arch in the world. Additionally, as with aspects of cost, by novating the

design WNSL were able to transfer the design risk to their contractor. However, it resulted in the client not taking a proactive role in resolving issues that arose on the project

- Poor expectation management: in respect of the UK government, the football/rugby fraternity, event organisers, spectators and the general public, perhaps it would have been more prudent for WNSL's to have opted for a longer construction phase, with the potential for delivering the stadium ahead of programme, rather than insisting on a very tight contract period

Political dimension

Already a complex project, in terms of its design and construction, Wembley was further hampered by being a multi-agency development, involving numerous public and private sector organisations (see appendix A).

Post-project outcome

The delivery issues encountered on the project attracted much media interest, initially predicting that they would damage London's prospect of hosting the 2012 Olympic Games, and then, once London had been awarded the staging rights to the games, that they would undermine the UK's ability to produce high-profile sporting venues. Calling for 'lessons to be learnt' from Wembley, the Conservative party demanded an official inquiry into the project. However, despite taking much longer to complete and costing considerably more than initially anticipated, and almost causing the insolvency of WNSL, the new stadium has generally been very well received by all its stakeholders (despite some initial issues over the quality of the pitch). The signature 'tiara' arch is '*instantly recognisable*' and visible across London. Moreover, Wembley received two prestigious industry awards in October 2007: the Institution of Civil Engineers (ICE) London's 2007 Merit Award for innovation, engineering quality and contribution towards sustainability, and the 'Sports and Leisure Project of the Year' award at the Builder and Engineer Awards.

At the time, the national stadium at Wembley was one of the largest redevelopment schemes in London, and, as with the Emirates Stadium in Islington and the Millennium Stadium in Cardiff, it became the catalyst for the regeneration of its surrounding area (and London overall), the objective being to turn the region into a sports and leisure zone of both national and international importance. Further, the redevelopment of the stadium has brought transport infrastructure improvements to the area, a benefit not only for spectators but also for the local inhabitants and commerce, and during major events the venue employs over 6000 people.

In August 2005, WNSL announced the establishment of a charitable trust through which 1% of the stadium's annual turnover would be used to encourage participation in sport and recreation at both a national and local level.

As a result of its participation in the project Multiplex suffered significant reputational damage. While asserting that the outcome of the project was the creation of a world-class facility, Multiplex has acknowledged that Wembley was one of the organisation's 'most challenging projects'. In addition to the losses incurred on the project, Multiplex entered into an Enforceable Undertaking with the Australian Securities and Investments Commission (ASIC) in December 2006, following an investigation into Multiplex's disclosures on the Wembley project. The settlement offer, completed in April 2007, resulted in Multiplex reimbursing eligible security holders a total of Au$20.3 million. Also, relating to disclosures on the project, a 'class action' was launched against Multiplex in December 2006. This was finally resolved when (Brookfield) Multiplex reached a conditional settlement in May 2010. Multiplex settled its claims on the project with WNSL in October 2006 (Multiplex, 2007).

Individual casualties of the project included Martin Tidd (Multiplex's Managing Director in Britain and Europe), who resigned in March 2007.

However, despite withdrawing from construction contracting in the UK in early 2006, Multiplex has since sought to re-enter the market, securing, for example, the contract to build the 288 m high Pinnacle Tower in London, for Arab Investments in August 2007. The £593 million project subsequently ran into difficulties. The contractor is currently (October, 2012) suing the developers for alleged breach of contract and non-payment of fees.

Following the acquisition of Multiplex by Brookfield Asset Management Inc. in December 2007, the company was delisted from the Australian Securities Exchange, becoming Brookfield Multiplex Limited in March 2008.

Acknowledgements

This case study was prepared by David Lowe from the following published sources.

Endnotes

1 For an overview of the legal cases arising from the Wembley stadium project see Thomas, D (2009) Lessons from the Wembley litigation. *Construction Law International*, **4**, 28–30

2 *Multiplex Constructions (UK) Ltd v Cleveland Bridge UK Ltd* [2006] Adj.L.R. 12/20. http://www.nadr.co.uk/articles/published/AdjudicationLawRep/Multiplex%20v%20Cleveland%20Bridge%20No2%202006.pdf (accessed 28 December 2012)

3 *Multiplex Constructions (UK) Ltd v Honeywell Control Systems Ltd* (No. 2) [2007] Adj.L.R. 03/06. http://www.nadr.co.uk/articles/published/ALR/Multiplex%20v%20Honeywell%20No2%202007.pdf (accessed 28 December 2012); Macpherson and Brennan (2007); Stewart (2007)

4 *PC Harrington Contractors Ltd v Multiplex Constructions (UK) Ltd* [2007] Adj.L.R. 11/30. http://www.nadr.co.uk/articles/published/ALR/Harrington%20v%20Multiplex%202007.pdf (accessed 28 December 2012)

5 *Multiplex Constructions (UK) Ltd v Mott MacDonald Ltd* [2007] EWHC 20 (TCC). http://www.bailii.org/ew/cases/EWHC/TCC/2007/20.html (accessed 28 December 2012)

References and sources

Anon (2000) Wembley hit by plans confusion. *BBC News*, 7 September. http://news.bbc.co.uk/sport1/hi/football/914739.stm (accessed 28 December 2012)

Anon (2002) Multiplex: no design changes at Wembley. *Building*, Issue 40. http://www.building.co.uk/news/multiplex-no-design-changes-at-wembley/1022270.article (accessed 28 December 2012)

Anon (2005) Stadium firm predicts £45m loss. *BBC News*, 30 May. http://news.bbc.co.uk/1/hi/business/4585533.stm (accessed 28 December 2012)

Anon (2006a) Wembley may miss FA Cup deadline. *BBC Sport*, 30 January. http://news.bbc.co.uk/sport1/hi/football/4661494.stm (accessed 28 December 2012)

Anon (2006b) Wembley dropped for FA Cup final, *BBC Sport*, 21 February. http://news.bbc.co.uk/sport1/hi/football/4731888.stm (accessed 28 December 2012)

Anon (2006c) Timeline: the new Wembley. *BBC Sport*, 21 February. http://news.bbc.co.uk/sport1/hi/football/4735072.stm (accessed 28 December 2012)

Anon (2006d) 'Steel the key' to extra time. *BBC News*, 21 February. http://news.bbc.co.uk/1/hi/uk/4735812.stm (accessed 28 December 2012)

Anon (2006e) Wembley crisis averted. *Contract Journal*, 5 April. http://www.contractjournal.com/Articles/Article.aspx?liArticleID=50803&PrinterFriendly=true (accessed 5 March 2009)

Anon (2006f) Multiplex QS takes the stand. *Building*, 5 May. http://www.building.co.uk/news/multiplex-qs-takes-the-stand/3067055.article (accessed 28 December 2012)

Anon (2006g) Wembley to open early next year. *The Guardian*, 19 October. http://www.guardian.co.uk/football/2006/oct/19/newsstory.sport10 (accessed 28 December 2012)

Anon (2007) Multiplex to return to UK contracting sector. *Contract Journal*, 5 September. http://www.contractjournal.com/Articles/2007/09/05/56160/multiplex-to-return-to-uk-contracting-sector.html (accessed 5 March 2009)

Anon (2008) Wembley cases to drag on into 2009. *Building*, Issue 11. http://www.building.co.uk/news/wembley-cases-to-drag-on-into-2009/3109321.article (accessed 28 December 2012)

Arup, Wembley Stadium Infrastructure. http://www.arup.com/projectmanagement/ (accessed 5 March 2009)

Austin, S (2006) Multiplex allays bankruptcy fears. *BBC Sport*, 21 February. http://news.bbc.co.uk/sport1/hi/football/4735922.stm (accessed 28 December 2012)

Baldock, (2000) Wembley compromised despite athletics deal. *Building*, 7 January. http://www.building.co.uk/news/wembley-compromised-despite-athletics-deal/2402.article (accessed 28 December 2012)

Bewsey, G (2006) No room for manoeuvre. *Construction Manager*, CIOB May. http://www.construction-manager.co.uk/story.asp?sectioncode=12&storycode=3066805 (accessed 5 March 2009)

Bill, T (2007) Are 'trophy' projects risky business? *Contract Journal*, 21 March. http://www.contractjournal.com/Articles/2007/03/21/54138/are-trophy-projects-risky-business.html (accessed 5 March 2009)

Blakely, R (2006) Building tensions revealed in Wembley case. *The Times*, 28 April. http://www.thetimes.co.uk/tto/business/industries/construction-property/article2166372.ece (accessed 28 December 2012)

Bose, M (2006a) Wembley could be the new Dome if legal dispute ends in stalemate. *The Telegraph*, 20 July. http://www.telegraph.co.uk/sport/football/2341026/Wembley-could-be-the-new-Dome-if-legal-dispute-ends-in-stalemate.html (accessed 28 December 2012)

Bose, M (2006b) Court battle looms over Wembley saga. *The Telegraph*, 31 July. http://www.telegraph.co.uk/sport/football/teams/england/2341925/Court-battle-looms-over-Wembley-saga.html (accessed 28 December 2012)

Bose, M (2006c) Carter the man to sort out Wembley. *The Telegraph*, 6 September. http://www.telegraph.co.uk/sport/football/2345112/Carter-the-man-to-sort-out-Wembley.html (accessed 28 December 2012)

Bose, M (2006d) Wrangle could stall Wembley until 2010. *The Telegraph*, 6 September. http://www.telegraph.co.uk/sport/football/teams/england/2345107/Wrangle-could-stall-Wembley-until-2010.html (accessed 28 December 2012)

Bose, M (2006e) Wembley to host 2007 FA Cup final. *The Telegraph*, 16 October. http://www.telegraph.co.uk/sport/football/2348115/Wembley-to-host-2007-FA-Cup-final.html (accessed 28 December 2012)

BovisLendLease.com (2000) *Bovis Lend Lease/Multiplex Appointed Preferred Contractor for New Wembley Stadium*, 23 February 2000. http://www.bovislendlease.com/llweb/bll/main.nsf/toprint/news_20000223?opendocument&print (accessed 5 March 2009)

Brent Council (2002) *Our Vision for a New Wembley*, Brent Council's Communications Unit. http://www.brent.gov.uk/regeneration.nsf/Files/LBBA-55/$FILE/Wembely%20vision.pdf (accessed 28 December 2012)

Brent Council (2007) *Wembley Regeneration Timeline*. Brent Council's Communications Unit. http://www.brent.gov.uk/regeneration.nsf/Wembley/LBB-72 (accessed 28 December 2012)

Broughton, T (2005) Multiplex pledges to finish Wembley a month early. *Building*, Issue 9. http://www.building.co.uk/news/multiplex-pledges-to-finish-wembley-a-month-early/3047727.article (accessed 28 December 2012)

Campbell, D (2006) Arch enemies meet in court. *The Observer*, 23 April. http://www.guardian.co.uk/wembley/article/0,,1759272,00.html (accessed 5 March 2009)

Carter, P (2002) *English National Stadium Review: Final Report October 2002*. The Stationery Office, London. http://webarchive.nationalarchives.gov.uk/+/http://www.culture.gov.uk/images/publications/NatStadiumreveiw.pdf (accessed 28 December 2012)

Chevin, D (2006) The bitter taste of success. *Building*, 9 June. http://www.building.co.uk/analysis/the-bitter-taste-of-success/3068593.article (accessed 28 December 2012)

Clark, P (2000a) Bovis may sue Multiplex over Wembley deal. *Building*, Issue 37. http://www.building.co.uk/news/bovis-may-sue-multiplex-over-wembley-deal/1417.article (accessed 28 December 2012)

Clark, P (2000b) Wembley stadium team plays on, despite doubts. *Building*, Issue 50. http://www.building.co.uk/news/wembley-stadium-team-plays-on-despite-doubts/1002288.article (accessed 28 December 2012)

Clark, P (2002) Tropus denies sour grapes claims over Wembley report. *Building*, Issue 21. http://www.building.co.uk/news/tropus-denies-sour-grapes-claims-over-wembley-report/1018827.article (accessed 28 December 2012)

Conn, D (2006) Lessons to be learned from Wembley's woes. *The Guardian*, 8 March. http://www.guardian.co.uk/sport/2006/mar/08/Olympics2012.politics (accessed 28 December 2012)

Cronin, S (2000) Multiplex slips in the back door at Wembley. *The Independent*, 17 September. http://www.independent.co.uk/news/business/news/multiplex-slips-in-the-back-door-at-wembley-699971.html (accessed 28 December 2012)

Cyril Sweett (2002) Report on Value for Money Analysis for Wembley National Stadium Ltd, Cyril Sweett, London

Daley, J (2006) Owner and contractor embark on war of words over Wembley delay. *The Independent*, 2 August. http://www.independent.co.uk/news/business/analysis-and-features/owner-and-contractor-embark-on-war-of-words-over-wembley-delay-410189.html (accessed 28 December 2012)

Denney, A (2003) A game of two halves. *The Culture Media and Sport Briefing*, Issue 2, National Audit Office, pp. 6–7. http://www.nao.org.uk/idoc.ashx?docId=9D16FA51-3B7D-43BE-91CC-E9E73B50087C&version=-1 (accessed 28 December 2012)

Department for Culture, Media and Sport (DCMS) (2003) *Wembley National Stadium Project: Into Injury Time*. The Stationery Office, London. http://www.publications.parliament.uk/pa/cm200102/cmselect/cmcumeds/843/843.pdf (accessed 28 December 2012)

Design Build Network. *Wembley Stadium, London, United Kingdom*. http://www.designbuild-network.com/projects/wembley/ (accessed 28 December 2012)

Eason, K (2006) Wembley runs for cover with decision to slow roof closure. *The Times*, 14 December. http://www.thetimes.co.uk/tto/sport/football/article2282009.ece (accessed 28 December 2012)

Fennell, E (2006) Wembley's lessons for 2012 – don't go grand, go simple. *The Times*, 18 April. http://www.thetimes.co.uk/tto/law/article2215643.ece (accessed 28 December 2012)

Football Association, *Press Releases*. http://www.thefa.com/TheFA/NewsFromTheFA/ (accessed 5 March 2009)

Football Association (2007) *Wembley Handover*, 10 March. http://www.thefa.com/TheFA/wembley/newsandfeatures/Postings/2007/03/WembleyHandover.htm (accessed 5 March 2009)

Froley, R (2006) Wembley National Stadium sceptical about September completion date. *Contract Journal*, 31 July. http://www.contractjournal.com/Articles/2006/07/31/51663/wembley-national-stadium-sceptical about-september-completion.html (accessed 5 March 2009)

Glackin M (2000) Wembley team set for showdown. *Building*, Issue 29. http://www.building.co.uk/news/wembley-team-set-for-showdown/738.article (accessed 28 December 2012)

Glancey, J (2007) We think it's all over ... *The Guardian*, 9 March. http://www.guardian.co.uk/uk/2007/mar/09/wembley stadium.football (accessed 28 December 2012)

Goodway, N (2005) Wembley walkout company collapses. The Evening Standard, 1 December. http://www.thisismoney.co.uk/news/article.html?in_article_id=405416&in_page_id=2 (accessed 9 March 2009)

Harris, N (2006) Football: Multiplex denies 'Wembley is sinking' claim. *The Independent*, 4 April. http://www.independent.co.uk/sport/football/news-and-comment/multiplex-denies-wembley-is-sinking-claim-472692.html (accessed 28 December 2012)

House of Commons Committee of Public Accounts (2004) *The English National Stadium Project at Wembley*. The Stationery Office, London. http://www.publications.parliament.uk/pa/cm200304/cmselect/cmpubacc/254/254.pdf (accessed 28 December 2012)

Infolink.com (2002) *Mutiplex given preferential treatment, Wembley committee told*. http://www.infolink.com.au/articles/DC/0C00DADC.aspx (accessed 9 March 2009)

Leitch, J (2007) Multiplex clear of Wembley woes – £220m profit unveiled. *Contract Journal*, 20 August. http://www.contractjournal.com/Articles/2007/08/20/55970/multiplex-clear-of-wembley-woes-220m-profit-unveiled.html (accessed 9 March 2009)

Macpherson, H and Brennan, P (2007) Wembley judge sends off NT case law. Construction Contractor, 27 June. http://www.infolink.com.au/n/Wembley-judge-sends-off-NT-caselaw-n765365 (accessed 26 August 2008)

Martinez, M (2006) Arch enemies and Armageddon: lessons learnt from Multiplex. *Quarterly Commercial Update*, Autumn, Trowers & Hamlins, pp. 2–3. http://www.trowers.com/uploads/Files/Publications/2007/T&H%20QCU%20Autumn%202006.pdf (accessed 28 December 2012)

McCulloch, C (2007) Multiplex wins another legal spat over Wembley stadium. *Building*, 5 December. http://www.building.co.uk/news/multiplex-wins-another-legal-spat-over-wembley-stadium/3101734.article (accessed 28 December 2012)

McCulloch, C and Schaps, K (2008) Multiplex UK arm reports 40% revenue fall. *Building*, 3 January. http://www.building.co.uk/news/multiplex-uk-arm-reports-40-revenue-fall/3103098.article (accessed 28 December 2012)

Morris, N (2002) Wembley – the venue of legends becomes a nightmare. *The Independent*, 22 May. http://news.independent.co.uk/uk/politics/article189432.ece (accessed 5 March 2009)

Morris, S (2004) Wembley building work row causes 'slippage'. *The Guardian*, 28 August. http://www.guardian.co.uk/uk/2004/aug/28/wembleystadium.stevenmorris (accessed 28 December 2012)

Multiplex (2005) Multiplex Group Press Release: Earnings guidance and Wembley update (30/05/2005). http://www.brookfieldmultiplex.com/about/news/news/?NewsID=250 (accessed 26 August 2008)

Multiplex (2007) Multiplex Group Annual Report 2007. http://www.brookfieldmultiplex.com/_uploads/documents/SITES%20annual%20reports/MultiplexGroupAnnualReport2007.pdf (accessed 26 August 2008)

Multiplex Group, *Press Releases*. http://www.multiplex.com.au/page.asp?partid=294 (accessed 5 March 2009)

National Audit Office (NAO) (2003) *The English National Stadium Project at Wembley*. The Stationery Office, London. http://www.nao.org.uk/idoc.ashx?docId=0ADAE6A8-43A7-4B27-BB6F-745AA1C3B0CB&version=-1 (accessed 28 December 2012)

Oliver, M (2002) Timeline: Wembley. *The Guardian*, 25 September. http://www.guardian.co.uk/uk/2002/sep/25/wembleystadium.football (accessed 28 December 2012)

Richardson, S (2006a) Wembley row is officially over. *Building*, 19 October. http://www.building.co.uk/news/wembley-row-is-officially-over/3075658.article (accessed 28 December 2012)

Richardson, S (2006b) Industry counts cost of Wembley stadium's 'appalling saga'. *Building*, 27 October. http://www.building.co.uk/news/industry-counts-cost-of-wembley-stadium's-'appalling-saga'/3076038.article (accessed 28 December 2012)

Richardson, S (2007a) Cleveland Bridge thrown Wembley lifeline. *Building*, 5 January. http://www.building.co.uk/news/cleveland-bridge-thrown-wembley-lifeline/3079383.article (accessed 28 December 2012)

Richardson, S (2007b) Multiplex loses first round of Mott MacDonald case. *Building*, 12 January. http://www.building.co.uk/news/multiplex-loses-first-round-of-mott-macdonald-case/3079716.article (accessed 28 December 2012)

Richardson, S (2007c) Not everyone's cheering. *Building*, 10 May. http://www.building.co.uk/analysis/not-everyone's-cheering/3086209.article (accessed 28 December 2012)

Richardson, S (2007d) Multiplex faces more Wembley costs. *Building*, 28 June. http://www.building.co.uk/news/multiplex-faces-more-wembley-costs/3089824.article (accessed 28 December 2012)

Richardson, S and Stewart, D (2007) Big setback for Cleveland Bridge. *Building*, 10 May. http://www.building.co.uk/news/big-setback-for-cleveland-bridge/3086386.article (accessed 28 December 2012)

Rogers, D (2006a) Multiplex develops new game plan. *Construction News*, 9 March. http://www.cnplus.co.uk/news/multiplex-develops-new-game-plan/383600.article (accessed 28 December 2012)

Rogers, D (2006b) Multiplex's new dawn. *Construction News*, 15 June. http://www.cnplus.co.uk/news/multiplexs-new-dawn/379999.article (accessed 28 December 2012)

Rookwood, D (2001) Wembley project 'scandalously inept'. *The Guardian*, 20 November. http://www.guardian.co.uk/sport/2001/nov/20/wembleystadium.whitehall (accessed 28 December 2012)

Ryland, J (2006) Goliath won. *Building*, 16 June. http://www.building.co.uk/professional/legal/goliath-won/3068872.article (accessed 28 December 2012)

Scott, M (2007) Wembley architect holds his breath as problem child takes first steps. *The Guardian*, 23 March. http://www.guardian.co.uk/football/2007/mar/23/sport.comment2 (accessed 28 December 2012)

Select Committee on Culture, Media and Sport (SCCMS) (2002a) Minutes of Evidence: *Supplementary memorandum submitted by the Football Association*, House of Commons, Publications and Records, 17 May. http://www.guardian.co.uk/football/2007/mar/23/sport.comment2 (accessed 28 December 2012)

Select Committee on Culture, Media and Sport (SCCMS) (2002b) Minutes of Evidence: *Memorandum submitted by Wembley National Stadium Limited*, House of Commons, Publications and Records, 20 May. http://www.publications.parliament.uk/pa/cm200102/cmselect/cmcumeds/843/2052106.htm (accessed 28 December 2012)

Select Committee on Culture, Media and Sport (SCCMS) (2002c) Minutes of Evidence: *Memorandum submitted by Tropus Limited*, House of Commons, Publications and Records, 20 May. http://www.publications.parliament.uk/pa/cm200102/cmselect/cmcumeds/843/2052102.htm (accessed 28 December 2012)

Select Committee on Culture, Media and Sport (SCCMS) (2002d) *Wembley National Stadium Project: Into Injury Time*, (SCCMS – Sixth Report), House of Commons, Publications and Records, 10 July. http://www.publications.parliament.uk/pa/cm200102/cmselect/cmcumeds/843/84303.htm (accessed 28 December 2012)

Smith, C (2000) *Wembley Stadium*. House of Commons Hansard Written Answers for 17 March 2000 (pt 9). http://www.parliament.the-stationery-office.co.uk/pa/cm199900/cmhansrd/vo000317/text/00317w09.htm (accessed 28 December 2012)

Smith, G (2006) Wembley Stadium, can we fix it? Er, well, sorry, no we can't, actually… *The Times*, 25 February. http://www.thetimes.co.uk/tto/sport/football/article2281727.ece (accessed 28 December 2012)

Sport England (2002) *The New Wembley Stadium press release*, 17 September 2002. www.sportengland.org/new_wembley.pdf (accessed 5 March 2009)

Sports Venue Technology, *Wembley Stadium, London, United Kingdom*. http://www.sportsvenue-technology.com/projects/wembley/index.html (accessed 5 March 2009)

Stewart, D (2007) Multiplex wins dispute with Honeywell, Building , 8 March http://www.building.co.uk/news/multiplex-wins-dispute-with-honeywell/3082715.article (accessed 28 December 2012)

Stone, S (2000) Football: completion of new Wembley faces delays. *The Independent*, 7 September. http://www.independent.co.uk/sport/football/news-and-comment/completion-of-new-wembley-faces-delays-700971.html (accessed 28 December 2012)

Wembley National Stadium Limited, *Building Wembley*. http://www.wembleystadium.com/brilliantfuture/thenewdesign/ (accessed 5 March 2009)

Wembley National Stadium Limited, *Press Releases*. http://www.wembleystadium.com/pressbox/pressReleases/ (accessed 5 March 2009)

Appendix A

Table A.10 Key stakeholders in the redevelopment of Wembley Stadium.

Stakeholder	Role
The Football Association	The governing body of football in England: responsible for promoting the game from professional to grass root levels. Wembley was one of two main investments undertaken by the FA, the other being the new National Football Centre. http://www.the-fa.org/
Wembley National Stadium Ltd	A wholly-owned subsidiary of the Football Association: responsible for the design, construction, finance and operation of the new stadium. http://www.wembleynationalstadium.co.uk/
Westdeutsche Landesbank (WestLB)	The largest provider of project finance in the world, WestLB was the main funding bank for the stadium
IMG	The Sales Agency for Wembley's premium seating
World Stadium Team	The project architects: a joint venture between Foster and Partners, and HOK Sport
Mott Stadium Consortium	Comprising: Mott MacDonald, Connell Wagner and Sinclair Knight Merz and Weidlinger Associates, the consortium provided civil and structural (C&S) engineering and mechanical and electrical (M&E) services on the project
Multiplex Construction	The main contractor: responsible for the design and construction of the stadium and in partnership with WNSL it also led the financing and commercial structuring of the new project
Subcontractors	Numerous specialist subcontractors were involved with the project. These included: AR Security (security and smoke alarms), Cleveland Bridge UK Limited, replaced by Hollandia, (steelwork), Emcor Drake & Scull (M&E), Griffiths McGee (demolition works), Honeywell Control Systems Ltd (communications and control systems), PC Harrington Contractors Ltd (concrete works), Phoenix Electrical (M&E), RMC UK (ready mixed concrete), Scanmoor (C & S engineering work), SGD Engineering (pipe work), Stadium Seating and Pel Stadium Seating (seat installation), Stent (piling), and Weatherwise (design and installation of the stadium's main roof)
Sport England	A Government agency (formerly The Sports Council) that leads the development of sport in England: responsible for administering the sports section of the national lottery in England. Sport England provided a £120 million Lottery grant used to fund the purchase of the existing Wembley Stadium
Department of Culture Media and Sport	Led by a Secretary of State (the sponsoring Department for Sport England), the Department contributed £20 million towards the project – used to provide non-stadium infrastructure improvements in the Wembley area
The Football League	Comprising of three leagues and 72 clubs. The Football league signed a staging agreement with WNSL to hold the Worthington Cup, Auto Windscreen Shield and three Nationwide play-offs at Wembley
The Rugby Football League	The governing body of Rugby League in England, founded in 1895, it also entered into a staging arrangement with WNSL. having held the Challenge Cup at Wembley since 1929
UK Athletics	The administrative body for athletics in the United Kingdom, supporting all levels of competition
London Borough of Brent	The planning authority for the Wembley area, Brent Council was also the licensing authority for the new stadium
The London Development Agency	The economic regeneration agency for the capital: the LDA contributed £21 million to the project and was instrumental in the Mayor and DTLR providing £14 million for station improvements

(Continued)

Table A.10 (*cont'd*).

English Heritage	The government agency responsible for the protection of historic buildings. As the existing Wembley Stadium was a Grade II listed building, English Heritage's approval had to be gained before demolition of any part of the existing Stadium could begin
Quintain Estates	Owners of the land adjoining the stadium complex. The development company has submitted plans to redevelop the site
The Wembley Stadium Residents Advisory Committee	A forum, established prior to the development of the new stadium, for representatives of the local residents associations. The committee met with WNSL, Brent Council and other key stakeholders approximately every 6 months to receive updates on the stadium's progress and to discuss matters of interest as they arose

Source: Adapted from Sport England (2002)

Appendix B

Table A.11 Media briefing by WNSL, Multiplex and the FA.

2004	
10 March	**WNSL: Wembley Stadium Update** The construction of the new Wembley Stadium remains five weeks ahead of schedule and on course to open early in 2006 … In the original construction schedule, it was actually planned that the Arch would be raised this summer, but the tremendous progress made since work began in September 2002 means that Multiplex have been able to move this date forward. Ashley Muldoon, Project Director for Multiplex Construction UK … said: *'The project is running ahead of schedule and Multiplex is confident that we will deliver it on time and to budget.'* The first match scheduled to take place at the new Wembley is the 2006 FA Cup Final, although it could open even earlier … Michael Cunnah, Chief Executive of WNSL, said: *'The new Wembley is an incredibly ambitious project, on a scale never seen before. It will be bigger and better than any other stadium in the world and to be ahead of schedule at the halfway stage of construction is a great achievement.'*
28 May	**WNSL: Wembley Stadium Arch – The Big Lift Begins** Ashley Muldoon, Project Director of Multiplex, said: *'We have made tremendous progress in the first 18 months of demolition and construction – so much so we are ahead of original construction schedule …'*
5 August	**WNSL: FA Statement: Wembley Clarification** … The stadium is on schedule and on budget. The arch has been raised and we are looking forward to Wembley opening its doors in May 2006 with the FA cup final. It will be the world's finest stadium.
10 August	**WNSL: Multiplex Statement** Multiplex has noted unattributed speculation in the media regarding substantial cost and time delays on the Wembley stadium project. Such speculation is untrue. Multiplex confirms that work at Wembley is continuing as scheduled. Multiplex is taking all steps necessary to mitigate the effects of Cleveland Bridge UK's (CBUK) recent decision to withdraw from the National Stadium Project … Multiplex remains hopeful that CBUK will comply with its legal obligation to take all steps necessary in ensuring that its decision will cause the least possible disruption to the orderly continuation of the project under Multiplex's management.
18 August	**MPX: Wembley Stadium Update** In light of recent press speculation, Multiplex Group advises that works on the Wembley Stadium project continue to progress at a rate which will permit Multiplex to complete the project ahead of program. Structural steel subcontractor Cleveland Bridge has

(*Continued*)

Table A.11 (*cont'd*).

	completed its works on site with the erection of the main arch, and Multiplex believes that a contractual dispute with Cleveland Bridge will not affect the program or impact upon Group earnings.
30 September	**WNSL: On this Day …** On this day two years ago work began on the new Wembley Stadium … Two years later the construction of the new Stadium is well underway and the raising of the iconic, 133 m high Arch in June marked a key stage in the construction of the new 'Venue of Legends,' signalling that the dream of a new national stadium is well on its way to being realised. The Stadium is on schedule to stage its first major event – The FA Cup Final in 2006.
2005	
13 May	**WNSL: In 365 Days Football is Coming Home!** **Wembley the 'Venue of Legends' opens 13th May 2006** Michael Cunnah, CEO of WNSL commented: *'Today marks a major milestone … To see the stadium that we have previously only seen in models and computer generated imagery come together is truly exciting. Although we face another 12 months of hard work we can now begin to see the light at the end of the tunnel.'* Ashley Muldoon, Project Director for Multiplex… said: *'Multiplex will handover the project on the contractual date. Multiplex will be then assisting WNSL as they prepare the stadium for licensing and operation for May 2006 for the FA cup. It is really exciting to think that this time next year the world will be able to see a world class Stadium in operation here at Wembley…'* [Ashley Muldoon, reiterated Multiplex's assertion that the stadium would be completed on time in a BBC News item broadcast on the same day (13 May 2005): http://news.bbc.co.uk/player/nol/newsid_4540000/newsid_4545900/4545987.stm?bw=bb&mp=wm&nol_storyid=4545987&news=1]
27 May	**WNSL: WNSL Statement** In response to Multiplex's recent announcement regarding Wembley Stadium. WNSL has issued the following statement: *'Multiplex have been in contact with us this morning and assured us that recent announcements will not, in any way, affect Wembley Stadium's construction schedule and that work on site continues as usual with the aim for the Stadium to open with The FA Cup Final in May 2006.'*
28/29 May	**MPX: Multiplex BBC interview** When questioned by a BBC reporter on whether Wembley would be completed on time, Martin Tidd (Multiplex's Managing Director in Britain and Europe) replied: *'I can absolutely guarantee that the FA Cup will be held here at Wembley and I can absolutely confirm that my seat is there and I'll be sat there.'*
30 May	**MPX: Multiplex Group - Earnings Guidance and Wembley Update** … an internal peer review… undertaken on the Wembley project… has highlighted that the productivity levels that had been assumed previously are not currently being achieved. The peer review has concluded that while Wembley is now expected to be completed by the end of March 2006, in time for the FA Cup Final to be played in May 2006, the costs associated with completion of this program are anticipated to have further increased which is expected to result in a loss on the project. Multiplex anticipates being able to commence a staged handover of the project to our client in January 2006. Multiplex estimates the loss in relation to Wembley to be £45 million (Au$109 million) excluding the Au$50 million Roberts Family indemnity, announced in February 2005.
31 May	**WNSL: 'Wembley Stadium is on track for The FA Cup Final next year** 'Michael Cunnah, Wembley Stadium's Chief Executive, confirmed in an interview on BBC Radio Five Live last night that… since Multiplex's release to the Australian stock exchange yesterday:

(*Continued*)

Table A.11 (*cont'd*).

	'From a Wembley Stadium perspective very little has changed. Unfortunately Multiplex… have had to disclose to their shareholders that they do expect a loss on the project. But they are still confirming that the stadium is going to be ready for The FA Cup Final. *The whole process of handing over such a large stadium doesn't happen overnight. There are over 2000 rooms in this building and there has to be a very long commissioning process to make sure that each aspect of the stadium is of the quality we require.'* … WNSL's contract with Multiplex is a fixed price contract meaning Multiplex is contracted to deliver the 90,000-seater stadium to a fixed specification for a fixed price The official opening of the stadium remains The FA Cup Final on May 13th, 2006 There will be a progressive handover of the stadium from Multiplex to WNSL from January 31st, 2006
2 November	**MPX: Multiplex 2005 AGM - Chairman's & CEO's Address** Our ongoing negotiations with Hollandia do not impact the program and I can confirm that completion of the Wembley project remains on target to permit the FA Cup Final to be played in May 2006 with a progressive handover from early next year. … there remain material residual risks in relation to adverse claims outcomes in respect of the client, the subcontractors and consultants on the project, and the risk of programme delays due to issues such as weather. In addition, there are risks in relation to liquidated damages, although this risk has been mitigated by an 'in-principle' agreement by the client to resolve various extensions of time entitlements to 31 March 2006… I would like to stress your Board remains confident that the project will be finished within the revised programme which will permit the FA Cup to be played at Wembley in May 2006.
9 December	**WNSL: Cunnah Reiterates Cup Final Aim** WNSL Chief Executive Michael Cunnah has today reiterated Wembley Stadium's aim to open with The FA Cup Final on May 13th, 2006. Following yesterday's comments from Football League Chairman Sir Brian Mawhinney, Cunnah told wembleystadium.com: *'Brian expressed his personal opinion and I respect his right to do so. However, our position is exactly the same as it was last week and the week before that. Multiplex continue to assure us that they will be in a position to handover the stadium on March 31st next year and we must respect their expert advice. It remains our aim to play The FA Cup Final on May 13th, 2006.'* [Michael Cunnah had previously commented on Multiplex's assurances in a BBC News item broadcast on 7 December 2005: http://news.bbc.co.uk/player/nol/newsid_4500000/newsid_4505800/4505890.stm?bw=bb&mp=wm&nol_storyid=4505890&news=1]
21 December	**MPX: Wembley Project – Further Information** … The project is currently tracking to a revised program which permits progressive handover of the stadium commencing in January 2006, with works substantially completed by end of March 2006. Some works that are not critical to delivering the operational status of the stadium continue past this date, but still permit the FA Cup Final to be played on May 13, 2006… There remains a material risk that these dates will not be achieved and the stadium will not be available for the FA Cup Final…
2006	
30 January	**MPX: Press briefing** Martin Tidd, the managing director of Multiplex UK, said that Multiplex and their subcontractors were: *'… fully committed to substantial completion by March 31.* *I need to restate there are residual risks that remain in terms of the stadium being able to hold the FA Cup Final in May.*

(*Continued*)

Table A.11 (*cont'd*).

	The substantial risks still remaining are subcontractor performance, design changes, industrial relations, the weather and final the integration and commissioning of systems. *My view would be a 70 per cent chance of completion for the Cup final.* *Work would have to run like a Swiss watch if the stadium was going to be completed according to schedule.* *It does not all come down to us at the end of the day. There's a lot of people needed to make sure the FA Cup comes off.* *We think it is too early to call whether it will be here or at Cardiff. We will be able to assess this in three weeks time.* *We have over 3,500 workers out there, there's tremendous pressure on the trade contractors. We are doing as much as we can to manage the trade contractors performance.* *There's still a risk, we are managing it, but there may be other contractors out there that might not see it through.* *We will not be a hostage to ransom and we are looking to our guys and the unions to manage things fairly.'*
30 January	**WNSL: Michael Cunnah's response** Michael Cunnah, WNSL chief executive, said his company would be working closely with Multiplex: *'We will continue to monitor progress on a daily basis and take their expert advice on how the stadium is progressing… Multiplex today said that in three weeks' time they will be able to advise whether the stadium will make March 31 completion. The FA will make the decision on where the Cup final will be held… The safety and quality of the stadium must remain the most important factors and will ultimately dictate the outcome of this decision.'*
31 January	**MPX: Wembley Update** Multiplex continues to target completion of the Project to enable the 2006 FA Cup Final to be played at Wembley. Multiplex has now completed certain initial aspects of the Project… Various areas of the Project that were previously targeted for handover by the end of January have not been completed. In these circumstances and as previously foreshadowed, there remains a material risk that the stadium will not be available for the FA Cup Final. However, assuming satisfactory performance in the month of February then it is expected that a progressive handover will be achieved so as to enable the stadium to be available for the 2006 FA Cup Final. Multiplex and its subcontractors are fully committed toward achieving this goal.
8 February	**WNSL: Wembley update** FA Director of Communications Adrian Bevington today stressed that no decision has been taken on whether this season's FA Cup Final will be played at the new Wembley Stadium or at the Millennium Stadium in Cardiff. Bevington said: *'I can categorically say that no decision has been taken on where The FA Cup Final will be played. We are giving Wembley every chance of being ready. We are in regular contact with the operating company WNSL and through them the builders Multiplex. There is a critical path that the builders have in place… We have taken the prudent measure of booking Cardiff as a back-up. It would have been naive not to do so. Whether or not it is ready for this season's Cup Final, the new Wembley will be a fantastic venue for Cup Finals and international matches for years and years to come. It will be one of the best stadiums in the world and one that everyone will feel immensely proud of.'*
21 February	**FA: Wembley statement** FA Chief Executive Brian Barwick today released the following statement regarding this year's FA Cup Final: *'I can confirm today that this season's FA Cup Final will be held at the Millennium Stadium in Cardiff on 13 May.* *I can also confirm that England's pre-World Cup games against Hungary and Jamaica – previously scheduled for Wembley – will now be held at Old Trafford.*

(Continued)

Table A.11 (cont'd).

	The FA has taken these decisions after meeting with Wembley Chief Executive Michael Cunnah and Multiplex's Managing Director Martin Tidd, following a site visit to the stadium yesterday. *The FA took the prudent step of booking the Millennium Stadium as a backup option to Wembley last August, therefore allowing us to extend this decision as long as possible and giving Multiplex every chance of making the 13 May date.* *It is clear that while they have over 3,500 people working hard on the site, Multiplex and indeed WNSL are unable to give us 100% certainty that the stadium will be completed in time for The FA Cup Final.* *Due to the magnitude of the Cup Final, we are not prepared to compromise or take any risks on the stadium being unable to stage such a significant event.* *We felt it was important to clarify the situation for everyone connected with the game in this country – especially for those teams and their supporters still in The FA Cup.* *Multiplex and WNSL remain committed to delivering the stadium at the earliest possible date.* *I would like to emphasise that WNSL have a fixed-price contract with constructors Multiplex. The total project cost remains £757 million.* *We should remember that when finished, Wembley will be world class and the stadium – with its magnificent arch – will host many Cup Finals and England internationals for decades to come.'* [Brian Barwick confirmed this statement in a BBC News item broadcast on the same day (21 February 2006): http://news.bbc.co.uk/player/nol/newsid_4730000/newsid_4736800/4736898.stm?bw=bb&mp=wm&nol_storyid=4736898&news=1 http://news.bbc.co.uk/player/nol/newsid_4730000/newsid_4736800/4736898.stm?bw=bb&mp=wm&nol_storyid=4736898&news=1]
21 February	**WNSL: FA Cup Final statement** Statement by Wembley Stadium Chief Executive, Michael Cunnah: *'The FA has confirmed today that The FA Cup Final will be held at The Millennium Stadium in Cardiff on May 13th and that England's matches against Hungary and Jamaica ahead of the World Cup – scheduled to be at Wembley Stadium – will be held at Old Trafford. Obviously I am as disappointed by this news…* *Multiplex have had over 3,500 construction workers on site in an effort to make The FA Cup Final. However given the magnitude of this event, WNSL and The FA require 100% certainty that the stadium will be ready in time to make this commitment. While Multiplex are confident of their progress they are not in a position to give us the 100% certainty that we require. Therefore we have had to inform The FA that the stadium cannot host the 2006 Cup Final.* *It is still too early to say exactly when the stadium will open. We will announce a definitive date once we have more certainty on when the stadium will be fully ready and finished to world-class specification.* *Our priority now is to work with Multiplex to finish the stadium to the highest standards – we have a once in a lifetime opportunity to build a world class stadium and we are working together to ensure both the quality of the stadium's facilities and the quality experience it offers our visitors.* *Although we share fans' disappointment at this news, the ultimate success of Wembley Stadium is not contingent on opening in time for The FA Cup Final. Wembley will be the greatest stadium in the world and will host the most exciting events in the sport and music calendar for many years to come.'*
21 February	**MPX: Wembley Update** Multiplex understands that the English Football Association ('the FA') is to transfer the 2006 FA Cup Final to its reserve venue, Millennium Stadium in Cardiff.

(Continued)

Table A.11 (*cont'd*).

	Whilst disappointed with the decision, Multiplex continues to work towards targeting completion of Wembley National Stadium at the earliest possible date. We understand that the FA has made this decision on the basis that it requires 100% certainty that the venue will be fully functional by 13 May 2006, the scheduled date for the 2006 FA Cup Final.
31 March	MPX: Wembley Update In relation to the Wembley National Stadium project, Multiplex advises as follows: Substantial completion of Contractor's works is now anticipated by the end of June 2006 (previously May 2006), based upon a review of the costs and benefits of any acceleration measures. Previous updates relating to programme have presumed an integrated approach with Client's works being completed simultaneously. Multiplex yesterday received advice from the client that it is now reconsidering whether it will or is able to continue to integrate its works, or whether it will commence these works post completion. Multiplex's client, WNSL, is solely responsible for determining when it starts and completes its works and also when and which events are hosted at the stadium. Substantial completion by the end of June will have certain works and certain activities such as commissioning and cleaning still to be completed after this date… … Multiplex has formally advised its client that Multiplex is entitled to substantial and legitimate extensions of time under the terms of its construction contract which will extend the contract completion date until at least September 2006. Not withstanding that it is Multiplex's belief that it has until at least September to complete, Multiplex is targeting substantial completion by end of June as noted above.
31 March	**WNSL: Wembley statement** Multiplex has given WNSL a revised construction schedule which now shows full practical completion by the end of September. Multiplex's statement anticipates 'substantial completion by the end of June' with key outstanding works up until the end of September. Wembley Stadium then requires a further period of two months to run test events (ramp-up) to ensure operational safety standards are met. Multiplex's revised schedule means WNSL has no alternative but to announce no major events will take place at the stadium for the remainder of this year. WNSL's main priority is to deliver a world-class stadium to the highest possible standards of both quality and safety. We will announce an opening date once we have 100% certainty from Multiplex. Wembley Stadium Chief Executive, Michael Cunnah, said: *'We share everyone's disappointment at Multiplex's announcement today of a further delay in the completion of the stadium. Their revised construction schedule leaves us with no other choice but to make this decision. However, this does not detract from our focus and determination to deliver a world-class stadium. We have an over-riding responsibility and duty to ensure that the stadium is completed to the highest standards of safety and quality.'*
1 May	MPX: Wembley Progress Update In relation to the Wembley National Stadium project, Multiplex advises as follows: Substantial completion of Contractor's works is anticipated by the end of June 2006… although it is not anticipated to finish commissioning, cleaning and other works… until September 2006. There are a number of critical works and activities that are the responsibility of our client, and the timing of the completion of these works is under the control of our client. The date for the opening of the stadium and the hosting of the first event will ultimately be a matter for our client. [Also, the statement also reiterated the following paragraphs: • Substantial completion by the end of June… (paragraph 4 – from the statement dated 31 March) • Multiplex has formally advised its client… (paragraph 4 – from the statement dated 31 March)]

(*Continued*)

Table A.11 (*cont'd*).

1 June	MPX: Wembley Progress Update Multiplex advises: Substantial completion is now anticipated on 13 July 2006 leaving certain works relating to the recent removal of the cranes and hoists and certain activities such as commissioning and cleaning to be completed after this date… Multiplex is targeting substantial completion in July as noted above, although it is not anticipated to finish commissioning, cleaning and other works… until September 2006. [Also, the statement also reiterated the following paragraphs: • Multiplex has formally advised its client… (paragraph 4 – from the statement dated 31 March and repeated in the statement dated 1 May) • There are a number of critical works and activities… (paragraph 2 – from the statement dated 1 May)]
3 July	MPX: Wembley Progress Update In relation to the Wembley Stadium project… Multiplex advises: … Multiplex remains confident that substantial completion will be achieved in July 2006 sufficient to enable us to achieve practical completion of our works at the end of September 2006. In order to achieve practical completion of Wembley Stadium certain works which are the responsibility of WNSL rather than Multiplex must first be completed; WNSL has not begun some of these works. Furthermore, once Practical Completion has been achieved, it is the responsibility of WNSL to carry out a series of critical works and activities in order to make the stadium operational for events. The timing and hosting of the first event is therefore entirely a matter for WNSL. Multiplex is of the view that the works for which WNSL is responsible may not finish by September 2006 and therefore the date of Practical Completion may be delayed. Such a delay should have no impact on Multiplex's project loss position Multiplex has formally claimed from WNSL lengthy extensions of time under the terms of its construction contract which will in effect extend the contract completion date. Multiplex has advised WNSL of extensions of time to October 2007.
3 July	**WNSL: Wembley Stadium statement** Following Multiplex's statement to the Australian stock exchange dated 3rd July, Wembley Stadium issued the following statement from its Chief Executive Michael Cunnah. '*Wembley Stadium firmly rejects Multiplex's statement and refutes any claims that they make against us. The delays experienced on the stadium have and continue to be under the complete control and responsibility of Multiplex. Although the stadium appears to be nearly finished, Multiplex has an important and substantial amount of work to complete, including the installation, testing and commissioning of mechanical and electrical systems.* *Wembley Stadium took the sensible decision in March to rule out all events at the stadium in 2006 as we were sceptical of Multiplex's ability to meet the various deadlines they have given us. The situation remains the same – Multiplex need to make more progress in delivering the stadium before we can name an opening date.*'
31 July	**WNSL: Wembley Stadium update** Wembley National Stadium Ltd (WNSL) today issued the following statement from Chief Executive Michael Cunnah. '*Wembley National Stadium Ltd's objective is to create a truly world class stadium of the highest standards of quality and safety. Because of their extensive experience in stadium construction outside the UK, Multiplex was appointed to design-build that stadium to a fixed cost. It is Multiplex's responsibility to deliver the stadium and it is WNSL's responsibility to ensure that they do.* *Unfortunately Multiplex has missed its original timetable of 30th January, 2006 as well as its own revised timetable of 31st March, 2006. Currently Multiplex states that it will complete the stadium by September 2006. It is WNSL's view that, at the current rate of progress,*

(*Continued*)

Table A.11 (*cont'd*).

	particularly in light of the extensive testing and commissioning regime that will be required, this is not likely to be achieved until later in the year. *Whilst the stadium is well on the way to being finished and looks magnificent from the outside, Multiplex still has major items to complete. For example Multiplex has yet to finish the stadium's roof, the extensive remedial works to the stadium's drainage network, the building management and life safety systems and the installation of approximately 10,000 of the stadium's 90,000 seats. It has always been our view that the stadium's delays are Multiplex's responsibility.* *Notwithstanding this, WNSL is geared up for opening the new stadium. We have recruited a team of over 6,200 people and the next step for them is to begin the final stages of training inside the stadium itself. We can only begin this final stage when Multiplex has sufficiently finished its works. At that point we will be able to enter the stadium to start the handover process, begin our final training and carry out limited works which include the installation of radio communications systems, mobile cash machines and national flags. These items will be completed at the same time as we prepare for the test events which will precede our first full capacity event.'*
1 August	MPX: Wembley Progress Update Multiplex advises the following progress update: The stadium works are now substantially complete, with the exception of the installation of the last remaining seats. The recent insolvency of the stadium seat subcontractor has caused unforeseen delays to the completion of the installation of the seats. Alternative seat supply arrangements are now in place, and the final seats will now be installed by early September 2006. As noted in previous progress updates, there are certain works that relate to the removal of temporary works, and works such as commissioning and cleaning that are required to achieve practical completion. These remaining works remain on programme and will complete prior to the date at which the stadium becomes operational. As noted in previous progress updates, there are a number of critical works and activities ('Client Works') that are the responsibility of the Client ('WNSL'), and the timing of the completion of these works is under the control of the Client. WNSL has not yet begun some of these Client Works and others which WNSL has begun are not complete. Other Client Works must be complete before the stadium can achieve all necessary operational licences and approvals. WNSL is responsible for obtaining approvals from various Authorities in order to ensure the stadium is fully operational for events. To obtain these approvals a series of test events will be required at the stadium to demonstrate to the Authorities that the integration of the safety and control facilities are working properly such that a 90,000 spectator event can take place. Unless and until WNSL hold and successfully complete this series of test events, and obtain the required approvals, Multiplex will be prevented from achieving practical completion. In the absence of a detailed programme having been provided by WNSL, Multiplex has conducted its own analysis of the programme to completion of the client works. This has been based on Multiplex's own assumptions of the current state of WNSL's preparedness. The analysis has been reviewed with an expert in the operation of major stadia and indicates that it is unlikely that the stadium will be able to hold a test event for 90,000 spectators before June 2007. Multiplex has advised WNSL of its entitlement to extensions of time to October 2007 and of its intention to pursue legal proceedings in that and other respects, including WNSL's acts and omissions which are preventing Multiplex's ability to achieve practical completion. Delays in the Client works give rise to entitlements in favour of Multiplex for extensions of time and financial entitlements.

(*Continued*)

Table A.11 (*cont'd*).

1 August	MPX: Multiplex statement to the press Multiplex spokesperson said: *'Our contract with the client clearly sets out the criteria required for us to achieve 'Practical Completion'… As WNSL have failed in their obligation to commence certain works critical to this criteria, the position has worsened with another month passing. We still have no clear idea from WNSL about when they intend to begin some of their works.'* (Quoted in *The Independent* – Daley, 2 August 2006)
1 August	**WNSL: Wembley Stadium statement** Wembley National Stadium Ltd (WNSL) today issued the following statement from Chief Executive Michael Cunnah. *'We have always sought to give a realistic appraisal of progress on site and following Multiplex's statement to its shareholders it is important that we clear up some issues.* *The contract with Multiplex has two critical future milestones which Multiplex have deliberately confused. Multiplex is required to hand WNSL a completed stadium which is defined in the contract as 'Practical Completion'. WNSL then has to work with Multiplex to finish certain works and to hold the various test events which will enable the safety certificate to be obtained and achieve 'Operational Completion', the point at which a fully-functioning stadium is delivered, capable of holding full-capacity events for 90,000 people.* *The latest date Multiplex had stated for PC was September, although we note that their latest statement does not actually give a new date or confirm the previous September date. Multiplex claim that they are substantially complete, however, we note again that Multiplex has yet to finish the stadium's roof, the extensive remedial works to the stadium's drainage network, the building management and life safety systems and the installation of approximately 10,000 of the stadium's 90,000 seats, or to hand over any of the c.3,400 spaces in the stadium for snagging.* *In the absence of a detailed programme of work from the contractor that we can rely on, we estimate that Multiplex will finish their work at some point late this year. When Multiplex do hand us the completed stadium, we have a detailed timetable of our works and test events which subject to co-operation from the contractor, should enable 'Operational Completion' to take place within two to three months of handover.* *We are still not in a position to name an opening event or rule any event in or out as we need Multiplex to make more progress on site. The bottom line for us is that we will be ready two to three months after they give us a finished stadium. If they can achieve that this year, then we will be able to open the stadium in early 2007. We hope that Multiplex will now devote all their energies to completing the stadium at the earliest opportunity as is their responsibility.'*
1 August	**WNSL: FA statement** Further to conversations with WNSL, The FA remains confident that Wembley Stadium can be ready for this season's FA Cup Final. This is of course dependent on when stadium constructors Multiplex meets all of its contractual obligations. WNSL believes it is unlikely that Multiplex will complete their work by September but they continue to expect practical completion to take place before the end of 2006. We do have to factor in that Multiplex have to date missed all key deadlines. However, once they do deliver the stadium they were contracted to hand over several months ago, WNSL will then be in a position to complete a detailed timetable of commissioning work, test events and any other remaining works. They cannot complete this work until the stadium is completed by Multiplex. We can also clarify that no plans have been made to book the Millennium Stadium in Cardiff for The FA Cup Final.

(*Continued*)

Table A.11 (*cont'd*).

17 August	MPX: Multiplex Group (MXG) Financial Results for the Year Ended 30 June 2006 In the UK, Wembley is now substantially complete and the Group is focusing on pursuing recoveries from third parties. During the year, the Group was successful in its claim against the former steel subcontractor at Wembley.
1 September	MPX: Wembley Progress Update Multiplex advises that there has been no material change to the project status from that reported on 1 August and 17 August 2006. As noted previously, works relating to the removal of temporary works and those such as commissioning and cleaning that are required to achieve practical completion are ongoing. These remaining works remain on programme and will complete prior to the date at which the stadium becomes operational. There are a number of critical works and activities ('Client Works') that are the responsibility of the Client ('WNSL') and the timing of the completion of these works is under the control of the Client. WNSL has not yet begun some of these Client Works and others which WNSL has begun are not complete.
5 September	**WNSL: WNSL statement to the press** Following the receipt of a 150 page formal claim from Multiplex, reported by the Daily Telegraph to be in the sum of £350m, a Wembley spokesman said:' *'It has come with a "without prejudice" letter, which means they do not believe it is a credible document which they could take to court… We would like them to take it to court because we believe we have a very strong position.'* (Quoted in *The Daily Telegraph* – Bose, 2006a)
5 September	**WNSL: WNSL statement to the press** A WNSL spokesman said: *'Multiplex have, from time to time, intimated various claims to us. We have refuted these claims and continue to believe that the delays to the construction of the stadium are delays that WSNL do not have responsibility for, such as the steel work and the installation of the roof.'* (Quoted in *The Daily Telegraph* – Bose, 2006b)
3 October	MPX: Wembley Progress Update Multiplex advises that there has been no material change to the project status from that reported on 1st September 2006. The installation of more than 90,000 seats is now complete… Multiplex is currently pursuing its entitlements against WNSL. An adjudication process has commenced in relation to a number of individual issues with many more adjudications and possible legal proceedings to follow. [Also, the statement also reiterated the following paragraphs: • As noted previously, works relating to the commissioning and cleaning… (paragraph 1 – from the statement dated 1 September) • There are a number of critical works and activities… (paragraph 2 – from the statement dated 1 May and repeated in the statements dated 1 June and 1 September)
16 October	MPX: Response to Media Speculation Re: Wembley Project In response to recent media speculation in the United Kingdom and Australia surrounding negotiations on the Wembley project dispute between Multiplex and the client, WNSL, Multiplex advises the following: Discussions in relation to a potential settlement between Multiplex and WNSL are progressing; and the commercial terms remain to be finalised. Multiplex will advise the market if and when any settlement with WNSL is completed.
19 October	**WNSL: Wembley Stadium Update** The parties involved in the Wembley Stadium project today announced that they have agreed a comprehensive settlement of all outstanding disputes. The agreement follows successful talks chaired by Lord Carter between The FA, Wembley National Stadium Limited (WNSL) and the constructors Multiplex. This agreement will avoid a lengthy and expensive legal dispute and all parties are committed to working together to ensure that the stadium opens for business as early as possible in 2007.

(Continued)

Table A.11 (*cont'd*).

	Multiplex Chief Executive Andrew Roberts said: *'We are very pleased to have reached agreement with WNSL and The FA to put our past differences behind us and put all our joint efforts into completing Wembley Stadium at the earliest opportunity.* 'The FA Chief Executive Brian Barwick said: *'This agreement secures the process of getting the new Wembley Stadium up, running and open to the public. We look forward to staging major events at the stadium next year and consider this agreement with Multiplex to represent the beginning of the end of the construction phase. Everyone's target is now to complete what will be the finest stadium in the world.* WNSL Chief Executive Michael Cunnah said: *'This is very welcome news for everyone involved with the stadium. This project was founded on the shared desire of WNSL and Multiplex to create a truly special stadium and it is very appropriate that we should enter the final stages of the project working together to get the stadium operational. Multiplex played an intrinsic part in getting this stadium project up off the ground and I am delighted to be working together to achieve our original vision'.'*
20 October	MPX: Wembley Stadium Settlement Multiplex is pleased to announce that it has resolved all variations and reached a comprehensive settlement of all disputes with its client at the Wembley Stadium project, WNSL. The agreement is conditional on a consent to be obtained by WNSL from its financiers. Multiplex and WNSL have agreed a streamlined process whereby the works and activities that are the responsibility of WNSL will be completed in parallel with other works. This should ensure that the stadium is able to complete its test events earlier than envisaged in previous progress updates. The financial effect of the agreement with WNSL does not alter the project loss position assumed in Multiplex's most recent annual results. As previously stated, this position remains dependent on the resolution of subcontractor claims and other third party recoveries.

Source: Adapted from FA, Multiplex and WSNL press releases, Bose (2006) (emphasis and [ ] added)

Case Study B
Terminal 5 (T5) Heathrow

Introduction

Variously described as 'enormous', 'epic', 'historic', 'huge' and 'massive', BAA plc (BAA)'s Terminal 5 (T5) programme at Heathrow Airport was at the time of its construction Europe's largest and most complex construction project costing £4.3 billion. It also promised to revolutionise UK construction project management practice, leading Brady *et al.* (2008) to classify the development as a 'megaproject'.[1]

T5 was conceived as London's and also the UK's primary entry point:

> 'T5 will provide a world-class gateway to the UK... Its airy, light and contemporary architecture should make the process of transition through the airport stress-free and enjoyable, supporting the fundamental design philosophy that the airport experience should delight the traveller.' (Steve West, Project Director – Pascall & Watson, quoted in Design Build Network)

Its main terminal building alone cost £1.4 billion and, at 396 m long by 158 m wide, is the UK's largest single-span building.

Commercial Management: theory and practice, First Edition. David Lowe.

> 'It is a spectacular piece of architecture and engineering that responds to the fact that this is London and the UK's primary entry point.' (Alan Lamond, Aviation Director – Pascall & Watson, quoted in Design Build Network)

Taking over 18-and-a-half years to develop (planning approval and its associated public inquiry lasted 10 years) T5 was delivered on time and to budget, becoming operational on 27 March 2008. And, despite some initial teething problems that included flight delays, technical hitches, baggage-handling issues, and protest action, it is deemed to have been one of the UK's most successful construction programmes.

Owned by BAA[2], T5 is operated by its subsidiary Heathrow Airport Limited, which has responsibility for the whole Heathrow Airport complex. Denationalised in 1986 and floated on the UK stock market in 1987 with a capitalisation of £1,225 million, BAA was at the beginning of the construction phase the world's largest commercial airport operator. In addition to Heathrow, BAA operated Aberdeen, Edinburgh, Gatwick, Glasgow, Southampton and Stansted airports in the UK, plus had interests in airports in Australia, Hungary, Italy and the US. In 2005 BAA generated a profit of £710 million, with 144.3 million passengers passing through BAA's seven UK airports (Heathrow alone accounting for 68 million) and, on average, 1700 daily aircraft departures.

At the start of the T5 construction phase, Heathrow Airport's wage bill contributed over £5 billion to the UK economy (the complex employing approximately 70 000 staff and indirectly supporting 250 000 jobs nationally). It was estimated that the T5 programme would protect a further 16 500 jobs at the airport.

In addition to operating airport facilities, BAA is a leading client and project management organisation with expertise in delivering total airport management contracts, having undertaken hundreds of capital projects, for example, the design and construction of Stansted Airport. BAA funds, develops and manages airports, and was both the client and project manager for T5.

Despite its considerable experience, previously BAA had neither developed integrated assets nor undertaken a programme on such a scale. T5 was a significant programme of construction works: phase 1 alone comprised 16 major interconnecting projects (all significant in their own right) and 147 sub-projects.

Independently, T5 is the fifth largest airport in Europe, providing extra terminal and aircraft parking facilities and increasing Heathrow's annual capacity by approximately 30 million passengers.

> 'The word "terminal" is mildly misleading. By itself it would be the fourth biggest airport in Europe. There would only be Heathrow, Schiphol in Amsterdam and Frankfurt bigger. And it comes complete with all this infrastructure, such as the new spur road off the M25 and extensions to the Heathrow Express and the Piccadilly Underground line.' (Tony Douglas, Managing Director T5, BAA, quoted in Leftly, 2004)

The T5 development was a genuine mixed-use scheme, including 23 000 m^2 of retail space (19 000 m^2 of which is located in prime airside sites), equating to over a third of that contained in the Bluewater shopping centre: it is the UK's third largest shopping venue based on market spend.

British Airways plc (BA) was, until the formation of International Airlines Group (AIG) in January 2011, the exclusive occupant of T5, having relocated its operations from other terminals at Heathrow with the aim of consolidating these activities within one building (this process commenced at 4 a.m. on 27 March 2008). Although not a contracting party in T5, BA participated in the key decision making through a joint planning team with BAA.

Requirement identification

Heathrow is the world's busiest international airport. Designed to cater for 45 million passengers per year, by the late 1990s passenger numbers at Heathrow (Terminals 1, 2, 3 and 4) exceeded 60 million; in 2004, for example, it handled 67 million passengers. Moreover, BAA estimated that this number could increase to 80 million per year by 2012. Consequently the airport operator began the process of developing a new terminal at Heathrow: T5. The aim of the project was to expand Heathrow's annual passenger capacity to 95 million. Following the official opening of T5C in June 2011, T5 has the capacity to handle 30 million passengers per year. In 2011 26.3 million passengers (on 184 616 flights) passed through the terminal (BAA, 2012).

BAA negotiated a sole tenancy agreement for T5 with BA, the terminal's primary airline operator utilising approximately 40% of Heathrow's capacity. While BA might have been the logical tenant for T5, BAA did consider various alternatives.

Table B.1 T5, Heathrow.

Capacity	30 million passengers per year (27 million phase 1, 3 million phase 2)
Cost	£4.3 billion
Construction started	September 2002
Construction completion	April 2008 (construction phase substantially completed September 2007)
Special features	Largest free-standing building – the waveform roof
Sponsor/Client	BAA plc
Architect	Concept architect Richard Rogers – Richard Rogers Partnership (now Rogers Stirk Harbour and Partners) Production architect Pascall & Watson
Structural engineer	Arup and others
Project manager	BAA plc
Cost consultants	Turner and Townsend/E C Harris
Main contractor	No main contractor – a framework agreement was used to appoint 60+ Tier-1 suppliers (see Appendix B for list of Tier-1 suppliers) Top 10 suppliers based on value of work: Laing O'Rourke, Amec M & E, Vanderlande, Balfour Beatty, Amec Civils, Rowen Structures, Morgan Vinci, MACE, Schmidlin UK, NTL Group

As BA generated significantly higher returns at Heathrow compared with other UK airports, it was eager to increase its capacity at the airport. At the time, however, its operations were divided between Terminal 4 and Terminal 1, with ensuing transfer issues between its long-haul and short-haul services. Exclusive occupancy of T5 would enable BA to relocate all its services within one terminal, simplifying flight transfers and significantly improving its operational efficiency. The move would also allow the airline operator to provide a world-class passenger experience through the introduction of innovative technologies (Booz Allen Hamilton, 2007). For BA, the efficient transfer, on schedule, of its operations to T5 was a strategic issue.

For BAA, the deal with BA generated extra passenger capacity at Heathrow with a guaranteed occupant for the new terminal. Further, it was anticipated that the single terminal model would lead to an increase in the appeal of Heathrow as a transfer hub, thereby increasing the number of passengers using the airport, while the relocation of BA's services in T5 would create spare capacity elsewhere on the terminal complex (Booz Allen Hamilton, 2007).

T5 was therefore developed specifically for a single airline occupant, with significant elements of the terminal designed to meet BA's particular requirements – for example, its baggage-handling system was designed to facilitate passenger transfer. As mentioned earlier, while BA was not a party to the T5 contract, the airline and airport operator cooperated for more than 10 years on the development, design and delivery of the terminal. An output of this collaboration approach was the introduction of an innovative 'self-service' passenger check-in and processing system (Booz Allen Hamilton, 2007).

Requirement specification

In defining the requirements of T5 BAA had to consider the varying and potentially conflicting interests of the numerous stakeholders on the project. These included:

- **Passenger requirements**: BAA was keen to '*delight the traveller*' and create a stress-free passenger experience. In particular, they were influenced by research into passenger behaviour, which had found that travellers exhibited elevated stress levels in the check-in area. A requirement of the terminal, therefore, was to the create the feeling of space:

  'A large hall creates an openness that increases the sense of well-being and avoids claustrophobia... Similarly, the screening process invokes concerns in some passengers, and so is located in the most open space possible.' (Alan Lamond, Aviation Director, Pascall & Watson, quoted in Design Build Network)

Further passenger requirements include:
- *accessibility*: public transport provision (rail and tube links) was a key requirement of the planning process. However, it was predicted that the impact of T5 would be to at least double the number of passengers arriving by road; therefore, improvements to the road infrastructure would be required, together with sufficient car-parking facilities. Adjacent hotel provision would also be required
- *efficient passenger processing*: for example, check-in, security clearance and baggage-collection systems, and especially the avoidance of queues
- *transit within the terminal*: direct, single-level access to conveniently located departure gates
- *clear signage*: visible route marking throughout the terminal and current flight information
- *refreshment and retail outlets*: aimed at satisfying the various requirements of the different passenger groups using the terminal. A further advantage of reducing the stress levels of passengers before they enter the terminal's retail section is that they are more likely to buy goods
- the provision of appropriate services and information for *'meeters and greeters'* (Rawlinson, 2008; Design Build Network, ND)

- **BA's requirements**: as the UK's primary 'full-fare' airline operator, BAA had specific design requirements. For example, as the terminal was to provide a world-class gateway to the UK, a signature building was necessary, reflecting BA's status. In addition, T5 had to provide a full range of services, such as exclusive 'first-class' and premium passenger facilities, baggage-handling systems to accommodate the relatively higher volumes and size of luggage of its users and a potentially higher level of security screening, plus a pier/gate service (Rawlinson, 2008; Design Build Network, ND)
- **Third-party requirements**: these included the specific needs of the security services, passport and customs control (the Border and Immigration Agency and Her Majesty's Revenue and Customs), and the UK Visa Services (the Foreign and Commonwealth Office). Since April 2008 these agencies have been combined within the UK Border Agency (Rawlinson, 2008)
- **BAA's requirements**: as the owner and operator of the terminal, BAA's concerns included revenue optimisation, through the provision of catering and retail space, etc., and achieving value-for-money from its investment in and performance of its fixed assets. It also needed to respond to the requirements of all its stakeholders, for example by considering the following brief requirements:
  - *value-for-money*: BAA's landing charges were fixed in advance by its regulator
  - *adaptability*: enabling BAA to respond flexibly to future business opportunities and needs, and to accommodate changing flight schedules and aircraft designs
  - *departure delay avoidance*: through the provision of intuitive transit routes through the departure and arrivals areas
  - *reliability*: of its operating systems
  - *efficiency*: in turning aircraft around by utilising well-organised apron services (Rawlinson, 2008)

In developing T5, BAA's principal objective was to make sure that the project was delivered on time and to budget. Additionally, as the 'client' and 'developer', they appreciated the need to consider the whole-life-cost implications of any design solution and the importance of creating an efficient, cost-effective and environmentally sound facility.

Solution selection

Site location

BAA's preferred location for T5 was a 260 ha site to the west of the existing Heathrow terminal complex.

Preliminary work on T5 started in 1989 following the appointment of the Richard Rogers Partnership (now Rogers Stirk Harbour and Partners) as concept architects via a design competition. However, it was not until February 1993 that BAA submitted its planning application. The proposed development provoked intense opposition from both local inhabitants and environmental pressure groups. This led to numerous protest rallies, and in August 2001 demonstrators occupied BAA's boardroom for five hours before being ejected.

Following two years of preparation, a public inquiry was instigated in May 1995, concluding four years later on 17 May 1999 – the longest recorded in the UK. An official inquiry report was presented to ministers in December 2000 and was finally approved in November 2001 by Stephen Byers, the Secretary of State for Transport. However, further objections and additional local approval requirements continued into 2003.

In total the process is thought to have cost BAA, the UK government and local authority objectors over £100 million (BAA's planning and consultation fees alone costing £63 million).

Site constraints

The T5 site was highly confined – constrained on all sides: to the north and south by two of the world's most heavily used runways, to the east by Heathrow's existing terminal buildings, with the M25, one of Europe's busiest motorways, running to the west. Additionally, as a result of the planning inquiry, BAA was required to comply with approximately 700 conditions. For example, before work could commence on the site BAA was obliged to:

- Undertake the UK's largest ever single-site archaeological dig: 80 archaeologists examined over 100 ha of the site
- Divert the Duke of Northumberland's and the Longford rivers which crossed the site, channelling them around the western perimeter of the airport

Additional conditions included:

- A requirement to extend the Heathrow Express rail and the Piccadilly Line underground links to T5
- The imposition of strict traffic restrictions around the site in terms of both volume and routes

Access to the site was therefore restricted to one key entry point. This was identified as a major risk at the beginning of the project, but the issue was mitigated by the introduction of a logistics plan. A further restriction was the imposition of a maximum building height of 39 m due to the airport's radar systems which limited working space to 43 m. The need to work in restricted airside zones imposed additional constraints, including the need to obtain security clearance for personnel and the potential for deliveries to be delayed. Other airport-specific constraints included:

- Compliance with stakeholder requirements, for example, the Department of Transport, HM Customs (now the UK Border Agency) and Special Branch, etc.
- Noise prevention and monitoring requirements
- Avoidance of electromagnetic interference and radar signatures
- Preventing disruption to passengers and aircraft using the existing terminals (Rawlinson, 2008)

Project scope

The programme to deliver T5 was divided into two phases. The first phase included:

- The main terminal building T5 (Concourse A) and one satellite terminal building (T5B) connected to T5 by an automated people mover
- 48 aircraft stands many of which are designed to accommodate the 550-seat Airbus A380, and taxiways
- A new air traffic control tower
- A public transport interchange: a dedicated rail station located under Concourse A comprising two platforms for the Piccadilly Line underground extension, two for the Heathrow Express extension and two for proposed rail links to the west
- Extensions to the Heathrow Express and London Transport's Piccadilly Line via 13 km of bored tunnels, a new railway and tube station
- Over 13.5 km of bored tunnelling (in nine sections) for drainage, road, underground and rail links
- A new spur road from T5 to the M25 motorway (to reduce the strain on the overburdened road network) and the diversion of the airport perimeter road
- The diversion of two rivers: the Duke of Northumberland's and the Longford
- A 605-bed hotel
- A four-storey car park with spaces for 3800 cars and 150 motorcycles
- Mechanical and electrical installations valued at £600 million

In addition, this phase included various additional structures that support the terminal, including de-icing stores, an energy centre, a fuel farm, ground pens and forward maintenance units.

Phase 2 included a second satellite terminal building (T5C), connected to T5 and T5B by an automated people mover, and an additional 12 aircraft stands.

With the completion of T5C in 2011, T5 increased Heathrow's annual passenger capacity by an additional 30 million.

Terminal design solution

The concept design for T5 was created by the Richard Rogers Partnership, who provided two on-site 'concept' guardians during the project delivery stage. Approximately 150 separate consultancy organisations were involved in the scheme, including production architects Pascall & Watson, structural engineers Arup and construction managers MACE (the main tier-1 designers/consultants are listed in Appendix A).

T5's design is considered to be cutting edge, incorporating, amongst others, retail space, baggage- and passenger-handling facilities, office accommodation and an integrated rail and tube interchange. Design features included:

- **Single-span terminal building**: at 396 m long by 158 m wide T5A (Concourse A) is the largest single-span terminal building in the UK, creating a distinctive landmark
- **High-rise configuration**: in comparison with other modern airport terminals, which have generally adopted a horizontal layout, T5 employs a comparatively high-rise format comprising four public levels above ground, plus three basement service floors. Passenger transit through the interchange zone is vertical, via lifts and escalators
- **'Loose-fit' configuration**: the building's shell and structure are independent of the four above-ground floor levels contained within it. In addition to the aesthetically pleasing open space that this creates, the arrangement offers long-term flexibility, as the internal space can be reconfigured at a later date, while retaining the building's envelope

  'As the nature of air travel changes over time, a fully flexible solution is needed to accommodate future needs...The loose-fit nature of the main terminal building (Concourse A) offers maximum flexibility because there are no columns to interrupt a huge floor space that stretches for over a quarter of a mile and is the size of 50 football pitches.' (Alan Lamond, Aviation Director, Pascall & Watson, quoted in Design Build Network)

- **Waveform roof**: formed without any intermediate columns, the 18 500 tonne single-wave effect roof is supported by 22 arched steel box girder rafters, 800 mm wide and up to 3.8 m deep, spaced at 18 m centres with 610 mm deep steel box section secondary beams spanning between the rafters. The roof covering was designed to be prefabricated into 'cassettes' to enable it to be lifted into position. The roof is supported by steel foot, arm and wing sections, connected via nodes, at 36 m creating the open-plan structure and providing adequate space for departure gates and their associated circulation space.

  'The natural curve of the roof will encourage the traveller to move towards the airside face, and at each decision point along the route to the aircraft, the next stage of the process will become obvious... For an arriving passenger, similarly structured design elements offer the same "flow-through" philosophy.' (Steve West, Project Director, Pascall & Watson, quoted in Design Build Network)

- **Track transit system**: T5 incorporates an automatic people mover (track transit system) to transfer passengers between the main terminal and its two satellite buildings
- **Pedestrian zone**: in keeping with BAA's desire to enhance the 'passenger experience', a 30 m pedestrian zone – the glass-roofed and landscaped Interchange Plaza – connects T5 to its car park, bus, coach and taxi facilities. Additionally, passenger transit through the terminal has been simplified and walking distances reduced compared with earlier terminals
- **High-speed baggage handling**: contained within a double-height basement area, T5 incorporates an 18 km long conveyor belt system, which BAA asserts is the world's most advanced baggage- handling system. The automated system is designed to sort, prioritise, and track up to 12 000 bags an hour. Luggage is then dispatched at 10 m/second to waiting aircraft, in driverless linear-induction-powered rail wagons. (Ironically, BAA claimed that the system would cut the instance of 'late bags' to below one in every thousand, thereby reducing flight delays and cancellations.)
- **Air traffic control tower**: at a height of 87 m, the new control tower is twice the height of its predecessor. Its radiused triangular design creates an optimal aerodynamic profile and affords its occupants a 360° view

Box B.1 Key statistics

T5 provides:

- T5A – the core terminal building (396 m × 176 m by 40 m high) with an continuous floor space of 300 000 m^2
- T5B – a satellite terminal building (442 m × 52 m by 19.5 high)
- Capacity for 30 million passengers per year
- 60 aircraft stands (12 stands in phase 2)
- 87-metre high control tower
- Over 13 km of tunnels (1.3 km, 8.1 m diameter airside road tunnel – 4.1 km, 2.9 m diameter storm water outfall tunnel)
- 18 km baggage system
- 6 km of twin rivers diversion
- 3800 space multi-storey car park
- 250 000 square feet of retail space
- 605 bed hotel
- 70% of the construction below the surface

However, despite a high degree of pre-construction design work, BAA recognised that there would still be a significant amount of design evolution during the delivery stage of the project:

> 'But one onerous result of the protracted public enquiry is that some commitments had to be made too early. For example, the maximum height of the building was established as a fixed commitment, but when (it) came to drawing up the detail we had to fit a big building into a very constrained space. Also, because of the extended gestation period, we have to live with design decisions taken ten years ago that we may not agree with today, although they were based on sound reasoning at the time. We are now delaying some procurement commitments for IT systems, so [that] we can install the latest technology, such as security screening equipment, at a future date.' (Alan Lamond, Aviation Director, Pascall & Watson, quoted in Design Build Network)

Development programme

T5 was initially planned to open at 4 a.m. on 30 March 2008, although this was subsequently revised to 4 a.m. on 27 March 2008. Crucially, the opening of the new terminal building was set to coincide with the introduction of the summer schedules by the airlines using Heathrow.

Phase 1 of T5's construction programme was scheduled to last five years and was divided into five key stages:

1. **Site preparation and enabling works**: included an archaeological survey, the diversion of services, site levelling, the removal of existing sludge lagoons and providing temporary and support facilities, for example, site roads, offices, compounds and building the logistics centres
2. **Groundworks**: included the major earthworks, substructure works – foundations and basements, drainage and rail tunnels
3. **Major structures**: included the construction of the main T5 terminal building (concourse A), the first satellite building (concourse B), the multistorey car park and ancillary structures
4. **Fit out**: included mechanical and electrical services installation, the baggage-handling system, a personal rapid transit system – an automated people mover (APM) and specialist electronic systems
5. **Implementation of operational readiness**: included ensuring that the facility was complete and operational, requiring all systems and procedures to have been tested and the workforce appropriately trained (Fullalove, 2004)

The key milestone dates for the T5 project are shown in Table B.2.

Table B.2 **T5 project milestones.**

1989	
	T5 design competition won by the Richard Rogers Partnership (now Rogers Stirk Harbour and Partners)
1993	
February	BAA submits planning application
1995	
May	T5 Public Inquiry commences
1999	
March	T5 Public Inquiry concludes
2001	
November	Planning consent granted by UK government
2002	
September	Work commences on site to construct T5
2003	
April	The archaeological dig draws to a close having found more than 80 000 artefacts
July	As part of the proactive approach to risk management, the trial erection of a full-size section of T5's roof is carried out at Thirsk in Yorkshire
August	A full-scale section of the baggage-handling system is built and tested
September	The trial erection of the roof is completed, leading to the identification and resolution of more than 140 issues, preventing a potential four-month delay
November	Designs for the new air traffic control tower are made public, while work commences on the main terminal building's steel superstructure
2004	
March	Tunnelling for the extension of the westbound Piccadilly Underground Line is completed 10 days ahead of schedule
April	The first section of the main terminal building's roof is successfully raised and four aircraft stands are handed over to BAA
May	The diversion of the 'twin rivers' – the Duke of Northumberland and Longford – is completed two weeks early. The scheme is applauded by the Environment Agency and Royal Parks Agency
July	T5 achieves one million consecutive man-hours without a reportable accident for the first time (it goes on to do this another nine times). The new control tower roof is lifted into position and the power supply to T5 is connected
September	The T5 extension is connected to the existing Heathrow Express tunnel
October	In one of the heaviest lifting operations ever undertaken at a UK airport, the cab section of the new control tower is transported 2 km across the airport
November	A further six aircraft stands are handed over to BAA
2005	
January	13.5 km of bored tunnel (comprising nine separate road, rail, underground and drainage tunnels) are completed on schedule
February	The mid-point of the T5 programme is reached. The UK's seventh longest road tunnel, the 1.3 km long 'airside road tunnel' (connecting T5 to the Central Terminal Area) is opened
March	The final (sixth) main terminal roof section is lifted into position. After a complex lifting procedure, the 87 m high control tower and 40 m high terminal roof successfully reach their full height
April	The main structure of the new air traffic control tower is completed
May	The twin river diversion scheme receives a CEEQUAL award for environmental excellence from the Institution of Civil Engineers
September	Alistair Darling, the Secretary of State for Transport, attends T5's topping-out ceremony, laying the final slab of concrete on the main building's departure floor
December	Ahead of schedule, T5 is made watertight and weatherproof following the completion of over 30 000 m^2 of glass façade. The M25 spur road is opened by the Secretary of State for Transport

(Continued)

Table B.2 (*cont'd*).

2006	
April	As part of its compatibility tests at Heathrow, the A380 is trialled at T5
July	The Duke of Edinburgh visits T5. T5 achieves two million consecutive man-hours without a reportable accident for the first time (it goes on to do this twice more)
September	An unexploded World War II bomb is discovered and safely dealt with. Six of T5's automated transit system vehicles arrive and are lowered down a 12 m shaft into position. After being closed for 20 months, the Piccadilly Underground Line connection to Terminal 4 is re-opened on schedule
November	Satellite terminal T5B nears completion
2007	
February	Including cycle paths and bridleways for the local residents, the Colne Valley landscaping project commences
April	The new air traffic control tower becomes fully functional and the integration of T5's local area network with Heathrow Airport is completed ahead of schedule
May	Cleaning of T5 begins and 40 semi-mature London plane trees are planted within the Interchange Plaza
July	Ahead of time, Heathrow Express and London Underground trains arrive at the six platform rail station to commence tests
August	T5 achieves one million consecutive man-hours without a reportable accident for the tenth time
September	Operational Readiness trials commence involving 15 000 volunteers
2008	
February	T5 achieves two million consecutive man-hours without a reportable accident for the third time
14 March	T5 is officially opened by Her Majesty the Queen
27 March	T5 becomes operational with BA flight BA026 from Hong Kong landing at 4:50 a.m.
2011	The second satellite terminal T5C opened

Source: Adapted from Anon (2004a) and BAA Terminal 5: Information

Table B.3 Based on a projected project cost of £3.725 billion, the costs can be broken down as below.

	£ million	£ million
Phase 1 (completion 2008)		
Buildings (including baggage systems)	1,350	
Airfield/civils	1,050	
Rail and tunnels	300	
Iver South and M25	275	
Other	125	3,100
Central contingency	200	3,300
Phase 2 (completion 2011)	425	
Terminal 5 (March 2002 prices)		3,725

Source: Stent (2003)

Project budget

The budget for T5 rose from £1.7 billion in 1996 to £4.2 billion at the start of construction in September 2002. A cost breakdown is provided in Table B.3 based on an overall budget of £3.725 billion set in March 2002. Despite criticisms from certain airlines that this figure was excessive, BAA considered the cost plan to be reasonable, basing it on recently completed airport developments such as Chep Lap Kok, Hong Kong's International Airport which opened in July 1998. Further, the budget was benchmarked by BAA's cost management team and endorsed by the Civil Aviation Authority (CAA) and BA. Moreover, research by BAA concluded that without the adoption of a fundamentally different delivery methodology the T5 project would exceed this budget by more than £1 billion.

Financing the project

T5 and its associated facilities were funded by BAA via debt finance, supported by landing charges determined on a five-yearly basis by its regulator the CAA. These charges were set at a level to ensure that BAA had an adequate income stream to fund the project (commencing at £6.48 per passenger in April 2003 the charges increased at the retail prices index [RPI] plus 6.5% in each following year).

> 'The regulator allows us a certain rate of return. But to satisfy our shareholders, we have to beat that.' (Matthew Riley, Commercial Director T5, BAA, quoted in Wolmar, 2005)

Significant completion delays or cost overruns on T5 would have impacted significantly on BAA's share value and caused major reputational damage. Additionally, CAA introduced a series of regulatory milestones – 'triggers' (penalties) based on the following stages:

1. Completion of the diversion of the twin rivers (2004–2005)
2. Completion of the T5 early release stands (2004–2005)
3. Handover of the T5 visual control room to National Air Traffic Services (NATS) (2005–2006)
4. T5A becoming weather proof (2006–2007)
5. T5B becoming weather proof (2006–2007)

Failure by BAA to meet the above timetable would have resulted in the charges levied at Heathrow Airport being reduced by 2% for each year a stage was behind schedule. The 'penalty', calculated on a monthly pro rata basis, was approximately £7 million per annum.

Project management

Most clients when faced with a project as complex and challenging as T5 would have adopted a familiar, well-established delivery approach. However, driven by a desire to reduce the costs of providing its airport facilities, BAA concluded that it could improve its project processes by adopting emergent project, risk and contract management methodologies.

In February 2001 BAA announced that it would project manage T5 itself.

> 'BAA has chosen to manage this programme and accept the risk rather than contract the programme out to a company to manage for us – a more common approach.' (Matthew Riley, Commercial Director T5, BAA, quoted in Wolmar, 2005)

as the complexity of the project precluded employing an external organisation:

> 'A project manager could not come into the T5 situation and pick up the pieces without having knowledge of the projects and all the stakeholders.' (Norman Haste, Project Director T5, BAA, quoted in Clark, 2001b)
>
> 'This is a fairly wide-ranging operation, which includes the airport, a transport interchange, retail operations and maintenance... BAA believes it has to manage the whole process and take the risk.' (Eryl Smith, Managing Director T5, BAA, quoted in Clark, 2001b)

Strategically, in order to deliver T5 successfully, BAA had to transform itself into the '*... most capable and sophisticated project management client in the UK*' (Brady et *al*. 2008).

As a result of retaining and managing risk, plus adopting a cost-reimbursable form of contract on T5 (a highly sophisticated approach), BAA required a large, highly proficient internal project management team to function as an 'intelligent client' on the project. Members of the BAA's 160-plus project management team took an active role in the management of each integrated (delivery) team. When making appointments to the core project management team, BAA selected individuals with experience gained on significant UK and international projects from both within and outside the construction sector.

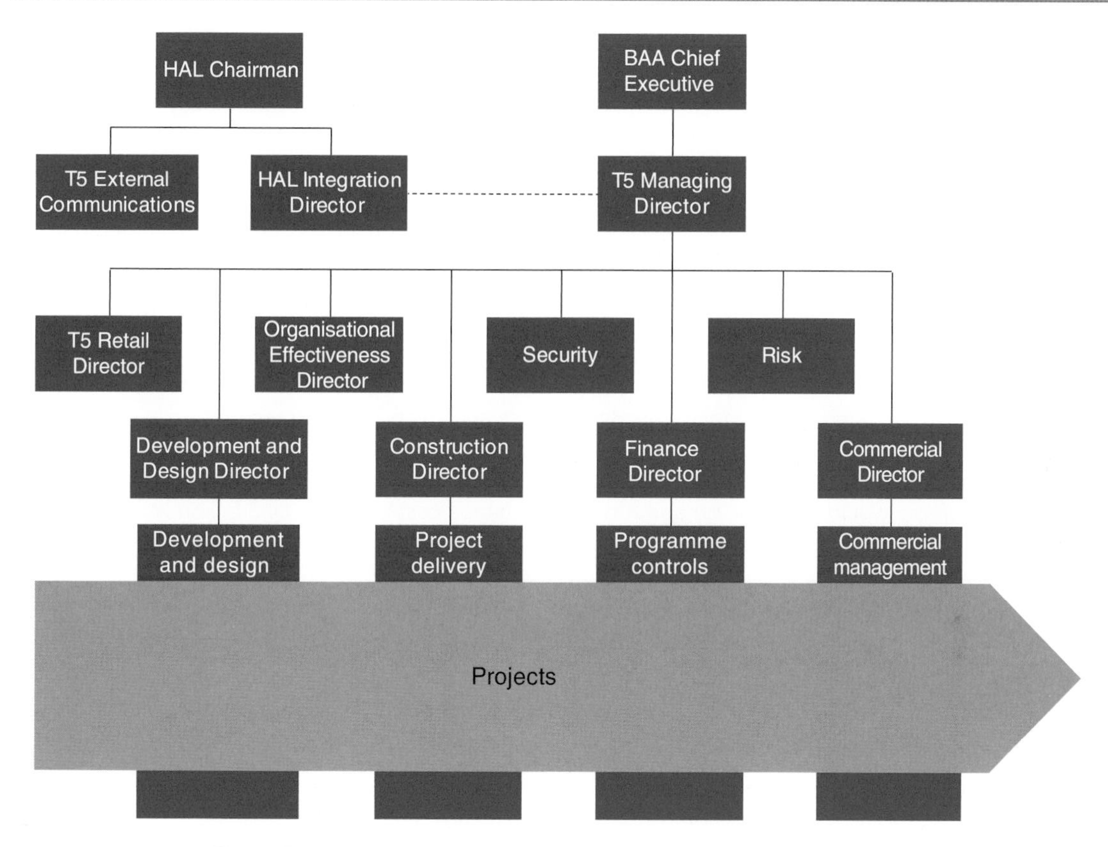

Figure B.1 T5 governance structure. HAL, Heathrow Airport Limited.

Committed 'high level project sponsorship and leadership' was also deemed to be vital to the success of the project; therefore, the project's managing director was an executive member of BAA's board of directors. Similarly, other board members provided support to and took an active interest in the project. Moreover, the board received frequent reports on the progress of the project. The governance structure adopted for T5 is shown in Figure B.1.

BAA considered project management to be '... *a tool of the risk management approach, not vice-versa*'.

Procurement strategy/contracting strategy

Development

The concept of an opportunity-based relational procurement/contract model surfaced in the 1990s when BAA developed its first-generation partnering agreements: CIPP (continuous improvement of the project process). Using these framework agreements, which incorporated integrated teamworking, BAA had entered into approximately 50 construction and consultancy agreements by the mid 1990s. The model agreement had resulted in enhanced project predictability and repeatability.

These early framework agreements were compliant with the European Union (EU) procurement directives, having been advertised in the OJEC (the *Official Journal of the European Community*[3]); suppliers were appointed with the expectation of being involved in the delivery of T5. As a result of the lengthy public enquiry into the development, however, the supply chain was unable to be mobilised as intended.

In part, BAA utilised the hiatus caused by the protracted T5 planning process to undertake a study of projects under construction both at Heathrow and elsewhere prior to commencing the T5 programme. BAA was keen to learn from previous major projects, particularly those that had failed in some aspect of delivery, so that it could avoid making the same mistakes and to inform its approach to procuring and managing T5.

'In the 1990s, BAA carried out research on large infrastructure projects in UK construction. We looked at the headlines of the outturn performance for things such as the Channel Tunnel, Jubilee Line extension, British Library, Scottish Parliament and West Coast Main Line.' (Matthew Riley, Supply Chain Director, BAA, quoted in Brass, 2008)

In addition, between 2000 and 2002, BAA analysed every UK construction project exceeding £1 billion in value constructed during the preceding 10 years, plus every international airport project completed in the previous 15 years. As part of this study BAA investigated project processes and organisation, and the influence of individual behaviour on project performance.

The research established that:

- Not one UK construction project (within the set parameters) had been delivered on time, on budget, safely or met its specified quality standards. Based on its analysis BAA predicted:

  '... if we were to apply the average outturn performance of those projects to Terminal 5, a five-year build programme would probably be about two years late, a cost target of £4.3 billion would probably be at least £1–£1.5 billion over budget, the quality would be variable and, statistically, 12 people would die on site. None of those consequences were acceptable to BAA... We were about to embark on making an investment that was going to add almost 50 per cent to the asset base of the company, yet if we applied some of those outturn performance metrics to T5, at best it would potentially run the risk of destabilising BAA financially. At worst it could have put us out of business.' (Matthew Riley, Supply Chain Director, BAA, quoted in Brass, 2008)

- None of the international airport projects studied had functioned as designed when initially opened. Again, based on the study, BAA predicted that if it used a similar approach it would take the T5 terminal three years to ramp up to its 30 million passengers a year capacity. BAA's aim was for T5 to achieve this over its first year of operation

BAA established that all the projects studied had experienced significant contractual and financial difficulties, and on each one the client had incurred immense reputational damage. As BAA operated within a regulated industry, with its income effectively fixed, it was particularly vulnerable to cost increases and, therefore, could not tolerate a similar outcome on T5. However, technical competence was not an issue on these projects.

The study concluded that failure on projects the size of T5 was due to two main reasons:

1. **Cultural confusion**: organisational and management issues arising from ill-defined project parameters and the failure of purchasers to appreciate the needs of supply chains
2. **The reluctance to acknowledge risk**: rather than identifying and apportioning risk appropriately at an early stage, traditional contracts generally sought to transfer risk to the supply chain, which often resulted in lengthy legal disputes when supplier performance did not meet the client's aspirations:

   '40 per cent of the cost of claims are the legal expenses.' (Matthew Riley, Commercial Director T5, BAA, quoted in Fullalove, 2004)

Alternatively, project success emanated from its culture and commercial milieu, and effective leadership was vital, particularly in environments subject to change. Further, the research established that current supplier 'behaviour', both positive and negative, was predominantly influenced by conditions of contract and anticipated profit margins.

Based on this research, BAA concluded that a step change in construction procurement best practice was essential if T5 was to be successfully delivered.

BAA's requirements

In selecting an appropriate procurement strategy, risk was a significant issue for BAA. Potential concerns included: the availability of resources, industrial relations, failure to manage the various interfaces and multiple suppliers required to deliver a project the size of T5, safety issues (a significant factor considering the project's location adjacent to two busy runways), restrictions imposed by the public inquiry, third-party dependencies (Highways Agency, London Underground, NATS [formerly, National Air Traffic Services Limited], Thames Water, etc.), and the impact of delay on the funding mechanism agreed with the CAA.

Similarly, the approach adopted had to be flexible and responsive to change. BAA recognised that over the duration of the contract its corporate needs and the requirements of airline operators were likely to change and many construction solutions could not be resolved before the appointment of contractors due to the rate at which technological advances were being made. Change management was therefore crucial in enabling the projects design to evolve during the project's delivery stage.

Main principles

BAA's objectives were:

- To create a project environment to support the delivery of T5
- To achieve and improve upon existing 'best-practice' performance in all stages of the project through the adoption of innovative and world-class procedure from a variety of industry sectors
- To initiate a blueprint for future major construction projects

To deliver these objectives BAA instigated a wide-ranging strategy to transform its own capabilities and those of its key suppliers built upon the principles of *Constructing the Team* (Latham, 1994) and *Rethinking Construction* (Egan, 1998), and on the success of its Heathrow Express Recovery project. However, BAA's contracting strategy went beyond any previous arrangement implemented in the UK, redefining its role as project client: participating fully in the delivery process; adopting a proactive approach to risk – managing the cause of risk, not its effect; and establishing a set of behaviours to facilitate innovative problem-solving.

The strategy embraced four key principles:

1. **The client always bears and pays for the risk**: irrespective of the contractual arrangement adopted it is impossible to transfer risk:

   'We realised that, to expose waste and manage the performance more efficiently, we would have to actively hold all the risk. We realised you cannot transfer corporate risks around that are so intrinsic to the success of your company; risks that relate to the City or to airlines or regulators or to your corporate citizenship. Those risks can't be transferred down a contract. You're kidding yourself if you think they can, because, in each of those examples we looked at, there were very few suppliers that went out of business as a consequence of those project failures. The risk ultimately comes back to the client organisation.' (Matthew Riley, Supply Chain Director, BAA, quoted in Brass, 2008)

2. BAA retained full liability for all project risks rather than acquiesce to the illusory transfer of risk to its suppliers:

   '… we had to have a strategy that was, at its highest level, BAA holding all the risk all the time, and in return we expected our suppliers to come together as partners and work in an integrated team or teams. They came together to deliver projects or products, and the financial consequences of risks were underpinned by insurance policies that BAA took out directly with the market, on a strictly no-fault basis.' (Matthew Riley, Supply Chain Director, BAA, quoted in Brass, 2008)

BAA's strategy, therefore, was to adopt a problem solving approach to risk, identifying its sources at an early stage and then assembling the best resources (integrated teams) to proactively manage them.

Box B.2 The underlying assumptions of BAA's contracting approach contrasted with conventional principles

T5 contracting (partnering)	Usual contracting (fixed price)
• Cannot transfer risk	• Transfer of risk
• Remain flexible	• Price in advance
• Integrated teams	• Profit at risk
• BAA holds the risk	• Penalties
• Active risk management	• Defined scope
• Reimburse properly incurred	• Employer's team
• Profit levels pre-agreed	• Skill and care
• Emerging pre-planned scope	• Compliance/remedies driven
• Single integrated team values	• Silos
• Exceptional performance	
• Goals/targets	
• Success driven	
• Liability	

Source: Doherty (2008)

3. **Suppliers' profit levels were predetermined and fixed**: BAA ring-fenced its suppliers' profits and incorporated a gain-share arrangement, whereby efficiency savings could be translated into higher margins
4. **Partners are worth more than suppliers**: BAA, implemented an 'integrated project team approach' embraced in a 'Delivery Team Handbook', which sought to create an appropriate environment for team-working, with the objective of:

   '... motivating, organising and generally getting the best out of the talented people working on the project.' (NAO, 2005)

These principles were incorporated into what was branded the 'T5 Agreement' – a bespoke project-specific contract developed by BAA to govern its relationships with tier-1 suppliers and designed to create a framework to deliver project success. A comparison of the underlying assumptions of BAA's contracting approach with conventional principles is provided in Box B.2.

Project procurement (contractor appointment)

Under the 'T5 Agreement', BAA entered into a direct contractual relationship with all its tier-1 suppliers (main suppliers, contractors and consultants), of which there were over 60. The agreement was used regardless of the type of service offered by the supplier or whether or not they would traditionally have been appointed as a subcontractor. (A list of the tier-1 suppliers as at May 2005 with their particular sphere of activity is provided in Appendix A.)

T5 consisted of 16 projects (each valued at between £50 million and £400 million), which in turn were divided into 147 sub-projects (comprising approximately 1000 work packages). Each sub-project was run by an integrated design and construction team, containing between 6 and 25 tier-1 suppliers, which was led by a BAA project manager.

Rather than specifying the precise work required or seeking a commitment from suppliers to carry out a specific amount of work, the T5 Agreement required tier-1 suppliers to provide a commitment as regards their capability and capacity to undertake work on the project. This allowed BAA to compile project teams from the collective expertise within the tier-1 suppliers – a process consistent with EU procurement law. Agreements were in place with many of the suppliers involved in the project from its inception.

Suppliers were then assigned to integrated teams when their capability was required to carry out discrete parts of the project based on plans of work.

Contract

The main objective of the agreement was to create a unique contract under which BAA retained all the risk relating to the project. Additionally, the contract needed to be flexible as BAA appreciated that their requirements would change during the course of the contract.

The contract is generally considered to be a balanced agreement, which facilitates appropriate relationships and behaviours. Drafted in a non-adversarial style, the negative and potentially confrontational aspects of traditional construction contracts were replaced by a commercial model and policy that created commercial tension without erecting commercial barriers. Although not explicitly derived from the NEC form, the two contacts have aspects of partnering and integrated working in common. The contract was designed to enable all participants to concentrate on:

- The root cause of problems – not their effects
- Working within integrated teams to deliver success in an uncertain environment
- The proactive management of risk rather than the avoidance of litigation (NAO, 2006a)

The structure of the agreement comprised:

- **The T5 Agreement (Delivery Team Handbook)**: contained the terms and conditions of contract
- **The supplement agreement**: identified the skill capability and capacity of the supplier and defined the likely scope of work on the programme
- **Functional execution plans**: detailed the requisite support to facilitate project delivery
- **Sub-project execution plans**: detailed the teams' plan of work
- **Work package execution plan**: detailed the work breakdown structure for each team member/individual supplier
- **Additional documents**: for example, the Commercial Policy, Core Processes and Procedures, Industrial Relations Policy and Programme Handbooks, which between them defined the commercial environment and generated the commercial tension (BAA documents, NAO, 2006a)

The T5 Agreement was supported by BAA's novel risk insurance policy.

Each tier-1 supplier was responsible for appointing, developing and managing their own supply chain (second- and lower-tier suppliers/subcontractors). However, BAA expected the contractual arrangements within the supply chain to conform to the principles of the T5 Agreement – for example, to avoid risks being transferred down the chain to those least able to carry them and to promote cooperative working methods.

To encourage this BAA stipulated how tier-2 suppliers were to be appointed and recommended the use of its modified version of the NEC Engineering and Construction Contract (ECC).

Additionally, BAA used a variety of NEC contracts, principally the Professional Services Contract to appoint approximately a further 150 consultants and other suppliers. In terms of value, these arrangements comprised around 10% of the overall project cost.

The contractual structure adopted on the project is represented diagrammatically in Figure B.2.

Risk management

A key component of the T5 Agreement, and a major departure from common practice, was the notion that BAA retained ownership of risk rather than seeking to transfer it, or more accurately transfer the economic implications of risk, to the numerous suppliers engaged on the project.

Following their review of major projects, BAA concluded that:

- The traditional model was ineffectual in the case of T5; many risks are unforeseeable before or during the bidding process and it is naive to behave as if they are:

  'The old game would be to go to the market with an incomplete understanding of what you want, ask for bids without understanding the inherent risks and then get bids from contractors that are designed

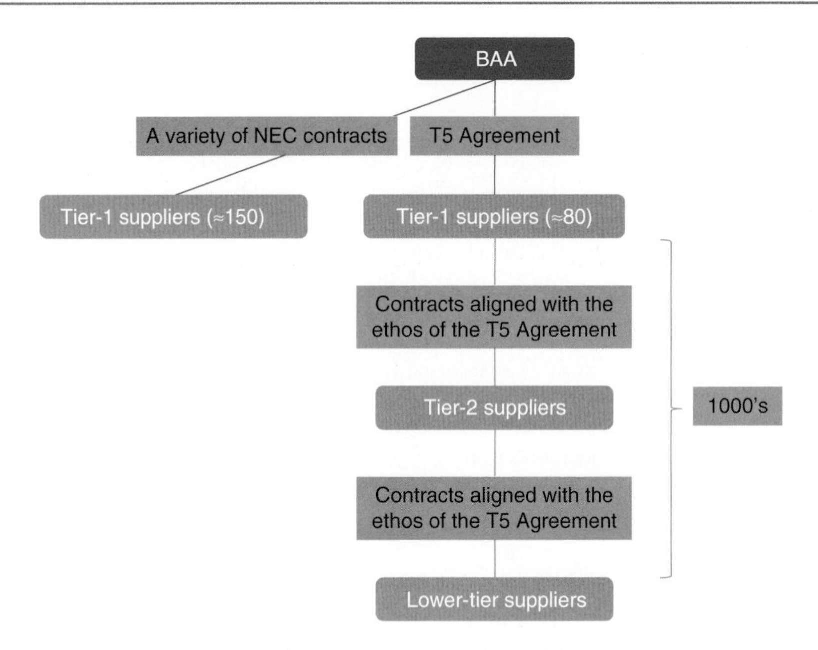

Figure B.2 Contractual structure adopted on T5. Source: Adapted from Ernst & Young LLP and Freshfields Bruckhaus Deringer (2005).

> to beat the competition rather than address the real risks.' (Tony Douglas, Managing Director T5, BAA, quoted in Simpkins, 2005)

- No construction company would be able to carry the financial liabilities generated by the £4.2 billion project
- Irrespective of how risks are apportioned, ultimately the purchaser always bears and pays for the risk

As a result BAA retained all risk on the project (eliminating it from the supply chain) and insured it, rather than requiring suppliers to include it in their prices. This approach, BAA maintains, obliges suppliers to acknowledge risk, while removing a common cause of the reallocation of capacity and talent to support legal action.

> 'We've redesigned the roles of the contractors. Instead of allocating risks we look at how to make risk visible and how to mitigate it. We make it quite clear that BAA holds all the risk all the time. We can then all work as an integrated group.... We choose key strategic partners and we fix the margin up front. Once the contractors know that they will make, say, 5% we can all work as an integrated group to clarify what is ambiguous – with total open-book transparency over costs.' (Tony Douglas, Managing Director T5, BAA, quoted in Simpkins, 2005)

The T5 contract embraced a proactive approach to risk management enabling the tier-2 suppliers (consultants and contractors) via integrated teams to concentrate on risk management, with the aim of mitigating potential risk, and technical delivery, with the objective of providing 'exceptional performance'.

> 'We removed one of the main commercial levers that UK construction is used to using, and that is making money from managing client risk. We were holding the commercial risk, and as a consequence our suppliers had no reason to hide, and we were able to demand complete transparency of plans where we felt it was appropriate to do so. We wanted all the best brains focused on good technical solutions, trying to actively look at managing the cause of potential risks – rather than all the best brains focused on protecting commercial positions and managing the consequences of failure after the

> event. That's a big shift in mindset and approach.' (Matthew Riley, Supply Chain Director, BAA, quoted in Brass, 2008)

Pre-emptive risk management is another tool used on the project. Integrated teams were responsible for identifying the '*root cause*' of each risk in a timely manner, assembling the most appropriate resources and managing the risk as effectively as possible. In some instances proposed solutions were trialled before their implementation to establish their effectiveness.

Acknowledging four levels of control risk, BAA sought to maintain all risks at levels 2 and 3 on the following scale:

1. *The solution exceeds the level of risk, and should be adjusted*
2. *The risk management solution is optimal*
3. *The risk is acceptably managed, but could be improved*
4. *The solution does not manage the risk acceptably* (NA0, 2005)

Established in the project's overall programme, risk control ratings formed the basis of objectives for individual team members.

Integrated teamwork

The T5 Agreement is also a partnering agreement between BAA and its tier-1 suppliers, which required the parties to adopt an integrated teamwork approach and demonstrate commitment and trust to the project. It also, provided a mechanism by which BAA could introduce talented individuals into the project.

BAA organised T5 delivery activities around integrated teams that were assembled to deliver customer products, not on the underlying supply organisations. Customer products were defined as operational facilities, not construction outcomes:

> 'we are creating an operating terminal, not a set of buildings'. (Quoted in NAO, 2005)

Examples include the transit of a passenger through the terminal (parking the car, checking in, passing through security, sitting in the lounge, shopping, embarking) or providing services for BA (offices, baggage systems, logistics centres, etc.).

The agreement enabled BAA to establish 'integrated teams' composed of the most appropriate personnel from BAA and their tier-1 suppliers, irrespective of who that might be, to meet the precise requirements of each of the numerous projects that comprised T5. BAA's vision was that all the suppliers involved on the project would function as a 'virtual company' that would work cooperatively to a common set of objectives and with a shared sense of values. Personnel from the tier-1 supplier organisations were expected to forego their corporate allegiances and share their knowledge, experience and information with individuals from other organisations and professions (some of whom may have formally been competitors of their parent company or lower-tier subcontractors). For the length of their task, individual team members were 'rebadged' and fully assimilated under the banner of T5 to ensure integration and foster a homogeneous culture on the project.

Each integrated team was responsible for its individual task and for working to the overall T5 project objectives.

The approach was designed to:

- Create highly effective 'integrated teams' that were committed to achieving T5 targets
- Place the interests of the T5 project above those of the individual suppliers
- Enable teams of experts to work collaboratively to:
  - identify potential problems and issues before design solutions were finalised and implementation stage commenced
  - find solutions as and when problems arose

  - ensure that the solutions derived were the most appropriate for the project as opposed to the most suitable for the suppliers
- Promote trust and reliance within the integrated team

Consequently, the integrated teams were able to add value, benefiting from additional time spent on creating innovative and safe solutions within the project's time, quality, cost and safety targets.

Collaborative project software

To facilitate open and timely communication within the integrated project teams, T5 utilised a collaborative project software package that provided access to the programme, scope of work and risk reports. The system was also credited with helping to reduce misunderstanding and delays. Additionally, the 'project flow' software system was used to compile and manage the demand for material and components, ensuring that they were delivered 'just-in-time'.

Culture change

Culture change was regarded as an essential, continuous facet of the project; the T5 Agreement specified the culture to be adopted on the project, allegedly the first time that this had been attempted within a construction contract. The behaviours BAA sought to engender on T5 comprised:

- ***Commitment***: *being seen to do what we say, challenging targets, value-focused, managing the supply chain and enabling processes and solutions*
- ***Teamwork***: *fully inclusive teams, integrated, co-located, share responsibility, share success, outcome (not problem) focused, helping and supporting, enabling individuals, managers as enablers, fully empowered, trained, celebrate achievements, and leading by example*
- ***Trust***: *partnering, cooperation, meaning what we say, respect, selected on merit, and treating our workforce as customers* (Ferroussat, 2007)

By relieving suppliers of the risk inherent in the project, BAA sought to avoid the development of an unproductive culture of blame and confrontation when the inevitable problems occurred on the project. Consequently, the T5 Agreement put a premium on collaborative working and the delivery of solutions and results.

BAA recognised that the T5 Agreement was a significant departure from the accepted norms within the sector, and in particular that UK suppliers were not generally familiar with the requirements of working in an open collaborative manner. The client, therefore, had to ensure that suppliers recognised the implications of the contract and that corporate objectives were aligned. To facilitate this T5 engaged an Organisational Effectiveness Director with responsibility for approximately 30 Change Managers. This team provided training and support in collaboration techniques and teamwork, and also supported teams facing particular challenges.

The culture of transparency and trust was established over the first 18 months to two years of the project; although BAA subsequently acknowledged that this requirement of the agreement had been particularly challenging for some suppliers:

> 'Some people get it straight away but some play the old game of putting all their intellectual effort into pushing margins and avoiding risks. They need to realise that the T5 Agreement fundamentally shifts the boundaries of how you handle the complexity of a project of this kind.' (Tony Douglas, Managing Director T5, BAA, quoted in Simpkins, 2005)

In practice, BAA had persistently to reinforce its philosophy of collaboration across the workforce due to changes in the supplier base over the evolution of the project. Scenario-based assessments were used to test the expected behaviours of individual team members.

Similarly, second- and lower-tier suppliers within the project supply chain were expected to adopt collaborative working methods.

Payment terms

The T5 Agreement is a cost-reimbursable form of contract, based upon an assumed common understanding of cost. Operating an 'open-book' policy, BAA had access to its tier-1 suppliers' books, allowing it to develop a comprehensive insight into its suppliers cost and organisational structures. This approach was at odds with the public sector norm, where firms generally used 'commercial confidentiality' as a means of withholding information. However, BAA believed that this entitlement was reasonable, as it had retained the risk liability on the project. The tier-1 suppliers received the prime cost of work undertaken in accordance with the CMA plus agreed overhead costs and their 'ring-fenced' profit margin.

The integrated project teams (which included BAA project managers) set cost targets, based on independently substantiated cost data derived from other projects, for the numerous work packages, then utilised a variety of payment methods to reimburse its supplier. Generally, however, suppliers were remunerated either on the substantiation of costs incurred (e.g. by means of receipts and wage slips, etc.) or via pre-agreed rates for specified work.

The agreement incorporated a gain-share provision; if the out-turn cost of a work package was less than the target cost then the difference (cost saving) was shared between the organisations that formed the integrated team. It also rewarded teams if they completed work packages within their time targets and achieved project milestones. These terms incentivised teams to collaborate and innovate, plus they were the only mechanism under the contract by which a supplier could improve its profitability.

BAA was fully aware of the inherent risk created by this approach: principally, that cost targets would be set too high, especially where innovative or complex solutions were proposed and for which there were limited or no relevant benchmark cost data. To mitigate this risk BAA:

- Used independent consultants to prepare detailed 'bottom-up' cost studies
- Maintained a competitive element within the framework by engaging more than one tier-1 supplier for a particular specialism or capability
- Retained the right to delay its 'approval to proceed'

Insurance

As with other elements of the T5 Agreement, BAA adopted an innovative approach to project insurance. It negotiated and purchased a single-premium, project-wide insurance policy from Swiss Re that covered BAA and the T5 supply chain for project risks, public liability, professional indemnity and marine cargo. Not only did the approach reduce the overall cost of insuring the project by removing waste and duplication within the supply chain, it also released value and supported the no-blame policy inherent in the agreement. Importantly, it placed the responsibility for risk with the party best able to manage it: the client BAA. As a result, BAA broke the cycle of risk transfer, electing to manage the project environment itself. The approach is set to become an exemplar for future construction projects, as insurers are enthusiastic for other clients to adopt a similar approach.

Incentive fund

Acknowledging profit to be a primary means of incentivising suppliers, BAA incorporated incentive payments to suppliers (integrated teams) that achieved exceptional performance within the T5 Agreement. Substandard performance was only penalised in terms of loss of profitability.

> 'By doing that you take away negativity, allow space for innovation and create any opportunity for people to perform at levels they haven't been allowed to before.' (Matthew Riley, Commercial Director T5, BAA, quoted in Mylius, 2005)

Incentive funds, therefore, were established for each team (derived from the savings made by BAA from retaining and insuring project risk themselves). The fund was designed to:

- Encourage teams to:
  - focus on risk management
  - collaborate, thereby generating the most effective means of achieving their target(s) and hopefully surpass the agreed plan
- Reward teams for completing tasks on time and under budget, and achieving the project's quality, safety and environmental targets
- Increase suppliers' profits provided they performed and responded to non-performance risks

As mentioned earlier, targets were established by integrated teams: therefore, incentive payments were shared equally among the tier-1 suppliers included within the team. Likewise, if the team failed in achieving their targets, then the incentive fund was used to make good any shortfall.

Performance requirements

Despite accepting the risk and advocating a no-blame culture, BAA did not provide suppliers with an 'open cheque book'. The T5 Agreement was predicated on 'best practice', which BAA verified by market-place benchmarking.

> 'The T5 agreement took technical competence largely for granted, and focused as much on the behavioural competence of both companies and people, which again was a unique feature.' (Matthew Riley, Supply Chain Director, BAA, quoted in Brass, 2008)

All tier-1 suppliers were, therefore, required to provide services to the highest standards possible as a minimum standard, although BAA later acknowledged that in practice this had not always been easy to define.

The agreement was viewed as enabler to deliver BAA's aspiration for their supply chain – 'exceptional performance' – which defined as follows:

> 'Exceptional performance is world class when it is better than anyone else has achieved so far.' (The T5 Agreement)

It also empowered the integrated team to evaluate, recommend and make decisions based on whole-life cost:

> 'Our experience all too often shows that, in keeping with customer demands, too many components and systems are removed before the end of their design life.'

Performance measurement

In order to capitalise on opportunities, frequent data collection and pre-emptive risk management were essential. Therefore, to meet the challenge of ongoing performance measurement, BAA undertook monthly performance management assessments and instigated procedures and controls to ensure transparency within the supply chain.

Remedies for poor performance

As mentioned earlier, the T5 Agreement advocated a 'no-blame' culture, enabling problems to be quickly identified and resolved as project success was in the interests of all parties. Under the agreement, for inadequate performance – for example, defective work – the supplier was required to perform the work again but no blame was incurred. However:

'If the [work] has to be done a second time, then the team will pay the cost with no profit margin. If it had to be redone a third time, then the cost would be down to the particular contractor.' (Matthew Riley, Commercial Director T5, BAA, quoted in Wolmar, 2005)

The cost of resolving poor performance, errors or defects was deducted from the incentive fund.

BAA implemented this approach as conventional liabilities – for instance, negligence – and defective workmanship would have been very difficult to establish on a project that utilised integrated teams.

Change management

BAA acknowledged that variations to both design and construction solutions would be inevitable due to the:

- Complexity of the project
- Rate of technological change
- Project's lengthy construction period (BAA's operational requirements for the terminal were liable amendment)

T5 utilised a five-stage approvals process, overseen by an '*intelligent client*' team, derived from the varying degree of risk encountered in the course of developing each project. A comparatively efficient decision-making process, BAA's practice echoed its integrative approach to design, procurement, logistics management and construction. The stages comprise:

A. **Project registration**: the identification and registration of an issue to be resolved
B. **Business case**: following on from Stage A – the development of the business case for the proposed solution including its cost and timing implications accurate to ±30%
C. **Design brief, or concept completion**: the preparation of the design brief for the selected solution including its cost and timing implications accurate to ±15%
D. **Approval to proceed**: whereupon the production design (solution) is required to be 80% complete
E. **Handover to operations**: handover of the solution for implementation to the relevant team (NA0, 2005)

A key element of the procedure was BAA's willingness to proceed to the next stage of the process before production designs had been finalised. Similarly, empowering the '*intelligent client*' team to make decisions based on both detailed information and experience enabled progress to be expedited.

Disputes

As a result of BAA selecting a cost-reimbursable contract, which incorporated pre-emptive risk management and integrated teams, and promoted a non-adversarial approach and no-blame culture, there have been no reported payment disputes on the project.

Asset supply

Construction work (or 'assembly' as termed on T5) commenced in September 2002.

The scale of the project was immense: at 396 m × 176 m × 40 m T5 is the UK's largest freestanding building - its construction taking 20 000 operatives 37 million man hours (overall more than 60 000 individuals were involved in the project working a total of 100 million man hours). In terms of the quantities of materials used, the project incorporated 1.2 million m^3 of concrete, 90 000 tonnes of structural steelwork, 150 000 tonnes of steel reinforcement, 30 000 m^2 of glass and 2300 km of cabling.

At its peak T5 employed approximately 6500 operatives on site, and BAA's average daily spend on the project was £3 million.

Logistics management

Constraints imposed by the restricted access to and working space within the site, prompted BAA to adopt an innovative master logistics plan. Based on its experience of operating the Heathrow Consolidation Centre (HCC), located adjacent to the airport, the strategy was credited with significantly reducing the impact of construction traffic (noise, pollution and congestion) on the local area. Two consolidation logistic centres were established to serve T5:

1. **Colnbrook Logistics Centre (CLC)**: this incorporated a railhead: the key route for delivery to and removal from the site of bulk items. CLC was also used for the prefabrication and assembly of components
2. **Heathrow South Logistics Centre**: controls manufactured products

These centres provided a holding area and buffer for material and components close to the site. The level of investment required to create these hubs was deemed appropriate due to the scale of the project and the severity of the access constraints and the restrictions imposed by the planning inquiry. According to BAA:

> 'No construction company would ever have put a logistics solution into their bid, and they would never have costed it.' (Tony Douglas, Managing Director T5, BAA, quoted in Simpkins, 2005)

Also, as a result of the restricted access and storage space associated with the site and the aspiration to increase the project's pace of delivery, BAA utilised a significant amount of off-site pre-assembly and 'just-in-time' delivery:

- **Pre-assembly of components off-site**: the assembly of T5 incorporated a significant amount of prefabrication and manufacturing in off-site factory-controlled environments. For example, prefabricated steel reinforcement sections for in-situ concrete works were pre-assembled in a purpose-built industrial building at the CLC. These were then delivered to site as required, off-loaded from the delivery vehicle and craned into position.

  > 'Rather than taking a week to build a column base we can do it in six hours on site... We've moved from 20% pre-assembled reinforcement to 80%.' (Colin Potts, T5 Substructures Production Leader, Laing O'Rourke, quoted in Pearson, 2004)

  Off-site prefabrication of the steel reinforcement had several benefits:

  > 'Factory production improves the quality of assembly, the factory is a safer working environment and production is quicker under factory conditions.' (Stuart Barr, Demand Fulfilment Manager, Laing O'Rourke, quoted in Pearson, 2004)

  Similarly, around 70% of the project's mechanical and electrical (M&E) installations were pre-assembled (compared with an industry average of 20%); AMEC, for example, developed a modular services system.

  Off-site prefabrication and assembly increased production levels on T5, improved site safety and resulted in a higher quality end-product (fewer defects). It also reduced the number of site-based operatives required from approximately 9000 to 6500, which had a positive impact on industrial relations:

  > 'We've modelled each worker population by skill... It's part of mitigating the risk of industrial action. These things are reasonably predictable in terms of dynamics. When the population of any particular trade reaches a critical mass of around 400 people they start to get a bit cocky!' (Tony Douglas, Managing Director T5, BAA, quoted in Simpkins, 2005)

However, BAA initially found that the UK construction sector did not possess the appropriate combination of skills (in prefabrication and off-site assembly) required by a complex project such as T5:

> 'We've had to turn to the offshore construction industry for the discipline and organisation to build things away from the T5 site.' (Tony Douglas, Managing Director T5, BAA, quoted in Simpkins, 2005)

- **'Just-in-time' delivery**: the adage 'deliver today what will be installed tomorrow' was embraced on T5. BAA utilised 'demand fulfilment' software (appropriated from the manufacturing and retail sectors), whereby contractors could 'pull' supplies as required on a 'just-in-time' basis. The technique required an integrated team to schedule its procurement needs six weeks in advance. These were initially delivered to the consolidation centres, which retained only a one-day reserve. Materials required by a particular team, enough for one day's work, were entered into the system the day before they were needed; they were then assembled into packages and dispatched overnight to the relevant location on site. Site deliveries through the sole access point were based on a 30-second cycle. In addition to delivering efficiency savings, the approach minimised waste, storage costs and the environmental impact on the surrounding area, and it also created a safer working environment

As part of its risk management approach, T5 also utilised cutting-edge technology, such as 3D computer modelling. A digital model of the terminal was created, enabling every component to be tested and coordinated, thereby reducing the likelihood of costly mistakes being made.

BAA's logistics and procurement methodology also led to a significant reduction in the levels of wastage on the site:

> 'Sir John Egan, in his report 'Rethinking Construction', put typical levels of waste at 30 per cent... I'm not prepared to put up with that! The challenge and opportunity is how you deal with waste and eliminate it.' (Tony Douglas, Managing Director T5, BAA, quoted in Simpkins, 2005)

T5 buy clubs

The collaborative working arrangements adopted on T5 enabled BAA and its tier-1 suppliers to further reduce costs by purchasing equipment, materials and services from tier-2 suppliers in bulk. For certain elements of the project, buy clubs were established, comprising teams of tier-1 suppliers with mutual supply chain, sourcing and logistics needs. They were considered to be instrumental in delivering value for money and minimising waste. Their role was to:

- Engender an appropriate commercial environment
- Decide upon the last possible dates for agreeing a design solution
- Enable planning on T5 to achieve 'exceptional performance' (Constructing Excellence, 2004)

An example is the M&E Buy Club, T5's and perhaps the UK construction industry's prototype. By combining their purchasing power M&E suppliers were able to procure cabling and other goods at a discount:

> '... on a couple of the packages, savings of up to 30% have been achieved.' (Matthew Riley, Commercial Director T5, BAA, quoted in Broughton, 2004)

As further example of the purchasing power of the M&E Buy Club, low-voltage electrical switchgear, which under normal arrangements would have cost around £22 million, was purchased for £15 million.

The ten core team members of the M&E Buy Club comprised: a team leader and a building services integration leader; five tier-1 M&E contractors (AMEC M&E, Balfour Kilpatrick, Crown House Engineering, Hotchkiss Ductwork and Novar Projects: Trend and Gent); and a TECHT team of cost consultants (E C Harris, Mott Green Wall, and Turner and Townsend):

> 'We have five first tier M&E suppliers working in different T5 projects, but essentially buying the same products and services. It seemed natural that we could better manage the purchasing for £600m of work by combining our suppliers' buying arrangements. We could also achieve better commercial terms. This unified approach prevents us from sending conflicting signals to the market and helps avoid competition between the 1st tier suppliers and the 2nd tiers. Standardising suppliers encourages commonality of products between the T5 projects and will bring maintenance benefits once T5 is up and running.' (Maya Jani, Team Leader T5 M&E Buy Club, BAA Supply Chain, quoted in Constructing Excellence, 2004)

BAA's requirement for total transparency in all financial transactions on the project is seen as the main driver of these savings.

Successfully applied to the M&E components of T5, the 'buy club' concept was also employed on the 'fit out' and 'systems' elements:

> 'This has many advantages. Pooled purchasing reduces prices by buying in greater volumes and also by gaining direct access to manufacturers to cut out wholesale mark-up. Then there's the market intelligence you gain by going through the procurement process with others. There are also administrative, logistical, environmental, commonality and quality advantages.' (Michael Puckett, T5 Building Services Integration Leader, BAA (MACE), quoted in Constructing Excellence, 2004)

The procurement process adopted by the buy clubs was designed by procurement experts from BAA, Mace and TECHT.

As mentioned earlier, the construction phase was subdivided into five key stages:

1. **Site preparation and enabling works**: the stage included the largest single-site archaeological dig in the UK; 80000 items were discovered including a 3000 BC hand axe, pots, buckets, flints and cups. Also, in order to divert the two rivers that crossed the site over 8.5 km of concrete river wall was constructed around its perimeter, of which around 5 km was formed using precast concrete sections in order to save time. The phase was completed three weeks ahead of schedule, on time and on budget
2. **Groundworks**: during this stage 9 million m^3 of earth were excavated, but due to poor weather conditions in the winter of 2002–2003 meant that the stage fell 14 weeks behind programme. However, due to a proactive approach to risk adopted on the project the substructure team were able to recover the slippage, handing over to the next team precisely on time
3. **Major structures**: construction of the distinctive waveform roof (the largest single-span roof in Europe) was a major component of this phase, due to its complexity and importance to the project as a whole. Its maximum height of 39 m, determined by the airport's radar system, was a challenge to both its design and construction; the height constraint ruled out the use of cranes. To overcome the problem, the roof's steel structure was designed to facilitate off-site fabrication; five 117 m × 56 m sections and one 117 m × 18 m section were delivered to site in 30 m sections, assembled on the ground and hoisted into position using strand jacking. As the majority of the operation took place at ground level, steel erectors spent less time at high levels, which made a positive impact on site safety.

   As part of the proactive risk management strategy adopted on the project, BAA commissioned the trial erection of one section of the roof structure in Yorkshire to assess the viability of the solution. According to BAA, the test cost £2.4 million. However, it:

   > '... saved three months work on the Heathrow site and taught us all kinds of things about tolerances and sequencing.' (Mike Forster, Development and Design Director. BAA, quoted in Wolmar, 2005)

   In April 2004, the first 2500 tonne steel roof section was successfully hoisted into place, taking a specialist team of 12 engineers 10 hours to complete. Overall, the erection of this section took a month longer than predicted; however, the team's response to this has been cited as an example of T5's problem-solving approach in action. As one of the suppliers involved in the exercise states:

   > 'Under a normal form of contract, there would have been an adversarial environment where you would end up defending yourself, but under the T5 agreement we all had the time and incentive to work together to find a solution.' (Peter Emerson, Managing Director, Severfield-Rowen, quoted in Broughton, 2004)

   To complete the roof a further five lifts were required.

   Developed by tier-1 supplier Hathaway Roofing Ltd, the roofs of both the main and satellite terminal buildings were clad using prefabricated roofing cassettes that incorporated the roof decking and both thermal and acoustic insulation. Each cassette measured 3 m × 6 m, which enabled delivery lorries to carry 10 cassettes per consignment. Over 3000 cassettes were installed on the main terminal building with a further 950 used on T5B. However, due to the originality of the approach, 12 prototype cassettes were manufactured and tested before the design solution was approved. By using pre-assembled components the building was watertight earlier, fewer roofers were required on site, and the associated reduction in man-hours worked at high level led to an improvement in health and safety standards. Moreover, the decision to construct the frame and envelope of the terminal building independently of its interior enabled the internal works to be undertaken in a waterproof environment, mitigating the risk of delay due to inclement weather and improving the quality of the works.

The main terminal building was declared weathertight in December 2005.

A prefabrication approach was also adopted for the construction of the air traffic control tower, which was assembled at the edge of the site, moved into position and raised in March 2005. The process enabled the structure to be constructed in a safer, more controlled environment with minor impact on the operation of the airport. However:

> 'Moving the top-heavy, 900t, 32m-high, cone-shape top cab section of the tower nearly 2km into position provided some unique engineering challenges. The solution was a specially designed temporary steel frame attached to the base to lower its centre of gravity and provide the stability needed to transport the tower safely. The 144-wheel flatbeds provided sufficient spread of the huge load to enable the team to move the tower across the airfield with no impact or damage.' (Nick Featherstone, Project Leader, Pascall & Watson, quoted in Design Build Network)

The tower was completed ahead of schedule and within budget

4. **Fit out**: included the installation of T5's baggage-handling system, one of the world's most complex and largest systems, incorporating 17km of conveyor belt and 27 X-ray machines
5. **Implementation of operational readiness**: T5 began a six-month programme of 'operational readiness' in September 2007 to enable the terminal to enter into service on 27 March 2008. During this phase of the project, all the processes and systems incorporated into T5 (the main terminal building and its first satellite building) were tested to ensure readiness for use and full integration and compatibility with Heathrow Airport's existing infrastructure. Over 15000 volunteers took part in 70 trials that checked every aspect of the terminal

Safety

Despite having a workforce, at its peak, of over 8000, T5's safety record exceeded the industry norm by a factor of four. During the assembly phase the site recorded one million consecutive man-hours without a reportable accident 15 times, and an impressive two million consecutive hours on three separate occasions. BAA attributed this achievement to the introduction of a safety programme that was aimed at changing site culture – Incident and Injury Free (IIF) – which everyone employed on the project had to undertake.

Project completion

On 1 September 2005, Alistair Darling, the Secretary of State for Transport attended a 'topping-out' ceremony to mark the completion of T5's building structure.

Generally, T5 is considered to have been delivered on time and within budget, but not without '*controversy*'. For example:

- The extension of the Piccadilly Line to provide a link to the new terminal required BAA to cooperate with London Underground. However, setbacks on the Jubilee Line extension caused some commentators to question whether similar delays would occur on T5, leading BAA to declare:

  > 'Any delays will not stop us from opening on time... We are planning for success, but we have mitigation plans in case anybody should let the side down.' (Mick Temple, Managing Director Heathrow Airport, BAA, quoted in Simpkins, 2005)

- BAA acknowledged that in October 2007 the M&E installations on the project were exceeding their target cost by approximately £56,000 an hour, prompting the instigation of efficiency savings. Similarly, the client was required to inject a further £90 million into the project to enable additional 'fit-out' operatives to be engaged

- Finally, in February 2008, BAA applied to the CAA for a suspension of Heathrow's regular service obligations in the period following T5's opening. This, allegedly, led BA to query the service levels they might expect at the new terminal

The T5 programme did not allow for 'soft commissioning' following the completion on the works; instead 'snagging issues' were resolved during the delivery phase:

> 'The complexity of integrating into the busiest international airport in the world is such that we are building plenty of room into the schedule to make sure that we do a world-class job. Most new airports take a few years after they open to work up to capacity. T5 will open immediately with 30 million passengers per year.' (Mick Temple, Managing Director, Heathrow Airport, BAA, quoted in Simpkins, 2005)

T5 was officially opened by HM the Queen on 14 March, following which Sir Nigel Rudd chairman of BAA optimistically stated:

> 'From every perspective, this is a landmark project and I am proud to think that Terminal 5 has become a model construction project, setting new, higher standards for an industry around the world.' (quoted in Wright 2008)

Although BAA did acknowledge at the time that:

> 'There is a small amount of work still left to do and the final tuning can only be achieved when the building comes under full load. For that we need a hot summer and a cold winter at full capacity. But this is all quite normal for a project of this size and we are confident we have done all we need to do.' (Andrew Wolstenholme, Capital Projects Director, BAA, quoted in Wright 2008)

Similarly and ironically, as it turned out, BA commented that:

> 'There will be some bedding down time but we do not think it will affect the customer experience in any way at all.' (David Noyes, Customer Services Director, BA, quoted in Wright 2008)

T5 became operational, as scheduled, at 4 a.m. on 27 March 2008. However, despite the six-month programme of 'operational readiness', by 2 p.m. BA had been obliged to withdraw 34 flights due to operating issues. These problems were initially attributed to 'teething problems'. During its first day:

- The high-tech baggage-handling system malfunctioned: early operational and logistical difficulties led to passengers, both departing from and arriving at the terminal, having to wait up to two hours to reclaim their luggage
- Workers at T5 had difficulties in accessing their car park (which also had insufficient spaces) and the terminal building, experiencing problems in passing through security
- Passengers were similarly affected: flights were delayed due to the check-in staff's lack of familiarity with T5's systems, resulting in the baggage-handling system becoming overloaded, which exacerbated the situation. Passengers were also critical of the confusing road signage and the payment arrangements at the car park. Additionally, an escalator broke down

During the first week T5 was open, over 200 flights were cancelled and around 20 000 items of luggage were mislaid. Ultimately, more than 300 flights were withdrawn and thousands of items of luggage misplaced. Also, instead of relocating 41 long-haul flights from T4 to T5 on 30 April 2008, BA delayed the move until October of that year.

The debacle of T5's opening, branded a 'disaster' by government ministers, is believed to have cost BA a minimum of £16 million and resulted in the resignations of Gareth Kirkwood, Director of Operations, and David Noyes, Director of Customer Services at BA. A further casualty included Mark Bullock, Chief Executive of Heathrow Airport, who left BAA 'with immediate effect' on 13 May. The incident also prompted the House of Commons Transport Select Committee to instigate an inquiry.

Initially, BA attributed their operational issues to a lack of '*staff familiarisation*'; subsequently, however, they were blamed on:

- **Failure of the baggage-handling system**: it was alleged that BAA had compromised its 'operational readiness' testing of the baggage-handling system, allocating only six months to the task, rather than the nine to 12 months its experts had recommended:

  'The baggage system at T5 required a degree of automation never achieved before. It is a huge, great spaghetti of mechanical systems. We decided we needed a year of operational readiness, doing nothing else but running operations through it.' (An unidentified source quoted in Stewart, 2008c)

- Similarly, Willie Walsh, Chief Executive of BA, attributed T5's calamitous opening to BA compromising its procedures to establish operational readiness following '... *delays in the building of the terminal*'. However, BAA countered this stating:

  'The period in the construction programme for operational readiness was six months, from the day we put a spade in the ground. We have been trialling the building since 17 September, and the baggage system has been operating throughout that time... The building was ready to go. The impact of human factors has created most challenges... Clearly, it's very disappointing we didn't get off to the glorious flying start we anticipated, but the key thing is there was nothing wrong with the building. We should not let this opening detract from the fact that we have raised the bar for construction and engineering in this country.' (Rob Stewart, Construction Director, BAA Capital Projects, quoted in Stewart 2008c)

- **Communication failures between BAA and BA**: speaking before a House of Commons Transport Select Committee on 10 July 2008, Colin Matthews, Chief Executive of BAA, acknowledged that it had been the failure of BAA and BA to work together effectively that had turned the opening of T5 from '... *what should have been an occasion of national pride into one of national humiliation*'. Similarly Unite, the trade union that represents employees at T5, criticised BAA for not adequately consulting its workforce before opening the terminal and for failing to respond appropriately to their concerns about the technology installed in it

As with the Wembley case study, the media were quick to exploit BAA and BA's catastrophe: describing the opening of T5 as 'shambolic', 'chaotic', a 'debacle', a 'fiasco', a 'shambles' and a 'national embarrassment'. However, unlike Wembley, T5 was delivered on time and to budget; despite this, the publicity generated following the opening of the terminal initially overshadowed the construction team's achievements. This was unfortunate, as it was widely anticipated that T5 would transform the public's perception of the UK construction sector.

In August 2008, BA launched a marketing campaign 'T5 is Working':

> 'The aim of the campaign is to communicate to the travelling public in an open and factual way that Terminal 5 is now working well.' (Katherine Whitton, BA's General Manager for Marketing Communications)

Further, BAA acknowledged that lessons had been learnt from the incident and that during its subsequent developments, valued at £4.54 billion in the period up to 2013, it would not make the same mistakes.

Reasons for success

Regardless of the difficulties encountered following the terminal's opening, the T5 delivery programme was a major success: realising its key design and construction objectives within budget and on time. Primarily, this achievement is generally considered to have been the result of BAA retaining the risk and undertaking a pivotal role in the management of the project. Additionally, the T5 delivery methodology introduced several innovative approaches and examples of best practice. These included, for example:

- **The contract form**: instead of requiring its suppliers to enter into fixed-price agreements (a more common approach), BAA utilised a bespoke cost-reimbursable contract, which ring-fenced the supplier's profit and incorporated a gain-share arrangement. An impartial agreement, the contract facilitated the development of relationships and behaviours:

  > 'Our contract actually states that we want different behaviours... That gives me a hook for what I do, which is incredibly well-grounded in the business. Basically we behave with our suppliers as we would want them to behave with us, and create a problem-solving environment rather than a finger-pointing one.' (Sharon Doherty, Organisational Effectiveness Director, BAA, quoted in Saunders, 2004)

  > and focused on the proactive management or risk.

  > 'T5 people always talk about the contract, it's what everything we do here is based on... Usually in this business, the contract tells you what to do when things go wrong, but our contract tells you what to do to make things go right.' (Andrew Wolstenholme, Construction Director T5, BAA, quoted in Saunders, 2004)

  Key features of this agreement considered to have supported the delivery of the project include:
  - a novel approach to risk allocation and management: the retention of risk by BAA, rather than transferring it through the supply chain to parties potentially unable to carry it, enabled project management to become a tool of risk management. As a result payments to suppliers were lower and the parties were free to concentrate on problem solving. The approach is recommended for use on complex projects where the client is unable to identify the inherent risks at an early stage
  - the procurement of a project-wide insurance policy by BAA, which reduced costs and prevented needless duplication
  - a commitment to partnering and integrated teamworking: this included the co-location of staff, open-book accounting and access to information held by all the parties. The approach led to a trusting and open working environment, reduced conflict, and enabled resources to be directed to project delivery. In particular it allowed the client to assemble teams to deliver T5 sub-projects based on the best combination of capabilities available within the tier-1 suppliers.

    > '... by integrating the design-and-build processes with our requirements, we avoid that adversarial system. If you look at what goes wrong with big construction projects, more than half the problems are people issues.' (Andrew Wolstenholme, Construction Director T5, BAA, quoted in Saunders, 2004)

- **Smart logistics and procurement strategies**: these included:
  - a wide-ranging logistics strategy that incorporated the establishment of two consolidation centres, the use of a dedicated railhead to deliver materials to site in bulk, and the application of 'demand fulfilment software' and 'just-in-time' delivery techniques
  - the extensive use of pre-assembly and off-site manufacture, prefabrication and testing led to a significant increase in productivity (of the order of 10–15%), and the use of off-site factories enabled BAA to circumvent regional labour shortages. Further significant cost savings were gained by undertaking off-site testing off the critical path, which also enabled the project to stay on programme
  - the creation of buy clubs enabled BAA and its tier-1 suppliers to reduce their costs by purchasing products and services from tier-2 suppliers in bulk. They were considered to be a key factor in delivering value for money and reducing waste
- **A single virtual project model**: this was supported by innovative IT software, and was used to pre-plan design and construction solutions. As BAA acknowledged:

  > 'At the outset we were really pushing the technology to its limits but as we have rolled out T5, software solutions capable of supporting these 3D design methodologies have increasingly come on line.' (Peter Rebbeck, former Technology Solutions Manager, BAA, quoted in Garrett, 2008)

- **The adoption of innovative design technologies**: this helped deliver a cheaper, faster and safer solution. As a BAA representative recognised:

  'If we had not used such an approach we would not have built the terminal as quickly and efficiently as we did.' (Dr Chris Millard, Head of Engineering for the T5 Programme and Head of Technology for BAA Capital Projects, quoted in Garrett, 2008)

- **Addressing the welfare of its workforce**: acknowledging that for the project to be delivered successfully, a culture centred on health and safety and respect for individuals would be essential, T5 established unparalleled levels of training, welfare, pay and safety for its entire workforce. BAA introduced an intensive induction and training programme, heavily focused on site safety. For example, it established, in conjunction with the CITB, an on-site health and safety test centre – a UK first. A condition of employment on T5, all employees had to successfully complete the CSCS test within three months. A further UK first was the provision of an on-site occupational health facility. Initiated to avoid overloading the local health services, the facility carried out thousands of pre-employment health checks and promoted health matters. In terms of salary levels, skilled operatives could earn in the region of £55,000 a year, provided they met their targets. (T5 also offered exceptionally high working conditions and facilities for its workforce.) These measures resulted in a relatively strike-free project, and also, in relation to site safety, they proved to be very successful; T5 had an exceptionally good safety record, particularly for a project of its scale, with no fatalities. Additionally, a major factor associated with the project's success, according to BAA, has been its use of highly skilled personnel

According to BAA, the approach saved it £1 billion, and it provided a platform for the tier-1 suppliers to demonstrate a transformational shift in their performance, with the potential of creating competitive advantage. Moreover, to date, no party involved in the project has resorted to litigation, in itself an important achievement and benefit.

However, while the project was successful – delivered on time and within budget – there have been suggestions that success may have come at the expense of the budget:

> 'Granted, they got great value from the money they spent, but it's still a lot of money. It will be delivered on budget, but the key to it is how high that budget is.' (A cost consultant close to BAA, quoted in Stewart, 2008a)

Examples of the lessons to be drawn from T5 are presented in Box B.3.

Closely examined by the construction sector, and the subject of public sector reviews (see for example: Ernst & Young LLP and Freshfields Bruckhaus Deringer, 2005; NAO, 2005, 2006a, 2006b) , the T5 model has been widely acclaimed as a genuine alternative procurement approach (for example, to PFI and PPP) for the delivery of large-scale infrastructure projects, although in many respects the T5 approach is the direct opposite of PFI particularly in its treatment of risk. Further, certain commentators suggested that the project's 'real legacy' would be its impact on the UK construction industry as a whole.

During the project BAA were keen to publicise the degree of interest shown in the T5 methodology from other industry sectors:

> 'We've had a lot of external interest because the T5 agreement is not only a totally new way of procuring a construction project, but is a whole new way of working... The T5 agreement could certainly be a marketable blueprint for other industries – but we've got the copyright on it.' (Matthew Riley, Commercial Director T5, BAA, quoted in Broughton, 2004)

Post-project outcome

During the course of the project BAA was the focus of a hostile takeover bid from a consortium led by Ferrovial, the Spanish construction and property group. Following months of uncertainty, the consortium purchased BAA on 6 June 2006 for between £10.1 and £10.3 billion, considered to be 50% above its market value. The purchase went ahead in the shadow of an Office of Fair Trading investigation into the

Box B.3 Key lessons learned

- The client always bears and pays for the risk
- Project management is a function of risk and opportunity management
- Contracts should facilitate a risk management approach:
  - recognising and allocating risk in advance
  - assigning risk and its associated management to the party best able to manage it
  - appointing proficient suppliers (partners) and then agreeing to working together to resolve project risk and determine an appropriate response

  The NAO acknowledged BAA's risk management approach to be an innovative solution to a challenging problem. However, the implications of adopting such an approach should not be underestimated:

  'If we say we hold all the risk all the time, it means all key decisions come back to BAA. The question is: were we geared up to do that from day one? The answer is no.' (Matthew Riley, Supply Chain Director, BAA, quoted in Brass, 2008)

- Recompensing suppliers using an 'open book' cost-reimbursable mechanism, coupled with determining their profit level in advance:
  - removes their motivation to indulge in post-award tactics to negotiate or prise more money from a contract at the expense of the client
  - fosters a problem-solving approach
  - underpins collaborative working
- All parties can gain significant benefits through the adoption of a collaborative, integrated working arrangement
- Relational confrontation can be reduced by creating an open and trusting working environment
- Board-level sponsorship and leadership are essential
- Off-site assembly and prefabrication can deliver buildings faster, more safely and with fewer defects
- The appointment of highly experience and exceptional personnel is crucial on major projects
- T5 has trialled a procurement and project management model that other organisations could apply. However, the approach should only be utilised by 'intelligent clients', that is, those who have the appropriate 'in-house' management capability.

  BAA, acknowledged that initially it had misjudged the number of personnel that would be required to manage the project:

  > 'A couple of years ago, we had just 60 people out on the site working with the various contractors to ensure that everything was progressing smoothly. Now we have three times that number, and it may well still not be enough.' (Matthew Riley, Commercial Director T5, BAA, quoted in Wolmar, 2005)

- Public sector procurement could benefit from the introduction of elements of the T5 approach. It has been suggested that if UK government adopted the T5 approach for its construction procurement programme it could annually save several hundreds of millions of pounds and numerous lives each year

However, despite its successful application it is unclear whether or not BAA will use the T5 model in the future:

'A potential downside of holding all the risk all the time and all the decisions coming back to the client is that you start to cloud accountability and get people performing roles they don't necessarily need to do... We're more agile now in our application of risk management. We focus on managing the high-level client risks that we're best placed to manage, and let our suppliers manage the risks that they're best placed to manage. The principles of T5 and active risk management still hold true, but we're making sure we place the right accountability back into the supply chain.' (Matthew Riley, Supply Chain Director, BAA, quoted in Brass, 2008)

Source: Adapted from NAO (2005); Brass (2008)

airport operator's virtual monopoly of passenger flights into and out of the London area (where it has a 92% share of the market) and its Scottish airports (where BAA has an 86% share); the issue was subsequently referred to the Competition Commission.

In order to finance the deal Ferrovial borrowed approximately £9 billion, saddling the company with substantial debt, although, the airport regulator, CAA, had made it clear that it would not sanction an increase in BAA's landing charges to service any debts incurred to finance a takeover. Ferrovial believed efficiencies could be made at BAA by reducing the company's overheads, outsourcing non-core activities and by changing its procurement practices. In line with this policy and to help reduce its debt, BAA sold World Duty Free (WDF) – its duty-free retail business – in March 2008 for £546.6 million. It also agreed a partial sale of its stake in the Airport Property Partnership, raising a further £309 million. Moreover, since the takeover several directors from BAA Capital Projects (Construction and Design Directors), Heathrow Airport (Chief Executive, Managing and Operations Directors) and BAA's board (Chief Executive, Finance and Corporate Affairs Directors), for example, have left the company, prompting the media to comment on the changing business culture within BAA. For example, an ex-BAA employee suggested that:

> 'The approach now is about finding deficiencies in the supply chain and cutting them out... It's a world away from the contracting approach of the [former chief executive John] Egan days right through to T5. I don't think we'll see that again at BAA.' (A former BAA Capital Projects employee quoted in Stewart, 2008a)

While a former BAA director was reported as saying:

> 'BAA should be asking itself what kind of company it is. Is this a company that is investing in the future infrastructure of Britain, or is it a company that is sacrificing its key concerns because of its debt structure?' (Quoted in Stewart, 2008a)

Over the next decade, BAA has a proposed investment programme valued at £9.3 billion, which includes the construction of Heathrow East, the refurbishment of Terminals 3 and 4 at Heathrow and, subject to planning approval, additional runways and terminal buildings at Stansted and Heathrow. Despite the scale of this investment, towards the end of January 2008 approximately 200 redundancies were made from BAA's in-house construction project management division, part of BAA Capital Projects. BAA described this as a process of 'simplification':

> 'If you look at the size of our organisation over the past two or three years, it has crept up to a level where we feel we can downsize... This is not axing 200 people to save money. This is actually getting a simplified process in order to make sure we are fit to deliver a £3.5bn programme at Heathrow.' (Andrew Wolstenholme, Capital Projects Director, BAA, quoted in Stewart, 2008a)

Simultaneously, it was reported that BAA would not be using the T5 Agreement/procurement approach to procure the second satellite building T5c. Rather, the operator appointed the contractor Carillion as 'complex build integrator' to deliver the terminal, under BAA's third-generation framework: 'Value in Partnership' frameworks agreement, on a two-stage design and construct basis. Under the agreement BAA transferred more risk onto its contractor than it had done previously under the T5 Agreement:

> 'BAA wants some accountability from its contractors. I'd be very surprised if it used something like construction management as it would be taking on all the risk again.' (A source close to the project quoted in Lane and Stewart, 2008)

The media interpreted these announcements as an indication that, under Ferrovial ownership, BAA was abandoning its 'groundbreaking' procurement methodology that had successfully delivered T5 and that the organisation was in danger of losing its reputation as the construction sector's most progressive client. BAA responded by stating:

> 'What we're doing with the next generation of frameworks is picking out the learning from T5 and acknowledging that one cap doesn't fit all our projects… The projects will be run very much in a T5 style, and we will handle risk in appropriately different ways… Perhaps BAA got too involved in some areas of integration where we should have stood back and let our first-tier suppliers take on more of that accountability. There are a lot of different ways of getting the same result.' (Andrew Wolstenholme, Capital Projects Director, BAA, quoted in Stewart, 2008a)

Finally, following an investigation into the UK airports operated by BAA, the Competition Commission (CC) concluded that there were 'competition problems' associated with all BAA's airports, which had led to 'adverse consequences' for both passengers and the airlines that used them. According to the CC the 'principal cause' of these problems was the 'common ownership' by BAA and recommended that BAA should sell some its airports. As a consequence BAA sold Gatwick Airport on 3 December 2009 for a reported £1.51 billion, followed by Edinburgh Airport in April 2012 for £807.2 million. Further, in August 2012 BAA decided to sell Stansted following the Competition Appeal Tribunal's judgment of 1 February 2012 that required them to dispose of the airport.

On 15 October 2012 BAA announced that Heathrow, Glasgow, Aberdeen, Southampton and Stansted Airports would operate under their own individual brand names and that the BAA group name would be no longer used.

Acknowledgements

This case study was prepared by David Lowe from the following published sources.

Endnotes

1 Flyvbjerg, B, Bruzelius, N and Rothengatter, W (2003) *Megaprojects and Risk: An Anatomy of Ambition*. Cambridge University Press, Cambridge

2 BAA announced on 15 October 2012 that the BAA group name would be no longer used and that its airports woud operate under their own individual brand names.

3 Now known as the Official Journal of the European Union (OJEU)

References and sources

Ahira, K (2008) Competition Commission wants BAA to sell two London airports. *Building*, 20 August. http://www.building.co.uk/news/competition-commission-wants-baa-to-sell-two-london-airports/3120724.article (accessed 28 December 2012)

AMEC (2006) *How Complex Projects Succeed: Lessons for the 2012 Olympics and Beyond*. AMEC debate, AMEC plc

Anon (1997) A framework for success? *Contract Journal*, 25 June. http://www.contractjournal.com/Articles/1997/06/25/22119/a-framework-for-success.html (accessed 30 January 2009)

Anon (2004a) T5 Supplement: Introduction: why T5 is the future of UK construction. *Building*, 27 May. http://www.building.co.uk/analysis/introduction-why-t5-is-the-future-of-uk-construction/3036113.article (accessed 28 December 2012)

Anon (2004b) T5 Supplement: On site. *Building*, 27 May. http://www.building.co.uk/on-site/3036119.article (accessed 28 December 2012)

Anon (2005) Terminal 5 reimbursable contract eliminates claims. *Chartered Institute of Building*, Englemere, 25 April. http://www.buildingtalk.com/news/tch/tch202.html (accessed 30 January 2009)

Anon (2008) Communication failures blamed for T5 fiasco at Heathrow. *Building*, 10 July. http://www.building.co.uk/story.asp?storycode=3118023 http://www.building.co.uk/news/communication-failures-blamed-for-t5-fiasco-at-heathrow/3118023.article (accessed 28 December 2012)

BAA (2004) *The T5 Agreement (*Fact sheet*)*, BAA. http://www.baa.com/assets/B2CPortal/Static%20Files/agreement.pdf (accessed 30 January 2009)

BAA (2012) *Heathrow Facts And Figures*. BAA Airports Limited, Hounslow. http://www.heathrowairport.com/about-us/facts-and-figures (accessed 25 September 2012)

BAA Terminal 5: Information. http://www.heathrowairport.com/assets/B2CPortal/Static%20Files/T5_Info_packnew.pdf) (accessed 30 January 2009)

BAA website http:/www.baa.com (accessed 28 December 2012)

Booz Allen Hamilton (2007) *Strengthening Airport and Airline Relations*. Presentation at 13th Annual US/Central Europe Airport Issues Conference, Munich, 12 November. www.aaae.org/products/_640_US_Central_Europe_Eurasia_Airport_Issues_Conference/downloads/ (accessed 25 September 2012)

Brady, T, Davies, A, Gann, D and Rush, H (2008) Learning to manage mega projects: the case of BAA and Heathrow Terminal 5. *Project Perspectives*, The Project Management Association Finland (PMAF), 32–39.

Brass, R (2008) Flying in formation. *Supply Management*, 13 March. http://www.supplymanagement.com/analysis/features/2008/flying-in-formation/ (accessed 28 December 2012)

Broughton, T (2004) T5 Supplement: Procurement. *Building*, 27 May. http://www.building.co.uk/analysis/procurement/3036122.article (accessed 28 December 2012)

Chevin, D (2008) That was then, this is now. *Building*, Issue 4. http://www.building.co.uk/analysis/that-was-then-this-is-now/3105173.article (accessed 28 December 2012)

Clark, P (2001a) Industry relief as Terminal 5 gets the go-ahead. *Building*, Issue 47. http://www.building.co.uk/news/industry-relief-as-terminal-5-gets-the-go-ahead/1013867.article (accessed 28 December 2012)

Clark, P (2001b) BAA to run £2bn Terminal 5 project itself. *Building*, Issue 7. http://www.building.co.uk/news/baa-to-run-£2bn-terminal-5-project-itself/1004103.article (accessed 28 December 2012)

Clark P (2001c) BAA to run £2bn Terminal 5 project itself. Building, Issue 7. http://www.building.co.uk/news/baa-to-run-£2bn-terminal-5-project-itself/1004103.article (accessed 28 December 2012)

Collie, B (2003) Terminal 5 Retail Perspective. *BAA Presentation*, 25 March. http://www.heathrowairport.com/assets/B2CPortal/Static%20Files/regulation_terminal5.pdf (accessed 25 September 2012)

Constructing Excellence (2004) *T5 Buy Club: How M&E contractors pool purchasing at Heathrow Terminal 5*. Constructing Excellence Limited. http://www.constructingexcellence.org.uk/pdf/case_studies/t5_buy_club_20040916.pdf (accessed 28 December 2012)

Design Build Network. *Terminal 5 Heathrow Airport*. London, United Kingdom. http://www.designbuild-network.com/projects/terminal-5-heathrow/ (accessed 28 December 2012)

Doherty, S (2008) *Heathrow's Terminal 5: History in the Making*. John Wiley & Sons, Chichester

Egan, J (1998) *Rethinking Construction: The report of the Construction Task Force*. The Stationery Office, London

Ernst & Young LLP and Freshfields Bruckhaus Deringer (2005) *Contract Management Study - Phase One*, Report for the Office of the PPP Arbiter. http://www.ppparbiter.org.uk/files/uploads/m_goodindustrypractice/200611101704_Contract%20Management%20Study%20-%20Phase%20One.PDF (accessed 25 September 2012)

Ferroussat, D (2007) *The Terminal 5 Project –Heathrow.*.BAA Heathrow. www.mod.uk/NR/rdonlyres/568AB0B4-0439-4437-8B75-9045FE189D9E/0/MODJuly2007Slides.pdf (accessed 25 September 2012)

Flyvbjerg, B, Bruzelius, N and Rothengatter, W (2003) *Megaprojects and Risk: An Anatomy of Ambition*. Cambridge University Press, Cambridge

Fullalove, S (2004) NEC helps BAA deliver Heathrow T5. *NEC Users' Group Newsletter*, Issue 30, August 2004. http://www.neccontract.com/news/article.asp?NEWS_ID=512 (accessed 28 December 2012)

Garrett, B (2008) Heathrow Terminal 5. *AEC Magazine*, 23 August. http://www.aecmag.com/index.php?option=com_content&task=view&id=253&Itemid=37 (accessed 28 December 2012)

Latham, M (1994) *Constructing the Team: Final Report Of of The Government/Industry Review of Procurement and Contractual Arrangements in the UK Construction Industry*. HMSO, London

Lane, T and Stewart, D (2008) BAA rejects T5 agreement for T5c. *Building*, 25 January. http://www.building.co.uk/news/baa-rejects-t5-agreement-for-t5c/3104726.article (accessed 28 December 2012)

Leftly, M (2004) T5 Supplement: the captain, speaking. *Building*, 27 May. http://www.building.co.uk/news/qs/the-captain-speaking/3036127.article (accessed 28 December 2012)

McKechnie, S (2004) Terminal 5 Roof, Heathrow Airport, London. *Nytheter om Stålbyggnad*, 12–18. http://www.sbi.se/uploaded/dokument/files/Art_Terminal%205%20Roof%20Heathrow%20Airport%20London.pdf (accessed 28 December 2012)

Mylius, A (2005) A game of two halves. *Supply Management*, 6 October. http://www.supplymanagement.com/analysis/features/2005/a-game-of-two-halves/ (accessed 28 December 2012)

NAO (2005) *Case Studies: Improving Public Services through Better Construction*. The Stationery Office, London. http://www.nao.org.uk/publications/0405/improving_public_services.aspx (accessed 28 December 2012)

NAO (2006a) *Case Study: BAA Terminal 5 Project*. NAO Defence Value for Money. http://www.nao.org.uk/our_work_by_sector/defence/defence_vfm/contracting_practices/idoc.ashx?docid=920dde14-0756-4219-b5eb-455fbe9ce572&version=-1 (accessed 28 December 2012)

NAO (2006b) *Ministry of Defence: Using the Contract to Maximise the Likelihood of Successful Project Outcomes*. The Stationery Office, London. http://www.nao.org.uk//idoc.ashx?docId=07409021-f83f-45ff-9afc-d94fa86fe11a&version=-1 (accessed 28 December 2012)

Pearson, A (2004) Offsite supplement: the big picture. *Building*, October. http://www.building.co.uk/analysis/the-big-picture/3041654.article (accessed 28 December 2012)

Rawlinson, S (2008) Cost model: airport terminals. *Building*, 1 August. http://www.building.co.uk/story.asp?sectioncode=661&storycode=3119327 (accessed 28 December 2012)

Richardson, S (2006) Up to 400 M&E workers face axe at T5. *Building*, 15 December. http://www.building.co.uk/news/up-to-400-me-workers-face-axe-at-t5/3078839.article (accessed 28 December 2012)

Saunders, A (2004) Project management: terminal velocity. *Human Resources*, 1 November.

Simpkins, E (2005) Terminal velocity. *RICS Business*, October, 16–20

Spring, M (2008) Quite a departure. *Building*, 1 February. http://www.building.co.uk/quite-a-departure/3105338.article (accessed 28 December 2012)

Stent, J (2003) *Terminal 5 Programme and Management*. BAA Presentation, 25 March. http://www.heathrowairport.com/assets/B2CPortal/Static%20Files/regulation_terminal5.pdf (accessed 25 September 2012)

Stewart, D (2008a) BAA the economy class client. *Building*, 1 February. http://www.building.co.uk/analysis/baa-the-economy-class-client/3105101.article (accessed 28 December 2012)

Stewart, D (2008b) T5 – where did it all go wrong? *Building*, 28 March. http://www.building.co.uk/analysis/t5-—-where-did-it-all-go-wrong?/3109927.article (accessed 28 December 2012)

Stewart, D (2008c) T5 baggage tests 'half as long as recommended'. *Building*, 4 April. http://www.building.co.uk/news/t5-baggage-tests-'half-as-long-as-recommended'/3110363.article (accessed 28 December 2012)

Stewart, D (2008d) BA delays moving long-haul services to T5. *Building*, 11 April. http://www.building.co.uk/news/ba-delays-moving-long-haul-services-to-t5/3111052.article (accessed 28 December 2012)

Stewart, D (2008e) Two BA directors leave after T5 chaos. *Building*, 15 April. http://www.building.co.uk/news/two-ba-directors-leave-after-t5-chaos/3111187.article (accessed 28 December 2012)

Stewart, D (2008f) Heathrow chief exec leaves BAA after T5 debacle. *Building*, 13 May. http://www.building.co.uk/news/heathrow-chief-exec-leaves-baa-after-t5-debacle/3113358.article (accessed 28 December 2012)

Teather, D (2006) Ferrovial lands BAA with final offer of £10.3bn. *The Guardian*, 7 June. http://www.guardian.co.uk/business/2006/jun/07/theairlineindustry.travelnews (accessed 25 September 2012)

Toms, M (2003) *The Regulatory Review; Now That It's All Over…* BAA Presentation, 25 March. http://www.heathrowairport.com/assets/B2CPortal/Static%20Files/regulation_terminal5.pdf (accessed 25 September 2012)

Topham, G (2012) BAA sells Edinburgh airport for £800 m, The Guardian, 23 April http://www.guardian.co.uk/business/2012/apr/23/baa-sells-edinburgh-airport-gip (accessed 28 December 2012)

Ward, A (2008a) T5 gets off to a faltering start. *Building*, 27 March. http://www.building.co.uk/news/t5-gets-off-to-a-faltering-start/3109874.article (accessed 28 December 2012)

Ward, A (2008b) BAA struggles to get T5 back on track. *Building*, 28 March. http://www.building.co.uk/story.asp?storycode=3109909

Wolmar, C (2005) Project management at Heathrow Terminal 5. *Public Finance*, 22 April. http://www.building.co.uk/news/baa-struggles-to-get-t5-back-on-track/3109909.article (accessed 28 December 2012)

Wolstenholme, A (2003) *Terminal 5 Project Delivery*. BAA Presentation, 25 March. http://www.heathrowairport.com/assets/B2CPortal/Static%20Files/regulation_terminal5.pdf (accessed 25 September 2012)

Wright, E (2008) Queen officially opens T5. *Building*, 14 March. http://www.building.co.uk/news/queen-officially-opens-t5/3108901.article (accessed 28 December 2012)

Appendix A

Table B.4 Quoted sources.

Individual	Company	Role
Stuart Barr	Laing O'Rourke	Demand Fulfilment Manager on T5
Sharon Doherty	BAA	HR and Organisational Effectiveness Director for Heathrow and T5 (2002 – 2007)
Tony Douglas	BAA	Managing Director of T5; Managing Director of Heathrow Airport (March – July 2006); Chief Executive Officer of Heathrow Airport (July 2006 – July 2007)
Peter Emerson	Severfield-Rowen	Managing Director
Nick Featherstone	Pascall & Watson	Project Leader
Mike Forster	BAA	Development and Design Director
Norman Haste	BAA	Project Director T5
Maya Jani	BAA	Team Leader T5 M&E Buy Club – BAA Supply Chain
Alan Lamond	Pascall & Watson	Aviation Director
Chris Millard	BAA	Head of Engineering for the T5 Programme and Head of Technology for BAA Capital Projects
David Noyes	BA	Head of Heathrow (December 2004 – October 2007); Director of Customer Services (October 2007 – April 2008)
Colin Potts	Laing O'Rourke	T5 substructures production leader
Michael Puckett	BAA (MACE)	Building Services Integration Leader T5
Peter Rebbeck	BAA	Former Technology Solutions Manager
Matthew Riley	BAA	Commercial Director T5; Supply Chain Director BAA (2006–2008)
Sir Nigel Rudd	BAA	Chairman (appointed September 2007)
Eryl Smith	BAA	Managing Director of T5 (1997–2002); Business Strategy Director for Heathrow Airport Ltd (2002–2005)
Rob Stewart	BAA	Construction Director – BAA Capital Projects (formerly Commercial Director T5)
Mick Temple	BAA/Heathrow Airport Limited	Managing Director of Heathrow Airport (left BAA November 2006)
Steve West	Pascall & Watson	Project Director
Andrew Wolstenholme	BAA	Project Director, T5 (2002–2007); Director, Capital Projects and T5 Project Director (May 2007 – September 2007); Capital Projects Director on BAA's Executive Committee (September 2007 – October 2008)

Source: BAA website (http://www.baa.com)

Appendix B

Table B.5 Key stakeholders in T5.

BAA plc (BAA)	Owner and operator of Heathrow Airport and six other UK airports. BAA was the client and project manager for Heathrow Terminal 5 (T5)
Heathrow Airport Ltd	The BAA subsidiary responsible for operating the whole Heathrow Airport complex (including T5)
British Airways plc	Sole occupier of T5; it was not contractually involved in the construction of T5. The trouble-free transfer of BA's operations to schedule was a key driver of the completion of T5
Tier-1 suppliers/contractors	
Principle works contractors	Bombardier Transportation (Holdings) USA Inc: track transit design and delivery services EDF Energy (Services) Ltd: high voltage power supply delivery Elliott Group: temporary accommodation/logistics Foster Yeoman Ltd: raw materials supplier Hathaway Roofing Ltd: roofing design and construction delivery services Hotchkiss Ductwork Ltd: ductwork services delivery Kone Escalators Ltd: escalator and passenger conveyor services delivery La Farge Cement UK: raw materials provider Laing O'Rourke Civil Engineering Ltd: civil construction infrastructure and logistics delivery Morgan/Vinci: deep bored tunnels delivery Novar Projects Ltd: design, supply and installation of building management systems NTL Group Ltd: communication infrastructure services delivery NTL UK: communication infrastructure Rowen Structures Ltd: building frame services delivery Rowen Structures Ltd: building structure Rugby Ltd: raw materials provider Schindler Management Ltd: lift services delivery Schmidlin (UK) Ltd: curtain walling service delivery Thyssenkrupp Airport Systems SA: loading bridge and apron services Ultra Electronics Ltd (Ferranti Air Systems Ltd): system integration services Vanderlande Industries Nederland B.V.: baggage systems Warings Contractors Ltd: fit-out contractor
Designers/consultants	Advantage Technical Consulting: software assurance services Alcatel Telecom Ltd: signalling, communications and control services AMEC Building and Facilities Services: building service delivery team AMEC Civil Engineering Ltd Axa Power: fixed ground power services delivery Balfour Beatty Rail Projects Ltd: track and tunnel (mechanical and electrical) rail design and delivery Bovis Engineering Ltd: planning supervisors Chapman Taylor: architectural retail design services CSE International Ltd: software assurance services DSSR: services (mechanical and electrical) design and engineering consultants E C Harris Group Ltd: cost consultant services Framework Archaeology JV (Wessex Archaeological and Oxford Archaeology): archaeological services delivery Halcrow Group Ltd: opportunity and risk management/strategic planning and design services

(Continued)

Table B.5 (*cont'd*).

	HOK International Ltd: architectural station design consultant, architectural production and brief development KBR (formerly Brown & Root): design of roadways and highways infrastructure Mace Ltd: production integration Mansell Construction Services Ltd: fit-out contractor and fixed links and nodes Mason Land Surveys Ltd: land surveying services Mott MacDonald Ltd: structural, tunnel and rail consultants Ove Arup & Partners Ltd: structural design consultants Parsons Brinkerhoff Ltd: project management and support services Pascal & Watson Ltd: production architectural consultant Richard Rogers Partnership: lead architectural consultant Taylor Woodrow Construction: project management services TPS Consult Ltd: design consultants for campus and airfield pavements Turner and Townsend Group: cost consultant services Warrington Fire Research Consultants: fire engineering services
Service providers	Compass Services (UK) Ltd: catering Duradimond Healthcare: occupational health service Menzies Aviation Group: bussing

Source: BAA website (http://www.baa.com)

Index

accelerated design-bid-build contracts 304
accountability 341
accounting
  definitions and conventions 19, 133, 143
  financial accounting information 136, 169–70
  management accounting information 41, 136, 150–151, 167, 169–70
  see also financial decision-making
accounting rate of return (ARR) 164
accruals 141–2
Achieving Excellence in Construction (AEC) 264–8, 277, 297, 309, 317
acquisitions 215–18
acumen 23–4, 59
adjustment 114
administration (insolvency) 226
administrative expenses 144
advance payments 191–2
adversarial dispute resolution 200–202
aerospace industry
  project-oriented organisations 7, 12–15
  risk and uncertainty 111–12, 116, 124
  strategy 96–7, 104
aggressive bidding tactics 327
Aircraft Carrier Alliance (ACA) 316
alliancing 25, 315–16
American Bar Association (ABA) 56
American Institute of Certified Public Accountants (AICPA) 55
American Marketing Association (AMA) 57
analogy estimating 324
anchoring 114
annual reports 138–9, 149, 170–171, 246
anti-competitive behaviour 218–20
antitrust legislation 218–22
approximate quantities 324
artificial neural networks (ANN) 354
assessment models 321
asset procurement *see* procurement
assets
  commercial management framework 294, 300
  definitions 140–142
  disposal 365–6
  financial decision-making 140–149, 160–163, 168
  football stadia 390–391, 406–7, 426
  Heathrow T5 programme 475
  specificity 65
Association for Project Management (APM) 8, 25, 28, 57–8, 109
auditors 139, 267
authority-compliance leadership style 84
availability 114
award of contract 293–4

BAE Systems 9, 13–14
balance sheets 133, 140–143
balanced functional/project matrix organisations 9–10
Balfour Beatty plc 142–8
bankruptcy 227
basic earnings per share 144–5
behavioural leadership theories 83, 84
behavioural perspective of risk 113–14, 126
benchmarking 361–2, 425–6
benefits realisation stage gate review 280, 364–5
Best Management Practice Portfolio 246–63
best value 336, 341
best-practice management 239–87
  Achieving Excellence in Construction 264–8, 277, 297, 309, 317
  case studies 267–8
  commercial management framework 271–7, 360–363
  contract management 268–77
  governance 242–6
  internal controls 243–6
  key performance indicators 240–242
  MoP framework 247, 256–7
  MoR framework 247, 258–60
  MoV framework 247, 260–263
  MSP framework 246, 253–6
  NAO/OGC good practice framework 271–5
  overview 19
  P3M3 framework 247, 257–8
  PRINCE2 framework 246, 247–53
  process improvement 246–7
  procurement 263–8, 357–8
  project success or failure 239–43
  risk management 243–5, 247, 255, 258–60, 266, 363
  support for frameworks 263
  see also OGC Gateway Process

Commercial Management: theory and practice, First Edition. David Lowe.

biases 113–14
bid-rigging 349
bidding
  bid cycle/process 291–2
  bid evaluation 336–40
  bid guarantees 197
  bid price strategy 347–57
  bid strategy/project screening stage gate review 331
  bid submission/adjudication stage gate review 355–6
  bidding models 353–5
  commercial management framework 45–6, 48, 326–7, 335–40, 347–57
  definitions 45–6
  transaction management 45–6, 48
bilateral governance 66
bill of quantities 194
blue ocean strategies 99–100
Boeing 96–7, 104
bonds 197–8
bookkeeping 133
borrowings 141–2
bounded rationality 66
boycotts 219
breach of contract 196–7
break-even analysis 154–6, 171
brief making 297
British Airports Authority Ltd *see* Heathrow T5 programme
British Standards Institution (BSI) 8, 28
BT Global Services 13, 15
budgets and budgeting
  best-practice management 273–4
  financial decision-making 157–60
  football stadia 385, 387, 404–5, 408, 416
  Heathrow T5 programme 463–4
build-own-operate-transfer (BOOT) contracts 305, 322
Building Schools for the Future (BSF) 311
business-to-business (b2b) exchanges
  changing nature of 3
  commercial management framework 291–2
  context for commercial management 7–17
  project-oriented organisations 7–17
business development 37, 40, 45, 50–51
business-to-government (b2g) exchanges 7
business justification stage gate review 279, 298–9, 325
business-level strategy 102
business management 37, 43–4
business transfer 228
buy clubs 477–9
buyers, definitions 24–5
buying and selling 24–6, 47–8

call-off contracts 316–17
capabilities 59–62, 103, 300
capability maturity model 60–62
capital goods 300
capital investment decisions 164–7
Carillion 16–17
cartel conduct 219
cash/cash equivalents 141–2
cash-flow management 160–163
cash-flow statements 133, 146–8
CATWOE mnemonic 396–7
change management
  best-practice management 250
  contracts and contracting 195–6
  Heathrow T5 programme 475
  key change drives 97–8
Chartered Institute of Management Accountants (CIMA) 55, 62–3
Chartered Institute of Marketing (CIM) 57, 62–3
Chartered Institute of Purchasing and Supply (CIPS) 56, 62–3
claims 198–200
classical contracting 66
Cleveland Bridge UK Limited 432–4
coercive power 82
cognitive perspective of risk 114–15, 126
collaboration
  b2b exchanges 3
  commercial management framework 312–15
  contracts and contracting 204–15
collaborative project software 472
Columbia accident 111–12, 116, 124
Combined Code on Corporate Governance 139
commencement of projects 191
Commerce and Industry Group 56
commercial activities
  commercial management framework 293–5, 323–4
  development of the commercial function 59–60
  examples in project-oriented organisations 6–7, 9–17
  purpose of the commercial function 26–7
commercial awareness 23–4, 59
commercial bid evaluation 337
commercial capabilities 59–62
commercial contracts 180
commercial function
  best-practice management 275–7, 360–363
  commercial activities 26–7, 59–60
  development of 58–70
  processes undertaken by 3, 7, 13–16, 36–53
  profit and value 27–9
  purpose of 26–33
  transmission of value through organisations 30–33
commercial function skills framework 275–7
commercial leadership 79–95
  conceptual framework of leadership 91–3
  contingency theories 83, 84–5
  definitions 79
  desirable qualities of leaders 90–91
  distributed leadership 83, 87–8
  implications for commercial management function 93
  key issues in leadership 81–2
  leadership and power 82
  leadership versus management 81–2
  learning 80, 83, 88–90
  nature of leadership studies 81
  overview 18
  style/behavioural theories 83, 84
  trait theories 83–4
  transformational leadership 83, 86–7
  trends in leadership literature 82–91

commercial management, definitions 3–4
commercial management framework 33–53
  best-practice management 271–7, 360–363
  bid cycle/process 291–2
  business development 37, 50–51
  case studies 366–7
  commercial activities 293–5, 323–4
  commercial strategies and tactics 288–377
  contract management 38–9
  corporate/business management role 37, 43–4
  deal creation stage 34, 35–6, 289, 293–4, 332–58
  development of the commercial function 58–70
  execution stage 34–5, 36, 289, 294–5, 359–67
  financial decision-making 37, 41–2
  industry sector reports 290–291
  intent stage 34, 35, 288, 292–3, 296–331
  legal, regulatory and governance 37, 42–3, 56
  negotiation 37, 44–5
  processes undertaken by the commercial function 36–53
  procurement 289–95, 298, 302–25, 332–45
  project and programme management 37, 49–50, 56–8
  relationship management 37, 39–41
  risk and opportunity management 37, 51–2, 57
  transaction management 37, 45–8, 56
  transaction process 289–95
commercial management maturity (CMM) 60–62
commercial managers, definitions 4–5
commercial practice 23–4, 55, 63, 70–71
commissioning 47
commitment 472
commitment management 5–6
communication failures 481
communication and learning 125
communications organisations 7, 13, 15–16
communityship 54
company voluntary arrangement (CVA) 226
compensation events 200
competency 91
competition
  best-practice management 267
  commercial management framework 291, 326, 353
  competitive advantage 27, 96–7, 98–100
  competitive dialogue 333
  competitor-related pricing systems 348
  contracts and contracting 218–22
completion dates 191–3
complex products and systems (CoPS) 300–301
compulsory competitive tendering (CCT) 300, 341
confidentiality agreements 224
conflict management 201
conjunctive event bias 114
consortia 25
construction/infrastructure industry
  best-practice management 240–241, 264–8
  commercial management framework 297, 306–15, 318–22
  contracts and contracting 195–7, 202–15
  financial decision-making 142–8
  project-oriented organisations 7, 16–17
  strategy 101, 104
  see also football stadia; Heathrow T5 programme
construction management 312
consultancy 303, 312
consultant advisers 303
contingency theories 83, 84–5
contract administration 361
contract lifecycle management (CLM) 39, 268
contract management
  best-practice management 268–77, 360–363
  commercial management framework 36, 37, 38–9, 55, 360–363
  contract administration 361
  definitions and components 5–7, 38–9, 174
  overview 20
  performance measurement 361–2
  professional bodies and associations 55
  relationship management 362–3
  risk management 363
contract management framework 39, 271–5, 361
contract notices 342
contract price adjustment 194
contracts and contracting
  best-practice management 269–70, 274–6
  bonds, guarantees and insurances 197–8, 212
  case studies 230–231
  choice of contract 185–8
  claims and compensation 198–200
  classical and neo-classical 66
  collaboration 204–15
  commercial contracts 180
  commercial management framework 31–3, 66, 293–5, 300–301, 303–25, 342–3
  competition and antitrust legislation 218–22
  complexity of contracts 180, 186
  contractual issues 174–81
  definitions 25, 175
  dispute resolution 200–203, 212–13
  employment rights on transfer of undertakings 228–9
  financial decision-making 160–161
  flexibility, clarity and simplicity 204
  football stadia 388–90, 394–5, 405–8, 423–4, 432–7
  freedom of information 225
  Heathrow T5 programme 465–70, 482, 484
  insolvency 225–8
  integrated 31–3
  intellectual property 209, 222–4
  labels and descriptions 178
  legal interpretation of contracts 178–9
  legal issues 19, 173–235
  letters of intent 177–8
  outsourcing 300–301, 323

overview 19
procurement/acquisition regulations 215–18
project-oriented organisations 6
relational 66–7, 68
remedies for breach of contract 196–7
roles, relationships and responsibilities 188–90
standard and model conditions of contract 178
strategy and type of contract 181–8
terms of contract 180
time, payment and change provisions 191–6, 231
transaction cost economics 66
understanding a contract 176
see also procurement; subcontracting
contribution to fixed costs 154
convertible payment terms 195
copyright 223
core value-proposition 104, 264
corporate culture 466, 472–3
corporate governance 139–40, 243–6, 422
corporate management 37, 43–4
corporate-level strategy 100–102
cost-based contracts 183–4, 194, 231
cost behaviour 151–4
cost determination 151–4
cost estimating 324, 350–351
cost models 324
cost-plus contracts 183, 194–5
cost prediction 156
cost-reimbursable contracts 184, 194, 482, 484
cost-related pricing systems 348
cost of sales 144–5
cost-volume-profit analysis 154–6, 171
country-club management style 84
creditors 141
critical reflectivity 80
cultural factors
commercial management framework 70
Heathrow T5 466, 472–3
risk and uncertainty 115–16
current assets/liabilities 141–2
customer relationship management (CRM) 40
deal creation
award 357
best-practice management 269
bid evaluation 336–40
bid price strategy 347–57
bid submission/adjudication stage gate review 355–6
commercial management framework 34–6, 289, 293–4, 332–58
investment decision stage gate review 345
multicriteria selection 337–9
overview 19
pre-award negotiations 343–5, 357
pre-qualification 334
procurement 332–45
proposition development 346–57
public sector procurement 341–2
supplier selection 335–40
debt relief orders (DRO) 227
decision to bid (d2b) 327–31
d2b models 330
d2b output 329
d2b process 327–8
factors influencing d2b decision 328–9
decision trees 123–4
defence commercial function (DCF) competencies 275–6
defence organisations 7, 12–15
defensive bidding tactics 327
delivery strategy stage gate review 279, 324–5
delivery system 104
demand-side transmission of value 30
design-bid-build contracts 303, 319, 322
design-build contracts 181, 187, 305, 311–12, 319, 320, 322
design-fund-build-operate (DFBO) contracts 305
design-manage contracts 305
design quality 267
diluted earnings per share 144–5
direct costs 151–2
directors 138
disclosure 139
discounted cash-flow method 164
disjunctive event bias 114
dispersed leadership 83, 87–8
dispute resolution
commercial management framework 363
contracts and contracting 200–203, 212–13
football stadia 391–3
Heathrow T5 programme 475
distributed leadership 83, 87–8
distribution costs 144
diversification 100–102
dividends 144
domain management 100
domain selection 100

early completion 192–3, 211
early-stage cost estimating 324
eAuctions 355
economy 29
effectiveness 29
efficiency 29
elemental estimating 324
Emirates Stadium, London 20, 399–409
asset supply 406–7
budget and financing 404–5, 408
commercial management framework 67, 304
development programme and milestones 385–7, 404
key features 381, 403
post-project outcomes 408–9
procurement 405–6
project completion 407
project scope and stadium design 402–3
relationship management 408
requirement identification 400
requirement specification 400
risk and uncertainty 127
site selection and constraints 400–402
transaction cost economics 67
emotional intelligence (EI) 83
employment rights 228–9
end users 25
endorsements 214
enterprise contract management (ECM) 39, 268
enterprise risk management (ERM) 126
environment 69–70
e-procurement 318–19

equity 140, 141–3
essential contract theory 68, 69
European Institute of Risk Management (EIRM) 57
exceptional items 144
exclusionary provisions 219
exclusive dealing 219
execution stage 34–5, 36, 289, 294–5, 359–67
  asset disposal/service termination 365–6
  benefits realisation stage gate review 364–5
  contract administration 361
  contract management 360–363
  implementing, delivering and fulfilling commitments 36, 359, 389
  maintenance of assets/ services 359–60
  operations review 364–5
  performance measurement 361–2
  post-award delivery 359–60
  post-contract stage gate review 364
  readiness for service stage gate review 364
  relationship management 362–3
  risk management 363
exit strategies 161–2
expansion 147, 169, 170
expectation management 437–8
expected utility (EU) theory 110, 112–13
expert power 82
express terms 180
external factors 186–7

Federal Acquisition Regulations System (FARS) 218
Ferrovial 101, 104
final certificate regimes 203
finance costs 144
financial accounting 136, 169–70
Financial Accounting Standards Board (FASB) 137
financial analysis 149
financial bid evaluation 337
financial decision-making 132–72
  annual reports 138–9, 149, 170–171
  balance sheets 133, 140–143
  best-practice management 280
  capital investment decisions 164–7
  cash-flow management 160–163
  cash-flow statements 133, 146–8
  commercial management framework 37, 41–2, 55
  cost determination and cost behaviour 151–4
  cost-volume-profit analysis 154–6, 171
  desirable qualities of financial information 135
  financial accounting information 136, 169–70
  financial analysis 149
  income statements 133, 143–6
  management accounting information 41, 136, 150–151, 167, 169–70
  nature of financial reporting environment 136–40
  nature and purpose of financial information 133–5
  overview 18–19, 132–3
  performance measurement 157–60, 167
  presentation of financial statements 133, 140–149
  professional bodies and associations 55
  relevance and reliability 135, 148, 168, 169
  user groups and information needs 134
financial environment 19
financial information 19, 41, 132–6, 150, 167
financial management 41–2
Financial Reporting Council (FRC) 139, 244–6
financial statements 133, 140–149
fitness for purpose 298
Five Forces model (Porter) 98
fixed costs 151, 154–6, 171
fixed-price contracts 183, 184–5, 191, 468
flexible budgets 157–9
football stadia 381–454
  asset supply 390–391, 406–7, 426
  budget and financing 385, 387, 404–5, 408, 416–20
  commercial management framework 67, 304
  development programme and milestones 385–7, 404, 415–18, 426–8, 444–54
  Emirates Stadium, London 20, 67, 127, 304, 381, 399–409
  key feature comparison 381
  key stakeholders 443–4
  media briefing by WNSL, Multiplex and the FA 444–54
  Millennium Stadium, Cardiff 20, 67, 127, 304, 381, 382–98
  overview 20
  post-project outcomes 397, 408–9, 438
  procurement 388–90, 405–6, 422–6
  project completion 393–4, 407, 428–30
  project management 387–90, 405–6, 420–426, 437
  project scope and stadium design 384–5, 402–3, 413–15
  relationship management 395–6, 408, 430–431
  requirement identification 383, 400, 411
  requirement specification 383, 400, 411
  risk and uncertainty 127
  site selection and constraints 383–4, 400–402, 411–13
  subcontracting 391–5, 432–6
  Wembley Stadium, London 20, 67, 127, 304, 381, 410–438
formal presentations 335
framework agreements/ arrangements 182, 316–17
framing effect 114
freedom of information 225
Fujitsu Services Limited 16
function-based organisations 9–10
function-led matrix organisations 9–10
function point analysis 324
future value 164

game theory 355
generic strategy concept 98–9
globalisation 3, 301
governance
  best-practice management 242–6, 272–3
  commercial management framework 37, 42–3, 63–7

contracts and contracting 186–7
financial decision-making 139–40
football stadia 422
Heathrow T5 programme 465
risk and uncertainty 108
growth 27
guaranteed maximum price (GMP) 187
guarantees 197–8

Harmon CFEM Facades (UK) Limited 338
health and safety 267, 479, 483
Heathrow T5 programme 455–91
asset supply 475
baggage-handling system failure 480–481
budget and financing 463–4, 485
buy clubs 477–9
change management 475
collaborative project software 472
commercial management framework 314–15
communication failures 481
contractual model 469–70
corporate culture 466, 472–3
development programme and milestones 461–3
incentive payments 473–4
insurance policies 473
integrated teamwork 471–2
key features 457, 461
key stakeholders 490–491
logistics management 475–7, 482
overview 20
payment terms 473
performance 474–5
post-project outcomes 483–6
procurement 465–75, 478, 482, 485–6
project completion 479–81
project management 464–75
project scope and terminal design 459–61, 482–3
reasons for success 481–3
requirement identification 456–7
requirement specification 457–8
risk management 469–71, 477–8, 482, 484–5
risk and uncertainty 466–8, 484–5
safety programmes 479, 483
site selection and constraints 458–9
heuristics 113–14
Honeywell Control Systems Ltd 434
horizontal integration 101

illusion of control 114
implementation cycle 24, 33–5, 71
implied contracts 338
implied terms 180
impoverished management 84
in-house project executives 303
incentive contracts 184–5, 210
incentive/disincentive (I/D) contracting 185
incentives
best-practice management 271, 273–4
football stadia 429
Heathrow T5 473–4
income statements 133, 143–6
incorporation by reference 180
Incoterms 221
indirect costs 151–2
industry dynamics 97, 99, 104
industry sector reports 290–291
information and communications technology (ICT) 7, 13, 15–16
innovation, market-led 99–100
insolvency 225–8
Institute of Commercial Management (ICM) 4–5
Institute of Risk Management (IRM) 57
Institute for Supply Management (ISM) 56
insurance 198, 212, 473
intangible non-current assets 141–2
integrated contracting 31–3
integrated procurement systems 304–12
integrated project teams (IPT) 8–9, 266
integrated supply teams (IST) 9, 266
integrated teamwork 471–2
intellectual property rights (IPR) 209, 222–4
intelligent customers 25
intent stage 34, 35, 288, 292–3, 296–331
assessment models 321
bid strategy/project screening stage gate review 331
business justification stage gate review 298–9, 325
commercial activities 323–4
decision to bid 327–31
delivery strategy stage gate review 324–5
disadvantages of inviting tenders 323
e-procurement 318–19
integrated procurement systems 304–12
management-orientated procurement systems 312–18, 321
opportunity identification and development 292, 325–6
procurement objectives 298
requirement identification/specification 296–7
selection criteria for procurement route 319–22
separated procurement systems 303–4
solution selection 299–301
supplier selection 321–3
supplier's bid strategy 326–7
supply chain management 302
tendering 293, 323
trends in procurement systems 319
internal controls 243–6
internal factors 186
internal rate of return (IRR) 164, 166–7
internal relationship management 40
International Accounting Standards Board (IASB) 133, 137
International Association for Contract and Commercial Management (IACCM) 4–7, 55
International Bar Association (IBA) 56
international commercial terms 221
International Federation of Accountants (IFAC) 55
International Federation of Purchasing and Supply Management (IFPSM) 56

International Financial Reporting Standards (IFRS) 137
International Project Management Association (IPMA) 8, 56
international trade law 220–221
inventory 141–2
investment
  best-practice management 280
  capital investment decisions 164–7
  financial decision-making 141–2, 147, 164–7
investment appraisal methods 164–7
investment decision stage gate review 280, 345
invitation to tender (ITT) 292, 338

joint ventures (JV) 25, 182, 346
just-in-time delivery 476–7

key performance indicators (KPI)
  best-practice management 240–242, 272
  commercial management framework 298
  contracts and contracting 193, 212
knowledge managers 89–90

Laing O'Rourke 17
Law Society 56
leadership
  best-practice management 255
  definitions 53–4, 79
  desirable qualities of leaders 90–91
  distributed 83, 87–8
  football stadia 437
  transformational 83, 86–7
  see also commercial leadership
leadership capability 93
learning
  commercial leadership 80, 88–90
  risk and uncertainty 125
  transaction cost economics 66
learning leadership 83, 90
learning organisations 89
legal issues
  commercial management framework 37, 42–3, 56, 338, 340
  contracts and contracting 19, 173–235
  professional bodies and associations 56
  subcontracting 432–6
legitimate power 82
letters of credit 221
letters of intent 177–8
liabilities 140, 141–2
limited liability companies 136–7
liquidated damages 193, 196
liquidation 226–7
logistics management 475–7, 482
London Underground 308–9
lowest price 335–6, 338, 343, 437

maintenance of assets/services 359–60
make-or-buy decisions 299–301
management
  commercial leadership 81–2
  contracts and contracting 189–90
  definitions 53–4
management accounting 41, 136, 150–151, 167, 169–70
management contracting 312
management control 157
management-orientated procurement systems 312–18, 321
Management of Portfolios (MoP) 247, 256–7
Management of Risk (MoR) 247, 258–60
Management of Value (MoV) 247, 260–263
Managing Successful Programmes (MSP) 246, 253–6
margin of safety 155
market governance 66
market-led innovation 99–100
market management 275
market-related pricing systems 348
marketing 51, 57
markets
  definitions 26
  market penetration and share 27
  market power 218–19
maturity models 60–62
measurement contracts 183
mergers 219
middle-of-the-road management style 84
milestone payments 192
Millennium Stadium, Cardiff 20, 382–98
  asset supply 390–391
  budget and financing 385, 387
  commercial management framework 67, 304
  development programme and milestones 385–7
  key features 381, 385
  post-project outcomes 397
  procurement 388–90
  project completion 393–4
  project management 387–90
  project scope and stadium design 384–5
  relationships management 395–6
  requirement identification 383
  requirement specification 383
  risk and uncertainty 127
  site selection and constraints 383–4
  subcontracting 391–5
mixed costs 151
model conditions of contract 178
monopolies 218–19
Monte Carlo analysis 122–3
most economically advantageous tender (MEAT) 333, 336, 338, 343
Mott MacDonald 435–6
multicriteria selection 337–9
multiple regression models 354

NASA 111–12, 116, 124
National Audit Office (NAO) 309–11
National Contract Management Association (NCMA) 5, 55
negotiated tendering 333
negotiation 37, 44–5, 343–5, 357
neo-classical contracting 66
net present value (NPV) 164–7, 171
neural network models 354
non-adversarial dispute resolution 200–202
non-compliant offers 349
non-current assets/liabilities 141–2
non-disclosure agreements (NDA) 224
nuclear power plants (NPP) 306–7
NuevoJoints 366–7

obligational contracting 66
off-site prefabrication 476
Office of Fair Trading (OFT) 349
Office of Government Commerce (OGC) 5, 28–9
OGC Gateway Process 239, 247, 265–7, 277–81, 298–9, 324–5, 331, 345, 355–6, 364–5
Olkiluoto 3 nuclear power plant 306–7
open source 224
open tendering 333
openness 312
operating income 144–5
operating profit 144
operations review stage gate review 280, 364–5
opportunism 65–6
opportunities and threats 108–10, 121–2, 125–6
opportunity development 292, 325–6
opportunity identification 292, 325
opportunity management 37, 51–2, 57
optimism bias 114
organisational interfaces 30–33
organisational structures 9–13
outsourcing 300–301, 323
overconfidence bias 114
overpricing 349

P3M3 framework 247, 257–8
package deals 305
parallel contracts 182
parenting advantage 101
partnering arrangements
  commercial management framework 309–11, 313–14
  contracts and contracting 182–3, 187, 205, 208, 227
  Heathrow T5 programme 468, 471–2
Partnerships for Schools (PfS) 311
passive bidding tactics 327
patents 223–4
path-goal theory of leadership 85
Pathclearer contract 180
payback period 164
payment
  best-practice management 273–4
  on completion 192
  Heathrow T5 programme 473
  provisions 191–5
PC Harrington Contractors Ltd 434–5
pension schemes 141–2
people development dimension 61–2
performance
  best-practice management 267, 273
  commercial management framework 353, 361–2
  contracts and contracting 197
  financial decision-making 157–60, 167
  Heathrow T5 programme 474–5
performance indicators (PI) 196
  see also key performance indicators
periodic payments 195
planned progress payments 192
Porter's Five Forces model 98
portfolio analysis 101
portfolio management 247, 256–7
positional negotiation 344
post-award activities 6
  best-practice management 269–70
  commercial management framework 38, 359–60
  football stadia 394
post-contract stage gate review 364
post-project outcomes
  football stadia 397, 408–9, 438
  Heathrow T5 programme 483–6
power and leadership 80–82
power perspective 68–9
power/interest matrix 392–3
power supply industry 306–7
pre-award activities 6
  best-practice management 269–70
  commercial management framework 38
  football stadia 394
  negotiations 343–5, 357
prepayments 141–2
pre-qualification 334
present value 164–7, 171
price-based contracts 183, 184
pricing
  bid price strategy 347–57
  commercial management framework 46
  price fluctuations 194
  price formulation process 350
  pricing policies and systems 152–3, 347–8
prime contracting 312, 322
PRINCE2 8
  benefits and limitations 252–3
  best-practice management 246, 247–53
  commercial management framework 25
  management levels, roles and responsibilities 248
  management projects 252
  principles 249
  process 250–251
  structure and elements 249–52
  techniques 251
  themes 249–50
principled negotiation 344
Private Finance Initiatives (PFI) 187, 305–11, 321
probability theory 353–4
probability trees 123–4
probability/impact/manageability matrix 122–3
process protocols 246–63
ProCost 321
ProCure21+ 318
procurement
  assessment models 321
  best-practice management 263–8, 357–8
  best value/value for money 336, 341
  bid evaluation 336–40
  commercial activities 293–5, 323–4
  commercial management framework 35, 46–8, 289–95, 298, 302–25, 332–45
  contracts and contracting 187, 215–18
  deal creation 332–45
  definitions 24–5, 298
  delivery strategy stage gate review 324–5
  disadvantages of inviting tenders 323
  e-procurement 318–19

procurement (*cont'd*)
European and UK regulations 342
football stadia 388–90, 405–6, 422–6
Heathrow T5 programme 465–75, 478, 482, 485–6
integrated procurement systems 304–12
intent stage 298, 302–25
investment decision stage gate review 345
key principles for good practice 333–4
lowest price 335–6, 338, 343
management-orientated procurement systems 312–18, 321
multicriteria selection 337–9
objectives 298, 315, 335
opportunity identification and development 292, 325–6
oversight 302–3
overview 19
pre-award negotiations 343–5
pre-qualification 334
procedures 333, 342
project-specific criteria 338–9
public sector procurement 341–2
selection criteria 319–21
separated procurement systems 303–4
sourcing strategy 35–6, 275–6, 293–4
supplier selection 321–3, 335–40
systems 303–25
time limits 342
trends in procurement systems 319
procurement cycle 24, 33, 48, 70–71
product-based planning 251
profit
allocation 349
commercial management framework 27–9
definitions 27–8
financial decision-making 144–5, 147
programme evaluation and review technique (PERT) 117
programme management
best-practice management 246–7, 253–6
commercial management framework 37, 49–50, 56–7
contracts and contracting 182
professional bodies and associations 56–7
project alliancing 315–16
project assurance 251
project-based organisations 9–10
project boards 248, 279
project briefs 297
project completion
football stadia 393–4, 407, 428–30
Heathrow T5 programme 479–81
project governance 42–3, 67, 174, 242–3
project-led matrix organisations 9–11
project management
best-practice management 246–53
commercial management framework 37, 49, 56–8, 302–3
contracts and contracting 175–7
football stadia 387–90, 405–6, 420–426, 437
Heathrow T5 programme 464–75
professional bodies and associations 56–8
Project Management Institute (PMI) 57
project management maturity (PMM) 60–62
project success or failure 239–43
project-oriented organisations 23–76
aerospace/defence organisations 7, 12–15
capability maturity model 60–62
case studies 67, 68
commercial activities 6–7, 9–13, 26–7, 59–60
commercial awareness 23–4
commercial management framework 33–53
communications/ICT organisations 7, 13, 15–16
components of commercial management 36–53
construction organisations 7, 16–17
context for commercial management 7–17
core capabilities and competencies 59–60
definitions 7–9
developing the commercial function 58–70
environment 69–70
management, leadership and communityship 53–4
organisational interfaces 30–33
organisational structures 9–13
overview 18
power perspective 68–9
professional bodies and associations 55–8, 62–3
profit and value 27–9
purpose of commercial function 26–33
relational contracting 66–7, 68
reporting lines 9–13
theoretical background 63–70
transaction cost economics 63–7, 69
trust 67–8
project organisation 266
project partnering 313
project risk management (PRM) 118–25
analysis phase 122–4
communication and learning 125
identification and categorisation phase 119–22
monitor and control phase 125
response phase 124
project-specific criteria (PSC) 338–9
proposed dividends 141–2
proposition development 346–57
bid price determination 347
bid submission/adjudication stage gate review 355–6
bidding and pricing tactics 349–50
cost estimation 350–351
eAuctions and game theory 355
factors influencing bid price decision 352
overview 19
price formulation process 350

pricing policies and systems 347–8
risk and uncertainty 352–3
setting the bid price 351–2
proposition identification 293
prospect theory 113
provisions for liabilities 141–2
Public Contracts Directive (2004/18/EC) 215–17
Public Finance Initiatives (PFI) 24
public limited companies (plc) 136–7
public–private partnerships (PPP) 5, 187, 305–11
Public Private Partnerships Programme (4Ps) 5
public sector procurement 341–2
public transport 308–9
purchasers and purchasing
claims 199
commercial management framework 47, 291, 326
definitions 175
obligations 190

qualifying projects 342
quality
best-practice management 250, 256
commercial management framework 298
contracts and contracting 231
quality reviews 251
quantitative analysis 122–3

rationalist perspective of risk 112–13, 126
readiness for service stage gate review 280, 364
rebates 196
receivership 226
recycling systems 385
referent power 82
regulatory issues 37, 42–3, 215–18, 342
related diversification 101
relational contracting 66–7, 68
relationship management
best-practice management 269–71, 273, 275, 362–3
commercial leadership 83
commercial management framework 37, 38, 39–41, 362–3
contracts and contracting 174
football stadia 395–6, 408, 430–431
re-measurement contracts 183
repayment guarantees 197
reporting lines 9–13
representativeness 114
request for proposal (RFP) 292, 293, 346
request for quotation (RFQ) 293, 346
request for tender (RFT) 293, 346
requirement identification
commercial management framework 292, 296–7
football stadia 383, 400, 411
Heathrow T5 programme 456–7
requirement specification
commercial management framework 292, 297
football stadia 383, 400, 411
Heathrow T5 programme 457–8
reserves 143
resource-based view (RBV) 102–3, 104, 299
restricted tendering 333
retained profit 144
retention guarantees 197
retention payments 195
return on assets (ROA) 28
return on capital employed (ROCE) 28
return on investment (ROI) 28
revenue 144–5
reward power 82
risk 108–31
behavioural perspective 113–14, 126
best-practice management 250, 272, 274
cognitive perspective 114–15, 126
commercial management framework 314, 352–3
contracts and contracting 174, 185–6, 189–90, 211
cultural perspective 115–16
definitions and descriptions 109–10
financial decision-making 170
football stadia 421
Heathrow T5 programme 466–8, 484–5
influencing factors 110–111
origins and development of risk management 116–17
overview 18
perspectives of risk 112–16, 126
project risk management 118–25
projects and risk 117–18
rationalist perspective 112–13, 126
threats and opportunities 108–10, 121–2, 125–6
tools and techniques 118–25
see also uncertainty
risk analysis 118, 122
risk breakdown structures 119–21
risk communication 125
risk identification 118, 119–22
risk management
best-practice management 243–5, 247, 255, 258–60, 266, 363
commercial management framework 37, 51–2, 57, 363
contracts and contracting 211
Heathrow T5 programme 469–71, 477–8, 482, 484–5
origins and development of 116–17
professional bodies and associations 57
project risk management 118–25
Risk Management Institution of Australasia Limited (RMIA) 57
risk monitoring and control 118, 125
risk registers 119–20
risk response 118, 124
Rolls Royce 9, 13, 14–15, 300
Royal Institution of Chartered Surveyors (RICS) 55

safety 230–231, 479, 483
sales 51, 57
Sarbanes–Oxley Act (2002) 108, 116, 246
schedule of measured work 194
schedule of rates 194
Scottish & Newcastle 180
Securities and Exchange Commission (SEC) 246
self-awareness 83
self-interest trust 66
self-management 83
sellers, definitions 25–6
selling 24–6, 47–8

semi-variable costs 151
sensitivity analysis 122
separated procurement systems 303–4
sequential contracts 182
sequential design 304
servant leadership 90
service-level agreements (SLA) 272, 301
service procurement *see* procurement
service providers 25
services
  commercial management framework 295, 300
  contracts and contracting 228
  termination 365–6
servitisation 3
share capital 143
share premium 143
shareholder value 164, 167
short-termism 170
single-stage tendering 335
site selection and constraints
  football stadia 383–4, 400–402, 411–13
  Heathrow T5 programme 458–9
social awareness 83
soft techniques 121
solution selection 299–301
sourcing strategy 35–6, 275–6, 293–4
sponsors 25
stakeholder management 39–40, 255
standard conditions of contract 178
start-up
  best-practice management 248, 250–251
  commercial management framework 62
  contracts and contracting 177
statement of comprehensive income 146
stepped costs 152
strategic agenda 100, 104
strategic partnering 313
strategy 96–107
  best-practice management 266, 278–9
  blue ocean strategies 99–100
  business-level strategy 102
  commercial strategies and tactics 19–20, 288–377
  competitive advantage 96–7, 98–100
  contracts and contracting 181–3
  corporate-level strategy 100–102
  definitions 97
  Heathrow T5 programme 465–8, 482
  key change drivers 97–8
  key components 97
  market-led innovation 99–100
  overview 18, 19–20, 96–7, 288–9
  Porter's Five Forces model 98
  resource-based view 102–3, 104
  risk and uncertainty 121
  see also commercial management framework
strategy implementation 18, 34–5, 96, 103, 296
strategy orientation 97, 103
style leadership theories 83, 84
subcontract management 360
subcontracting
  commercial management framework 295, 301
  football stadia 391–5, 432–6
  legal issues 177, 178, 231, 432–6
  project-oriented organisations 6
subjective expected utility (SEU) theory 110, 112–13
submission date extensions 349
superficial method 324
supernormal profit 98
supervision of suppliers 190
supplier relationship management (SRM) 40
suppliers
  best-practice management 269–70, 274–5
  claims 198–200
  commercial management framework 291, 314–15, 321–3, 326–7, 335–40, 354
  definitions 175
  obligations 190
  payment systems 192–3
  supplier selection 321–3, 335–40
supply chain integration 209
supply chain management (SCM) 40, 46–8, 302
supply-side bidding and implementation cycles 19, 288, 291
supply-side transmission of value 30
surety bonds 197
suspension of projects 191
sustainability 267, 403
synergies 100, 101

tactical pricing 350
tactical withdrawal of bids 349
tangible non-current assets 141–2
target-incentive contracts 194–5
target price 183
tax liabilities 141–2, 144–5, 168
team management 84
teaming agreements (TA) 346
teamwork 471–2
technical bid evaluation 337
technical/engineered system techniques 121–2
tender guarantees 197
tendering
  deal creation stage 333–6, 338, 341, 343
  football stadia 437
  intent stage 292–3, 323
  transaction management 45–6, 48
term contracting 182, 317–18
Terminal 5 (T5) *see* Heathrow T5 programme
termination of contract 197
threats and opportunities 108–10, 121–2, 125–6
time extensions 198–200
time limits 342
time provisions 191, 231
time value of money 164
trade payables 141–2
trade receivables 141–2
trade secrets 224
trait theories of leadership 83–4
transaction cost economics (TCE)
  commercial management framework 63–7, 69, 299
  contracts and contracting 174, 186
transaction management 37, 45–8, 56
transaction-specific governance 66–7
transactional leadership 86

transfer of an undertaking (TUPE) 228–9
transformational leadership 83, 86–7
transparency
  contracts and contracting 212
  financial decision-making 139
  Heathrow T5 programme 472, 477
trilateral governance 66
trust
  commercial management framework 67–8, 312, 314
  Heathrow T5 programme 472
  transaction cost economics 67–8
Turnbull report (ICAEW, 1999) 108, 116, 244–6
turnkey contracts 181, 305, 306–7
two-stage tendering 335

uncertainty
  commercial management framework 65–6, 352–3, 358
  contracts and contracting 185–6
  definitions 110
  influencing factors 110
  origins and development of risk management 117
  overview 8
  perspectives of risk 112–13, 115
  project risk management 121–3, 125
  transaction cost economics 65–6
  see also risk
unit method 324
unrelated diversification 101
Utilities Contracts Directive (2004/17/EC) 215–17
utility 29

value
  commercial management framework 27–33
  core value-proposition 104, 264
  definitions 28–9
  strategy 98–104
  transmission through an organisation 29–33
  value management 28–9, 195, 247, 260–263, 266
  value opportunity assessments 272
value for money (VfM)
  best-practice management 271
  commercial management framework 29, 336, 341
  football stadia 425–6
value-at-risk (VaR) 110
variable costs 151, 154–6, 171
variation of price 194
vertical integration 66, 101
Vickrey Auction 336
Victorian Government Purchasing Board (VGPB) 5
VINCI 137–8
Vodafone plc 139
voluntary arrangements 227

wealth creation 137, 170
Wembley Stadium, London 20, 410–438
  asset supply 426
  budget and financing 416–20
  commercial management framework 67, 304
  development programme and milestones 415–18, 426–8, 444–54
  key features 381, 416
  key stakeholders 443–4
  media briefing by WNSL, Multiplex and the FA 444–54
  post-project outcomes 438
  procurement 422–6
  project completion 428–30
  project management 420–426, 437
  project scope and stadium design 413–15
  relationship management 430–431
  requirement identification 411
  requirement specification 411
  risk and uncertainty 127
  site selection and constraints 411–13
  subcontracting 432–6
  transaction cost economics 67
whole supply network management 276
whole-life costing 29, 266–7